W9-BQT-908

Mastering
Mathematica

Programming Methods
and Applications

Second Edition

LIMITED WARRANTY AND DISCLAIMER OF LIABILITY

ACADEMIC PRESS ("AP") AND ANYONE ELSE WHO HAS BEEN INVOLVED IN THE CRE-
ATION OR PRODUCTION OF THE ACCOMPANYING CODE ("THE PRODUCT") CANNOT AND DO
NOT WARRANT THE PERFORMANCE OR RESULTS THAT MAY BE OBTAINED BY USING THE
PRODUCT. THE PRODUCT IS SOLD "AS IS" WITHOUT WARRANTY OF ANY KIND (EXCEPT AS
HEREAFTER DESCRIBED), EITHER EXPRESSED OR IMPLIED, INCLUDING, BUT NOT LIMITED
TO, ANY WARRANTY OF PERFORMANCE OR ANY IMPLIED WARRANTY OF MERCHANTABILI-
TY OR FITNESS FOR ANY PARTICULAR PURPOSE. AP WARRANTS ONLY THAT THE CD-ROM
ON WHICH THE CODE IS RECORDED IS FREE FROM DEFECTS IN MATERIAL AND
FAULTY WORKMANSHIP UNDER THE NORMAL USE AND SERVICE FOR A PERIOD OF NINETY
(90) DAYS FROM THE DATE THE PRODUCT IS DELIVERED. THE PURCHASER'S SOLE AND
EXCLUSIVE REMEDY IN THE EVENT OF A DEFECT IS EXPRESSLY LIMITED TO EITHER
REPLACEMENT OF THE CD-ROM OR REFUND OF THE PURCHASE PRICE, AT AP'S SOLE DIS-
CRETION.

IN NO EVENT, WHETHER AS A RESULT OF BREACH OF CONTRACT, WARRANTY OR TORT
(INCLUDING NEGLIGENCE), WILL AP OR ANYONE WHO HAS BEEN INVOLVED IN THE CRE-
ATION OR PRODUCTION OF THE PRODUCT BE LIABLE TO PURCHASER FOR ANY DAMAGES,
INCLUDING ANY LOST PROFITS, LOST SAVINGS OR OTHER INCIDENTAL OR CONSEQUENTIAL
DAMAGES ARISING OUT OF THE USE OR INABILITY TO USE THE PRODUCT OR ANY MODIFICA-
TIONS THEREOF, OR DUE TO THE CONTENTS OF THE CODE, EVEN IF AP HAS BEEN ADVISED
OF THE POSSIBILITY OF SUCH DAMAGES, OR FOR ANY CLAIM BY ANY OTHER PARTY.

Any request for replacement of a defective CD-ROM must be postage prepaid and must be accompanied by the
original defective CD-ROM, your mailing address and telephone number, and proof of date of purchase and purchase price. Send such requests, stating the nature of the problem, to Academic Press Customer Service, 6277
Sea Harbor Drive, Orlando, FL 32887, 1-800-321-5068. AP shall have no obligation to refund the purchase
price or to replace a CD-ROM based on claims of defects in the nature or operation of the Product.

Some states do not allow limitation on how long an implied warranty lasts, nor exclusions or limitations of incidental or consequential damages, so the above limitations and exclusions may not apply to you. This Warranty
gives you specific legal rights, and you may also have other rights which vary from jurisdiction to jurisdiction.

THE RE-EXPORT OF UNITED STATES ORIGIN SOFTWARE IS SUBJECT TO THE UNITED STATES
LAWS UNDER THE EXPORT ADMINISTRATION ACT OF 1969 AS AMENDED. ANY FURTHER SALE
OF THE PRODUCT SHALL BE IN COMPLIANCE WITH THE UNITED STATES DEPARTMENT OF
COMMERCE ADMINISTRATION REGULATIONS. COMPLIANCE WITH SUCH REGULATIONS IS
YOUR RESPONSIBILITY AND NOT THE RESPONSIBILITY OF AP.

Mastering
Mathematica

Programming Methods
and Applications

Second Edition

John Gray
University of Illinois
Urbana, Illinois

ACADEMIC PRESS

San Diego London Boston
New York Sydney Tokyo Toronto

This book is printed on acid-free paper. ∞

Copyright © 1998, 1996, 1994 by Academic Press
All rights reserved.
No part of this publication may be reproduced or
transmitted in any form or by any means, electronic
or mechanical, including photocopy, recording, or
any information storage and retrieval system, without
permission in writing from the publisher.

Mathematica is a registered trademark of Wolfram Research, Inc.
The Mathematica 3.0 Logo is a trademark of Wolfram Research, Inc., Champaign, Illinois, and is used by Academic Press
publishing company under License. Use of the Logo unless pursuant to the terms of a license granted by Wolfram
Research, Inc. or as otherwise authorized by law is an infringement of the trademark. The Publication is independently
published by Academic Press publishing company. Wolfram Research, Inc. is not responsible in any way for the contents
of this publication.
Unix is a registered trademark of AT&T.
Macintosh is a trademark of Apple Computer, Inc.
NeXT is a trademark of NeXT, Inc.
Sun Sparc is a trademark of Sun Microsystems, Inc.
MS-DOS is a registered trademark of Microsoft Corporation.

ACADEMIC PRESS
525 B Street, Suite 1900, San Diego, CA 92101-4495, USA
1300 Boylston Street, Chestnut Hill, MA 02167, USA
http://www.apnet.com

Academic Press Limited
24–28 Oval Road, London NW1 7DX, UK
http://www.hbuk.co.uk/ap/

Library of Congress Cataloging-in-Publication Data

Gray, John W.
    Mastering Mathematica : programming methods and applications /
John W. Gray. — 2nd ed.
        p.   cm.
    Includes bibliographical references and index.
    ISBN 0-12-296105-6
    1. Mathematica (Computer file). 2. Mathematics—Data processing.
I. Title.
    QA76.95.G68   1997
    510'.285'53042—dc21   97-30329
                                                                    CIP

Printed in the United States of America
97 98 99 00 01 IP 9 8 7 6 5 4 3 2 1

Contents

Part III: *Mastering Knowledge Representation in* Mathematica

Preface

There are three distinct levels of competence that are relevant to the use of *Mathematica*, all of which are addressed in this book. They provide the headings for its three main divisions:

Mathematica as a Symbolic Pocket Calculator
Mathematica as a Programming Language
Knowledge Representation in *Mathematica*.

Much of this material grew out of a course in mathematical software that has been taught at the University of Illinois at Urbana-Champaign almost every semester since 1987. The first edition of this book presented this material in a form that was accessible to anyone interested in programming in *Mathematica*. This second edition contains essentially the same material, with hundreds of small and large changes, ranging from better formatting to significantly faster and better algorithms. One intriguing discovery was that Dot could be redefined in the context of object-oriented programming to mimic its use in Java, leading to the observation that much of *Mathematica* is, in fact, object-oriented. Extensive use is made of the new text-formatting capabilities of *Mathematica*, and this entire book has been produced in *Mathematica*, rather than a word processor, as was done previously.

The course itself is intended for upper division and graduate students in mathematics, mathematics education, engineering, and the sciences, and its purpose is to teach students how to do their own mathematics using symbolic computation programs. The emphasis then and now is on how to take known, but rather vaguely described, mathematical results and turn them into precise algorithmic procedures that can be executed by a computer. In this way, the range of known examples of a given procedure is extended and insight is provided into more complex situations than can be investigated by hand. There is a vast difference between "understanding" some mathematical theory and actually implementing it in executable form. Our main goal is to provide tools and concepts to overcome this gap. Naturally, there is nothing new about finding computer implementations of mathematical theories and efforts in this direction have been going on for 40 years. What is new is that *Mathematica* makes it possible for "ordinary" people, who are not computer professionals, to join in these efforts on an equal basis. There are innumerable opportunities in our highly technological society for such developments, ranging from theoretical mathematical questions in group theory or graph theory, through optimization routines in econometrics to intensely practical questions such as predicting results in tournaments or calculating docking orbits for satellites.

Perhaps the most important contribution of *Mathematica*, particularly in its notebook interface versions, is the way in which it has empowered mathematicians, engineers, scientists, teachers, and students to take advantage of these opportunities.

"Empowered" is the key word here, for there are a number of other general purpose symbolic computation programs, such as Macsyma, Reduce, Derive, Maple, and Axiom, as well as dozens of special purpose symbolic programs. The main difference between these general purpose programs and *Mathematica* lies in their archaic approach to programming. Their languages are Pascal-like; i.e., imperative languages based on the language of while-programs. For many people, programming in such a language is drudgery. Everything is broken down into such tiny steps and the built-in facilities are so meager that there seems to be no place to exercise insight and ingenuity. *Mathematica*, on the other hand, supports four distinct styles of programming, functional programming, rule-based programming, imperative programming, and object-oriented programming, and its built-in facilities are so incredibly rich that nearly any algorithmic, mathematical thought has an almost direct expression in it. There is another seemingly small difference which is actually an important aspect of empowerment. The "arcane" knowledge possessed by professional programmers frequently consists in knowing what key strokes will accomplish their desired end; i.e., which abbreviations, acronyms, or whimsical terms will cause the computer to do what is desired. Symbolic computation programs are large and have many built-in commands (in the current version of *Mathematica* there are more than 1500 names). It would be very difficult to try to remember that many abbreviations. It would even be very difficult to find them in a manual if they were alphabetized as abbreviated, as they are in Macsyma and Maple. Instead, *Mathematica* writes out almost all terms in full, and this makes a tremendous difference in ease of learning to use the language. Finally, the notebook interface is an order of magnitude improvement over any of the previous ways in interacting with a symbolic computation program. It is the thing that empowers people to produce documents containing embedded active mathematics in a very simple way. The previous edition of this book was a collection of notebooks which were exported into a word processor. The current edition makes use of all of the new formatting and editing capabilities of Version 3.0 and later to produce everything in *Mathematica*. Since notebooks and their contents are directly programmable, things like the index and table of contents (including their hyperlinks) are made by *Mathematica* programs.

The first part of the book is concerned with *Mathematica*'s use as a symbolic pocket calculator and requires almost no mathematical sophistication, except in certain sections (for instance the one on differential equations). Essentially, "buttons" are pushed to see what happens. The second part treats programming in *Mathematica*. In these first two parts, there is a practice section and a section of exercises at the end of almost every chapter. The practice sections address the question "What should I do first?" Faced with a new program, how do you get it to do anything? Here, just try out what's in the practice sections. The exercises are extremely important. It is only after trying to do something yourself that you are motivated to learn the various ways that it can be done. Answers to selected exercises are given, sometimes in great detail, at the end of the book. A number of exercises are repeated from chapter to chapter, each time asking for a more sophisticated answer. Similarly, answers may be given in several forms, starting with crude programs that just barely work and leading to elegant, brief programs that display their outputs in graphical form. Once button pushing and programming have been mastered, the problem then is to use this knowledge to develop some part of mathematics in detail. The third part of the book is devoted to examples of how to do this.

Considering the contents in more detail, Part I is devoted to using *Mathematica* as a symbolic pocket calculator. Chapter 1 does just this. Chapter 2 investigates the three ways of interacting with *Mathematica*, and and discusses some of the capabilities of the new notebook interface. Chapter 3 looks in more detail at numerical calculations and solving equations. Both algebraic and differential equations are considered here, and a whole mini-course in differential equations is included, mostly in exercises, because experience has shown that this very dramatically demonstrates how much can be done by such a symbolic program. Chapter 4 is concerned with built-in graphics; i.e., how to make pictures without programming. If all you want is a simple picture with a certain amount of customizing, this chapter shows you how to make it.

In Part II, we turn to the real concern of the book, which is using *Mathematica* to program mathematics. Chapter 5 discusses the *Mathematica* language, and then we see in Chapters 6, 7, 8, and 9 that *Mathematica* is capable of four styles of programming: functional programming, rewrite programming, imperative programming, and object-oriented programming.

i) The functional aspects of the language are explained in Chapter 6, with functional programming itself, via "one-liners" as the main topic. LISP is a typical functional programming language, but the actual functionality available to the *Mathematica* programmer is many times that to be found in LISP, thanks to the very many built-in functions that are immediately usable.

ii) Rule-based programming is studied in Chapter 7. *Mathematica* actually works by systems of rewrite rules and the *Mathematica* programmer can freely create and use his or her own systems of rules. This distinguishes it from traditional programming languages, which normally have no such features.

iii) Imperative programming is treated in Chapter 8, where we present several examples of imperative programs from Pascal and C and show how to translate them into *Mathematica* programs. This is an important skill since many thousands of such programs have been published and they serve as a source for precise statements of algorithms. In our examples, there is first a direct translation of the program into *Mathematica*, and then a translation of the purpose, rather than the form, of the program into a style that expresses its mathematical content in a much more direct and "mathematical" form. The possibility of writing such programs is one of the things that makes *Mathematica* such an attractive language.

iv) Chapter 9 turns to the topic of object-oriented programming. *Mathematica* is able to shed a piercing ray of light onto this most confusing of all programming methodologies for several reasons: first, because the objective extension of *Mathematica* is written in top-level code and hence can be examined to see how it works. We do not actually carry out this examination in detail, but just show, through carefully chosen examples, how it is possible to create active data objects that know how to respond to messages. Second, *Mathematica* is interactive, so classes and objects are immediately available for experimentation, without any intervening linking and compilation steps. Third, the entire *Mathematica* language can be used to write methods and interact with objects. For all of these reasons, *Mathematica* will surely become the prototyping tool par excellence for object-oriented programming.

Chapter 10 is concerned with graphics primitives; i.e., how to make pictures with programming. If you want a fully customized picture in which you control all elements of the final result, this chapter shows how to do it. Finally, Chapter 11 studies the language from a more technical point of view. Packages, which are a technique for, so to speak, engraving a body of code in stone, are treated here. They are the appropriate mechanism for adding functionality to *Mathematica*. No program can possibly contain all of the mathematical procedures that a mathematician, scientist, engineer, economist, etc. could want. It is very easy to extend *Mathematica* for one's own use, but if you want to supply new functions for others to use, then common courtesy and concern for others demands that the code for these functions should be carefully organized and protected from accidentally interfering with or being interfered with by other code. Packages are exactly the mechanism for doing this. Several more technical questions involved with evaluation of expressions and the process of substitution are also treated here. Along the way we provide a simple implementation of the lambda calculus—an abstract, theoretical, functional programming language.

The point of becoming fairly fluent in writing short programs is to be able to then use this facility in developing your own mathematics. Part III consists of some topics that have interested me, often because of student interest. Chapter 12 on Polya's Pattern Inventory [Polya], began with a student project by Kungmee Park. Chapter 13 was inspired by material from an early version of Skiena's book on Discrete Mathematics [Skiena]. Graph theory is such an obvious topic for computer implementation that one has to be careful not to get carried away with seeing how one's own particular concerns manifest themselves there. We decided to program it all in object-oriented style, using Maeder's package, **Classes**. It also serves in a minor way to introduce the capabilities of the package **Notation**. Chapter 14, concerning differentiable mappings, builds on a problem set that comes earlier in the book. A direct attack on this problem set usually results in confusion, as the answers show. Once everything is treated from a more abstract and systematic point of view, the calculations become clear. Chapter 15 extends the treatment of differentiable mappings to consider the analysis of critical points of functions and the developments in differential geometry that are required to study minimal surfaces.

One brief comment on the notation used here. Built-in *Mathematica* operations all begin with capital letters. Everything that is defined in this book starts with a lower case letter, so there should never be any question whether some operation is built-in or user defined. (I strongly support the suggestion that only employees of Wolfram Research, Inc. are allowed to define operations starting with capital letters, and in the finest Quaker tradition, I even have my doubts about some of them.) Inputs and outputs are shown in monospaced fonts as they appear in Notebook implementations on machines where bold face fonts are available. Thus, a typical interaction looks like:

```
Expand[ (1 + x) ^6 ]
```

$1 + 6x + 15x2 + 20x3 + 15x4 + 6x5 + x6$

In the earlier edition of this book, outputs were frequently edited to make them look nicer on the page, but in Version 3.0 and later, this is no longer necessary. The standard reference for everything concerning *Mathematica* is Stephen Wolfram, The *Mathematica* Book, 3rd ed. (Wolfram Media/Cambridge University Press, 1996).

As previously mentioned, the kind of material in this book has been taught at the UIUC nearly every semester since 1987. Furthermore, it has been the subject of four week-long summer workshops during the summers of 1991–95 sponsored by the Office of Continuing Engineering Education of the UIUC under its Illinois Software Summer School program. The students in these courses have contributed a great deal to the final form of this book, both locally and globally. Locally, they have frequently come up with better ways to do something than anything I could think of, and globally they have kept the entire organization of the book in flux, finding out what works educationally and what doesn't. Anybody concerned with elementary aspects of *Mathematica* is bound to be influenced by Nancy Blachman's book [Blachman] and anybody concerned with more advanced aspects will be equally influenced by Roman Maeder's book [Maeder 1]. I owe Roman especial thanks for everything he taught me about symbolic programs. Finally, I thank my son, Theodore Gray, for his patience and constant help and advice in dealing with all aspects of *Mathematica* and my wife, Eva Wirth Gray, for carefully proofreading and improving much of the book.

How to Use the CD-ROM

The CD-ROM disk that comes with this book can be used with Macintosh, MS-DOS, Next, and Unix computers. It contains the complete contents of the book as *Mathematica* notebooks. These are ASCII text files that *Mathematica* displays as notebooks that look exactly like the contents of the book. The only difference is that all input cells are active and can be evaluated or changed as desired. Furthermore, the hyperlinks in the Table of Contents and the Index are active and can be followed just by clicking on them.

To use the CD-ROM, insert the disk in your CD-ROM player. Open up the icon on the desktop if necessary to see a folder labeled "Mastering *Mathematica*". You have two choices for proceeding.

1. Simply move the folder to your hard disk. (It is about 10 megabites of material.) You can then open the individual files in *Mathematica* as usual. If you open the Index or Table of Contents file, then clicking on a hyperlink there will open the appropriate chapter with the relevant cell selected.

2. The more adventurous choice is to place the folder where it can be accessed by the *Mathematica* Help Browser. To do this, open the *Mathematica* 3.0 files folder, where you will see a folder named AddOns. Open this to see yet another folder named Autoload. Put the entire folder "Mastering *Mathematica*" in the Autoload folder. (On multi-user Unix systems, you can either locate it in .*Mathematica*/AddOns/Autoload for your own use or have your systems administrator put it in the general *Mathematica* directory for everyone to use. The only thing that is actually autoloaded is a one-sentence init.m file.) After moving the folder, restart the *Mathematica* front end and choose "Rebuild Help Index" from the Help menu. This will incorporate the book into the Help Browser, and include its index in the Master Index section. This has two consequences:

   a) If you click on the radio button "Add-ons" in the Help Browser, it will include Mastering *Mathematica* as one of the options in the left-hand column. Clicking on its name there puts the parts of the book in the second column. Clicking on a part number puts the chapters in that part in the third column. Clicking on a chapter name puts the sections of the chapter in the fourth column. Clicking on one of them brings up the material contained in the section in the bottom part of the Help Browser. Alternatively, if you click on "Index" in the second column, then the Index is opened in the bottom part of the Help Browser. Clicking on an item there opens the relevant chapter as a notebook as in 1.

b) If you click on the radio button "Master Index" in the Help Browser, then click on a letter, such as A, and then scroll way down to "Attributes" and click on it, you will discover that "Mastering *Mathematica*" has been added to the general index. (Alternatively, type "Attributes" in the top line of the Help Browser and click on "Go To" to get there directly.) There will be (at least) three main headings, Build-In Functions, Mastering *Mathematica*, and The *Mathematica* Book. Each of these has a listing for "Attributes" and for its subitems. Clicking on hyperlinks there opens the relevant document in the bottom part of the Help Browser. The "Back" button in the upper right corner of the Help Browser takes you back to the index so you can explore further.

Mastering
Mathematica

Programming Methods
and Applications

Second Edition

I

Mastering
Mathematica *as a*
Symbolic Pocket
Calculator

1

CHAPTER

A Quick Trip through Elementary Mathematics

Anything you can do I can do better.

1 Opening Remarks

On the simplest level, *Mathematica* is just a glorified pocket calculator, with over 1500 "buttons" to "push". We will begin our study of the language by looking at just this aspect of it. There are all kinds of different buttons:

Kind of button	Examples
Arithmetic button	+, ×, −, /, ^
Special functions	**Sin, Cos, BesselJ**, etc.
Algebraic manipulations	**Expand, Factor**, etc.
Calculus routines	**D, Integrate, Series, Limit**, etc.
Solutions of equations	**Solve, NSolve, DSolve**, etc.
Linear algebra	**Det, Eigensystem**, etc.
Graphics routines	**Plot, Plot3D, ListPlot**, etc.

The first chapter provides an introduction to this very rich world by examining various parts of mathematics in the order in which they are usually introduced in school, starting from grade school arithmetic and running through advanced mathematics.

2 Grade School Arithmetic

By grade school arithmetic, we mean the study of numbers: integers, fractions, decimals and for completeness, complex numbers, but no symbols. Naturally, *Mathematica* has very refined facilities for treating all kinds of numbers in a precise and flexible way.

2.1 Basic Operations

When you first begin a *Mathematica* session, you should start out with some ridiculously simple calculation to check that the program is working, and to start the kernel if you are working in an interface mode. For instance:

In[1]:= **2 + 2**

Out[1]= 4

Input to *Mathematica* is shown in a boldface, equispaced font (Courier bold) and output is shown on the next line in a plain face, equispaced font (Courier plain). We consider grade school arithmetic to consist of addition, subtraction, multiplication, division, and exponentiation by integers. *Mathematica* can of course deal with bigger numbers than one usually works with by hand, so our examples will be correspondingly bigger than those you worked in the third grade. Let us try adding two 32 digit numbers.

In[2]:= **9172584429161413285761749248879 +
11773984116181554151698259468319**

Out[2]= 10349982840779568700931575175098

The answer comes back almost immediately, provided the kernel has been loaded with the preceding simple example. This addition could be carried out by hand with a certain amount of diligence, but probably a mistake would be made somewhere in the middle, which would then be difficult to find.

To make the problem a bit more challenging, insert minus signs in the middle of each of the summands, so that both addition and subtraction are involved.

In[3]:= **9172584429161413 − 2857617492488779 +**
       1177398411618155 − 4151698259468319

Out[3]= 3340667088822470

This can still be checked by hand, but the chances of error have gone up even more. To make the problem considerably more interesting, insert multiplication signs, indicated by spaces (or if desired by stars "*****"), in the middle of each of the preceding numbers.

In[4]:= **91725844 29161413 − 28576174 92488779 +**
       11773984 11618155 − 41516982 59468319

Out[4]= −2300273380507712

Note that multiplication takes precedence over addition and subtraction; that is, it is carried out first. It would take a great deal of time and diligence to check this computation by hand. There would be 256 multiplications of 8 digit numbers by single numbers, 4 additions of 8 rows of shifted 8 digit numbers, two more additions of 16 digit numbers to combine the positive and negative parts, and one subtraction. Alternatively, one can see that the first two products more or less cancel each other and that the fourth product is bigger than the third, so it is at least correct that the answer is negative.

Now create an almost impossible problem by inserting division signs, indicated by "**/**", in the middle of each of the preceding numbers.

In[5]:= **9172/5844 2916/1413 − 2857/6174 9248/8779 +**
       1177/3984 1161/8155 − 4151/6982 5946/8319

Out[5]= $\dfrac{7350539986062779909393943317}{3103373239800909095133051120}$

The answer still comes back almost instantaneously, but it is now a very large fraction. Note again that division takes precedence over multiplication, addition, and subtraction. Scarcely anybody would have the patience to try to do this calculation by hand and the chance of getting the correct answer must be close to 0.

Finally, insert exponent signs, indicated by ^ (i.e., **2^3** becomes **8**) in the middle of each of the preceding numbers.

In[6]:= **91^72/58^44 29^16/14^13 − 28^57/61^74 92^48/87^79 +**
       11^77/39^84 11^61/81^55 − 41^51/69^82 59^46/83^19

Out[6]= 31745395910427015424122195845563477740000970246847733627941999928⋮
        338497859784646755116561543400621264063846134917252325396788466
        522842824867832405837422750938502050672183172721393407603551319
        925686571948614529058427180978598246276955407583874339698435285
        039883731856824540003347387132673295110938865885196582719679672

1307863711889107574179380718916562364143844417070546362963897049020983805601858539703025222849749860260343327309865707870554158502874174644210964820747820124788471924756672050128278556529741544416347514937017249072683449144980063513339017144493115611788167274251157568473158855886852962951587029186202531828038300512151492826011581670293103808919110094099640490485346886733620498239052276655331845074957456 89 /

349002642126839797438754826702298177750663465486510044255898897⸳7065800978182945791294150770622373094208145134516106857379149⸳24958932008508785181515805631222570064209941911812217349089555⸳305322942895097438157143293948976131416985052431168459049311721⸳02188159684603153608661739617865351563860497508986299957304168⸳87416056167265168496410110701230744053904380067875045180460353⸳47550692092638560775838457404167009801711141735181880916790660⸳72319667911976405761076290385036085546558525014229258545016826⸳30569779189670597431818307664867615051597851579211406792819615⸳34083043462211974010937987017414673962430835963451689569012893⸳9807959592510799138792778235904

This calculation takes a noticeable length of time. Note that exponentiation takes precedence over all the other arithmetic operators. The single slash in the middle of the output indicates division because the numerator and the denominator each require many lines. The back slashes at the ends of the lines just represent line breaks and have no mathematical meaning. Surely nobody could do this calculation by hand, and we have no effective way, other than repeating it, perhaps in a different program, to know if it is correct or not.

This sequence of computations shows a general property of symbolic mathematics programs. They will do all of the usual operations that one does by hand much more rapidly and much more reliably than a person can. In addition, they will carry out calculations that are beyond the possibility of even the most determined human being. Nevertheless, they won't do everything. The preceding example was deliberately arranged to end up with 2 digit exponents since, had the exponents been larger, the calculation would have taken too long. Starting with two 64 digit numbers would have led to 4 digit numbers raised to 4 digit exponents. We got tired of waiting for such a result to return and aborted the calculation.

Of course *Mathematica* is perfectly able to deal with larger exponents. For instance, using two-dimensional positional notation (see Chapter 2) chosen from the palette:

In[7]:= 3^{10}

Out[7]= 59049

We can now find the 10th root of this result.

In[8]:= $\sqrt[10]{\%}$

Out[8]= 3

Here, % refers to the previous output. We repeat these last two calculations replacing 10 by 100 and then by 1000.

In[9]:= 3^{100}

Out[9]= 515377520732011331036461129765621272702107522001

In[10]:= $\sqrt[100]{\%}$

Out[10]= 3

In[11]:= 3^{1000}

Out[11]= 13220708194808066368904552597521443659654220327521481676649203⁚
82268285973467048995407783138506080619639097776968725823559509⁚
54582100618911865342725257953674027620225198320803878014774228⁚
96484127439040011758861804112894781562309443806156617305408667⁚
44905061781254803444055470543970388958174653682549161362208302⁚
68563778582290228416398307887896918556404084898937609373242171⁚
84635993869551676501894058810906042608967143886410281435038564⁚
87471658320106143661321731027689028552200001

In[12]:= $\sqrt[1000]{\%}$

Out[12]= 3

In the Exercises you are asked to try 3^{10000} for yourself.

2.2 Factoring Integers

Integers can be factored quickly if they are not too large. (Too large means generally more than 30 digits.)

In[13]:= **FactorInteger[44261662123343986901383109450 03]**

Out[13]= {{37, 1}, {173, 1}, {2143, 2}, {150568994203431074347, 1}}

FactorInteger writes the prime factors of an integer in the form of a list of pairs. The first entry in a pair is the prime factor and the second entry is the number of times it occurs in the factorization. Thus our number is equal to

$$37^1 \ 173^1 \ 2143^2 \ 150568994203431074347^1$$

We can check that the last number here really is a prime number using the built-in predicate **PrimeQ**. (Predicates are functions that return the value True or False.)

In[14]:= **PrimeQ[150568994203431074347]**

Out[14]= True

2.3 Real Numbers

The number 3^{1000} that we calculated above has very many digits. We can find out just how many by converting it to a real number in scientific notation.

In[15]:= **N[3^{1000}]**

Out[15]= $1.322070819480807 \times 10^{477}$

N[anything] finds the numerical value of "anything" expressed as a floating point number in scientific notation by showing a 6 digit number, with 1 digit to the left of the decimal point, times a suitable power of 10 (as soon as the number requires 7 or more digits for its expression). Integer arithmetic such as was used in the first section is done with infinite precision; that is, all relevant digits are shown and no approximations are made. All calculations involving integers and fractions remain in integer or fractional form with all digits shown. Numbers are converted to approximate real values only if **N** is explicitly used.

Square roots can be calculated by using the square root function **Sqrt[-]** or the square root palette item. We try out the two forms together in a list, indicated by the curly brackets.

In[16]:= $\{$ **Sqrt[9]**, $\sqrt{9}\}$

Out[16]= $\{3, 3\}$

Note that in the linear form, the square root function must be typed in exactly as it appears with a capital letter and square brackets. **sqrt[9]**, **Sqrt(9)** and **sqr(9)** all don't work. Square brackets are always used for function application and all built-in operations begin with a capital letter. Try another example.

In[17]:= $\{$ **Sqrt[10]**, $\sqrt{10}\}$

Out[17]= $\{\sqrt{10}, \sqrt{10}\}$

Since 10 is an integer and the square root of 10 is not, the function remains unevaluated. However, its numerical value as an approximate real number can be found to as many decimal places as desired.

In[18]:=**N[Sqrt[10], 40]**

Out[18]= 3.162277660168379331998893544432718533720

This gives the numerical value of the square root of 10 to 40 decimal places. In all occurrences, **N** can take a second argument indicating how many significant digits are desired. (See Chapter 3 for the exact meaning of the second argument.) A single real number containing a decimal point in an arithmetic expression contaminates the entire numerical calculation and turns everything into real numbers.

In[19]:= **1.0** $+ \dfrac{1398}{1434} + \dfrac{21582}{4323} - \dfrac{8935}{9602}$

Out[19]= 6.03673

Pi, or the symbol π from the palette, denotes the mathematical constant π. We can calculate it to any desired number of decimal places, depending of course on the amount of computer memory that is available and the length of time we are willing to wait. The following calculation is almost instantaneous.

In[20]:=**N[π, 500]**

Out[20]= 3.14159265358979323846264338327950288419716939937510582097494459230781640628620899862803482534211706798214808651328230664709384460955058223172535940812848111745028410270193852110555964462294895493038196442881097566593344612847564823378678316527120190914564856692346034861045432664821339360726024914127372458700699

606315588174881520920962829254091715364367892590360011330530541
882046652138414695194151160943305727036575959195309218611738191
326117931051185480744623799627495673518857527248912279381830111
9491

2.4 Complex Numbers

Complex numbers are written in the form a + b I, where I, or the symbol *i* from the palette, is the square root of –1. For instance:

In[21]:= $(6 + i)^5$

Out[21]= 5646 + 6121 I

The number here is actually a Gaussian integer (the real and imaginary parts are integers). They are closed under addition, multiplication, and exponentiation by ordinary integers. As before, the fifth root should take us back to where we started.

In[22]:= $\sqrt[5]{\%}$

Out[22]= 6 + I

Try another example.

In[23]:= $(2 + 5 i)^{12}$

Out[23]= -86719879 + 588467880 I

In[24]:= $\sqrt[12]{\%}$

Out[24]= $(-86719879 + 588467880 \ I)^{1/12}$

In[25]:= N[%]

Out[25]= 5.33013 + 0.767949 I

Clearly, the 12th root of 2 + 5 I to the 12th power is not the same as 2 + 5 I. In the exercises, you are asked to investigate this situation more carefully.

2.5 Number Types in Mathematica

The following table, constructed in *Mathematica,* shows the kinds of number types that are available in *Mathematica*. We have divided them into "Real types" in the left-hand column and "Complex types" on the right. More general types are to the right and down in the table.

```
Integers      Gaussian Integers

Rationals     Gaussian Rationals

Reals         Complexes
```

A Gaussian rational number is a quotient of Gaussian integers. It can always be represented as a complex number with rational real and imaginary parts. Gaussian rationals are closed under arithmetic operations. For example:

$$\text{In[26]:=} \quad \frac{3 + 5\,i}{2 + 4\,i}$$

$$\text{Out[26]=} \quad \frac{13}{10} - \frac{I}{10}$$

Any arithmetic calculation is carried out in the least general type that is common to all of the arguments of the calculation. For instance, the sum of a rational number and a Gaussian integer is a Gaussian rational number.

$$\text{In[27]:=} \quad \frac{1}{2} + (3 + 5\,i)$$

$$\text{Out[27]=} \quad \frac{7}{2} + 5\,I$$

The function **N[]** converts any number to the type at the bottom of its column. Built-in numerical functions such as **Sqrt[]** are usually only evaluated if the type of the answer matches the type of the arguments. For example,

$$\text{In[28]:=} \quad \sqrt{10\,.}$$

$$\text{Out[28]=} \quad 3.16228$$

3 High School Algebra and Trigonometry

Virtually every computer program and every person who has been to school is able to handle numbers in some way. The first step upwards in mathematical sophistication comes with the introduction of variables and symbolic constants. Most programming languages and many people never take this step. Those programs that do are called symbolic computation programs. The place where this happens in school is in high school algebra, which consists of manipulating algebraic expressions, solving linear and quadratic equations in one variable, and possibly solving systems of linear equations. The crucial new ingredient is the inclusion of symbols representing constants or variables.

3.1 Manipulating Algebraic Expressions

One of the main strengths of *Mathematica* lies in its facilities for symbolic manipulation of mathematical expressions involving symbolic constants and variables.

3.1.1 The first example

Start by entering an algebraic expression with symbolic components.

In[1]:= $(x + y)^2 + 7 (3 + x) (x + y)$

Out[1]= $7 (3 + x) (x + y) + (x + y)^2$

Nothing has been done to this expression except that the order of the two summands has been rearranged according to *Mathematica*'s own notion of what should come first. The reason that nothing was done is that nothing was asked for. If something is asked for, then another form of the expression will be displayed. For instance:

In[2]:= **Expand [%]**

Out[2]= $21 x + 8 x^2 + 21 y + 9 x y + y^2$

Expand does exactly what you would expect. It distributes multiplication over addition until all terms are monomials. Here it has expanded the previous expression, referred to by **%**. A person would have no difficulty in carrying out this expansion by hand. We could now raise this new expression to the fourth power.

In[3]:= %^{4}

Out[3]= $(21 x + 8 x^2 + 21 y + 9 x y + y^2)^4$

Again, nothing happened because nothing was asked for other than this expression itself. Presumably, we meant to expand this expression as well.

In[4]:= **Expand [%]**

Out[4]= $194481 x^4 + 296352 x^5 + 169344 x^6 + 43008 x^7 + 4096 x^8 +$
$777924 x^3 y + 1222452 x^4 y + 719712 x^5 y + 188160 x^6 y + 18432 x^7 y +$
$1166886 x^2 y^2 + 1926288 x^3 y^2 + 1188054 x^4 y^2 + 324576 x^5 y^2 +$
$33152 x^6 y^2 + 777924 x y^3 + 1407672 x^2 y^3 + 941976 x^3 y^3 +$
$276948 x^4 y^3 + 30240 x^5 y^3 + 194481 y^4 + 444528 x y^4 + 354564 x^2 y^4 +$
$119952 x^3 y^4 + 14721 x^4 y^4 + 37044 y^5 + 52920 x y^5 + 24696 x^2 y^5 +$
$3780 x^3 y^5 + 2646 y^6 + 2352 x y^6 + 518 x^2 y^6 + 84 y^7 + 36 x y^7 + y^8$

This produces a large expression containing many terms, each of which is a monomial in **x** and **y**. It would be quite difficult to do this expansion by hand, but it is humanly possible. We can check the number of terms using the **Length** function.

In[5]:= **Length [%]**

Out[5]= 35

Now, factor this large expression. (The command **Factor** is reserved for algebraic expressions. To factor integers use **FactorInteger**.) The expression we want to factor is now two outputs back, so we have to use **%%** to refer to it.

In[6]:= **Factor [%%]**

Out[6]= $(x + y)^4 (21 + 8 x + y)^4$

A quick visual check shows that this agrees with the factored form of our first expression, raised to the fourth power. It would be virtually impossible for a person to find this factorization by hand without knowing where the expression being factored came from. Note that *Mathematica* does not know this either. Human beings are very bad at factoring polynomials in more than one variable, but there is a very efficient machine algorithm for the same purpose. Finally, for completeness, note that there is a case in which **Expand** does not do the expected thing.

In[7]:= **Expand** $\left[\sqrt[3]{x\,y} \right]$

Out[7]= $(x\,y)^{1/3}$

Thus, **Expand** does not distribute fractional powers over products. Instead, one has to use **PowerExpand**.

In[8]:= **PowerExpand [%]**

Out[8]= $x^{1/3}\,y^{1/3}$

3.1.2 Another example

Type in a rational expression; that is, a quotient of polynomials. Note that we have to carefully bracket the numerator and denominator (actually, bracketing the denominator is sufficient here) to get the correct expression, using round brackets, which are reserved just for the purpose of grouping terms. (Try this expression without the outer brackets on top and on the bottom.) This time we give it a name, **exp**, to use in later calculations by typing **exp** = "the expression" (i.e., "=" is used for what is called *assignment* in some computer languages).

In[9]:= **exp** $= \dfrac{(\mathbf{x} - \mathbf{1})^{2}\,(\mathbf{2} + \mathbf{x})}{(\mathbf{1} + \mathbf{x})\,(\mathbf{x} - \mathbf{3})^{2}}$

Out[9]= $\dfrac{(-1 + x)^{2}\,(2 + x)}{(-3 + x)^{2}\,(1 + x)}$

Let's see what **Expand** does to this. Now we refer to **exp** by name rather than using %.

In[10]:= **Expand [exp]**

Out[10]= $\dfrac{2}{(-3 + x)^{2}\,(1 + x)} - \dfrac{3\,x}{(-3 + x)^{2}\,(1 + x)} + \dfrac{x^{3}}{(-3 + x)^{2}\,(1 + x)}$

If **Expand** is applied to a quotient of polynomials, it just expands the numerator and writes each term over a separate copy of the (unexpanded) denominator. There is a command that will expand both numerator and denominator.

In[11]:= **ExpandAll [exp]**

Out[11]= $\dfrac{2}{9 + 3\,x - 5\,x^{2} + x^{3}} - \dfrac{3\,x}{9 + 3\,x - 5\,x^{2} + x^{3}} + \dfrac{x^{3}}{9 + 3\,x - 5\,x^{2} + x^{3}}$

Now we can put these back together in expanded form to get what we may have wanted in the first place.

In[12]:= **Together[%]**

Out[12]= $\dfrac{2 - 3\,x + x^3}{9 + 3\,x - 5\,x^2 + x^3}$

Together just writes fractions over a common denominator. Here is another form of **exp**.

In[13]:= **Apart[exp]**

Out[13]= $1 + \dfrac{5}{(-3 + x)^2} + \dfrac{19}{4\,(-3 + x)} + \dfrac{1}{4\,(1 + x)}$

Apart carries out a partial fractions decomposition of a quotient of polynomials. **Factor** takes us back to the original form of the expression in which both numerator and denominator are factored.

In[14]:= **Factor[%]**

Out[14]= $\dfrac{(-1 + x)^2\,(2 + x)}{(-3 + x)^2\,(1 + x)}$

Finally, we can ask *Mathematica* how it thinks **exp** should be written.

In[15]:= **Simplify[exp]**

Out[15]= $\dfrac{(-1 + x)^2\,(2 + x)}{(-3 + x)^2\,(1 + x)}$

Simplify looks at all possible ways of writing **exp** and returns the one which it thinks is the simplest, based on a certain number of possible rewritings. There is another such general simplification command, **FullSimplify**, that is useful for more complicated expressions. Finally, if we just want to look at the numerator and denominator of **exp** separately, they are given by the commands:

In[16]:= **Numerator[exp]**

Out[16]= $(-1 + x)^2\,(2 + x)$

In[17]:= **Denominator[exp]**

Out[17]= $(-3 + x)^2 (1 + x)$

One of the hardest things to do in any symbolic algebra program is to get the program to display an expression in the form that you want, rather than the form that it wants to give you. The only way to do this is to become thoroughly familiar with the commands that are available and the ways there are to apply them. We shall see some more complicated ways to simplify expressions in Chapter 3.

3.1.3 Yet another example

Type in another expression in expanded form.

In[18]:= **newexp = Expand[(3 + 2 x + y)^3]**

Out[18]= $27 + 54 x + 36 x^2 + 8 x^3 + 27 y + 36 x y + 12 x^2 y + 9 y^2 + 6 x y^2 + y^3$

The following command lets us concentrate on how **x** occurs in the expression.

In[19]:= **Collect[newexp, x]**

Out[19]= $27 + 8 x^3 + 27 y + 9 y^2 + y^3 + x^2 (36 + 12 y) + x (54 + 36 y + 6 y^2)$

Collect[expression, variable] tries to write **expression** as a polynomial in **variable** (here equal to **x**) whose coefficients are expressions in any other variables that are present. The ordering of the output is somewhat unfortunate. Basically, it consists of all of the terms not involving **x** followed by decreasing powers of **x**. This consistent scheme is ruined by putting **x^3** before anything involving **y**. The powers of **y** in the coefficients, however, are arranged in increasing order. But, if we collect coefficients of **y**, then the ordering is just what we want.

In[20]:= **Collect[newexp, y]**

Out[20]= $27 + 54 x + 36 x^2 + 8 x^3 + (27 + 36 x + 12 x^2) y + (9 + 6 x) y^2 + y^3$

It is possible to specify the order of symbols by using the operation **$StringOrder**, but we won't go into that here. It is also possible to collect in two variables simultaneously, but in this case nothing new happens.

In[21]:= **Collect[newexp, {x, y}]**

Out[21]= $27 + 8 x^3 + 27 y + 9 y^2 + y^3 + x^2 (36 + 12 y) + x (54 + 36 y + 6 y^2)$

The following two commands produce the coefficient of **x** in **Collect[newexp, x]** and the highest power of **y** in **newexp**.

In[22]:= **Coefficient[newexp, x]**

Out[22]= $54 + 36\,y + 6\,y^2$

In[23]:= **Exponent[newexp, y]**

Out[23]= 3

3.2 Solving Equations

Manipulating expressions is subsidiary to the main purpose of symbolic programs. Nearly everything that such a program does can be characterized as solving some kind of an equation. The simplest kinds are algebraic equations in one or more variables. *Mathematica* has a very powerful built-in equation solver. Equations are indicated by double equals signs, written ==. (Recall from above that a single equals sign, =, is used for assignment.)

3.2.1 A single equation in one variable

The syntax for solving the equation **2 x - 3 == 5** for the variable **x** is as follows:

In[24]:= **Solve[2 x - 3 == 5, x]**

Out[24]= $\{\{x \rightarrow 4\}\}$

The answer, **x** equals 4, is presented as a list (indicated by the outer curly brackets, which are reserved for lists) of solutions. In this case, there is only one solution, which is itself a list consisting of a *replacement rule*. A replacement rule is an expression of the form $x \leftarrow n$. The meaning is that if **x** is replaced in the equation by the value **n** to the right of the arrow, then the equation is satisfied. To actually carry out the substitution of **4** for **x** in the left-hand side of the equation, one uses "**/.**" which stands for the command **ReplaceAll**. (See Chapter 7 for a thorough discussion of rules.)

In[25]:= **2 x - 3 /. x -> 4**

Out[25]= 5

Quadratic polynomials are treated in exactly the same way.

In[26]:= **Solve[x² - 4 x - 8 == 0, x]**

Out[26]= $\left\{\left\{x \to 2\left(1 - \sqrt{3}\,\right)\right\}, \left\{x \to 2\left(1 + \sqrt{3}\,\right)\right\}\right\}$

Clearly the two substitutions here consist of the values given by the usual quadratic formula. Actually, *Mathematica* can be made to display the general formula just by asking for the solution of a generic quadratic equation with symbolic coefficients.

In[27]:= **Solve[a x² + b x + c == 0, x]**

Out[27]= $\left\{\left\{x \to \dfrac{-b - \sqrt{b^2 - 4\,a\,c}}{2\,a}\right\}, \left\{x \to \dfrac{-b + \sqrt{b^2 - 4\,a\,c}}{2\,a}\right\}\right\}$

So, *Mathematica* has given us the usual formula for solving quadratic equations. Let's try a fourth degree polynomial equation in the variable **x** involving a symbolic constant **a**.

In[28]:= **Solve[x⁴ - 7 x³ + 3 a x² == 0, x]**

Out[28]= $\left\{\{x \to 0\}, \{x \to 0\}, \left\{x \to \dfrac{1}{2}\left(7 - \sqrt{49 - 12\,a}\,\right)\right\}, \left\{x \to \dfrac{1}{2}\left(7 + \sqrt{49 - 12\,a}\,\right)\right\}\right\}$

The result is four exact solutions for **x** in terms of **a**. In this case $x = 0$ is a double root since **x²** is a factor of the equation.

3.2.2 Simultaneous equations in more than one variable

The syntax for a single equation in one variable is **Solve[equation, variable]**. The general form for the arguments of **Solve** consists of a list of equations followed by a list of variables to be solved for. For instance, the general case of two linear equations in variables **x** and **y** has coefficients **a, b, c, d** on the left-hand side and constants **e** and **f** on the right. This gives the general solutions of such a 2 × 2 system.

In[29]:= **Solve[{a x + b y == e, c x + d y == f}, {x, y}]**

Out[29]= $\left\{\left\{x \to -\dfrac{-d\,e + b\,f}{-b\,c + a\,d}, \; y \to -\dfrac{c\,e - a\,f}{-b\,c + a\,d}\right\}\right\}$

The solution is unique, so it consists of a list with one entry consisting of a list of two substitutions, one for each of **x** and **y**.

3.2.3 Exact, closed form solutions

Mathematica can deal with much more complicated equations. Here is a system consisting of a second degree and a third degree polynomial in two variables.

In[30]:= **Solve[{x^3 + y^3 == 1, x^2 + y^2 == 1}, {x, y}]**

Out[30]= $\{\{x \to 0, y \to 1\}, \{x \to 0, y \to 1\}, \{x \to 1, y \to 0\},$

$\{x \to 1, y \to 0\}, \{x \to \dfrac{-5\,I + \sqrt{2}}{4\,I + \sqrt{2}}, y \to \dfrac{1}{2}\left(-2 + I\,\sqrt{2}\right)\},$

$\{x \to \dfrac{5\,I + \sqrt{2}}{-4\,I + \sqrt{2}}, y \to \dfrac{1}{2}\left(-2 - I\,\sqrt{2}\right)\}\}$

The result this time is a list of six solutions, each solution consisting of a list of two substitutions, one for each of **x** and **y**. Note that two of the solutions occur with multiplicity 2.

   Mathematica can give us a picture of this pair of equations, but we have to use a command that is found in one of the packages rather than built in to the kernel. Such packages have to be loaded before they can be used. This is done as follows:

In[31]:= **Needs["Graphics`ImplicitPlot`"]**

Use this on the pairs of equations whose simultaneous solution we found above.

In[32]:= **ImplicitPlot[{x^3 + y^3 == 1, x^2 + y^2 == 1}, {x, -2, 2}];**

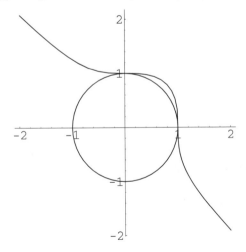

In this remarkable picture, we can see the two double solutions at (0, 1) and (1, 0). The two complex solutions of course are not in the picture.

3.2.4 An impossible equation

However, not all polynomial equations, even in one variable, have exact solutions.

In[33]:= $\mathbf{Solve[1 + 8\,x^3 + x^5 - 2\,x^6 + 4\,x^7 == 0,\ x]}$

Out[33]= $\left\{\left\{x \to \dfrac{1}{4}\left(1 - I\sqrt{3}\right)\right\}, \left\{x \to \dfrac{1}{4}\left(1 + I\sqrt{3}\right)\right\},\right.$

$\qquad \left\{x \to \text{Root}[1 + 2\,\#1 + \#1^5\ \&,\ 1]\right\},$

$\qquad \left\{x \to \text{Root}[1 + 2\,\#1 + \#1^5\ \&,\ 2]\right\}, \left\{x \to \text{Root}[1 + 2\,\#1 + \#1^5\ \&,\ 3]\right\},$

$\qquad \left.\left\{x \to \text{Root}[1 + 2\,\#1 + \#1^5\ \&,\ 4]\right\}, \left\{x \to \text{Root}[1 + 2\,\#1 + \#1^5\ \&,\ 5]\right\}\right\}$

Here, we gave *Mathematica* a seventh degree equation in **x**. It found two solutions, which left a fifth degree equation to be solved. It is well known from the theory of equations that equations of degree 4 or less have exact, closed form solutions in terms of roots of expressions constructed from the coefficients. However, as Galois showed, for equations of degree 5 or more, there need be no such solution. These solutions are expressed in the form: x → Root[equation, n]. In the expression, #1 stands for a generic (first) variable and n counts which root is being described; thus, the equation could be written more legibly as 1 + 2 x + x^5 = 0. This is a typical example of a fifth degree equation which has no exact, closed form solution. See Chapter 3 for more details about **Root**. Of course, a polynomial equation can be solved for all of its roots by numerical methods. The effect of **N[]** here is to find all seven of these roots.

In[34]:= $\mathbf{N[\%]}$

Out[34]= $\{\{x \to 0.25 - 0.433013\,I\}, \{x \to 0.25 + 0.433013\,I\}, \{x \to -0.486389\},$

$\qquad \{x \to -0.701874 - 0.879697\,I\}, \{x \to -0.701874 + 0.879697\,I\},$

$\qquad \{x \to 0.945068 - 0.854518\,I\}, \{x \to 0.945068 + 0.854518\,I\}\}$

3.3 Trigonometry

Hardly anybody thinks that trigonometry is their favorite subject. Pocket calculators have eliminated the extensive tables and interpolation formulas that previously were the bane of trying to use actual values of trigonometric functions. Modern programs let us calculate values to any desired precision and make arbitrarily detailed plots of these values. All of the standard trigonometric functions are found as built-in operations. If they are given real arguments, they return real values, just like an ordinary pocket calculator.

In[35]:= $\mathbf{Sin[1.3]}$

Out[35]= 0.963558

The normal, Standard form for this input is **Sin[1.3]**. However, if we type this in a new cell, select it, and then choose the Menu item **Cell ▷ ConvertTo ▷ TraditionalForm**, then we get an input that looks like the usual mathematical notation. We use this procedure on the next several cells.

In[36]:= **sin(1.3)**

Out[36]= 0.963558

Mathematica also knows about their complex values for complex arguments, which is more than most pocket calculators know. For instance, consider a product of a cos and a tan. (The space indicates multiplication.)

In[37]:= **cos(3.2 + 5.1 *i*) tan(0.4 + 3.7 *i*)**

Out[37]= -4.8548 - 81.8002 I

Furthermore, the built-in **Plot** command lets us make pictures of trigonometric functions.

In[38]:= **sinplot = Plot[sin(*x*), {*x*, 0, 2 π}]**

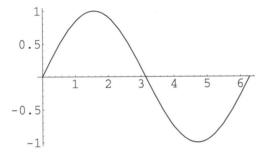

Out[38]= - Graphics -

Plot takes two arguments, the first being a numerical function of one variable and the second being a list of a special form called an *iterator*. (The same form was used earlier in **Implicit-Plot**.) The iterator, **{x, 0, 2 π}**, means that the variable **x** is to take values between **0** and **2** π. Note that the output consists of the term -Graphics-, while the picture is an extra, side effect of the command. *Mathematica* knows how to deal with plots of singular functions as well; for instance:

In[39]:= **tanplot = Plot[tan(*x*), {*x*, 0, 2 π}]**

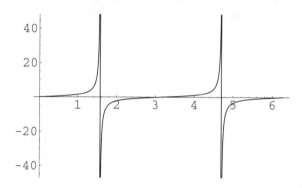

Out[39]= - Graphics -

Mathematica has decided on its own to show values only up to about 44. We'll see later how to increase or decrease that if desired. The function **Show** takes the names of a number of pictures and combines them in the same drawing, which is why we gave names to the preceding plots. It adjusts the scales of the drawing so they fit together correctly.

In[40]:= **Show[sinplot, tanplot]**

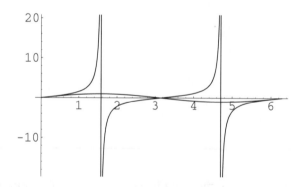

Out[40]= - Graphics -

Notice how *Mathematica* has decreased the maximum **y** values that are shown in order to see what is happening to the sin curve.

Another way to see two plots together is to use **GraphicsArray**, which takes a list (actually a matrix) of names of graphics objects and creates a new graphics object consisting of all of the individual graphics objects scaled to the same size. **Show** displays this in a rectangular array.

In[41]:=`Show[GraphicsArray[{sinplot, tanplot}]];`

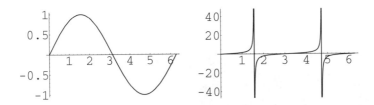

Trigonometric functions can also be used to make interesting three-dimensional plots as well. The syntax is the obvious extension of the two-dimensional case.

In[42]:= **Plot3D[sin(x) sin(3y), {x, −2, 2}, {y, −2, 2}]**

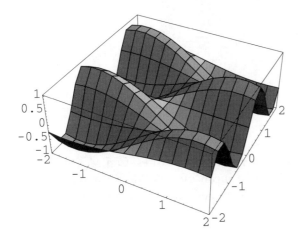

Out[42]= **-** SurfaceGraphics **-**

Here is another way to illustrate the same function.

In[43]:= **ContourPlot[sin(*x*) sin(3 *y*), {*x*, −2, 2}, {*y*, −2, 2}]**

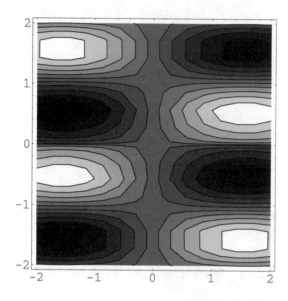

Out[43]= **-** ContourGraphics **-**

Of course, there is more to trigonometry than just pictures. Can *Mathematica* prove trigonometric identities? It depends on what you mean by this. Modifications of the **Expand** and
Factor functions we used earlier will handle many cases of simplifying trigonometric expressions. For instance:

In[44]:= **Expand[cos^2(*x*) + sin^2(*x*), Trig → True]**

Out[44]= 1

In[45]:= **Factor[tan(2 *x*), Trig → True]**

Out[45]= $\dfrac{2 \cos(x) \sin(x)}{(\cos(x) - \sin(x))\,(\cos(x) + \sin(x))}$

Here we have converted the output cell to **TraditionalForm** as well. This output can be
improved by using **ExpandAll**.

In[46]:= **ExpandAll[%]**

$$Out[46]= \frac{2\cos(x)\sin(x)}{\cos^2(x) - \sin^2(x)}$$

The extra arguments to **Expand** and **Factor** are called *optional arguments*. They are an important feature of *Mathematica* operations. Trigonometric identities can be checked by *Mathematica*.

However, proving trigonometric identities should mean that it is possible to check an identity like $\frac{\cos z}{1+\cos z} = \frac{\sin z}{\sin z + \tan z}$. *Mathematica* is not able to make substitutions and turn the left-hand side into the right-hand side by itself, which is what you might mean by proving such an identity. However, you can subtract the right-hand side from the left-hand side and use **Simplify**, hoping that the result will be 0. (**Simplify** also takes an optional argument for trigonometric simplification, but the default value is **True**, so we don't have to specify it explicitly.)

In[47]:= **Simplify** $\left[\dfrac{\cos(z)}{\cos(z) + 1} - \dfrac{\sin(z)}{\sin(z) + \tan(z)} \right]$

Out[47]= 0

Identities that are surprisingly complex can be handled this way.

4 College Calculus, Differential Equations, and Linear Algebra

College mathematics means calculus to most people, and that is what most people expect symbolic computation programs to do. As soon as early symbolic computation programs could do anything at all, it was realized that symbolic integration posed a major challenge. Symbolic differentiation is very simple (we'll use it to illustrate different styles of programming) but there are still aspects of integration which have no easy answer. The first commercially successful symbolic computation program, Macsyma, grew out of these early efforts in the 1960s to teach a program to integrate, first as well as an MIT freshman, then as well as an MIT graduate, and finally as well as the most knowledgeable expert. Humans integrate functions mostly by experience. Fortunately, there is an algorithm, the Reisz algorithm, that is very well suited to integration by computers. Essentially it guesses the form of the answer and then finds the precise result by means of undetermined coefficients. The more kinds of functions that are available in the guessing stage, the more functions that can be successfully integrated. Current efforts to complete this endeavor center around the treatment of situations where the form of the answer depends on the values of symbolic parameters in the integrand.

4.1 Integration, Differentiation, Series, and Limits

Mathematica of course carries out the standard operations of calculus in symbolic form. The command to find the antiderivative, or indefinite integral, of f[x] with respect to x is Integrate[f[x], x]. Using a palette, or converting a cell with such an expression, gives the usual notation for an integral. For instance:

$$\text{In[1]:= } \mathbf{int} = \int \frac{x}{1 - x^3} \, dx$$

$$\text{Out[1]= } -\frac{\text{ArcTan}\left[\frac{1+2\,x}{\sqrt{3}}\right]}{\sqrt{3}} - \frac{1}{3}\,\text{Log}[-1 + x] + \frac{1}{6}\,\text{Log}[1 + x + x^2]$$

The output here is in the standard Mathematica output form. If this cell is converted to **TraditionalForm** using the Menu item, then the result appears as follows:

$$-\frac{\tan^{-1}\left(\frac{2\,x+1}{\sqrt{3}}\right)}{\sqrt{3}} - \frac{1}{3}\log(x-1) + \frac{1}{6}\log(x^2 + x + 1)$$

Note that the answer omits the constant of integration that all freshmen are told is required.

Differentiation is the inverse operation to integration. It is one of the few commands that are abbreviated in Mathematica, being denoted just by D. Thus, D[f[x], x] means df(x)/dx. A good way to check the operation of integration is to differentiate the result, so differentiate the previous integral. The Standard syntax for the derivative of the preceding expression, int, with respect to x is D[int, x]. If such a cell is converted to **TraditionalForm**, then a partial differentiation operator is used instead of the usual "d".

$$\text{In[2]:= } \frac{\partial \mathbf{int}}{\partial x}$$

$$\text{Out[2]= } -\frac{1}{3\,(-1 + x)} + \frac{1 + 2\,x}{6\,(1 + x + x^2)} - \frac{2}{3\left(1 + \frac{1}{3}\,(1 + 2\,x)^2\right)}$$

This doesn't look like the function we started with, but, after simplification, we get back the original expression.

$$\text{In[3]:= } \mathbf{Simplify[\%]}$$

$$\text{Out[3]= } \frac{x}{1 - x^3}$$

Higher order derivatives are given by `D[f[x], {x, n}]`, where **n** is the order. Again, **TraditionalForm** gives the expected expression for a second derivative. Thus, the second derivative of `int` is

In[4]:= **Simplify** $\left[\dfrac{\partial^2 \text{int}}{\partial x^2} \right]$

Out[4]= $\dfrac{1 + 2\,x^3}{\left(-1 + x^3\right)^2}$

The expression `Integrate[f[x], {x, a, b}]` gives the definite integral of `f[x]` with respect to **x** from **a** to **b**. Similarly, `NIntegrate` finds numerical values of definite integrals of functions even if there is no closed form for their indefinite integral. We continue converting such inputs to **TraditionalForm**.

In[5]:= $\displaystyle\int_0^\pi \sin(x)\,d\,x$

Out[5]= 2

In[6]:= **NIntegrate**$[\sin(\sin(x)), \{x, 0, \pi\}]$

Out[6]= 1.78649

The command `Series[f[x], {x, a, n}]` finds the first **n** terms of the Taylor's series expansion of `f[x]` about the point **a**.

In[7]:= **Series**$[e^{-x} \sin(2\,x), \{x, 0, 6\}]$

Out[7]= $2\,x - 2\,x^2 - \dfrac{x^3}{3} + x^4 - \dfrac{19\,x^5}{60} - \dfrac{11\,x^6}{180} + O[x]^7$

The command `Limit[f[x], x -> a]` finds the limit of `f[x]` as **x** approaches **a**.

In[8]:= $\displaystyle\lim_{x \to 0} \dfrac{\sin(x) - \tan(x)}{x^3}$

Out[8]= $-\dfrac{1}{2}$

4.2 *Calculus of several variables*

Mixed derivatives are easily calculated. Start with some expression in **x** and **y**. We don't need to see it repeated as output so we suppress the output by following the definition with a semicolon.

In[9]:= **exp** = $x^3 \sin(y^4)$;

The mixed partial derivative of **exp** with respect to **x** and then **y** is given by using the same symbol **D** that is used for ordinary derivatives, with an extra argument for the second variable. The following cell is made by converting **D[exp, x, y]** to **TraditionalForm**.

In[10]:= $\dfrac{\partial^2 \exp}{\partial x\, \partial y}$

Out[10]= $12\, x^2\, y^3\, \mathrm{Cos}\,[y^4]$

Now differentiate twice with respect to **x** and three times with respect to **y**. Standard notation is **D[exp, {x, 2}, {y, 3}]**. Converting as usual gives

In[11]:= $\dfrac{\partial^5 \exp}{\partial x^2\, \partial y^3}$

Out[11]= $144\, x\, y\, \mathrm{Cos}\,[y^4] - 384\, x\, y^9\, \mathrm{Cos}\,[y^4] - 864\, x\, y^5\, \mathrm{Sin}\,[y^4]$

Just as **D** denotes ordinary or partial differentiation, **Integrate** denotes single or multiple integration. Converting **Integrate[E^(-2x) Cos[y], x, y]** gives

In[12]:= $\displaystyle\int\int e^{-2x}\cos(y)\,dy\,dx$

Out[12]= $-\dfrac{1}{2}\, E^{-2x}\, \mathrm{Sin}\,[y]$

Multiple definite integration uses two (or more) iterators. The Standard notation **Integrate[·E^(-2x) Cos[y], {x, 0, Pi/4}, {y, 0, x}]** gives

In[13]:= $\displaystyle\int_0^{\frac{\pi}{4}}\int_0^x e^{-2x}\cos(y)\,dy\,dx$

Out[13]= $\dfrac{1}{5} - \dfrac{3\, E^{-\pi/2}}{5\, \sqrt{2}}$

Notice that the second integration in the linear notation is performed first; that is, this result is the same as the iterated integral:

In[14]:= `Integrate[Integrate[E^(-2x) Cos[y], {y, 0, x}],`
          `{x, 0, Pi/4}]`

Out[14]= $\dfrac{1}{5} - \dfrac{3 \, E^{-\pi/2}}{5 \sqrt{2}}$

All of this converting from **StandardForm** to **TraditionalForm** can be eliminated by choosing the Menu items **Cell ▷ Default Input Format Type ▷ TraditionalForm** and **Cell ▷ Default Output Format Type ▷ TraditionalForm.** But then one has to learn how to actually type **TraditionalForm** input from the keyboard. This will be discussed briefly in Chapter 2.

4.3 *Differential Equations*

From one point of view, mathematics education is a long line of development leading from counting to differential equations. It is differential equations that allow the prediction of the future, so they are a crucial ingredient in everything from ballistics to bridge construction, automobile controls to economic forecasts and weather prediction. The ultimate test of a symbolic computation program is how it deals with them. Integration, of course, is a special case of solving a differential equation, namely, one of the form **y' = expression**. Again, Mac-Syma was the first symbolic program to easily handle a large class of differential equations. It tried to recognize the form of an equation and then apply a solution method adapted to that form. It told you as it went along what methods it was trying. As with integration, the more functions that are available to use in the answers, the more equations that can be solved. *Mathematica*'s current capacities, which are astonishing, are based on hypergeometric functions.

The command to solve differential equations is **DSolve**. Here is a typical second order, linear, non-homogeneous differential equation.

In[15]:= `diffeq1 = y''[x] - 5 y'[x] + 6 y[x] == 2 E^x;`

Differentiation is indicated by primes, and it is necessary to include the independent variable **x** in the expression for the dependent variable **y[x]**. The syntax for a single differential equation is **DSolve[equation, dependent variable, independent variable].** Thus:

In[16]:= `DSolve[diffeq1, y[x], x]`

Out[16]= $\{\{y[x] \rightarrow E^x + E^{2\,x}\, C[1] + E^{3\,x}\, C[2]\}\}$

Constants of integration are called `C[1]` and `C[2]` here.

Mathematica can also handle certain non-linear equations, even with symbolic constants.

In[18]:= **diffeq2 = y'[x] + a x y[x]^2 == 0;**

In[19]:= **DSolve[diffeq2, y[x], x]**

Out[19]= $\left\{\left\{y[x] \rightarrow \dfrac{2}{a\,x^2 - 2\,C[1]}\right\}\right\}$

If we give *Mathematica* a Bessel's differential equation, it recognizes it immediately.

In[20]:= **diffeq3 = x^2 y''[x] + x y'[x] + x^2 y[x] == 0;**

In[21]:= **DSolve[diffeq3, y[x], x]**

Out[21]= $\left\{\left\{y[x] \rightarrow \text{BesselJ}\left[0, \sqrt{x^2}\,\right] C[1] + \text{BesselY}[0, x]\,C[2]\right\}\right\}$

Here `BesselJ[0, x]` and `BesselY[0, x]` are the usual zeroth order Bessel functions. *Mathematica* knows all about these functions, as well as all the other usual functions that arise in physics and engineering. For instance, we can plot both of them together by giving them as a list to the **Plot** command.

In[22]:= **Plot[{BesselJ[0, x], BesselY[0, x]}, {x, 0, 10}]**

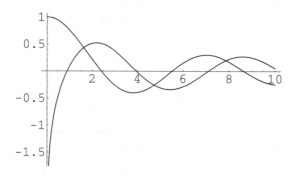

Out[22]= - Graphics -

Even if a differential equation cannot be solved exactly, it may be possible to solve it numeri-cally. There is a built-in function **NDSolve** to do this. It works with systems of equations, and equations have to be added specifying the initial conditions. Here is an example.

```
In[23]:= diffeqSystem =
             {x'[t] == -y[t] - x[t]^2,
              y'[t] == 2 x[t] - y[t],
              x[0] == y[0] == 1};
```

In the **NDSolve** command, we have to give the system of equations, the dependent variables (here **x** and **y**) and the range of the independent variable (here **t**).

```
In[24]:= solution = NDSolve[diffeqSystem, {x, y}, {t, 0, 10}]

Out[24]= {{x → InterpolatingFunction[{{0., 10.}}, <>],
            y → InterpolatingFunction[{{0., 10.}}, <>]}}
```

The answer is expressed in terms of Interpolating Functions for **x** and **y** as functions of **t**. We can use these to find individual values of the solution at some point, e.g., **t** = 3, by substituting the interpolating functions for **x** and **y**.

```
In[25]:= {x[3], y[3]} /. solution

Out[25]= {{-0.139737, -0.517751}}
```

It is much more interesting to plot the solution using the built-in command for plotting a parametric curve.

```
In[26]:= ParametricPlot[Evaluate[{x[t], y[t]} /. solution],
                 {t, 0, 10},
                 PlotRange -> All]
```

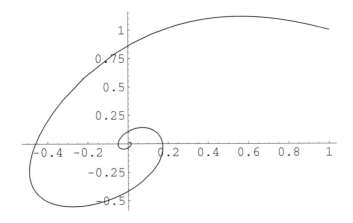

```
Out[26]= - Graphics -
```

The reason for **Evaluate** in this command will be explained later.

4.4 Lists

Lists are a very important built-in data type in *Mathematica*. They are used for themselves and to represent vectors and matrices. As we have seen, lists are indicated by curly brackets.

In[27]:= **{a, b, c}**

Out[27]= {a, b, c}

A convenient way to construct a list whose elements are given by some mathematical formula is to use the **Table** command.

In[28]:= **Table[Expand[(1 + x)^n], {n, 1, 8}]**

Out[28]= $\{1 + x, 1 + 2 x + x^2, 1 + 3 x + 3 x^2 + x^3, 1 + 4 x + 6 x^2 + 4 x^3 + x^4,$
$1 + 5 x + 10 x^2 + 10 x^3 + 5 x^4 + x^5, 1 + 6 x + 15 x^2 + 20 x^3 + 15 x^4 + 6 x^5 + x^6,$
$1 + 7 x + 21 x^2 + 35 x^3 + 35 x^4 + 21 x^5 + 7 x^6 + x^7,$
$1 + 8 x + 28 x^2 + 56 x^3 + 70 x^4 + 56 x^5 + 28 x^6 + 8 x^7 + x^8\}$

The command **TableForm** will display this in a nicer format.

In[29]:= **TableForm[%]**

Out[29]//TableForm=
$1 + x$
$1 + 2 x + x^2$
$1 + 3 x + 3 x^2 + x^3$
$1 + 4 x + 6 x^2 + 4 x^3 + x^4$
$1 + 5 x + 10 x^2 + 10 x^3 + 5 x^4 + x^5$
$1 + 6 x + 15 x^2 + 20 x^3 + 15 x^4 + 6 x^5 + x^6$
$1 + 7 x + 21 x^2 + 35 x^3 + 35 x^4 + 21 x^5 + 7 x^6 + x^7$
$1 + 8 x + 28 x^2 + 56 x^3 + 70 x^4 + 56 x^5 + 28 x^6 + 8 x^7 + x^8$

A list of numbers in sequence can also be constructed by the **Range** command.

In[30]:= **Range[5, 20]**

Out[30]= {5, 6, 7, 8, 9, 10, 11, 12, 13, 14, 15, 16, 17, 18, 19, 20}

There are many operations that take lists as arguments; for instance:

In[31]:=**Permutations[{a, b, c}]**

Out[31]= {{a, b, c}, {a, c, b}, {b, a, c}, {b, c, a}, {c, a, b}, {c, b, a}}

In[32]:=**Flatten[%]**

Out[32]= {a, b, c, a, c, b, b, a, c, b, c, a, c, a, b, c, b, a}

4.5 Vectors

Vectors do not appear in *Mathematica* as a separate data type but are represented as lists. For instance, the dot product of two vectors is given by typing a dot between the vectors.

In[33]:=**{x, y, z} . {a, b, c}**

Out[33]= a x + b y + c z

Vectors can be added and multiplied by scalars in the usual way.

In[34]:=**{a, b, c} + {1, 2, 3}**

Out[34]= {1 + a, 2 + b, 3 + c}

In[35]:=**4 {a, b, c}**

Out[35]= {4 a, 4 b, 4 c}

4.6 Matrices

One reason for the "unreasonable effectiveness" of mathematics in science is the observation that many phenomena can be described quite effectively in linear terms. Linear algebra is the part of mathematics that deals with this. There are large and important Fortran and C programs that deal with numerical linear algebra, and *Mathematica*'s facilities in this direction, although effective, are no substitute for these packages. However, one of the main purposes of symbolic programs is to deal with symbolic linear algebra; for instance, matrices with symbolic rather than numeric entries.

Matrices also do not appear in *Mathematica* as a separate data type. Rather, they are represented as lists of lists. For instance:

In[36]:= {{1, 2, 3}, {4, 5, 6}, {7, 8, 9}}

Out[36]= {{1, 2, 3}, {4, 5, 6}, {7, 8, 9}}

The commands **TableForm** and **MatrixForm** display the output as a two-dimensional table.

In[37]:= **TableForm[%]**

Out[37]//TableForm=

1	2	3
4	5	6
7	8	9

Alternatively, matrices can be input in rectangular form by using a menu item or palette items. For instance, choosing the menu item **Input ▷ Create Table/Matrix/Palette...** brings up a dialogue box. Clicking the **Matrix** button and accepting the default numbers of rows and columns gives an input cell with the following structure.

$$\begin{pmatrix} \square & \square & \square \\ \square & \square & \square \\ \square & \square & \square \end{pmatrix}$$

Clicking on the first small box here selects for entering an expression. The Tab key steps through the other positions.

As with lists themselves, matrices can be constructed by the **Table** command when the entries are given by some mathematical formula. Here is the 3 × 3 Hilbert matrix.

In[38]:= **matrix = Table$\left[\dfrac{1}{i + j - 1},\ \{i, 1, 3\},\ \{j, 1, 3\}\right]$**

Out[38]= $\left\{\left\{1, \dfrac{1}{2}, \dfrac{1}{3}\right\}, \left\{\dfrac{1}{2}, \dfrac{1}{3}, \dfrac{1}{4}\right\}, \left\{\dfrac{1}{3}, \dfrac{1}{4}, \dfrac{1}{5}\right\}\right\}$

In[39]:= **matrix//TableForm**

Out[39]//TableForm=

1	$\frac{1}{2}$	$\frac{1}{3}$
$\frac{1}{2}$	$\frac{1}{3}$	$\frac{1}{4}$
$\frac{1}{3}$	$\frac{1}{4}$	$\frac{1}{5}$

Instead of the prefix form **TableForm[]**, we have used the suffix form of function application **//TableForm** here.

If we think of **matrix** as a matrix, then we can carry out matrix operations on it.

In[40]:= **Inverse[matrix]**

Out[40]= {{9, -36, 30}, {-36, 192, -180}, {30, -180, 180}}

Matrix multiplication is also represented by a dot, so the following calculation checks that the preceding result is the inverse of **matrix**.

In[41]:=**% . matrix // TableForm**

Out[41]//TableForm=

1	0	0
0	1	0
0	0	1

Given a matrix, we can calculate its eigenvalues by the usual procedure of solving its characteristic polynomial. Recall that the characteristic polynomial of a matrix is the determinant of the matrix given by subtracting **x** from each diagonal entry of the original matrix. We'll use this procedure to find the eigenvalues of **matrix**.

In[42]:=**matrix - x IdentityMatrix[3]//TableForm**

Out[42]//TableForm=

$1 - x$	$\frac{1}{2}$	$\frac{1}{3}$
$\frac{1}{2}$	$\frac{1}{3} - x$	$\frac{1}{4}$
$\frac{1}{3}$	$\frac{1}{4}$	$\frac{1}{5} - x$

IdentityMatrix[n] is the n × n identity matrix, as one might expect. Multiplying it by **x** gives a matrix with **x**'s on the main diagonal and **0**'s elsewhere. Subtracting the resulting matrix from **matrix** gives the desired result (since subtraction of matrices of the same size subtracts corresponding entries). Next, we calculate the determinant of this matrix by using the command **Det** (which is another of the rare abbreviations in *Mathematica*).

In[43]:= **Det[%]**

Out[43]= $\dfrac{1}{2160} - \dfrac{127\,x}{720} + \dfrac{23\,x^2}{15} - x^3$

Actually, there is a built-in command to find the characteristic polynomial.

In[44]:=**CharacteristicPolynomial[matrix, x]**

Out[44]= $\dfrac{1}{2160} - \dfrac{127\,x}{720} + \dfrac{23\,x^2}{15} - x^3$

Since the coefficients of **matrix** are rational numbers, these calculations yield a polynomial with rational coefficients. This third degree polynomial has an exact solution in terms of roots of these coefficients, but the resulting answer fills a whole screen, so we content ourselves with numerical approximations to the roots. There is a special command to find numerical solutions of equations.

In[45]:=**NSolve[% == 0, x]**

Out[45]= $\{\{x \to 0.00268734\}, \{x \to 0.122327\}, \{x \to 1.40832\}\}$

These are the eigenvalues of **matrix** by definition. Of course, there is also a built-in function to calculate the eigenvalues of a matrix. It gives the results in a different order and different form.

In[46]:=**Eigenvalues[N[matrix]]**

Out[46]= $\{1.40832, 0.122327, 0.00268734\}$

One can also calculate the exact eigenvalues. We'll just look at the first one, which is chosen by the **[[1]]** following the command to calculate the eigenvalues without the **N[]**.

In[47]:=**eigen1 = Eigenvalues[matrix][[1]]**

Out[47]= $\dfrac{23}{45} + \dfrac{6559}{180\left(517148 + 405\,I\,\sqrt{89799}\right)^{1/3}} + \dfrac{1}{180}\left(517148 + 405\,I\,\sqrt{89799}\right)^{1/3}$

This appears to have a non-trivial complex component. Finding its numerical value shows otherwise.

In[48]:=**N[eigen1]**

Out[48]= $1.40832 + 0.\,I$

The command **Eigenvalues** also works for matrices with symbolic entries. For instance, try a general 2 × 2 matrix.

In[49]:=**Eigenvalues[{{a, b}, {c, d}}] // Simplify**

Out[49]= $\left\{ \frac{1}{2} \left(a + d - \sqrt{a^2 + 4bc - 2ad + d^2} \right), \frac{1}{2} \left(a + d + \sqrt{a^2 + 4bc - 2ad + d^2} \right) \right\}$

5 *Graduate School*

Most of the entries in the following long list of built-in functions would not be encountered in a typical undergraduate mathematics course.

```
AiryAi, AiryAiPrime, AiryBi, AiryBiPrime, ArithmeticGeometricMean,
BernoulliB, BesselI, BesselJ, BesselK, BesselY, Beta,
BetaRegularized, Catalan, ChebyshevT, ChebyshevU, ClebschGordan,
CoshIntegral, CosIntegral, DivisorSigma, EllipticE, EllipticExp,
EllipticExpPrime, EllipticF, EllipticK, EllipticPi, EllipticTheta,
EllipticThetaC, EllipticThetaD, EllipticThetaN,
EllipticThetaPrime, EllipticThetaS, Erf, Erfc, Erfi, EulerE,
ExpIntegralE, ExpIntegralEi, FresnelC, FresnelS, Gamma,
GammaRegularized, GegenbauerC, GroebnerBasis, HermiteH,
HypergeometricPFQ, HypergeometricPRQRegularized, HypergeometricU,
Hypergeometric0F1, Hypergeometric0F1Regularized,
Hypergeometric1F1, Hypergeometric1F1Regularized,
Hypergeometric2F1, Hypergeometric2F1Regularized, InverseJacobiCD,
InverseJacobiCN, InverseJacobiCS, InverseJacobiDC,
InverseJacobiDN, InverseJacobiDS, InverseJacobiNC,
InverseJacobiND, InverseJacobiNS, InverseJacobiSC,
InverseJacobiSD, InverseJacobiSN, InverseWeierstrassP,
JacobiAmplitude, JacobiCD, JacobiCN, JacobiCS, JacobkDC, JacobiDN,
JacobiDS, JacobiNC, JacobiND, JacobiP, JacobiSC, JacobiSD,
JacobiSn, JacobiSymbol, JacobiZeta, JordanDecomposition,
LaguerreL, LatticeReduce, LegendreP, LegendreQ, LerchPhi,
LogGamma, LogIntegral, LUBackSubstitution, LUDecomposition,
MoebiusMu, NBernoulliB, Pochhammer, PolyGamma, PolyLog,
PseudoInverse, QRDecomposition, Resultant, RiemannSiegelTheta,
RiemannSiegelZ, SchurDecomposition, SimplifyGamma,
SimplifyPolyGamma, SinhIntegral, SinIntegral, SixJSymbol,
SphericalHarmonicY, StirlingS1, StirlingS2, ThreeJSymbol,
WeierstrassP, WeierstrassPPrime, Zeta
```

In reading over this list, what strikes one is the preponderance of functions from physics, number theory, and algebraic geometry. There are a few general operations like **GroebnerBasis** or **LUDecomposition**, but mainly these functions serve as a substitute for specialized tables, just as the more common operations **Sin**, **Cos**, etc. are substitutes for tables of constant everyday use. The moral is that, unless your use of mathematics is restricted to the kinds of operations sketched in this chapter or to the specialized functions mentioned here, you will have to program the mathematics you want yourself. Fortunately, *Mathematica* has a powerful, highly developed programming language that permits programs to be written in a variety of styles. These programming facilities are the main subject of the second part of this book.

6 Practice

In learning any language, one of the most important things is to practice simple phrases and statements until they become second nature. In an interpreted programming language, this means typing simple commands into the language until you are thoroughly familiar with simple aspects of the syntax of the language, such as using brackets correctly, typing functions with capital letters, separating variables with commas, etc. Here are a few things to try for practice. Whenever it makes sense, try converting the input and/or output cells to **TraditionalForm**.

```
 1.  2 + 2
 2.  2 - 2
 3.  3 5
 4.  4 / 8
 5.  3.14159 + 2.3456
 6.  3^10
 7.  3.14159^10
 8.  (3 + 2 I)^10
 9.  Pi
10.  N[Pi]
11.  N[Pi, 100]
12.  Sqrt[2]
13.  Sqrt[2.0]
14.  N[Sqrt[2], 20]
15.  Sin[2]
16.  Sin[2.0]
17.  E^(Pi I)
18.  N[E^Pi > Pi^E]
19.  1 + 2 3
20.  (1 + 2) 3
21.  1 / 2 - 3
22.  1 / (2 - 3)
```

```
23. 3^10000
24. %^(1/10000)
25. 2/5 + 3.0/8
26. Random[]
27. Round[N[23^(2/3)]]
28. Ceiling[N[23^(2/3)]]
29. Plot3D[Sin[x y], {x, 0, Pi}, {y, 0, Pi}]
30. PSPrint[%] (in Unix systems)
31. Eigenvalues[{{a, b, 1}, {-b, 2, -a}, {b, 0, -a}}]
32. D[x^2, x]
33. D[x^2 y^3, x, y]
34. D[x^2, {x, 2}]
35. D[x^2 y^3, {x, 2}, {y, 3}]
36. Integrate[x^2, x]
37. Integrate[x^2 y^3, x, y]
38. Integrate[Sin[x], {x, 0, Pi}]
39. Integrate[Sin[x] y, {x, 0, 2 Pi}, {y, 0, 2}]
40. Integrate[Sin[x] y, {x, 0, Pi}, {y, 0, x}]
41. Integrate[Sin[x] y, {x, 0, y}, {y, 0, 2}]
42. Series[ Exp[-x] Sin[2x], {x, Pi I, 6} ]
43. Table[i^3, {i, 1, 10}]
44. m = Table[1 / (i + j), {i, 1, 3}, {j, 1, 3}]
45. m . m//TableForm
46. m . Inverse[m]//TableForm
47. {{a, b}, {c, d}} + {{1, 2}, {3, 4}}
48. {{a, b}, {c, d}} - {{1, 2}, {3, 4}}
49. {{a, b}, {c, d}} {{1, 2}, {3, 4}}
50. {{a, b}, {c, d}} / {{1, 2}, {3, 4}}
```

7 Exercises

Find the inputs and outputs to *Mathematica* that solve the following problems.

1. i) Factor the polynomial $1 - x^{10}$.

ii) Investigate the factors of polynomials of the form $1 - x^n$ for n between 1 and 10 by making a suitable table.

2. Use *Mathematica* to verify the following trigonometric identities. (Hint: subtract the right-hand side from the left-hand side.)

(i) $\dfrac{1 - \cos(2\,t)}{\cos(2\,t) + 1} = \tan^2(t)$

(ii) $(\cot(t) + \csc(t))^2 = \dfrac{\cos(t) + 1}{1 - \cos(t)}$

(iii) $\dfrac{(\cos^3(t) + \sin^3(t))}{\cos(t) + \sin(t)} = 1 - \sin(t)\cos(t)$

3. Use *Mathematica* to calculate the following integrals. In each case differentiate the result to check the answer if possible. Use **Simplify**, **Factor**, **Together**, etc. wherever it seems appropriate.

$$\int \frac{x^2 + 5}{x^5 + x^4 - x - 1}\, dx$$

$$\int \frac{\sqrt{x^2 - 1}}{x^6}\, dx$$

4. a) Convince *Mathematica* to display the expression
   $(a + b)\,((c + d\,x)\,x + e\,x^2)$ in the following forms:

   i) $a\,c\,x + b\,c\,x + a\,d\,x^2 + b\,d\,x^2 + a\,e\,x^2 + b\,e\,x^2$

   ii) $(a + b)\,c\,x + (a\,d + b\,d + a\,e + b\,e)\,x^2$

   iii) $(a + b)\,x\,(c + d\,x + e\,x)$

   b) Let $\exp = \left(\frac{ab+cd}{ab-cd}\right)^2 + \left(\frac{ab-cd}{ab+cd}\right)^2$. Convince *Mathematica* to display exp as

   i) $\dfrac{2\,(a^4\,b^4 + 6\,a^2\,b^2\,c^2\,d^2 + c^4\,d^4)}{a^4\,b^4 - 2\,a^2\,b^2\,c^2\,d^2 + c^4\,d^4}$

   ii) $\dfrac{2\,(a^4\,b^4 + 6\,a^2\,b^2\,c^2\,d^2 + c^4\,d^4)}{(-(ab)+cd)^2\,(ab+cd)^2}$

5. Graph the conic section $9\,x^2 + 4\,x\,y + 6\,y^2 = 1$.

6. Find all integer values of n between 0 and 5 such that *Mathematica* can evaluate the following integral. Hint: make a table. Note: the answer is different in Versions 2.2 and 3.0, providing a good illustration of the new power of the *Mathematica* integrator.

$$\int \frac{\left(1 - \frac{1}{u}\right)^{4/3}}{u^n} \, du$$

Use differentiation to check that the values it does find are correct. Hint: subtract the integrand from the derivative of its integral and use **Factor**.

7. Same problem as number 5 for the following family of integrals.

$$\int x^{2\,n+2} \sqrt{4x^{2\,n} - 1} \, dx$$

Show that the case n = 3 can be integrated by a substitution. Check your result.

8. Find the numerical value of the integral of $\sin(x^3) \, / \, \cos(x^3)$ from 0 to 1 in three different ways. The third way should use a series approximation. (Hint: look up **Normal**.) How many terms in the series are required to get six digit accuracy?

9. Evaluate the double integral:

$$\int_0^3 \int_0^y x \sqrt{y^3 + 16} \, dx \, dy$$

10. Let

$$expr1 = \frac{6x^5 + x^3}{2(1 - x^3)}$$

i) Differentiate expr1.

ii) Simplify the result of i).

iii) Integrate the result of ii).

iv) Show that the answer to iii) is correct.

11. We saw in the text that $((2 + 5 \text{ I})^{12})^{(1/12)} \neq 2 + 5 \text{ I}$. What is the precise relationship between these two numbers?

12. Evaluate

$$\lim_{x \to \frac{\pi}{2}} \frac{\cos(x) - \cot(x)}{\left(x - \frac{\pi}{2}\right)^3}$$

13. i) Consider the matrix

$$A = \begin{pmatrix} 1 & 2 & 3 \\ 4 & 5 & 6 \\ 7 & 8 & 9 \end{pmatrix}$$

Find the exact values and the numerical values of the eigenvalues and eigenvectors of A. Display the answers as a table in which the first column has the eigenvalues and the second column has the corresponding eigenvectors. (Hint: look up commands starting with **Eigen**. Also, consider **Transpose**.) Display your answers in a nice, readable form.

ii) The transpose of the matrix of eigenvectors of A determines the coordinate transformation that diagonalizes A. Use this to check the results of part i).

14. Same problem as 13 for the matrix

$$B = \begin{pmatrix} 1 & 4 & 3 \\ 4 & 2 & 3 \\ 3 & 3 & 1 \end{pmatrix}$$

The exact values here are very large, but given enough time, *Mathematica* is able to find and check them.

15. In the September 1991 issue of the Notices of the AMS, David Stoutemyer proposed some tests for symbolic computation programs. Here are two of them

a) Over what range of values does *Mathematica* give a continuous antiderivative for $\frac{1}{2+\cos x}$. Hint: make a plot of the antiderivative. Does *Mathematica* give the correct answer for the definite integral of the function from 0 to 2π?

b) Use the expression $\frac{x^3+2x^2+3x+2}{x^3+4x^2+5x+6}$ to show that simplification does not commute with substitution. Hint: the numerator and denominator have a common factor.

16. In the September 1990 issue of the Notices of the AMS, Barry Simon described the results of submitting test problems to several symbolic computation programs – Derive, Macsyma, Maple, and *Mathematica*. Here are modified versions of some of the problems

a) Try inverting the $n \times n$ Hilbert matrix (just change 3 to n in the definition) for larger values of n. For n about 10, it is still possible to look at the result. Simon asks for $n = 20$. Don't display the result, but do show a check that the answer is correct.

b) Find the symbolic sum of i^p for i from 1 to n. Do this for p equal to various small values (e.g., 3, 5). You have to use a package to do this, so execute the statement **Needs["Algebra`SymbolicSum`"]** first. Simon asks for the value when $p = 30$

c) Differentiate $x^{10} \cos(x^5 \log(x))$ with respect to x and then integrate the result.

d) Here is the Van Der Monde matrix of size 3.

$$\begin{pmatrix} 1 & 1 & 1 \\ x(1) & x(2) & x(3) \\ x[1]^2 & x[2]^2 & x[3]^2 \end{pmatrix}$$

Define a function that constructs the Van Der Monde matrix of size n.

i) Simon's problem is: factor the determinant of the Van Der Monde matrix of size 6

ii) How many terms are there in the unfactored form of the Van Der Monde determinant of size n?

iii) How many symbols are there in each term? How many symbols in the entire determinant? (Don't forget about spaces and + and - signs.)

iv) Assume that there are eighty symbols per line, that *Mathematica* breaks expressions only at + and - signs where possible, and that there are 50 lines per page. How many pages are needed to display Van Der Monde of size 6? of size 10?

e) Factor the integer 236789456789432678.

17. Construct the matrix $\mathbf{mat} = \begin{pmatrix} 0 & 4 & 0 & 0 & 0 \\ 1 & 0 & 3 & 0 & 0 \\ 0 & 2 & 0 & 2 & 0 \\ 0 & 0 & 3 & 0 & 1 \\ 0 & 0 & 0 & 4 & 0 \end{pmatrix}$ in *Mathematica*.

i) Find the eigenvalues and eigenvectors of **mat**.

ii) Find the matrix **S** which diagonalizes **mat** is such a way that in the resulting diagonal matrix the entries on the main diagonal are decreasing. Check your result. Hint: look up commands starting with **Eigen**. Look up **Transpose** and **Reverse**.

CHAPTER 2

Interacting with Mathematica

Mathematica est omnis divisa in partes tres.

1 The Different Aspects of Mathematica

There are three distinct aspects to working with *Mathematica*, as indicated in the following tables.

Aspects	Explanation	Things to master
The kernel	The kernel is a very large C program that deals with inputs and returns outputs by means of two processes: calling hardwired C code to do various computations and using rewrite rules to reduce expressions to normal form. (These will be explained in great detail in later chapters.)	What the commands are that the kernel recognizes and how to use them.
Notebook front end	The Notebook front end is a graphical user interface with the kernel which is now supported by most computers. They all present essentially the same appearance to the user. These front ends provide facilities for editing and organizing text and sending inputs to the kernel for evaluation. The kernel sends the results back to the front end, which is then responsible for displaying the outputs in an appropriate form, including displaying graphics in place. Documents developed in the Notebook front end can be printed exactly as they appear on the screen.	How to most efficiently make use of the many facilities of the notebook front end to create interesting and useful documents in *Mathematica*.

| Packages | Packages are small, or not so small, programs written in the *Mathematica* programming language that extend the functionality of the kernel. Even though the kernel recognizes over 1500 commands, these do not begin to cover all of the operations that are needed in various parts of mathematics, science, engineering, commerce, etc., so the program is also supplied with 148 packages organized into 13 directories (or folders), containing over 2000 additional commands and constants that supply some of the other desired operations. Many other packages are available through *MathSource*. | How to understand and use the programs that are available in packages. |

The kernel is functionally the same on all platforms, but details of the Notebook front end may vary slightly from one computer to another. However, individual notebooks are completely portable. This second edition of the book was produced entirely in *Mathematica*. Packages are normally not written as notebooks so they can be used on any computer, whether or not it has a notebook interface. When you get the program, you also receive a rather substantial book describing the current versions of the packages. These packages have to be deliberately loaded, as was illustrated in the first chapter by the `ImplicitPlot` package, in order to use the operations contained in them.

It is easy to create your own extensions to the kernel. In fact, the ease with which such extensions can be created is an important way to distinguish between various symbolic computation programs. There are two ways to write extensions in *Mathematica*. One is to write notebooks containing detailed discussions of the topics being treated along with examples, graphics, etc., all implemented in the very flexible *Mathematica* programming language. The other, more formal way, which is suitable for code intended for use by others, is to write your own packages using the supplied packages as models. See also [Maeder 1] and Chapter 10, Section 2, of this book.

2 Interacting with the Kernel

Inputs are typed in from a keyboard, typically in an input cell in a notebook, or at a command line in a raw kernel. (They can also be read in from a file; see Chapter 8.) It is unnecessary to end an input with any particular symbol. Carriage returns can be used so that a single input can extend over many lines. However, be careful to make line breaks in such a way that the material before the break is not a complete *Mathematica* expression. If it is, and you are working in a raw kernel, *Mathematica* will try to evaluate it. If it is unable to do so or if there is nothing to be done, then the input will be returned in unevaluated form. In a notebook, nothing is sent to the kernel until Enter or Shift-Return is typed. Even so, complete expressions will be evaluated separately. After a shorter or longer time, the evaluated form of the input will be returned as an output. In a notebook, if the output is not what was desired, then the input can be edited in place and reevaluated. The inputs and outputs are numbered consecutively and provide a temporal ordering, which may differ from the spatial ordering in a notebook because of reevaluations or because new inputs were inserted between previous ones.

2.1 Help Facilities in the Kernel

The kernel provides help facilities via **?** and **??** commands. For instance, to find out about the **Plot** command use the following:

In[1]:= **?Plot**

```
Plot[f, {x, xmin, xmax}] generates a plot of
    f as a function of x from xmin to xmax. Plot[{f1,
    f2, ... }, {x, xmin, xmax}] plots several functions fi.
```

This tells us what the arguments to **Plot** should look like and what the command does. To get more information use the following form.

In[2]:= **??Plot**

```
Plot[f, {x, xmin, xmax}] generates a plot of
    f as a function of x from xmin to xmax. Plot[{f1,
    f2, ... }, {x, xmin, xmax}] plots several functions fi.

Attributes[Plot] = {HoldAll, Protected}
```

```
Options[Plot] = {AspectRatio -> GoldenRatio^(-1),
  Axes -> Automatic, AxesLabel -> None,
  AxesOrigin -> Automatic, AxesStyle -> Automatic,
  Background -> Automatic, ColorOutput -> Automatic,
  Compiled -> True, DefaultColor -> Automatic,
  Epilog -> {}, Frame -> False, FrameLabel -> None,
  FrameStyle -> Automatic, FrameTicks -> Automatic,
  GridLines -> None, ImageSize -> Automatic, MaxBend -> 10.,
  PlotDivision -> 30., PlotLabel -> None, PlotPoints -> 25,
  PlotRange -> Automatic, PlotRegion -> Automatic,
  PlotStyle -> Automatic, Prolog -> {}, RotateLabel -> True,
  Ticks -> Automatic, DefaultFont :> $DefaultFont,
  DisplayFunction :> $DisplayFunction,
  FormatType :> $FormatType, TextStyle :> $TextStyle}
```

This tells us in addition that **Plot** has two attributes and 30 options. Three of these, Image-Size, FormatType, and TextStyle are new to Version 3.0. Both attributes and options are under the control of the user and we will discuss how to use them in great detail later. One can use * as a wild card in requests for information. To see all of the commands starting with **B**, use:

In[3]:= **?B***

Background	BlankForm	Button
Backward	BlankNullSequence	ButtonBox
BaseForm	BlankSequence	ButtonCell
Baseline	Block	ButtonContents
Before	Bold	ButtonData
Begin	BoldItalic	ButtonEvaluator
BeginPackage	Bottom	ButtonExpandable
Below	BoxData	ButtonFrame
BernoulliB	BoxDimensions	ButtonFunction
BesselI	Boxed	ButtonMargins
BesselJ	BoxForm	ButtonMinHeight
BesselK	BoxFrame	ButtonMnemonic
BesselY	BoxMargins	ButtonNote
Beta	BoxRatios	ButtonNotebook
BetaRegularized	BoxRegion	ButtonSource
BinaryGet	BoxShift	ButtonStyle
BinaryOp	BoxSizeAdjustments	Byte
Binomial	BoxStyle	ByteCount
Blank	Break	

The fact that there are more than twice as many such commands in Version 3.0 as there were in Version 2.2 is accounted for by all the commands starting with Box or Button. To see all of the commands containing the word **List**, use:

In[4]:= **?*List***

CoefficientList	Listable	MessageList
CompletionsListPacket	ListContourPlot	MonomialList
ComposeList	ListDensityPlot	NestList
FactorList	Listen	ReadList
FactorSquareFreeList	ListInterpolation	RecordLists
FactorTermsList	ListPlay	ReplaceList
FindList	ListPlot	SampledSoundList
FixedPointList	ListPlot3D	TrigFactorList
FoldList	ListQ	$MessageList
List		

The only new item here is TrigFactorList.

2.2 A Quick Overview of Definitions in Mathematica

There are three kinds of *definitions* in *Mathematica*, and a great deal can be done just using the simplest aspects of these forms.

    i) Assignment statements

    ii) Function definitions

    iii) Recursive (function) definitions

2.2.1 Assignments

An *assignment statement* assigns a value to some symbol (or expression). It is given by a single equals sign "=". For instance, to assign to **r** the product **x** times **y** use the syntax

In[5]:= **r = x y**

Out[5]= x y

From now on, whenever *Mathematica* encounters **r** in the evaluation of an expression, **r** is replaced by its value **x y**.

In[6]:= **5 + 2 r + 3 r^2**

Out[6]= $5 + 2 \, x \, y + 3 \, x^2 \, y^2$

One can also mimic arrays by assigning values to expressions of the form **w[i]**. For example:

In[7]:= **w[2] = 1 + 2 a**

Out[7]= $1 + 2 \, a$

Again this is used whenever possible.

In[8]:= **w[1] + b w[2]**

Out[8]= $(1 + 2 \, a) \, b + w[1]$

2.2.2 Function definitions

The other kind of definition is function definition. This is usually specified by "colon equals", i.e., "**:=**". On the lefthand side there is an underscore "**_**" preceded by some symbol, e.g., "**x_**". This should be read as "a pattern named **x**". The righthand side then specifies what is to be done with **x**. For instance:

In[9]:= **f[x_] := x^2**

This defines **f** to be the squaring operation; i.e., **f** applied to "anything" is replaced by "anything squared".

In[10]:= **f[3] + f[a+b] + f[anything]**

Out[10]= $9 + \texttt{anything}^2 + (a + b)^2$

See Chapter 6 for a thorough discussion of functional programming and Chapter 7 for the precise meanings of = and **:=**.

2.2.3 Recursive functions

Patterns as described earlier can be used to define functions recursively; i.e., the function being defined can also appear on the right-hand side of the definition. For instance, we can construct our own factorial function by the following two rules.

In[11]:= `fac[n_] := n fac[n-1]`

In[12]:= `fac[1] = 1;`

To ask *Mathematica* what it has learned about our factorial function, use the same command as for built-in functions.

In[13]:= `?fac`

> Global`fac
>
> fac[1] = 1
>
> fac[n_] := n * fac[n - 1]

It can be used just like any other function.

In[14]:= `fac[20]`

Out[14]= 2432902008176640000

We can even ask *Mathematica* to show us how it uses these rules to calculate values of our factorial function.

In[15]:= `Trace[fac[4]]`

Out[15]= {fac[4], 4 fac[4 - 1],
          {{4 - 1, 3}, fac[3], 3 fac[3 - 1], {{3 - 1, 2}, fac[2],
            2 fac[2 - 1], {{2 - 1, 1}, fac[1], 1}, 2 1, 2}, 3 2, 6}, 4 6, 24}

This says that to calculate **fac[4]**, first calculate **4 fac[4 - 1]**. Next calculate **4 - 1**, which is **3**, so calculate **fac[3]**. But this requires **3 fac[3 - 1]**, etc., until one reaches **fac[1]**, which is given to be **1**. Then these results have to be multiplied together. So **2*1** is the same as **1*2**, which is **2**, **3*2** is the same as **2*3** which is **6**, and finally **4*6** is **24**.

2.2.4 Recursive programming viewed as rewrite rules

Instead of thinking of "**:=**" definitions as defining functions, which might be recursive, we can think of them as rewrite rules that say that anything that matches the lefthand side should be rewritten as the righthand side. For instance, let us program our own logarithm rules. First just consider the rule that says that the logarithm of a product is the sum of the logarithms of the factors.

In[16]:=`log[x_ y_] := log[x] + log[y]`

This is not a definition of a logarithm function but rather a rule for rewriting expressions containing "log". Now we can use this rewrite rule.

In[17]:=`log[a b c^2 d]`

Out[17]= $\log[a] + \log[b] + \log[c^2] + \log[d]$

The rule has been applied several times to reduce the original expression to this form, but it is not quite what we wanted. Apparently, *Mathematica* does not recognize that c^2 is the same as c*c so we need a second rule to handle this case also.

In[18]:=`log[x_ ^ n_] := n log[x]`

Try the example again.

In[19]:=`log[a b c^2 d]`

Out[19]= $\log[a] + \log[b] + 2\log[c] + \log[d]$

This is what we wanted. We use **?** again to check what *Mathematica* knows about our logarithm function.

In[20]:= `?log`

    Global`log

    log[(x_) * (y_)] := log[x] + log[y]

    log[(x_) ^ (n_)] := n * log[x]

Thus, we have given two different rules for expressions containing "log" that do different things depending on the form of the argument to log. See Chapter 7 for a thorough discussion of programming with rewrite rules.

2.3 Some General Observations

2.3.1 Different forms of expressions

Type in an expression.

In[21]:=**expr = (2 - 3 x^2)/(a + Sin[3]) + Integrate[Tan[x^3], x]**

Out[21]= $\int \mathrm{Tan}\,[x^3]\,dx + \dfrac{2-3\,x^2}{a+\mathrm{Sin}\,[3]}$

We can get back the input form of this expression as an output if we want it.

In[22]:= **InputForm[expr]**

Out[22]//InputForm=
Integrate[Tan[x^3], x] + (2 - 3*x^2)/(a + Sin[3])

It has a **TraditionalForm**, as we saw in Chapter 1,

In[23]:= **TraditionalForm[expr]**

Out[23]//TraditionalForm=
$$\dfrac{2-3\,x^2}{a+\sin(3)} + \int \tan(x^3)\,dx$$

as well as an **OutputForm**.

In[24]:= **OutputForm[expr]**

Out[24]//OutputForm=

$$\mathrm{Integrate}\,[\mathrm{Tan}\,[x^3],\ x] + \dfrac{2 - 3\ x^2}{a + \mathrm{Sin}\,[3]}$$

StandardForm is the current default form for outputs.

In[25]:= **StandardForm[expr]**

Out[25]//StandardForm=
$$\int \mathrm{Tan}\,[x^3]\,dx + \dfrac{2-3\,x^2}{a+\mathrm{Sin}\,[3]}$$

These forms will be discussed more in the next section. We can also output the TeX form, the Fortran form, or the C form if they are needed.

In[26]:= **TeXForm[expr]**

Out[26]//TeXForm=
```
\int \tan ({x^3})\Mfunction{\,}dx +
   {\frac{2 - 3\,{x^2}}{a + \sin (3)}}}
```

In[27]:= **FortranForm[expr]**

Out[27]//FortranForm=
```
Integrate(Tan(x**3),x) + (2 - 3*x**2)/(a + Sin(3))
```

In[28]:= **CForm[expr]**

Out[28]//CForm=
```
Integrate(Tan(Power(x,3)),x) + (2 - 3*Power(x,2))/(a + Sin(3))
```

These could then be copied and pasted into a TeX document, a Fortran program, or a C program.

2.3.2 Kinds of brackets

There are four kinds of brackets that are used in *Mathematica*. Just for fun, we will use *Mathematica* to construct a table of these kinds of brackets and what they are used for.

In[29]:= **{{"Brackets", "Usage"}, {"---------", "------------"},**
    {"[ ]", "function application"},
    {"{ }", "lists"}, {"( )", "grouping"},
    {"[[ ]]", "part extraction"}} // TableForm

Out[29]//TableForm=
```
Brackets            Usage

---------           ------------

[ ]                 function application
{ }                 lists
( )                 grouping
[[ ]]               part extraction
```

3 Interacting with the Notebooks Front End

Notebooks provide many facilities for the user.

i) The most dramatic of these is the ability to edit inputs in place and reevaluate them without losing control of the sequence in which things have been evaluated.

ii) The next obvious thing is the hierarchical organization of the cells in a notebook which provide a very convenient outlining facility and enable one to hide those parts of a notebook that are not being worked on.

iii) Graphics and text can be intermixed and printed exactly as they appear on the screen.

iv) Finally, the text and formula formatting capabillities give an unprecedented control over the appearance of both online and printed documents.

3.1 Help Facilities in the Front End

3.1.1 The Help Browser

The Menu selection **Help ▷ Help...** brings up the **Help Browser**, an immensely useful program which attempts to explain all aspects of *Mathematica*, including itself. To get started, press the radio button **Getting Started**, near the top of the window. This includes an item "Using the Help Browser" that explains how to use the **Help Browser** itself. The button **The Mathematica Book** takes one to a hypertext version of the complete book by that name. As a practical matter, the most important radio button in the Help Browser is the **Built-in Functions** button. For instance, press it and then click on the entries in the four boxes below the button to find out about any of the operations in *Mathematica*. Thus, selecting **Algebraic Computation**, **Formula Manipulation** and **FactorTerms** in turn leads to an explanation of how to use this particular operation. The explanation includes hyperlinks to the description of this operation in **The Mathematica Book**. Notice the **Back** button in the top right corner to return from a hyperlink. The second most useful radio button is the one labeled **Add-ons**. Here are found the contents of the standard packages discussed in the next section. The CD-ROM version of this book can be added to the items under this button. Furthermore, its index can be merged with the **Master Index** so that references to all documents indexed there can be found in the same listing.

3.1.2 Other help facilities

i) Type the beginning of some command, highlight it and select **Completion Selection** from the **Input** menu. (Note that it has a command key equivalent.) If there is only one possible completion of your partial command, that will be made. If there are several, a scrollable dialogue box will appear showing all possible completions.

ii) Type a command or the beginning of a command and select **Make Template** from the **Input** menu. The complete command will be shown with sample arguments, giving the same information as the first part of a **?** command to the kernel.

iii) The *Mathematica* program comes with many files covering all aspects of the program. They can be found under the **Getting Started** and **Other Resources** buttons in the **Help Browser**.

4 Producing Documents Using the Front End

The entire second edition of this book was produced in *Mathematica*. It is not possible in a brief section here to describe all of the facilities for formatting documents that are found in the front end. Instead, we will try to point out those places where it is not entirely obvious how to proceed to get the desired result. Most of the document formatting is done by using items in menus. Detailed information about all of the menu items can be found under the **Other Resources** button in the **HelpBrowser**. We concentrate on those having to do with document formatting by explaining what we did when various situations arose. Nearly all the information here that is not found in The *Mathematica* Book comes from Theodore Gray. His book with Neal Soiffer will contain a great deal more information about these matters, as well as code automating many procedures such as making an index.

4.1. The Format Menu – WYSIWYG

The first thing everybody tries is using the palettes to produce some integrals or sums involving Greek letters, etc., just to see how it works. When you print such an experiment, you discover that the printed version is not identical to the screen version. What's going on? This is controlled by the menu items

Format ▷ Screen Style Environment

Format ▷ Printing Style Environment

both of which lead to the four choices

Working - Presentation - Condensed - Printout

However, the default for **Screen Style** is **Working**, while the default for **PrintingStyle** is of course **Printout**. The experiment shows that these are different. The **Screen Style** can be set to **Printout** to achieve a WYSIWYG working environment, but then the screen appearance is bad; letters may be blurred and symbols may be too small to show up easily on the screen. A solution is to keep the default values and just convert to the **Printout Screen Style** to check things like page breaks, which are found under

Format ▷ Show Page Breaks.

Suppose you are not happy with the default styles for the **Working** or **Printout** environments. Maybe you don't like the font family or the font sizes or the spacing between cells, etc. How do you change these things? The procedure is very different from what was done in Version 2. First choose the menu item

Format ▷ Edit Style Sheet

This brings up a dialogue box where you should choose **Import Private Copy**. This opens up a new notebook called "Definitions for <file name>", which contains a bewildering collection of Sections labeled "Styles for Whatever" or "Whatever Styles". As an example, pick **Styles for Headings**. Opening it leads to a collection of subcells named by the usual kinds of heading cells. Choose **Section** and open it, displaying four subsubcells called

Section - Section/Presentation - Section/Printing - Section/Condensed.

It is not possible to go into these cells to edit them. Instead, just select one of them, say the **Section** one which is actually **Section/Working** in the previous nomenclature. Then, while the Definitions window is still the top window, make the menu selection

Format ▷ Option Inspector...

This brings up another Notebook called **Option Inspector**, containing seven items called **"Whatever" Options**. If you know exactly which option you want to change, you can type its name in the bar at the top and then click on **LookUp** to go directly to it. Otherwise, browsing through the inspector will eventually lead you to all possible changes. For instance, opening **Cell Options ▷ Display Options** leads to a list containing **CellMargins**. Opening it shows a list of four items:

left - right - bottom - top

Clicking on one of the numbers shows that you can edit it by typing in some new number. Type Return to record your change. This change will then happen for all **Section** cells in your notebook and will be visible immediately.

Other options can only take fixed values. For instance, opening **Formatting Options ▷ Font Options** leads to an item **FontFamily** whose default value is "Times". Holding down the mouse button on the box to the right brings up a list of all fonts in the system. Scrolling down this list and highlighting one of the fonts will change the font family to whatever is selected. The change should be visible immediately in the affected notebook.

In these procedures, we have used the **Options Inspector** to modify the style sheet for the notebook. Another way to accomplish the same goal is to use the **Options Inspector** on cells in the notebook itself. At the top of the **Options Inspector** is a line **Show option values for Selection**, and **Selection** is part of a dropdown menu that includes **Selection's Style**. If this is chosen, then any modifications have the same effect as it they were made to the style sheet; i.e., they affect all cells with the same style.

In producing this book, I of course wanted all the chapters to have the same styles. The idea was to set the styles for Chapter 1 and then use the same styles in all the other chapters. At the time of writing this, there was only one way to do this, although other procedures may work by the time this appears in print. The method is as follows. Having used **Import Private Copy** as above to fix the desired styles for Chapter 1, select all the cells (use **Edit ▷ SelectAll**) in the Styles Notebook and copy them to a new notebook. Save it under some name like MyStyles. Then go to **Format ▷ StyleSheet ▷ Other...** in the original notebook, which brings up a dialogue box asking one to select a style sheet. Choose MyStyles to make it the style sheet for Chapter 1. If you now choose **Format ▷ Edit Style Sheet**, it should bring up MyStyles. Follow this same procedure, starting with **Format ▷ StyleSheet ▷ Other...**, for the later chapters to make MyStyles the style sheet for them.

4.1.3 Show Expression

Select a cell and choose **Format ▷ Show Expression**. What you then see is something of the form

$$Cell[- - -].$$

This is an aspect of the most dramatic change in Version 3; namely, notebooks and the cells in them are just *Mathematica* expressions. **Show Expression** toggles between the expression for a cell and its displayed form. To find out about cells, highlight Cell and use

Help ▷ Find In Help....

4.2 The File Menu

4.2.1 Printing

Once the styles are adjusted properly, it's time for a test printout. After looking at a few pages, you will want to change the page margins and the headers and footers. These are found under **Printing Settings**. The **Printing Options** dialogue box takes care of margins, but the **Headers and Footers** item is mysterious. How do you get the headers and footers to have the proper text? For instance, the **Right Aligned** box says

Cell[TextData[{OptionValueBox["FileName"]}]]

while you would like it to say, perhaps,

Part I • Symbolic Pocket Calculator

in some appropriate size and font. The way to achieve this is to type what you want in a notebook somewhere and use **Format ▷ Font**, etc. to format it as desired. Then use **Format ▷ Show Expression** to turn it into a Cell expression. Copy and Paste that into the **Headers and Footers** dialogue box.

4.2.2 Other items

File ▷ Palettes is a walking menu showing palettes in addition to the ones that open automatically when the program is started. To add your own palette to this list, you must first make a palette and turn it into a notebook using **File ▷ Generate Notebook From Palette** and then add that notebook to the directory *Mathematica* 3.0: SystemFiles: FrontEnd: Palettes. The item **File ▷ Notebooks** shows the names of recently used notebooks. Under the heading **Modify This Menu...** there is a sophisticated control mechanism to determine which files appear there.

4.3 The Edit Menu

The **Preferences** submenu now just opens the **Options Inspector**. Items that used to be found under this menu are now there, but they are sometimes hard to find. For instance, to get a **Real-time scroll bar**, type "scroll" in the entry box and click on Lookup. This takes you to **Window Options ▷ Scrolling Options** where you find **RealTimeScrolling**. To get automatic timing of evaluations, go to **Notebook Options ▷ Evaluation Options ▷ EvaluationCompletionAction**. The kernel stack size also used to be changed here. Now you have to go to **Edit ▷ Stack Preferences** in the MathKernel program itself.

The most useful item under **Edit** is **Check Balance** with its keyboard equivalent of **Command b**. Put the cursor anywhere in the content of an input cell and type **Command b**. The smallest string between two brackets will be selected. Continue pressing **Command b** and progressively longer strings will be selected. This is how you find unbalanced brackets.

The items under **Edit ▷ Motion** seem to me to be intended for emacs users since they are similar to emacs operations. Those under **Edit ▷ Expression Input** give the keyboard equivalents to **Palette** items, as well as other common operations like adding rows or columns to a matrix, or adding an inline cell to a text cell. Using them as well as other keyboard equivalents allows very many things to be done from the keyboard. All key equivalents can be found in the file

Mathematica 3.0 Files ▷ SystemFiles ▷ FrontEnd ▷ TextResources
▷ KeyEventTranslations.tr.

4.4 The Cell Menu

4.4.1 Displaying cells

This particular file looks as if one could add one's own key equivalents if desired.

The first group of items here is concerned with various ways of formatting text. As we remarked in Chapter 1, if you are already familiar with the standard way to enter *Mathematica* expressions, then the easiest way to get expressions in traditional form is to type them in standard form, select the cell and then choose **Cell ▷ Convert To ▷ Traditional Form**, or use its key equivalent. If you want to be able to type things in traditional form and have them evaluated correctly, then you have to choose **Cell ▷ Default Input FormatType ▷ Traditional-Form**. Then, for instance, expressions like sin(1.3) give the correct answer. As remarked above, to see the Cell expression for a cell, use **Format ▷ ShowExpression**. This item perhaps belongs under the **Cell** menu.

4.4.2 Other items

Everything here is useful. **Cell Grouping ▷ Automatic grouping** is wonderful when it does exactly what you want. If it doesn't, this is where you turn it off.

I use both a monochrome and a color monitor, so I frequently choose **Rerender Graphics** to change a monochrome picture to a colored picture. However, resizing the picture accomplishes the same end. **Animate Selected Graphics** is here with its keyboard equivalent **Command y**.

4.5 The Input Menu – Numbering Headings

The items I use most are the **3D ViewPoint Selector**, the **Create Table/Matrix/Palette...** item and the **Create Automatic Numbering Object...** item. This last item is what created the section numbers in this book. To use it, as explained by Theodore Gray, bring up the dialogue box and select **Section** in the Counter Name box. Click on **Copy**, which puts the counter on the Clip Board. Paste this where the number should go in the first of your Section cells. Add a period and a space. Then Copy the number, period, and space and Paste this in all the other Section cells, which will then be numbered correctly.

To number the subsections, Copy the section number and period without the space and Paste it where the number should go in the first Subsection cell. Go back to **Create Automatic Numbering Object...** and select **Subsection** in the Counter Name box. Click on **Copy** and then Paste the result right after the period following the section number in the first Subsection box. Add a period and space and then Copy the section number, period, subsection number, period, and space. Paste this in all other Subsection cells, which will then magically be numbered correctly.

Repeat this whole process in the obvious way to number Subsubsection cells. These numbers automatically update themselves if new sections, subsections, etc. are added. (Perhaps at some later date there will be an item **Number Heading Cells** with the option **Automatic**.) Finally, to remove the Cell Dingbats, go to the **Options Inspector** for the Style Sheet under **Cell Options ▷ Display Options ▷ Cell Dingbat** and change the option to None.

4.6 The Find Menu – Making an Index

Most of the items here are standard. Two important items that perhaps also belong in the Cell menu are **Show Cell Tags** and **Make Index...**. Indexes are made from cell tags. You can use **Add/Remove Cell Tags...** to create these cell tags. However, it is more direct to use **Format ▷ Show Expression** to see the Cell expression for a cell and add the optional argument

CellTags -> {"whatever1", "whatever2", . . . }.

If **Show Cell Tags** is chosen, then these tags appear at the beginning of each cell so they can easily be checked for correctness.

Once cell tags have been added for index items, it is very simple to make the actual index. First, select **Format ▷ Show Page Breaks** and then just choose the item **Make Index...** and check that **Page References** is selected. Click on **OK**, and the index is made and placed on the Clipboard. Go to the end of your document and create a Text cell there and paste the index into it.

4.7 Mathematical and Two-Dimensional Input

4.7.1 Mathematical input

The *Mathematica* front end now has very refined capabilities for producing typeset-quality documents both on the screen and in print. The simplest way to access these capabilities in through palettes, but there cannot be enough palettes to contain all 569 named characters to be found on pages 1266 through 1312 of The *Mathematica* Book. This listing can also be found via the **Help Browser** by choosing The *Mathematica* Book and then

Reference Guide ▷ Listing of Named Characters.

The basic keyboard input form for these characters is \[Whatever]. For instance, the little triangle in the preceding line is made by substituting RightTriangle for Whatever. If you do this, then the symbol replaces the text as soon as the closing bracket is typed. This one happens to also be an operator that can be used as input. For example:

In[30]:= **RightTriangle[p, q]**

Out[30]= p ▷ q

Many of the named characters have aliases that are given in the listing. A standard form for most of these aliases is ⦂whev⦂, where whev is some acronym suggestive of Whatever. The symbol ⦂ stands for the Escape key. For instance, π is given by substituting Pi for Whatever in the basic input form, as well as being aliased by ⦂p⦂, which is typed as ESC p ESC. Instead of p, one can also use pi in this form.

4.7.2 Two-dimensional input

Another issue is the usual two-dimensional form of fractions, exponentials, radicals, integrals, sums, products, etc. Many people are so accustomed to forms like $\int_a^b f(x)\,dx$ that it comes as a surprise that an expression like **Integrate[f[x], {x, a, b}]** means (in the sense of denotes) exactly the same thing. Of course they have different connotations. Thus, the $\int_a^b$ symbol suggests a summation from a to b and the f(x)dx suggests a product of a value of f(x) times an infinitesimal dx representing an infinitesimal volume, whereas the second form has been stripped of all intensive qualities. But, for the purposes of symbolic computation, they are interchangeable. Here we are just concerned with the mechanics of producing these richly resonant forms. Fortunately, there is a long discussion in Section 1.10 of The *Mathematica* Book showing exactly how to do this. To find it in the **Help Browser**, press the radio button for The *Mathematica* Book and then choose **Practical Introduction ▷ Input and Output in Notebooks**.

5 *Using Packages*

The current version of *Mathematica* has 16 directories (or folders) of packages with names like **Calculus** and **Graphics**. **Calculus** contains 11 notebooks, a **Master** file, a **Common** directory, and a **Kernel** directory, while **Graphics** contains 23 items. A very convenient way to find out about packages is to use the **Help Browser**. Just click on the radio button **Add-Ons** there, and then on **Standard Packages**. The browser will show you the names of all the packages. If you then click on **Calculus**, for example, it will show you a list of all the Notebooks in this directory. If you click on one of the package names, e.g., **FourierTransform**, then the browser will display a very instructive, detailed account of the contents of this package and how to use them. The amount of mathematical knowledge contained in these descriptions of the standard packages is phenomenal. Just be sure you have set aside plenty of memory for the front end if you want to look at more than one of them. A brief introduction to each of the packages can be found under **AddOns ▷ Loading Packages**. There are two ways to load one of these notebooks into a session of *Mathematica:* by a **Get** command, abbreviated by **<<**, or by a **Needs** command. A command with **<<** uses the actual name of the file. You may get a dialogue box if your system can't locate the file. This can be avoided by using a complete path name for the file. The **Get** command takes either the exact file name or the context name. If you use

<<Polyhedra.m

then, depending on the machine, you may or may not succeed in loading the file. On a Macintosh, a dialogue box appears saying the file can't be found. If you use

<<:Graphics:Polyhedra.m

there is a better chance of succeeding. The strange form with back ticks,

<<Graphics`Polyhedra`

now seems to be the most reliable, working just like the **Needs** command. This particular package enables one to display regular and stellated polyhedra. Using **Needs** directly is system independent.

In[31]:=**Needs["Graphics`Polyhedra`"]**

Note the quotation marks and the back ticks after **Graphics** and **Polyhedra**. This is actually a context name rather than a file name. Contexts will be explained when we study packages in Chapter 11, Section 2.

5.1 Graphics and Geometry Packages

Start with some examples from graphics packages. Here is an example from the **Polyhedra** package.

In[32]:= **Show[Graphics3D[GreatDodecahedron[]]]**

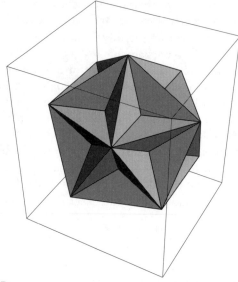

Out[32]= - Graphics3D -

Actually, each folder in the **Package** subdirectory has a file called **Master.m**. If this package is loaded, then all of the commands in all of the other packages in the folder are made available.

In[33]:= **Needs["Graphics`Master`"]**

The following command will list all of the packages that have now been made available.

In[34]:= **$ContextPath**

Out[34]= {Graphics`ThreeScript`, Graphics`SurfaceOfRevolution`,
        Graphics`Spline`, Graphics`Shapes`, Graphics`PlotField`,
        Graphics`PlotField3D`, Graphics`ParametricPlot3D`,
        Graphics`MultipleListPlot`, Graphics`Legend`,
        Graphics`ImplicitPlot`, Graphics`Graphics`,
        Graphics`Graphics3D`, Graphics`FilledPlot`,
        Graphics`ContourPlot3D`, Graphics`ComplexMap`,
        Graphics`Common`GraphicsCommon`, Graphics`Colors`,
        Graphics`Arrow`, Graphics`ArgColors`, Graphics`Animation`,
        Graphics`Master`, Graphics`Polyhedra`,
        Geometry`Polytopes`, Global`, System`}

To find out what is in the **Shapes** package, one can use the following command.

In[35]:= **Names["Graphics`Shapes`*"]**

Out[35]= {AffineShape, Cone, Cylinder, DoubleHelix, Helix, MoebiusStrip,
        RotateShape, Sphere, Torus, TranslateShape, WireFrame}

Thus, this package makes available 11 more operations in _Mathematica_.

5.2 Miscellaneous Packages

The directory **Miscellaneous** contains many useful and interesting constants.

In[36]:= **Needs["Miscellaneous`Master`"]**

For instance, the package **Units** has 241 scientific and common units and converts between them. (See also the package **SIUnits**.) E.g.,

In[37]:= **{Convert[27 BTU, Calorie], Convert[0.5 Gallon, Teaspoon]}**

Out[37]= {6803.91 Calorie, 384. Teaspoon}

Similarly, the package **ChemicalElements** has all 106 elements together with some 25 operations to manipulate them.

In[38]:= **{HeatOfVaporization[Xenon], ElectronConfigurationFormat[Zinc]}**

Out[38]= $\left\{ \dfrac{12.65 \, \text{Joule Kilo}}{\text{Mole}}, \ 1s^2 \ 2s^2 2p^6 \ 3s^2 3p^6 3d^{10} \ 4s^2 \right\}$

The package **PhysicalConstants** has exactly what its name suggests.

In[39]:= {AccelerationDueToGravity, ThomsonCrossSection}

Out[39]= $\left\{ \dfrac{9.80665 \, \text{Meter}}{\text{Second}^2}, \ 6.65224 \times 10^{-29} \, \text{Meter}^2 \right\}$

In[40]:= Convert[AccelerationDueToGravity, Feet / Second^2]

Out[40]= $\dfrac{32.174 \, \text{Feet}}{\text{Second}^2}$

The package **Music** has absolute and relative frequencies.

In[41]:= TableForm[{MeanMajor, PythagoreanMajor, Fflat3},
            TableSpacing - <{1, 1}]

Out[41]//TableForm=

0	193.2	386.3	503.4	696.6	889.7	1082.9	1200
0	204	408	498	702	906	1110	1200
329.628							

5.3 Math Source

MathSource is a resource maintained by Wolfram Research, Inc. The most convenient way to access it is via the World Wide Web at

http://www.wolfram.com/MathSource/

At the time of writing, it contains over 100,000 pages of material, some of which are produced in house by Wolfram Research, Inc., and some of which are contributed by users of the program. There are short programs, long programs, programs written in a very naive style, and programs written in a very sophisticated style, as well as other documents, examples, etc. Before embarking on a project of your own, it would seem wise to check to see what is available there.

6 Saving Work to Be Reused

6.1 Notebook Front Ends

If you work in a notebook front end, then saving work to be reused seems like a simple matter. Just put the work you want to save in a separate notebook and save the notebook under some convenient name using the menu selection in the **File** menu. However, this notebook may contain many other things besides the operations you have defined to carry out certain tasks, and you can arrange things so that just these operations will be evaluated when you start up *Mathematica* again and open this notebook. Just select each of the cells containing the important definitions and give them the attribute **Initialization Cell**, found under **Cell ▷ Cell Properties**. When such a notebook is reopened, a dialogue box appears asking if you want to evaluate the initialization cells. You can either answer Yes at this point, or answer No and wait until later when the same thing can be accomplished using the menu item **Evaluate Initialization** found under **Kernel ▷ Evaluation**.

6.2 Saving Definitions in a File

It is very easy to save a series of definitions in a file to be loaded directly into *Mathematica* at some later time. First make the definitions. For example:

```
In[42]:= ff[x_] := x^2
```

```
In[43]:= fff[x_] := x^3
```

```
In[44]:= ffff[x_] := x^4
```

Try out these functions just to see how they work.

```
In[45]:= {ff[2], fff[2], ffff[2]}
```

```
Out[45]= {4, 8, 16}
```

The command to save these definitions in a file named "sessionHistory" is just

```
In[46]:= Save["sessionHistory", ff, fff, ffff]
```

Now clear the definitions of **ff**, **fff**, and **ffff**, so that they no longer work.

```
In[47]:= Clear[ff, fff, ffff]
```

Check that they don't work.

In[48]:= **{ff[2], fff[2], ffff[2]}**

Out[48]= {ff[2], fff[2], ffff[2]}

The command **Get["filename"]** reads in a file.

In[49]:= **Get["sessionHistory"]**

There is no output, but the definitions have been read into *Mathematica* and they now work again.

In[50]:= **{ff[2], fff[2], ffff[2]}**

Out[50]= {4, 8, 16}

One can also ask to see exactly what is in the file.

In[51]:= **! ! sessionHistory**

```
ff[x_]  := x^2

fff[x_]  := x^3

ffff[x_]  := x^4
```

However, this output is a Print cell and does not result in the definitions being evaluated.

7 Practice

1. **Needs["Graphics`Polyhedra`"]**

2. **Show[Graphics3D[Icosahedron[ ]]]**

3. **?D***

4. **?*Plot***

5. **?A***

8 Exercises

1. Write rewrite rules for a function **logb[x]** that reverse the rules given for **log[x]**. Note that in writing these rules it is necessary to use **^:=** instead of **:=** for reasons that will be explained in Chapter 7.

2. Make a three-dimensional plot from a different view point. First evaluate

$$\texttt{Plot3D[Sin[x] Cos[y], \{x, 0, Pi\}, \{y, 0, Pi\}]}$$

to see the picture from the default viewpoint. Then change the command to

$$\texttt{Plot3D[Sin[x] Cos[y], \{x, 0, Pi\}, \{y, 0, Pi\},]}$$

and place the cursor just after the last comma. Go to the **Input** menu and select the **3D View-Point Selector**. Use the cursor to drag the outline box to a new orientation and click on the **Paste** button. Your command will now look like

$$\texttt{Plot3D[Sin[x] Cos[y], \{x, 0, Pi\}, \{y, 0, Pi\},}$$
$$\texttt{ViewPoint->\{1.927, -2.501, -1.216\}]}$$

Evaluate this new graphics command and compare with the original picture.

3. Make a plot of the curve y = x - cos x for x between 0 and 1. Try to use **Solve** to find the value of x where y = 0. Plot the curve over smaller and smaller ranges to try to estimate this value. Or, look at **Input ▷ Get Graphics Coordinates...** Another method is to start with some value and repeatedly apply **Cos** to it and compare the values found. Alternatively, type **?FixedPoint** to find out about this function and use it to find the solution.

4.   i) Type **Be** leaving the cursor just after the **e** and select **Complete Selection** from the **Input** menu. Choose **BesselY**.

   ii) Now, with the cursor after **BesselY**, choose **Make Template** from the Input menu.

5. In the **Help Browser**, click on the **Getting Started** radio button. Select **Using *Mathematica*** and investigate some of the items there.

Here are some suggestions from Wolfram Research, Inc., in course notes by Paul Abbot, of things to do with the front end.

6. Select all of the **Subsubsection** cells in this notebook by holding down the Option key and selecting one of them. Copy them and paste them into another new notebook. Select all of them by choosing **Select All** in the **Edit** menu. and choose **Convert To ▷ Pict** from the **Cell** menu. While they are still selected, choose **Cell Grouping ▷ Group Cells**, also in the **Cell** menu. (Note that **Manual Grouping** has to be chosen for this to work.) Now close the group of cells and choose **Animate Selected Graphics**, also in the **Cell** menu. Use the controls that appear in the lower left–hand corner of the window to control the speed, or drag the horizontal scroll bar to view the cell names one at a time.

7. Use a drawing program such as MacDraw to produce a PICT graphics. Copy and Paste it into a cell in *Mathematica* and use **Cell ▷ Convert To ▷ InputForm** to produce a *Mathematica* input cell yielding the same graphics. Give a name to the Graphics item in this drawing. Evaluate the cell and compare the result with the original graphics item. Load the package **Graphics`Graphics`** and get information on the command **TransformGraphics**. Apply **TransformGraphics** to your named graphics item using some function like **Sin** for the second argument, and **Show** the result.

CHAPTER 3

More about Numbers and Equations

1 Introduction

At the heart of any symbolic computation program lie its abilities to deal in different ways with equations of all kinds. The possible ways include exact and approximate numerical solutions and exact symbolic solutions. The kinds can be linear, polynomial, algebraic, and transcendental equations in one or more variables, as well as ordinary and partial differential equations involving one or several unknown functions of one or more variables. There are many subtle questions which we only have space to dwell on briefly in introducing the reader to this very rich world that *Mathematica* makes available to users.

2 Numbers

2.1 Precision and Accuracy

There are two important measures attached to numbers in *Mathematica*, precision and accuracy. The definitions are:

```
Precision[x] = the total number of significant digits in x

Accuracy[x]  = the number of significant
   decimal digits to the right of the decimal point in x
```

Here are some simple examples, presented as a list of inputs. The answer will be a corresponding list of outputs. We will use this format frequently to save space.

In[1]:= `{{Precision[10], Accuracy[10]},`
    `{Precision[3/5], Accuracy[3/5]},`
    `{Precision[314.159], Accuracy[314.159]}}`

Out[1]= $\{\{\infty, \infty\}, \{\infty, \infty\}, \{16, 13\}\}$

It is clear that infinite precision numbers like integers and rational numbers should have **Precision** equal to ∞. Presumably having **Accuracy** also equal to ∞ suggests an infinite number of zeros to the right of the decimal point. But why the value 16 and 13 for 314.159, rather than 6 and 3? This is because of the way that real numbers are handled by default. They use the built-in machine-level floating-point arithmetic. For any specific machine the number of digits can be accessed by the command

In[2]:= `$MachinePrecision`

Out[2]= `16`

This result is for a Macintosh. Unix workstations usually have the same machine precision. Commands that start with **$** have values or effects concerned with the environment in which *Mathematica* is running or the way in which it works. For instance:

In[3]:= `{$Version, $TimeUnit, $RecursionLimit}`

Out[3]= $\left\{\text{Power Macintosh 3.0 (October 5, 1996)}, \dfrac{1}{60}, 256\right\}$

This shows that I am using the Macintosh Version 3.0 of *Mathematica* from October 5, 1996, that the minimal unit of time on my machine is 1/60 of a second, and that a recursive program will carry out 256 steps before stopping and asking if I want to continue. Anyway, **$Machine-Precision** equal to 16 means that all machine-level real numbers are treated as though they have 16 significant digits. For real numbers with specified precision, things work correctly.

In[4]:= `sq3 = N[`$\sqrt{30}$`, 25]`

Out[4]= `5.477225575051661113456970`

In[5]:= `{Precision[sq3], Accuracy[sq3]}`

Out[5]= `{25, 24}`

However, calculations with numbers of specified precision can result in values that have a different precision. Thus, for instance, start with the square root of 30 calculated with precision 50.

In[6]:= `N[`$\sqrt{30}$`, 50]`

Out[6]= 5.4772255750516611345696978280080213395274469949980

In[7]:= **{Precision[%], Accuracy[%]}**

Out[7]= {50, 49}

If we calculate the 25th power of $\sqrt{30}$, we get

In[8]:= **N$\left[\sqrt{30}, 50\right]^{25}$**

Out[8]= $2.9108222368310298450168547834141086869980593454 \times 10^{18}$

In[9]:= **{Precision[%], Accuracy[%]}**

Out[9]= {49, 30}

One digit of precision has been lost by raising $\sqrt{30}$ to the 34th power, starting with $\sqrt{30}$ to 50 significant digits

In[10]:= **N$\left[\left(\sqrt{30}\right)^{25}, 50\right]^{34}$**

Out[10]= $5.977688167482936639858090991639370366691225981888 \times 10^{627}$

In[11]:= **{Precision[%], Accuracy[%]}**

Out[11]= {48, -579}

Another digit of precision has been lost. The negative accuracy value means that the significant digits start 579 places to the left of the decimal point. Note that

In[12]:= **627 - 579**

Out[12]= 48

However, given input of sufficient precision, in Version 3.0, the function **N** itself will try to achieve the requested precision by increasing the precision of the input. In Version 2.2, **N[something, 20]** meant start with 20 digits of precision for **something** and see what you get. In Version 3.0, it means try to achieve 20 digits of precision in the answer by increasing the precision of **something** if necessary. Here is an example from [Sofroniou] (which, incidentally, contains a great deal of information about the new numeric capabilities of *Mathematica*). First generate a rational approximation to π. (See Section 2.2 later.)

In[13]:= **q = Rationalize[π, (0.1)^{100}]**

Out[13]= $\dfrac{39437283434272590306994370980763234507447310 2456264}{12553277201361201519554317372950508261618601 2726141}$

Now compare this with π, using first machine precision and then 20 digits precision.

In[14]:= **{N[π - q], N[π - q, 20]}**

$MaxExtraPrecision::meprec :
  $MaxExtraPrecision = 50.` reached while evaluating
  $- \dfrac{39437283434 \ll 28 \quad 473102456264}{12553277201 \ll 28 \quad 186012726141} + \pi$. Increasing the value
  of $MaxExtraPrecision may help resolve the uncertainty.

Out[14]= $\{0., 0. \times 10^{-83}\}$

The first calculation, done in machine precision, gives the value 0. The second, in trying to achieve 20 digits of precision, runs into the default limit of 50 extra digits of precision in the input. One can increase the value of **$MaxExtraPrecision** and try again.

In[15]:= **$MaxExtraPrecision = 100;**

In[16]:= **N[π - q, 20]**

Out[16]= $5.752043867880545649 \times 10^{-102}$

Machine-precision numbers being stored as 16 digit numbers even when fewer are displayed affects certain calculations. For instance, suppose we want to make a table of approximations to π with the values of **Sin** of those approximations to show the values approaching 0. The following attempt fails.

In[17]:= **Table[{N[π, n], Sin[N[π, n]]}, {n, 1, 5}] // TableForm**

Out[17]//TableForm=

3.	0.
3.1	0.
3.14	0.
3.142	0.
3.1416	0.

We get what appear to be increasingly accurate approximations to π, but the values of **Sin** are all the same. The reason is that in the left-hand column we are just being shown fewer digits of π at the beginning. However, consider the following construction, which uses **ToString** to turn a number into a string, i.e., something which has no numerical value. It then uses **ToEx-**

pression to turn it back into a number. Along the way all of the hidden digits get lost and what we see is what we get.

In[18]:=**piApprox = ToExpression[ToString[N[π, 5]]]**

Out[18]= 3.1416

In[19]:=**N[piApprox, 10]**

Out[19]= 3.1416

This means that **piApprox** is really **3.1416000000....** Using this, we can make the desired table.

In[20]:=**Table[{ ToExpression[ToString[N[π, n]]],**
            N[Sin[ToExpression[ToString[N[π, n]]]], 11] },
          {n, 1, 5}] // TableForm

Out[20]//TableForm=

3.	0.14112000806
3.1	0.041580662433
3.14	0.0015926529165
3.142	-0.00040734639894
3.1416	$-7.3464102068 \times 10^{-6}$

There is an interesting number called **$MachineEpsilon** which is "the smallest machine-precision number which can be added to 1.0 to give a result not equal to 1.0".

In[21]:= **$MachineEpsilon**

Out[21]= 2.22045×10^{-16}

Adding it to 1.0 doesn't appear to change the value.

In[22]:=**1.0 + $MachineEpsilon**

Out[22]= 1.

However, comparing this with 1 shows that there is a difference.

In[23]:=**% - 1**

Out[23]= 2.22045×10^{-16}

Other interesting machine numbers are the biggest and smallest ones.

In[24]:= **{$MaxMachineNumber, $MinMachineNumber}**

Out[24]= $\{1.79769 \times 10^{308}, 2.22507 \times 10^{-308}\}$

2.2 *Inverses to* $\mathrm{N}[\,]$

There are three commands that convert real numbers into integers: **Floor**, **Ceiling**, and **Round**. They behave exactly as might be expected.

In[25]:= **{Floor[3.5], Ceiling[3.5], Round[3.5]}**

Out[25]= $\{3, 4, 4\}$

What is more interesting is to convert real numbers into rational numbers. We saw in Chapter 1 that **N[]** converts integers and rational numbers (real or complex) into floating-point reals or complexes. If a second argument is given, then it converts them into reals or complexes with a specified precision. An inverse operation should convert reals or complexes into rational numbers or integers (real or complex as the case may be). There are operations in *Mathematica* that do exactly this. **Rationalize** converts decimal numbers into rational numbers.

In[26]:= **Rationalize[3.456 + 1.234 I]**

Out[26]= $\dfrac{432}{125} + \dfrac{617\,I}{500}$

The result is not very interesting. It has just regarded the decimals as fractions whose denominator is an appropriate power of 10. In general, this result will be reduced to lowest terms, so it might look more interesting without really being so. **Rationalize** becomes much more interesting when it, like **N**, is given a second argument which represents the intended accuracy of the rational approximation to the real or complex number.

In[27]:= **Rationalize[3.456 + 1.234 I, 0.001]**

Out[27]= $\dfrac{235}{68} + \dfrac{58\,I}{47}$

To check the accuracy of this, just subtract it from the original number to see that it is accurate to three decimal places.

In[28]:= **3.456 + 1.234 I - %**

Out[28]= $0.000117647 - 0.0000425532$ I

Let's try to find an approximation to π.

In[29]:=**Rationalize[N[π], 0.001]**

Out[29]= $\dfrac{355}{113}$

In fact, let's find many approximations to π.

In[30]:=**Table[Rationalize[N[π],(0.1)^n], {n, 1, 12}]**

Out[30]= $\left\{ \dfrac{22}{7}, \dfrac{22}{7}, \dfrac{355}{113}, \dfrac{355}{113}, \dfrac{355}{113}, \dfrac{355}{113}, \dfrac{104348}{33215}, \right.$

$\left. \dfrac{104348}{33215}, \dfrac{104348}{33215}, \dfrac{312689}{99532}, \dfrac{1146408}{364913}, \dfrac{5419351}{1725033} \right\}$

Surely a curious result! There is no best approximation to π whose denominator has two or four digits because $\frac{22}{7}$ is a better approximation than any fraction whose denominator has two digits and $\frac{355}{113}$ is better than any one whose denominator has four digits. To check the accuracy of $\frac{355}{113}$, just calculate the difference.

In[31]:= **N[π]** $- \dfrac{355}{113}$

Out[31]= -2.66764×10^{-7}

For another approach to rationalizing real numbers, see Chapter 11, Section 6.3.1.

2.3 *Working with Fixed Precision*

It is possible to specify the form in which *Mathematica* displays floating-point numbers. For instance:

In[32]:=**NumberForm[N[π, 35], NumberSeparator** ←" ", **DigitBlock** ←5]

    Out[32]//NumberForm=
  3.14159 26535 89793 23846 26433 83279 5029

The two optional arguments to the command **NumberForm**, indicated by the ←, mean that we want digits before and after the decimal point divided into groups of five, separated by spaces. The following might be more appropriate for large integers.

In[33]:= `NumberForm[3`^{24}`, NumberSeparator ↵",", DigitBlock ↵3]`

  Out[33]//NumberForm=
  282,429,536,481

If we are going to work frequently with 35 decimal places, we could define a function that formats such numbers for us.

In[34]:= `n36[x_] :=`
           `NumberForm[N[x, 36],`
               `NumberSeparator  ↵" ", DigitBlock  ↵5]`

The following command will now apply n36 to every output.

In[35]:= `$Post = n36`

  Out[35]//NumberForm=
  n36

In[36]:= `Sqrt[3]`

  Out[36]//NumberForm=
  1.73205 08075 68877 29352 74463 41505 87237

In[37]:= `Precision[%]`

  Out[37]//NumberForm=
  36.00000 00000 00000 00000 00000 00000 000

Notice that this doesn't affect non-numerical expressions.

In[38]:= `a + b`

  Out[38]//NumberForm=
  a + b

One way to turn off the post-processing of all outputs is to redefine $Post as nothing, using a period.

In[39]:= `$Post = .`

Large floating-point numbers are usually displayed in scientific notation. To see all of the digits to the left of the decimal point, use **AccountingForm**.

In[40]:= {3.24^{24}, AccountingForm[3.24^{24}], N[3.24^{24}, 20]}

Out[40]= {1.79094 × 10^{12}, 1790936736361., 1.790936736360974 × 10^{12}}

2.4 Different Bases

All of the numbers we have discussed up to now have been written in base 10, but *Mathematica* can deal with numbers in different bases and convert values between different bases. The following illustrates how to convert a decimal integer to various other bases.

In[41]:= {BaseForm[12345678, 2], BaseForm[12345678, 15],
        BaseForm[12345678, 36]}

Out[41]= {101111000110000101001110$_2$, 113cea3$_{15}$, 7clzi$_{36}$}

Thus, any base up to 36 is acceptable (since there are 10 ordinary digits and 26 letters to use to represent the extra digits). Decimal real numbers, fractions, and complex numbers can also be converted to other bases.

In[42]:= {BaseForm[1234.5678, 15], BaseForm[1234.5678, 36],

        BaseForm[$\frac{3}{4}$, 2], BaseForm[1234 + 5678 I, 36]}

Out[42]= {574.87b5$_{15}$, ya.kfv$_{36}$, $\frac{11_2}{100_2}$, ya$_{36}$ + 4dq$_{36}$ I}

To convert numbers in a specified base back to decimal numbers, use the following form.

In[43]:= {2^^101111000110000101001110,
        15^^113cea3, 36^^7clzi}

Out[43]= {12345678, 12345678, 12345678}

2.5 Fun with Factor

FactorInteger was discussed in Chapter 1. If we give it a prime number like 2 as argument, the results are uninteresting.

In[44]:=`FactorInteger[2]`

Out[44]= $\{\{2, 1\}\}$

However, if **FactorInteger** is told to use Gaussian integers (complex numbers with integer real and imaginary parts) in its factorizations via an optional second argument, then the results are much more interesting.

In[45]:=`FactorInteger[2, GaussianIntegers -> True]`

Out[45]= $\{\{-I, 1\}, \{1 + I, 2\}\}$

Check that the product of these Gaussian integers does equal 2.

In[46]:=`(-I) (1 + I)^2`

Out[46]= 2

But are the entries prime numbers? We can check using the predicate (= function that returns the value `True`, or `False`) **PrimeQ** which tests if a number is prime or not.

In[47]:=`{PrimeQ[-I], PrimeQ[1 + I]}`

Out[47]= $\{$False, True$\}$

No, **-I** is not a Gaussian prime. Actually, it is a unit (= a number that divides 1). The Gaussian integers have four units, 1, -1, I, and -I. Factorizations into primes in the Gaussian integers are unique up to multiplication by units. For instance, `1 - I` is also a Gaussian prime and obviously, `(1 + I) (1 - I) = 2;` but `(1 - I) = (-I) (1 + I)` so everything is OK. The following amusing use of **PrimeQ** for Gaussian integers appeared in the *Mathematica* One-Liners column in the *Mathematica* Journal, Vol. 1, No. 4, Spring 1991. It illustrates all Gaussian primes of the form a + b I, where a and b are less than or equal to 50.

In[48]:=`Table[`
`     If[PrimeQ[a + b I], 1, 0], {b, 0, 50}, {a, 0, 50}];`

In[49]:=`ListDensityPlot[%];`

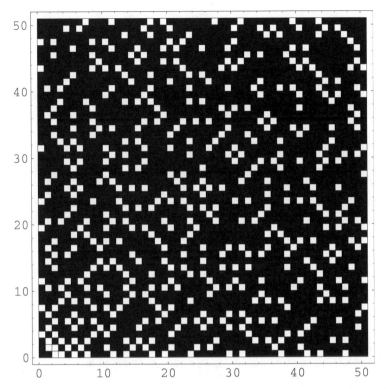

The white squares here are the Gaussian primes. What can be said about the distribution of such primes?

2.6 *The N functions*

There are five numerical functions in *Mathematica* starting with **N**; namely, **NDSolve**, **NInte-grate**, **NProduct**, **NSolve**, and **NSum**. In addition, there are four more in the package **Numer-icalMath`NLimit`** called **NIntegrateInterpolatingFunction** , **NLimit**, **NResidue**, and **NSeries**. Each of them is a numerical version of the symbolic, exact command without the **N**; i.e., **DSolve**, **Integrate**, **Product**, **Solve**, **Sum**, **Limit**, **Residue**, and **Series**. As a general rule, try the exact command first. If that fails, by returning the input unevaluated, or by returning a partial result, or by never returning, then try the corresponding **N** command. In all cases there are in fact four possible ways to get a result; in the case of **Integrate**, one can try **Integrate[]**, **N[Integrate[]]**, **NIntegrate[]**, and **Integrate[N[]]**. The advantage to using **N[command[]]** is that **N[]** takes a second argument specifying the precision, but in this case, if **Integrate[]** fails, so will **N[Integrate[]]**, whereas **NIntegrate[]** may very well succeed.

Fortunately, or unfortunately, except for **NSolve**, all of these functions have several optional arguments, which complicates their use, but gives us a better chance to get an accurate answer. These actually refine the single optional second argument to **N**. For instance:

In[50]:= **Options [NIntegrate]**

Out[50]= {AccuracyGoal → ∞, Compiled → True, GaussPoints → Automatic,
        MaxPoints → Automatic, MaxRecursion → 6, Method → Automatic,
        MinRecursion → 0, PrecisionGoal → Automatic,
        SingularityDepth → 4, WorkingPrecision → 16}

WorkingPrecision determines how accurately the integrand is evaluated in approximating the integral. This has the same effect as giving a second argument to **N[]**. PrecisionGoal means how precise the answer should be. By default, Automatic means that it is 10 digits less than WorkingPrecision. AccuracyGoal similarly sets the desired accuracy of the answer. These same three options are available in **NDSolve**, **NProduct**, and **NSum**. For more information about the use of these options, see [Skeel].

3 Solving Algebraic Equations

3.1 One Variable

3.1.1 Solutions of equations in one variable

The standard format for solving an equation is **Solve[equation, variable]**, as we have seen in Chapter 1.

In[1]:= **Solve [x^2 + 3 x == 2, x]**

Out[1]= $\left\{\left\{x \to \frac{1}{2}\left(-3 - \sqrt{17}\right)\right\}, \left\{x \to \frac{1}{2}\left(-3 + \sqrt{17}\right)\right\}\right\}$

(The "variable" here does not have to be a symbol. See 4.7.2 later.) The output is a list of rules. However, there is another form of **Solve** that gives its result in a different form.

In[2]:= **Roots [x^2 + 3 x == 2, x]**

Out[2]= $x == \frac{1}{2}\left(-3 - \sqrt{17}\right) \mid\mid x == \frac{1}{2}\left(-3 + \sqrt{17}\right)$

The result here is a pair of equations for the values of **x**, separated by **| |**, which means **Or** in *Mathematica*. **Solve** is the same as **Roots** followed by **ToRules**.

In[3]:= **{ToRules[%]}**

Out[3]= $\left\{\left\{x \to \frac{1}{2}\left(-3 - \sqrt{17}\right)\right\}, \left\{x \to \frac{1}{2}\left(-3 + \sqrt{17}\right)\right\}\right\}$

We would like to check that the answer is correct. One way to do that is to substitute the values for **x** into the left-hand side of the equation and see if the results equal the right-hand side. This is done by using **/.** indicating application of the rules, followed by **%**, referring to the previous output, which consists of a list of rules.

In[4]:= x^2 **+ 3 x /. %**

Out[4]= $\left\{\frac{3}{2}\left(-3 - \sqrt{17}\right) + \frac{1}{4}\left(-3 - \sqrt{17}\right)^2, \frac{3}{2}\left(-3 + \sqrt{17}\right) + \frac{1}{4}\left(-3 + \sqrt{17}\right)^2\right\}$

It's hard to see if this is right or not, so we **Simplify** it.

In[5]:= **Simplify[%]**

Out[5]= $\{2, 2\}$

This may seem rather mysterious. The important thing is that the **/.** in the form **expression /. rules** means: use the rules to change the expression by replacing the occurrences of the left-hand sides of the rules in **expression** by the right-hand sides. For instance,

In[6]:= **a /. a -> 5**

Out[6]= 5

I read something like this as "a, where a gets the value 5." So, I read **/.** as "where". (Actually, in *Mathematica*, **/.** stands for **ReplaceAll**.) Rules can be applied simultaneously by putting them in a list. For instance, calculate the value of **x y** where **x** gets the value 2 and **y** gets the value 3.

In[7]:= **x y /. {x -> 2, y -> 3}**

Out[7]= 6

If the **rules** part of in **expression/.rules** is a list of lists, then the result is a list of modified expressions, one for each substitution in the list. For instance,

In[8]:= **x y /. {{x -> 2}, {x -> 3}}**

Out[8]= $\{2 y, 3 y\}$

   Another way to check equations is to substitute the answers in the equation itself, and that is the format we will use in looking at a number of equations. First, give a name to the equation. The output here is suppressed by following the input with a ";".

In[9]:= **equation1 = x^2 + 3 x == 2;**

Now, solve the equation, giving a name to the solution.

In[10]:= **solution1 = Solve[equation1, x]**

Out[10]= $\left\{\left\{x \to \frac{1}{2}\left(-3-\sqrt{17}\right)\right\}, \left\{x \to \frac{1}{2}\left(-3+\sqrt{17}\right)\right\}\right\}$

Finally, substitute the solution in the equation, simplifying the result.

In[11]:= **Simplify[equation1 /. solution1]**

Out[11]= {True, True}

We got a list of two values of True since **solution1** is a list of two solutions. This also tells us something more about ==. It behaves somewhat like a predicate. For instance,

In[12]:= **{2 == 2, 2 == 3, a == b}**

Out[12]= {True, False, a == b}

If *Mathematica* can determine that the left-hand side of == does or does not equal the right-hand side, then it returns the value True or False as appropriate. Otherwise, it leaves the result unevaluated. This is exactly what one wants for an equation; i.e., a "predicate" that asks if the two sides are the same, but leaves them unevaluated if there are variables without values on one side or the other.

   Now let's try a more complicated example. Experience shows that the solution to the following equation is a very large expression consisting of three different values. To save space, we just look at the third one by typing **[[3]]** after **Solve**, which picks out the third entry in the list of solutions. It is often a good idea to put in an extra simplification step in solving equations, so we will always include that, using the postfix form of function application **//**.

In[13]:= **equation2 = x^3 + 34 x + 1 == 0;**

In[14]:= **solution2 = Solve[equation2, x][[3]] // Simplify**

Out[14]= $\left\{x \to \dfrac{1}{2\,6^{2/3}}\left(68\,\left(1-I\,\sqrt{3}\right)\left(\dfrac{3}{-9+\sqrt{471729}}\right)^{1/3}-\right.\right.$

$\qquad\left.\left.\left(1+I\,\sqrt{3}\right)\left(-18+2\,\sqrt{471729}\right)^{1/3}\right)\right\}$

This is quite complicated looking, but *Mathematica* is able to substitute it back into the equation and simplify the result to see that it is correct, providing we use **FullSimplify** instead of **Simplify** (which sufficed in Version 2.2). Calculations that require **FullSimplify** tend to take a long time.

In[15]:= **equation2 /. solution2 // FullSimplify**

Out[15]= True

(Try this without the **FullSimplify** at the end.) As a general policy, you should never believe the result of a symbolic computation program unless you can find some way to check the result. For instance, what is one to think about the calculations of π to 100 decimal places or the value of 3^{1000}? The second one can be checked by taking the 1000th root, which is an independent calculation, but the only way to check the calculation of π is to compare it with some other similar calculation by a different program. If we take the previous equation and complicate it by adding some symbolic constants, then the answer will become much larger.

In[16]:= **equation3 = x^3 + a x^2 + b x + 2 == 0;**

In[17]:= **solution3 = Solve[equation3, x][[3]] // Simplify**

Out[17]= $\left\{x \to \dfrac{1}{12}\left(-4\,a+\left(2\,I\,2^{1/3}\left(I+\sqrt{3}\right)\left(a^2-3\,b\right)\right)\right/\right.$

$\qquad\left(-54-2\,a^3+9\,a\,b+3\,\sqrt{3}\,\sqrt{108+8\,a^3-36\,a\,b-a^2\,b^2+4\,b^3}\,\right)^{\wedge}$

$\qquad(1/3)-2^{2/3}\left(1+I\,\sqrt{3}\right)\left(-54-2\,a^3+9\,a\,b+\right.$

$\qquad\left.\left.3\,\sqrt{3}\,\sqrt{108+8\,a^3-36\,a\,b-a^2\,b^2+4\,b^3}\,\right)^{\wedge}(1/3)\right)\right\}$

Again, *Mathematica* can deal with checking this.

In[18]:= **equation3 /. solution3 // Simplify**

Out[18]= True

One can of course replace the symbolic values by actual numbers in the solution.

In[19]:= **solutionAb1 = solution3 /. {a -> 3, b -> 2} // Simplify**

Out[19]= $\left\{ x \to \dfrac{1}{12} \left(-12 + \dfrac{6\left(1 + I\sqrt{3}\right)}{\left(27 - 3\sqrt{78}\right)^{1/3}} - 2^{2/3}\left(1 + I\sqrt{3}\right)\left(-54 + 6\sqrt{78}\right)^{1/3} \right) \right\}$

Does this result agree with the solution of the equation where the substitution is made before solving it?

In[20]:= **solutionAb2 =**
        Solve[equation3 /. {a -> 3, b -> 2} , x][[3]]//Simplify

Out[20]= $\left\{ x \to \dfrac{1}{6} \left(-6 + \left(1 - I\sqrt{3}\right)\left(27 - 3\sqrt{78}\right)^{1/3} + \dfrac{3^{1/6}\left(3\,I + \sqrt{3}\right)}{\left(9 - \sqrt{78}\right)^{1/3}} \right) \right\}$

In Version 2.1, these two solutions looked quite different, while in Version 2.2 they came out to be identical. Now in Version 3 they are different again. To check that they are really the same, we can take their difference and see if it simplifies to 0. Ordinary **Simplify** doesn't work (try it), but **FullSimplify** does.

In[21]:= **((x /. solutionAb1) - (x /. solutionAb2)) // FullSimplify**

Out[21]= 0

Next, let's look at an equation that cannot be solved exactly.

In[22]:= **equation4 = x^5 + 5 x + 1 == 0;**

In[23]:= **solution4 = Solve[equation4, x]**

Out[23]= $\{\{x \to \text{Root}[1 + 5\,\#1 + \#1^5 \,\&, \, 1]\},$
        $\{x \to \text{Root}[1 + 5\,\#1 + \#1^5 \,\&, \, 2]\}, \{x \to \text{Root}[1 + 5\,\#1 + \#1^5 \,\&, \, 3]\},$
        $\{x \to \text{Root}[1 + 5\,\#1 + \#1^5 \,\&, \, 4]\}, \{x \to \text{Root}[1 + 5\,\#1 + \#1^5 \,\&, \, 5]\}\}$

This is our first encounter with a new aspect of Version 3. If the solutions of an equation cannot be found exactly, then they are expressed in terms of Root objects of the form **Root[function, n]**. Here **function** is some pure function (as explained in Chapter 6) in which the variable has been replace by a **#1** with the whole expression followed by **&**, and **n** refers to a particular root in some numbering of the roots. *Mathematica* does know that these strange expressions are solutions of the corresponding equation (although the check fails without the simplification step).

In[24]:= **equation4 /. solution4 // Simplify**

Out[24]= {True, True, True, True, True}

Such solutions can always be converted to numerical form.

In[25]:= **solution4n = N[solution4]**

Out[25]= {{x → -0.199936},
{x → -1.0045 - 1.06095 I}, {x → -1.0045 + 1.06095 I},
{x → 1.10447 - 1.05983 I}, {x → 1.10447 + 1.05983 I}}

If we try to check the numerical solutions, then the check fails.

In[26]:= **equation4 /. solution4n // Simplify**

Out[26]= {False, False, False, False, False}

We have to try harder, since we expect that these really are approximate solutions. We could substitute the values of **solution4n** in the left-hand side of **equation4** and see if we get the right-hand side, i.e., **0**.

In[27]:= **equation4[[1]] /. solution4n**

Out[27]= {-1.30104 × 10^{-17}, 8.88178 × 10^{-16} + 8.88178 × 10^{-16} I,
8.88178 × 10^{-16} - 8.88178 × 10^{-16} I, -2.66454 × 10^{-15} + 2.66454 × 10^{-15} I,
-2.66454 × 10^{-15} - 2.66454 × 10^{-15} I}

These are all tiny numbers, so **Chop** should eliminate them. As long as we believe that these tiny results are artifacts of the solution algorithm used by *Mathematica* (and all other such programs), we are probably justified in using **Chop**. (Of course, it is trivial to write an equation which genuinely has such a tiny solution.)

In[28]:= **Chop[%]**

Out[28]= {0, 0, 0, 0, 0}

It's reassuring to see five zeros. Another way to proceed is to find the numerical solutions to greater accuracy.

In[29]:= **solution4nn = N[solution4, 17]**

Out[29]= {{x → -0.19993610217122000},
{x → -1.0044974557968355 - 1.0609465064060406 I},

$\{x \to -1.0044974557968355 + 1.0609465064060406 \, I\}$,
$\{x \to 1.1044655068824455 - 1.0598296691525201 \, I\}$,
$\{x \to 1.1044655068824455 + 1.0598296691525201 \, I\}\}$

Now the check proceeds without difficulty. (Any precision greater than 16 works.)

In[30]:= **equation4 /. solution4nn**

Out[30]= {True, True, True, True, True}

Question: Should we believe this result more than the previous one? The following is faster and more efficient if one knows that the best that can be achieved is a numerical solution.

In[31]:= **NSolve[equation4, x, WorkingPrecision -> 17]**

Out[31]= $\{\{x \to -1.0044974557968355 - 1.0609465064060406 \, I\}$,
$\{x \to -1.0044974557968355 + 1.0609465064060406 \, I\}$,
$\{x \to -0.1999361021712200\}$,
$\{x \to 1.1044655068824455 - 1.0598296691525201 \, I\}$,
$\{x \to 1.1044655068824455 + 1.0598296691525201 \, I\}\}$

The check proceeds without difficulty.

In[32]:= **equation4 /. %**

Out[32]= {True, True, True, True, True}

3.1.2 Transcendental equations

Mathematica can deal with certain equations containing transcendental functions applied to the variable. It always gives a warning that it may not find all solutions.

In[1]:= **equation5 = Cos[x]^2 + 2 Cos[x] + 4 == 0;**

In[2]:= **solution5 = Solve[equation5, x]**

    Solve::ifun : Inverse functions are being
        used by Solve, so some solutions may not be found.

Out[2]= $\left\{\left\{x \to -\text{ArcCos}\left[-1 - I \sqrt{3}\,\right]\right\}, \left\{x \to \text{ArcCos}\left[-1 - I \sqrt{3}\,\right]\right\},\right.$
$\left.\left\{x \to -\text{ArcCos}\left[-1 + I \sqrt{3}\,\right]\right\}, \left\{x \to \text{ArcCos}\left[-1 + I \sqrt{3}\,\right]\right\}\right\}$

In[3]:= **equation5 /. solution5 // Simplify**

Out[3]= {True, True, True, True}

Mathematica used to find only two solutions to this equation, but now it knows that **Cos** is an even function and hence there are four solutions. Here is another example that only began working in Version 2.1.

In[4]:= **equation6 = 2^x == 8;**

In[5]:= **solution6 = Solve[equation6, x]**

Out[5]= {{x → 3}}

But not all such equations can be solved so easily.

In[6]:= **Solve[Cos[x] == x, x]**

```
Solve::tdep :
  The equations appear to involve transcendental functions
    of the variables in an essentially non-algebraic way.
```

Out[6]= Solve[Cos[x] == x, x]

The message tells the whole story. There is no way we can hope to "solve" equations like this exactly. Instead, numerical methods are required. Newton's method is the obvious one, which is implemented in the **FindRoot** command. We ask it to find a root near **x** = 0.5.

In[7]:= **FindRoot[Cos[x] == x, {x, 0.5}]**

Out[7]= {x → 0.739085}

Of course, if you ask something impossible, **FindRoot** may also give up.

In[8]:= **FindRoot[Sin[x] == 2, {x, 1}]**

```
FindRoot::cvnwt : Newton's method failed to
    converge to the prescribed accuracy after 15 iterations.
```

Out[8]= {x → -10.3883}

The problem is that **Sin[x]** is always between -1 and +1 for real arguments and so it can never equal 2. However, if **x** is allowed to take on complex values, then there is no problem. We tell *Mathematica* this by giving a complex seed. Again we set the **WorkingPrecision** high enough (namely, 1 more than `$MachinePrecision`) so that the subsequent check succeeds.

```
In[9]:= FindRoot[Sin[x] == 2, {x, 1 + I},
            WorkingPrecision -> 17]
```

Out[9]= {x → 1.5707963267948966 + 1.3169578969248167 I}

```
In[10]:= Sin[x] == 2 /.%
```

Out[10]= True

3.1.3 Example of an equation with an exact solution that isn't found

Consider the following special sixth degree equation.

```
In[11]:= equation7 = x⁶ - 9 x⁴ - 4 x³ + 27 x² - 36 x - 23 == 0;
```

```
In[12]:= solution7 = Solve[equation7, x]
```

Out[12]= {{x → Root[-23 - 36 #1 + 27 #1² - 4 #1³ - 9 #1⁴ + #1⁶ &, 1]},
          {x → Root[-23 - 36 #1 + 27 #1² - 4 #1³ - 9 #1⁴ + #1⁶ &, 2]},
          {x → Root[-23 - 36 #1 + 27 #1² - 4 #1³ - 9 #1⁴ + #1⁶ &, 3]},
          {x → Root[-23 - 36 #1 + 27 #1² - 4 #1³ - 9 #1⁴ + #1⁶ &, 4]},
          {x → Root[-23 - 36 #1 + 27 #1² - 4 #1³ - 9 #1⁴ + #1⁶ &, 5]},
          {x → Root[-23 - 36 #1 + 27 #1² - 4 #1³ - 9 #1⁴ + #1⁶ &, 6]}}

But we can give a solution ourselves.

```
In[13]:= solution77 = {x → ∛2 + √3 }
```

Out[13]= {x → $2^{1/3} + \sqrt{3}$ }

```
In[14]:= equation7 /.solution77 // Simplify
```

Out[14]= True

3.1.4 A funny equation

Sometimes *Mathematica* can solve strange equations.

In[15]:= **equation8 = $\sqrt{1-x}$ + $\sqrt{1+x}$ == a**

Out[15]= $\sqrt{1-x}$ + $\sqrt{1+x}$ == a

In[16]:= **solution8 = Solve[equation8, x]**

Out[16]= $\left\{\left\{x \to -\frac{1}{2}\sqrt{4a^2 - a^4}\right\}, \left\{x \to \frac{1}{2}\sqrt{4a^2 - a^4}\right\}\right\}$

However, it's not able to do anything about checking this solution by itself. Even **FullSimplify** can't deal with this.

In[17]:= **equation8 /. solution8 // FullSimplify**

Out[17]= $\left\{\sqrt{1 - \frac{1}{2}\sqrt{-a^2(-4+a^2)}} + \sqrt{1 + \frac{1}{2}\sqrt{4a^2-a^4}} == a,\right.$

$\left.\sqrt{1 - \frac{1}{2}\sqrt{-a^2(-4+a^2)}} + \sqrt{1 + \frac{1}{2}\sqrt{4a^2-a^4}} == a\right\}$

Here is some magic, using pure functions as discussed in Chapter 6 together with local patterned rewrite rules as discussed in Chapter 7, which shows that at least the squares of the two sides are the same.

In[18]:= **PowerExpand[**
       Map[Expand[#^2]&,
           equation8 /. solution8, {2}]/.
       Sqrt[x_] Sqrt[y_] :> Sqrt[Simplify[x y]]]

Out[18]= {True, True}

3.2 Simultaneous Equations – Groebner Bases

The same methods work for several equations in several variables. In Chapter 1 we looked at linear equations with symbolic constants and also higher order equations. They are checked in exactly the same way.

In[1]:= **equations9 = {a x + b y == 1, x - y == 2};**

In[2]:= **solution9 = Solve[equations9, {x, y}]**

Out[2]= $\left\{\left\{x \to -\frac{-1-2b}{a+b}, y \to -\frac{-1+2a}{a+b}\right\}\right\}$

In[3]:= **equations9 /. solution9 // Simplify**

Out[3]= {{True, True}}

Here is a more complicated pair of non-linear equations related to the system we investigated in Chapter 1.

In[4]:= **equations10 = {x^2 + y^2 == 13, x^3 + y^3 == 9};**

We suppress the output completely by ending the **Solve** command with a ";". The answer fills a whole screen and the calculation takes a noticeable length of time.

In[5]:= **solution10 = Solve[equations10, {x, y}];**

We can still check that the answer is correct and see that six solutions were in fact found.

In[6]:= **equations10 /. solution10 // Simplify**

Out[6]= {{True, True}, {True, True}, {True, True},
          {True, True}, {True, True}, {True, True}}

Instead of giving a list of equations, one can give a list of left-hand sides "equals equals" to a list of right-hand sides. This time we solve them numerically.

In[7]:= **equations11 = {x^2 + y^2, x^3 + y^3} == {13, 9};**

In[8]:= **solution11 = NSolve[equations11, {x, y}]**

Out[8]= {{x → -3.23205 - 1.98649 I, y → -3.23205 + 1.98649 I},
          {x → -3.23205 + 1.98649 I, y → -3.23205 - 1.98649 I},
          {x → -2.30688, y → 2.77098}, {x → 2.77098, y → -2.30688},
          {x → 3. - 1.58114 I, y → 3. + 1.58114 I},
          {x → 3. + 1.58114 I, y → 3. - 1.58114 I}}

As before, there are six solutions since Bezout's theorem says that a curve of degree 3 and a curve of degree 2 intersect in six points. If we add a symbolic constant, then the solution of this kind of system really takes somewhat longer. We look at just the first exact solution.

In[9]:= **equations12 = {x^2 + y^2 == 13 a^2, x^3 + y^3 == 9 a^3};**

In[10]:= **solution12 = Solve[equations12, {x, y}][[1]]//Simplify**

Out[10]= $\left\{ x \rightarrow \left(3 - I \sqrt{\dfrac{5}{2}} \right) a, \ y \rightarrow \left(3 + I \sqrt{\dfrac{5}{2}} \right) a \right\}$

Let us investigate how *Mathematica* goes about solving such systems of equations. The idea is to "diagonalize" the equations just as is done for linear equations, except that now the equations will be polynomial ones. The goal is to end up with an equation in just one of the variables. The resulting set of equations is called a Groebner basis for the original equations. (Actually, it is a basis of a particular form for the polynomial ideal spanned by the original equations.) There is a built-in command to find such a basis.

In[11]:= **gBasis = GroebnerBasis[equations12, {x, y}]**

Out[11]= $\{-2116\,a^6 + 507\,a^4\,y^2 - 18\,a^3\,y^3 - 39\,a^2\,y^4 + 2\,y^6,$
$2116\,a^4\,x + 2116\,a^4\,y - 351\,a^3\,y^2 - 169\,a^2\,y^3 + 18\,a\,y^4 + 26\,y^5,$
$169\,a^4 - 9\,a^3\,x - 9\,a^3\,y + 13\,a^2\,x\,y - 26\,a^2\,y^2 + 2\,y^4,$
$9\,a^3 - 13\,a^2\,x + x\,y^2 - y^3, \ -13\,a^2 + x^2 + y^2\}$

The first entry in this list of five equations involves only **y**, so we can try to solve it. We will just look at the second solution.

In[12]:= **solutionY = Solve[gBasis[[1]] == 0, y][[2]]//Simplify**

Out[12]= $\left\{y \to \left(3 + I\,\sqrt{\dfrac{5}{2}}\right) a\right\}$

This agrees with the value we found for **y** in **solution12**, so let's try to find the corresponding value of **x**. In the remaining equations in the Groebner basis, the second one involves just the first power of **x**, so we can substitute in the value we just found for **y** in it and solve for **x**.

In[13]:= **solutionX = Solve[(gBasis[[2]]/.solutionY) == 0, x] // Simplify**

Out[13]= $\left\{\left\{x \to \left(3 - I\,\sqrt{\dfrac{5}{2}}\right) a\right\}\right\}$

These values for **x** and **y** are exactly what we found earlier in the direct solution.

3.3 Simultaneous Equations – FindRoot

If the equations are not multivariate polynomials, then **Solve** and even **NSolve** may fail. For instance, the following system is one of Simon's challenge problems in the *Notices of the AMS*, Sept. 1991.

In[14]:= **equations13 =**
   {Sin[x] + y^2 + Log[z] == 7,
           3 x + 2^y -   z^3 == -1,
              x + y +   z   == 5};

In[15]:=**NSolve[equations13, {x, y, z}]//Simplify;**

We have suppressed the output because it is 15 screens of apparent nonsense. Attempting to check the output results in many more pages of nonsense. So this is not the way to try to solve this problem. However, **FindRoot** tries to find a solution, given seed values for the variables, for any system of equations in any number of variables. Here are the only two solutions we have been able to find.

In[16]:=**solutions13 =**
     {FindRoot[equations13, {x, 1}, {y, 1}, {z, 1},
              WorkingPrecision->17],
      FindRoot[equations13, {x, 0}, {y, 0}, {z, 2},
              WorkingPrecision->17]}

Out[16]= {{x → 0.5990537566405673, y → 2.3959314023778168,
       z → 2.0050148409816158}, {x → 5.100412729886776,
       y → -2.6442371270278302, z → 2.5438243971410540}}

In[17]:=**equations13 /. solutions13**

Out[17]= {{True, True, True}, {True, True, True}}

3.4 Matrix Equations

Two vectors or matrices are "equals equals" providing corresponding entries are the same. In particular, this means that we can write matrix equations. First define a coefficient matrix, a variable vector, and a right-hand side vector.

In[18]:=**A = {{3, 1}, {2, -5}};**
     X = {x, y};
     B = {7, 8};

In[21]:=**equations14 = A . X == B;**

In[22]:=**solution14 = Solve[equations14, {x, y}]**

Out[22]= $\left\{\left\{x \rightarrow \dfrac{43}{17}, y \rightarrow -\dfrac{10}{17}\right\}\right\}$

In[23]:=`equations14 /. solution14 // Simplify`

Out[23]= `{True}`

3.5 *Indexed Variables*

The **Table** command can also be used to construct equations and lists of variables.

In[24]:=`equations15 =`
`        Table[2 a[i] + a[i - 1] == a[i + 1], {i, 5}]`

Out[24]= `{a[0] + 2 a[1] == a[2], a[1] + 2 a[2] == a[3],`
`        a[2] + 2 a[3] == a[4], a[3] + 2 a[4] == a[5], a[4] + 2 a[5] == a[6]}`

In[25]:=`solution15 =`
`        Solve[equations15, Table[a[i], {i, 5}]]//Simplify`

Out[25]= $\left\{\left\{a[1] \rightarrow \frac{1}{70} (-29 a[0] + a[6]),\right.\right.$

$\quad a[2] \rightarrow \frac{1}{35} (6 a[0] + a[6]), a[3] \rightarrow \frac{1}{14} (-a[0] + a[6]),$

$\quad \left.\left. a[4] \rightarrow \frac{1}{35} (a[0] + 6 a[6]), a[5] \rightarrow \frac{1}{70} (-a[0] + 29 a[6])\right\}\right\}$

In[26]:=`equations15 /. solution15 // Simplify`

Out[26]= `{{True, True, True, True, True}}`

3.6 *Complete Solutions*

Besides **Solve** and **Root**, there is another command to solve equations that is particularly useful for equations with symbolic constants where the form of the answer may depend on relations between the constants. If we use **Solve** to solve a generic quadratic equation, we get the usual high school formula.

In[27]:= `Solve[a x² + b x + c == 0, x]`

Out[27]= $\left\{\left\{x \rightarrow \frac{-b - \sqrt{b^2 - 4 a c}}{2 a}\right\}, \left\{x \rightarrow \frac{-b + \sqrt{b^2 - 4 a c}}{2 a}\right\}\right\}$

However, this is clearly wrong if, for instance, **a** is zero. The full story is given by the command **Reduce**.

In[28]:= **Reduce[a x^2 + b x + c == 0, x]**

Out[28]= $a \neq 0$ && $x == \dfrac{-b - \sqrt{b^2 - 4\,a\,c}}{2\,a}$ $||$ $a \neq 0$ && $x == \dfrac{-b + \sqrt{b^2 - 4\,a\,c}}{2\,a}$ $||$

$\quad c == 0$ && $b == 0$ && $a == 0$ $||$ $b \neq 0$ && $x == -\dfrac{c}{b}$ && $a == 0$

This output uses the logical operators $||$ for "**Or**" and && for "**And**". Here **And** takes prece-dence over **Or**. The output means that the possible solutions are: **a** is not zero, and there are the usual two high school solutions; or all three of **a**, **b**, and **c** are zero, in which case there is no restriction on **x**, or **b** is not zero but **a** is zero and **x == $-\frac{c}{b}$**.

   Equations can also be given as logical combinations of equalities instead of as lists. We use **equations9** as an example.

In[29]:= **equations9a = a x + b y == 1 && x - y == 2;**

In[30]:= **solution9a = Solve[equations9a, {x, y}] // Simplify**

Out[30]= $\left\{\left\{x \to \dfrac{1 + 2\,b}{a + b},\ y \to \dfrac{1 - 2\,a}{a + b}\right\}\right\}$

Checking this solution requires a little bit of magic. Try **Simplify** and even **FullSimplify** to see what happens.

In[31]:= **Map[Together, equations9a /.solution9a, {3}]**

Out[31]= {True}

Now there is just one value of True since **equations9a** is the conjunction of the two equa-tions.

3.7 *Eliminating Variables*

Solve can also take a third argument which is a "variable" or list of "variables" that should be eliminated from the solution.

In[32]:= **equations16 = {x == 1 + 2 a, y == 9 + 2 x a};**

First solve for **x** and **y** (in terms of **a**).

In[33]:= **Solve[equations16, {x, y}]**

Out[33]= $\{\{y \rightarrow 9 + 2\, a + 4\, a^2, x \rightarrow 1 + 2\, a\}\}$

Next, solve for **x** and **a** (in terms of **y**).

In[34]:= **Solve[equations16, {x, a}]**

Out[34]= $\{\{x \rightarrow \frac{1}{2}\,(1 - \sqrt{-35 + 4\,y}\,), a \rightarrow \frac{1}{4}\,(-1 - \sqrt{-35 + 4\,y}\,)\},$
$\{x \rightarrow \frac{1}{2}\,(1 + \sqrt{-35 + 4\,y}\,), a \rightarrow \frac{1}{4}\,(-1 + \sqrt{-35 + 4\,y}\,)\}\}$

Now solve for **x**, eliminating **y**.

In[35]:= **Solve[equations16, x, y]**

Out[35]= $\{\{x \rightarrow 1 + 2\, a\}\}$

Then solve for **x** eliminating **a**.

In[36]:= **Solve[equations16, x, a]**

Out[36]= $\{\{x \rightarrow \frac{1}{2}\,(1 - \sqrt{-35 + 4\,y}\,)\}, \{x \rightarrow \frac{1}{2}\,(1 + \sqrt{-35 + 4\,y}\,)\}\}$

Finally, eliminate **a** from the equations.

In[37]:= **Eliminate[equations16, a]**

Out[37]= $y == 9 - x + x^2$

There is much more to be said about solutions of equations, but they are not our main concern so we will leave that for other books to treat in depth. See, e.g., The *Mathematica* Book, Chapter 3.4.

3.8 Working Modulo a Prime Number

The operation **Factor** has an optional argument, **Modulus** ←n, which gives factorizations modulo **n**. Here is a well known example of a polynomial that has no factorizations over the reals, but factors modulo p for all primes p. In the following, **Prime[n]** means the **n**th prime number. Look up **TableForm** and its options in The *Mathematica* Book.

```
In[38]:= TableForm[
     Table[
        {Prime[n], Factor[x^4 + 1, Modulus    ←Prime[n]]},
        {n, 1, 10}],
       TableHeadings   ←{None, {"prime", "factorization"}},
       TableSpacing    ←{0, 6}]
```

Out[38]//TableForm=

prime	factorization
2	$(1 + x)^4$
3	$(2 + x + x^2) \ (2 + 2\,x + x^2)$
5	$(2 + x^2) \ (3 + x^2)$
7	$(1 + 3\,x + x^2) \ (1 + 4\,x + x^2)$
11	$(10 + 3\,x + x^2) \ (10 + 8\,x + x^2)$
13	$(5 + x^2) \ (8 + x^2)$
17	$(2 + x) \ (8 + x) \ (9 + x) \ (15 + x)$
19	$(18 + 6\,x + x^2) \ (18 + 13\,x + x^2)$
23	$(1 + 5\,x + x^2) \ (1 + 18\,x + x^2)$
29	$(12 + x^2) \ (17 + x^2)$

Solve also works modulo a prime number. The condition on the modulus is added as another equation. If we use the preceding equation, then **Solve** finds solutions in case the left-hand side factors into linear terms. The following slightly mysterious command gives just these solutions for the first 20 primes.

```
In[39]:= sol = Select[Table[
         Solve[{x^4 + 1 == 0, Modulus == Prime[n]}, x],
         {n, 1, 20}], # ≠ {} &]
```

Out[39]= { { {Modulus → 2, x → 1}, {Modulus → 2, x → 1},
            {Modulus → 2, x → 1}, {Modulus → 2, x → 1} },
          { {Modulus → 17, x → 2}, {Modulus → 17, x → 8},
            {Modulus → 17, x → 9}, {Modulus → 17, x → 15} },
          { {Modulus → 41, x → 3}, {Modulus → 41, x → 14},
            {Modulus → 41, x → 27}, {Modulus → 41, x → 38} } }

It is simple to check these solutions.

```
In[40]:= Mod[x^4 + 1, Modulus] == 0 /. sol
```

Out[40]= { {True, True, True, True},
          {True, True, True, True}, {True, True, True, True} }

4 Solving Ordinary Differential Equations

There are (at least) five ways to approach ordinary differential equations in *Mathematica*: **DSolve**, **NDSolve**, **RungeKutta** methods (in a package), series solutions (by hand), and Laplace transform methods in the package **LaplaceTransform.m.** There used to be a package called **DSolve.m** that extended the functionality of the built-in **DSolve**, but all of its routines–in a greatly extended form–are now autoloaded and used by the built-in **DSolve**. These operations continue to be under intensive development and so the problems that can be solved and the forms of their solutions are a moving target. There are large differences among Versions 2.0, 2.1, 2.2, and 3.0. Everything here is from Version 3.0.

4.1 DSolve

Many simple differential equations can be solved by the built-in operation **DSolve**.

4.1.1 A linear, first order differential equation

DSolve works in two different ways. Consider a simple example and its solution.

In[1]:=**diffEq1 = y'[x] + y[x] == 1;**

In[2]:=**solution1 = DSolve[diffEq1, y[x], x]**

Out[2]= $\{\{y[x] \rightarrow 1 + E^{-x} C[1]\}\}$

The solution contains an arbitrary constant denoted by C[1]. If we try to check this solution in the same way we checked algebraic equations, it doesn't work.

In[3]:=**diffEq1 /.solution1**

Out[3]= $\{1 + E^{-x} C[1] + y'[x] == 1\}$

The trouble is that **y'** is not calculated so the solution cannot be verified. We could work around this by calculating **y'[x]** ourselves:

In[4]:=**solution1' = D[solution1, x]**

Out[4]= $\{\{y'[x] \rightarrow -E^{-x} C[1]\}\}$

Then the check succeeds if we make substitutions for both **y[x]** and **y'[x]**.

In[5]:=**diffEq1 /.solution1 /.solution1'**

Out[5]= {{True}}

However, there is a better way to do this using the other form of **DSolve**. The only difference is that instead of using **y[x]** as the second argument, one uses **y**.

In[6]:=**newsolution1 = DSolve[diffEq1, y, x]**

Out[6]= {{y → (1 + E$^{-\#1}$ C[1] &)}}

What has happened is that the **x** has been replaced by #1 and the whole expression on the right-hand side now ends with an ampersand, &. This syntax indicates a *pure function* in *Mathematica*. Pure functions will be explained in great detail in Chapter 6. The important thing here is that it gives a value for **y** rather than **y[x]**, so this substitution works for **y'** as well. Thus:

In[7]:=**y' /.newsolution1 // Simplify**

Out[7]= {(E$^{-\#1}$ (−1) Log[E]) C[1] &}

The check now proceeds exactly as in the algebraic case.

In[8]:=**diffEq1 /.newsolution1**

Out[8]= {True}

Initial conditions are given as additional equations to be satisfied. For instance:

In[9]:=**diffEq2 = {y'[x] == a y[x], y[0] == 1};**

In[10]:=**solution2 = DSolve[diffEq2, y, x]**

Out[10]= {{y → (E$^{a\,\#1}$ &)}}

The check now verifies both the differential equation and the initial condition.

In[11]:=**diffEq2 /.solution2**

Out[11]= {{True, True}}

4.1.2 A non-linear first order equation

Mathematica can solve non-linear first order equations, finding two solutions in this case.

In[12]:=**diffEq3 = y[x] y'[x] == 1;**

In[13]:=**solution3 = DSolve[diffEq3, y, x]**

Out[13]= $\left\{\left\{y \to \left(-\sqrt{2}\ \sqrt{\#1 + C[1]}\ \&\right)\right\}, \left\{y \to \left(\sqrt{2}\ \sqrt{\#1 + C[1]}\ \&\right)\right\}\right\}$

From now on, checks are omitted unless there is some difficulty in carrying them out.

4.1.3 Linear equations with constant coefficients

In principle, *Mathematica* will solve arbitrary order linear equations with constant coefficients, provided it can solve the associated indicial equation. For the special case of second order equations, there are three kinds of solutions for an equation of the form y''[x] + b y'[x] + c y[x] == 0 depending on the roots of the indicial equation z^2 + b z + c == 0. We treat the case of complex roots and leave the others to the exercises. In this case, the solution consists of a sin and a cos term times an exponential function. We also make a picture which shows the exponentially increasing oscillations. Note that it is necessary to assign values to the arbitrary constants to make such a plot.

In[14]:=**diffEqComplex = y''[x] - 2 y'[x] + 5 y[x] == 0;**

In[15]:=**solutionComplex = DSolve[diffEqComplex, y, x]**

Out[15]= $\left\{\left\{y \to \left(E^{\#1}\ C[2]\ Cos[2\ \#1] - E^{\#1}\ C[1]\ Sin[2\ \#1]\ \&\right)\right\}\right\}$

In[16]:=**Plot[Evaluate[y[x] /.solutionComplex /.{C[1]->1, C[2]->1}],
          {x, 0, Pi}];**

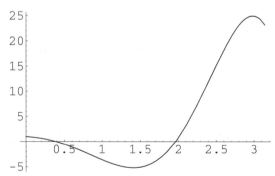

Consider also a simple higher order example where the coefficients are chosen by expanding an algebraic product.

In[17]:=**Expand[Product[x - i, {i, 1, 5}]]**

Out[17]= $-120 + 274\,x - 225\,x^2 + 85\,x^3 - 15\,x^4 + x^5$

In[18]:= `diffEq4 = y'''''[x] - 15 y''''[x] + 85 y'''[x] -`
           `225 y''[x] + 274 y'[x] - 120 y[x] == 0;`

In[19]:= `solution4 = DSolve[diffEq4, y, x]`

Out[19]= $\{\{y \to (E^{\#1}\,C[1] + E^{2\,\#1}\,C[2] + E^{3\,\#1}\,C[3] + E^{4\,\#1}\,C[4] + E^{5\,\#1}\,C[5]\,\&)\}\}$

Just for fun, at this stage, to produce the preceding equation without copying the coefficients, use the form

In[20]:= `Table[Derivative[i][y], {i, 0, 5}] .`
        `CoefficientList[Expand[Product[x - i, {i, 1, 5}]], x]`

Out[20]= $-120\,y + 274\,y' - 225\,y'' + 85\,y^{(3)} - 15\,y^{(4)} + y^{(5)}$

On the other hand, an equation like the following, which is only of the third degree and can be solved, in earlier versions of *Mathematica* resulted in an answer which was too complicated for a human to comprehend. The form of the answer is now completely different, being expressed in terms of Root objects. This equation will be investigated numerically later.

In[21]:= `diffEq5 = y'''[x] + y''[x] + y'[x] + a y[x] == 0;`

In[22]:= `solution5 = DSolve[diffEq5, y, x]`

Out[22]= $\left\{\left\{y \to \left(E^{\mathrm{Root}[a + \#1 + \#1^2 + \#1^3 \&,\, 1]\,\#1}\,C[1] + \right.\right.\right.$
$\left.\left.\left. E^{\mathrm{Root}[a + \#1 + \#1^2 + \#1^3 \&,\, 2]\,\#1}\,C[2] + E^{\mathrm{Root}[a + \#1 + \#1^2 + \#1^3 \&,\, 3]\,\#1}\,C[3]\,\&\right)\right\}\right\}$

A check of this answer proceeds without difficulty. Looking at it we can see that the result is just a sum of three exponentials (as is to be expected) and the only real complication is hidden in the description of the roots of the characteristic equation. We can, for instance, plot the solution by giving values to the arbitrary constants and to the constant **a**.

In[23]:= `solution51 =`
        `y[x] /. solution5[[1, 1]] /. Table[C[i] -> i, {i, 1, 3}]`

Out[23]= $E^{x\,\mathrm{Root}[a + \#1 + \#1^2 + \#1^3 \&,\, 1]} + 2\,E^{x\,\mathrm{Root}[a + \#1 + \#1^2 + \#1^3 \&,\, 2]} + 3\,E^{x\,\mathrm{Root}[a + \#1 + \#1^2 + \#1^3 \&,\, 3]}$

This result depends on both **a** and **x**, so we plot the real part of this as a function of two variables.

In[24]:= `Plot3D[Evaluate[Re[solution51]], {x, 1, 3}, {a, -1, 1}];`

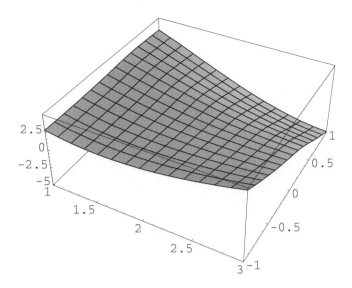

4.2 Some Typical First Order Differential Equations

DSolve can deal with very many common first and second order differential equations. These examples and the problems in the exercises, together with the series solution examples and the Laplace transform examples, constitute a mini-course in ordinary differential equations.

4.2.1 Exact equations

Here are three equations that that are made exact by integrating factors, which can be solved directly.

In[25]:= **diffEqInt1 = (x y[x] + x²) y'[x] + y[x]² + 3 x y[x] == 0;**

In[26]:=**solutionInt1 = DSolve[diffEqInt1, y, x]//Simplify**

Out[26]= $\left\{\left\{y \to \left(\dfrac{-\#1^2 - \sqrt{\#1^4 + 2\, C[1]}}{\#1}\; \& \right)\right\}, \left\{y \to \left(\dfrac{-\#1^2 + \sqrt{\#1^4 + 2\, C[1]}}{\#1}\; \& \right)\right\}\right\}$

This first one could be solved completely. The second one is only solved implicitly, so we solve for **y[x]** rather than for **y**.

In[27]:= **diffEqInt2 = (2 x y[x] - E$^{-2\,y[x]}$) y'[x] + y[x] == 0;**

In[28]:= **solutionInt2 = DSolve[diffEqInt2, y[x], x]**

Out[28]= $\text{Solve}[E^{2\,y[x]}\,x - \text{Log}[y[x]] == C[1], \{y[x]\}]$

The equation to be solved that is given as the output here is one of the usual ways to present solutions to differential equations, as relations between x and y rather than finding y as a function of x. In this case, the solution is in the form of x expressed as a function of y. Checking this result requires a certain amount of experimentation. First find the derivative of $y[x]$.

In[29]:= $\textbf{solution' = Solve[D[solutionInt2[[1]], x], y'[x]]}$

Out[29]= $\left\{\left\{y'[x] \to -\dfrac{E^{2\,y[x]}\,y[x]}{-1 + 2\,E^{2\,y[x]}\,x\,y[x]}\right\}\right\}$

In[30]:= $\textbf{solx = Solve[solutionInt2[[1]] /. y[x] - <foo, x][[1]]}$

Out[30]= $\{x \to E^{-2\,foo}\,(C[1] + \text{Log}[foo])\}$

Then, essentially substitute the values for x and $y'[x]$ in the differential equation. We replace $y[x]$ everywhere by the nonsense symbol foo because it has to be treated as independent of x here.

In[31]:= $\textbf{diffEqInt2 /. solution' /. y[x] - <foo /. solx // Simplify}$

Out[31]= $\{\text{True}\}$

The third one can also be solved exactly.

In[32]:= $\textbf{diffEqInt3 = y'[x] == }\dfrac{3\,\textbf{y[x]}^2 + \textbf{x}^2}{2\,\textbf{x}\,\textbf{y[x]}}\textbf{;}$

In[33]:= $\textbf{solutionInt3 = DSolve[diffEqInt3, y, x]}$

Out[33]= $\left\{\left\{y \to \left(-\sqrt{-\#1^2 + \#1^3\,C[1]}\ \&\right)\right\}, \left\{y \to \left(\sqrt{-\#1^2 + \#1^3\,C[1]}\ \&\right)\right\}\right\}$

4.2.2 Generalized homogeneous equations

In this case, DSolve finds 30 rather complicated solutions. We look at just the first solution in terms of $y[x]$.

In[34]:= $\textbf{diffEqGenHom1 = x (y[x]}^2\textbf{ - 3 x) y'[x] + 2 y[x]}^3\textbf{ - 5 x y[x] == 0;}$

In[35]:= $\textbf{solutionGenHom1 = DSolve[diffEqGenHom1, y[x], x][[1]]}$

Out[35]= $\{y[x] \to -\sqrt{\text{Root}[}$
$$-169\, x^2\, C[1]^{65} + 130\, x\, C[1]^{65}\, \#1 - 25\, C[1]^{65}\, \#1^2 + x^{52}\, \#1^{15}\, \&, 1]\}$$

Notice that the polynomial in the **Root** object has coefficients that depend on **x**. It is easily checked as in 4.1.1 that this answer is correct. However, if we try to solve for **y**, then we get the following result.

In[36]:= **solution2GenHom1 = DSolve[diffEqGenHom1, y, x][[1]]**

Out[36]= $\{y \to (-\sqrt{\text{Root}[-169\, \#1^2\, C[1]^{65} + }$
$$130\, \#1\, C[1]^{65}\, \#1 - 25\, C[1]^{65}\, \#1^2 + \#1^{52}\, \#1^{15}\, \&, 1]\, \&)\}$$

Here, #1 has been syntactically substituted for **x** in the coefficients, resulting in the #1 being captured by the #1 that is there already. A check shows that this is wrong.

In[37]:= **diffEqGenHom1 /. solution2GenHom1**

Out[37]= False

4.2.3 A Riccati equation

The exact solution is found with no difficulty.

In[38]:= **diffEqRic = (1 - x^2) y'[x] == 1 - (2 x - y[x]) y[x];**

In[39]:= **solutionRic = DSolve[diffEqRic, y, x]**

Out[39]= $\left\{\left\{y \to \left(\dfrac{1 + \#1\, C[1] - \#1\, \text{Log}\left[\frac{\sqrt{1+\#1}}{\sqrt{-1+\#1}}\right]}{C[1] - \text{Log}\left[\frac{\sqrt{1+\#1}}{\sqrt{-1+\#1}}\right]}\, \&\right)\right\}\right\}$

4.2.4 Bernoulli's equation

The last general type equation we look at here is a Bernoulli's equation

In[40]:= **diffEq6 = x^2 y'[x] - y[x]^3 + 2 x y[x] == 0;**

In[41]:= **solution6 = DSolve[diffEq6, y, x] // Simplify**

Out[41]= $\left\{\left\{y \to \left(-\dfrac{\sqrt{5}\ \sqrt{\#1}\ C[1]^{5/2}}{\sqrt{-\#1^5 + 2\,C[1]^5}}\ \&\right)\right\},\ \left\{y \to \left(\dfrac{\sqrt{5}\ \sqrt{\#1}\ C[1]^{5/2}}{\sqrt{-\#1^5 + 2\,C[1]^5}}\ \&\right)\right\}\right\}$

4.2.5 A different kind of equation

In[42]:= `DSolve[y'[x] == 1/(x y[x] + 1), y[x], x]`

Out[42]= $\text{Solve}\left[-\dfrac{1}{2}\ E^{-\frac{1}{2}y[x]^2}\left(-2\,x + E^{\frac{y[x]^2}{2}}\ \sqrt{2\,\pi}\ \text{Erf}\left[\dfrac{y[x]}{\sqrt{2}}\right]\right) == C[1],\ \{y[x]\}\right]$

This can be checked using the method for **diffEqInt2**.

4.2.6 A harder equation

In[43]:= `DSolve[y'[x] == a y[x]^3 + b x^(-3/2), y[x], x]`

Out[43]= $\text{Solve}\Big[$

$\qquad \text{Log}[x] - 2\,\text{RootSum}\Big[2\,b + \sqrt{x}\ \#1 + 2\,a\,x^{3/2}\ \#1^3\ \&,\ \dfrac{\text{Log}[-\#1 + y[x]]}{1 + 6\,a\,x\,\#1^2}\ \&\Big] ==$

$\qquad C[1],\ \{y[x]\}\Big]$

This cannot be checked using the method for **diffEqInt2** because the step involving a solution for **x** cannot be carried out.

4.3 Some Typical Second Order Differential Equations

4.3.1 An exact second order differential equation

In[1]:= `diffEqEx2 = y''[x] + x y'[x] + y[x] == 0;`

In[2]:= `solutionEx2 = DSolve[diffEqEx2, y, x]`

Out[2]= $\left\{\left\{y \to \left(E^{-\frac{\#1^2}{2}}\ C[1] + E^{-\frac{\#1^2}{2}}\ C[2]\ \text{Erfi}\left[\dfrac{\sqrt{\#1^2}}{\sqrt{2}}\right]\ \&\right)\right\}\right\}$

4.3.2 Bessel's equation

First, the zeroth order equation.

In[3]:= `diffEqBessel = y''[x] +` $\dfrac{y'[x]}{x}$ `+ y[x] == 0;`

In[4]:= `solutionBessel = DSolve[diffEqBessel, y, x]`

Out[4]= $\left\{\left\{y \to \left(E^{-\frac{\text{Log}[\#1]}{2} - I\,\#1}\left(\dfrac{E^{I\,\#1}\ \text{BesselK}[0,\ I\,\#1]\ C[1]}{\sqrt{\pi}} +\right.\right.\right.\right.$
$\left.\left.\left.\left. E^{I\,\#1}\ \text{BesselI}[0,\ I\,\#1]\ C[2]\right)\sqrt{\#1}\ \&\right)\right\}\right\}$

If the equation is multiplied through by x^2 as we did in Chapter 1, Section 4.3, then the answer comes out in terms of **BesselJ** and **BesselY** as one expects. Nevertheless, *Mathematica* is able to check this solution using **FullSimplify**. Try just **Simplify** here to see how much more **FullSimplify** knows than **Simplify**.

In[5]:= `diffEqBessel /. solutionBessel // FullSimplify`

Out[5]= {True}

Now try the **n**th order Bessel's equation.

In[6]:= `diffEqBesseln = x² y''[x] + x y'[x] + (x² - n²) y[x] == 0;`

In[7]:= `solutionBesseln = DSolve[diffEqBesseln, y, x]`

Out[7]= $\left\{\left\{y \to \left(\text{BesselJ}\left[-n,\ \sqrt{\#1^2}\ \right]C[1] + \text{BesselJ}\left[n,\ \sqrt{\#1^2}\ \right]C[2]\ \&\right)\right\}\right\}$

This time, even **FullSimplify** is unable to complete the check of this solution, which we know is correct anyway.

4.3.3 Variation of parameters

Here is a non-homogeneous second order linear differential equation.

In[8]:= `diffEqVar1 = (x² + 1) y''[x] + 2 x y'[x] +` $\dfrac{3}{x^2}$ `== 0;`

In[9]:= **solutionVar1 = DSolve[diffEqVar1, y, x]**

Out[9]= $\left\{\left\{y \to \left(\frac{1}{2} \text{ArcTan}[\#1] \ (-3 \ I + C[1]) + \right.\right.\right.$

$\quad\quad \frac{1}{2} \text{ArcTan}[\#1] \ (3 \ I + C[1]) + C[2] + 3 \ \text{Log}[\#1] -$

$\quad\quad \left.\left.\left. \frac{1}{4} \ I \ (-3 \ I + C[1]) \ \text{Log}[1 + \#1^2] + \frac{1}{4} \ I \ (3 \ I + C[1]) \ \text{Log}[1 + \#1^2] \ \& \right)\right\}\right\}$

And here is a non-linear equation involving a power of **y'[x]**.

In[10]:= **diffEqNonL = y''[x] + y[x] y'[x]^3 == 0;**

In[11]:= **solutionNonL = DSolve[diffEqNonL, y, x];**

There are three lengthy solutions, which are suppressed. A check proceeds without difficulty.

4.3.4 The Legendre equation

Mathematica is able to solve and check the general second order Legendre differential equation.

In[12]:= **diffEqLeg = (1 - x^2) y''[x] - 2 x y'[x] + n (n - 1) == 0;**

In[13]:= **solutionLeg = DSolve[diffEqLeg, y, x]**

Out[13]= $\left\{\left\{y \to \left(C[2] + \frac{1}{2} \ (-n + n^2 + C[1]) \ \text{Log}[-1 + \#1] + \right.\right.\right.$

$\quad\quad \left.\left.\left. \frac{1}{2} \ (-n + n^2 - C[1]) \ \text{Log}[1 + \#1] \ \& \right)\right\}\right\}$

4.3.5 A different equation

In[14]:= **diffEqDif = y''[x] + x y'[x] + Exp[-x^2] y[x] == 0;**

In[15]:= **solutionDif = DSolve[diffEqDif, y, x]**

Out[15]= $\left\{\left\{y \to \left(C[2] \ \text{Cos}\left[\sqrt{\frac{\pi}{2}} \ \text{Erf}\left[\frac{\#1}{\sqrt{2}}\right]\right] - C[1] \ \text{Sin}\left[\sqrt{\frac{\pi}{2}} \ \text{Erf}\left[\frac{\#1}{\sqrt{2}}\right]\right] \ \& \right)\right\}\right\}$

4.3.6 A differential equation *Mathematica* can't solve

On this equation, *Mathematica* just gives up immediately.

In[18]:= **diffEqBad2 = x^2 y''[x] + 2 x y[x] - 1 == 0;**

In[19]:= **solutionBad2 = DSolve[diffEqBad2, y, x]**

Out[19]= $\text{DSolve}[-1 + 2 \text{ x y}[x] + x^2 \text{ y}''[x] == 0, y, x]$

4.3.7 Partial differential equations

DSolve can also solve partial differential equations. Explicit notation has to be used for partial derivatives, but everything else is the same. Consider first an unknown function **u[x, y]** of two variables subject to the following condition.

In[20]:= **pde1 = x^2 (D[u[x, y], x] - D[u[x, y], y]) - (u[x, y] - x - y)^2 == 0;**

One can solve for either **u[x, y]**, or **u**. The second is easier to check.

In[21]:= **sol1 = DSolve[pde1, u, {x, y}]**

Out[21]= $\left\{ \left\{ u \rightarrow \left(\dfrac{-2\ \#1 - \#2 + \#1^2\ C[1][\#1 + \#2] + \#1\ \#2\ C[1][\#1 + \#2]}{-1 + \#1\ C[1][\#1 + \#2]}\ \& \right) \right\} \right\}$

Here #1 is the first variable **x** and #2 is **y**. Note that C[1] is now an unknown function, rather than a constant. The check is exactly the same as before.

In[22]:= **pde1 /. sol1 // Simplify**

Out[22]= $\{\text{True}\}$

Next, consider an unknown function **w[x, y, z]** of three variables subject to the following condition.

In[23]:= **pde2 = x^2 D[w[x, y, z], x] +**
        y^2 D[w[x, y, z], y] + z^2 D[w[x, y, z], z] - x y z == 0;

In[24]:= **sol2 = DSolve[pde2, w, {x, y, z}]**

Out[24]= $\left\{ \left\{ w \rightarrow \left(\left(\#1\ \#2\ \#3 \left(\#2\ (\#1 - \#3)\ \text{Log}\left[\dfrac{\#1}{\#2}\right] + (-\#1 + \#2)\ \#3\ \text{Log}\left[\dfrac{\#1}{\#3}\right] \right) \right) \middle/ \right. \right. \right.$
        $\left. \left. ((\#1 - \#2)\ (\#1 - \#3)\ (\#2 - \#3)) + \right. \right.$
        $\left. \left. C[1]\left[-\dfrac{1}{\#1} + \dfrac{1}{\#2}, -\dfrac{1}{\#1} + \dfrac{1}{\#3}\right]\ \& \right) \right\} \right\}$

Check the solution.

In[25]:= **pde2 /. sol2 // Simplify**

Out[25]= {True}

The unknown function C[1] satisfies the "homogeneous" part of the equation. We can extract these parts of the expressions by using the **Part** operation. Thus, the "homogeneous" part is

In[26]:= **pde2 [[1, {2, 3, 4}]]**

Out[26]= $z^2\, w^{(0,0,1)}\, [x,\ y,\ z] + y^2\, w^{(0,1,0)}\, [x,\ y,\ z] + x^2\, w^{(1,0,0)}\, [x,\ y,\ z]$

and the function C[1] is

In[27]:= **Evaluate[sol2 [[1, 1, 2, 1, 2]]] &**

Out[27]= $C[1]\, \left[-\dfrac{1}{\#1} + \dfrac{1}{\#2},\ -\dfrac{1}{\#1} + \dfrac{1}{\#3} \right]\,$ &

The following substitutes C[1] in the 'homogeneous" part.

In[28]:= **pde2 [[1, {2, 3, 4}]] == 0 /.**
        {w - <(Evaluate[sol2 [[1, 1, 2, 1, 2]]] &)} // Simplify

Out[28]= True

4.4 NDSolve

The fastest way to get a numerical solution is to use **NDSolve**. In order to use this it is necessary to give enough initial conditions to ensure a well-determined, numerical answer.

In[29]:= **ndiffEq = y'''[x] + y''[x] + y'[x] + 2 y[x] == 0;**

In[30]:=**numSolution =**
        NDSolve[{ndiffEq, y[0] == y'[0] == y''[0] == 1},
            y, {x, 0, 10}]

Out[30]= {{y → InterpolatingFunction[{{0., 10.}}, <>]}}

In[31]:=**Plot[Evaluate[y[x]/. numSolution], {x, 0, 10}];**

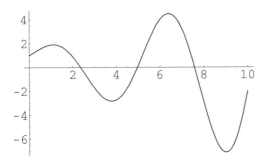

It is interesting to investigate the sense in which this is a solution. Following a suggestion of J. Keiper, evaluate the left-hand side of the equation for this solution.

In[32]:=**lhs[x_] := ndiffEq[[1]]/. numSolution**

The point is that interpolating functions can be differentiated, so this is itself another interpolating function.

In[33]:=**lhs[x]**

Out[33]= {2 InterpolatingFunction[{{0., 10.}}, <>][x] +
    InterpolatingFunction[{{0., 10.}}, <>][x] +
    InterpolatingFunction[{{0., 10.}}, <>][x] +
    InterpolatingFunction[{{0., 10.}}, <>][x]}

Hence, this function can be plotted. Near 0, the result seems to be very bad, but elsewhere, the difference of the left-hand side from 0 is less than 0.003.

In[34]:=**Plot[Evaluate[lhs[x]], {x, 0, 10}];**

4.4.1 A planetary orbit

A more challenging problem is to determine the trajectory of a mass in the gravitational field caused by a very large mass at the origin. This is described by a pair of differential equations:

$$d^2x/dt^2 = -x / r^3, \quad d^2y/dt^2 = -y / r^3 \text{ where } r = (x^2 + y^2)^{(1/2)}$$

NDSolve requires that we also specify initial conditions for **x** and **y** and their derivatives at **t** = 0.

In[35]:= **orbit =**

$$\text{NDSolve}\Big[\Big\{ \mathbf{x''[t]} == \frac{-\mathbf{x[t]}}{\left(\sqrt{\mathbf{x[t]}^2 + \mathbf{y[t]}^2}\right)^3},$$

$$\mathbf{y''[t]} == \frac{-\mathbf{y[t]}}{\left(\sqrt{\mathbf{x[t]}^2 + \mathbf{y[t]}^2}\right)^3},$$

$$\mathbf{x[0]} == 1, \quad \mathbf{x'[0]} == 0.2,$$

$$\mathbf{y[0]} == 0, \quad \mathbf{y'[0]} == 1.25\Big\},$$

$$\{\mathbf{x, y}\}, \{\mathbf{t, 0, 45}\}\Big]$$

Out[35]= {{x → InterpolatingFunction[{{0., 45.}}, <>],
            y → InterpolatingFunction[{{0., 45.}}, <>]}}

In[36]:= **ParametricPlot[Evaluate[{x[t], y[t]}/.orbit], {t, 0, 45}, Aspect-Ratio Automatic];**

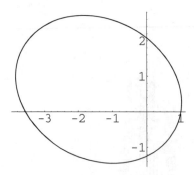

As is to be expected, the picture shows that the result is an ellipse with one focus at the origin.

4.4.2 Two equal masses

A still more challenging problem is that of two bodies of equal mass acting under mutual gravitational attraction. If the bodies have coordinates (x_1, y_1) and (x_2, y_2), then the equations are essentially the same as before, except expressed in terms of the differences $(x_2 - x_1)$ and $(y_2 - y_1)$. One gets four second order equations, which require eight initial conditions. We start the bodies off located symmetrically with respect to the origin, the left one moving down and the right one moving up. Note that most of the space is just entering the differential equations and the initial conditions.

```
In[37]:= twoorbits =
     NDSolve[
```

$$\left\{x1''[t] == -\frac{x1[t] - x2[t]}{\left(\sqrt{(x1[t] - x2[t])^2 + (y1[t] - y2[t])^2}\right)^3},\right.$$

$$y1''[t] == -\frac{y1[t] - y2[t]}{\left(\sqrt{(x1[t] - x2[t])^2 + (y1[t] - y2[t])^2}\right)^3},$$

$$x2''[t] == -\frac{x2[t] - x1[t]}{\left(\sqrt{(x1[t] - x2[t])^2 + (y1[t] - y2[t])^2}\right)^3},$$

$$y2''[t] == -\frac{y2[t] - y1[t]}{\left(\sqrt{(x1[t] - x2[t])^2 + (y1[t] - y2[t])^2}\right)^3},$$

```
     x1[0] == 1,  x1'[0] == 0, y1[0] == 0,  y1'[0] == 0.3,
     x2[0] == -1, x2'[0] == 0, y2[0] == 0,  y2'[0] == -0.3},
     {x1, y1, x2, y2}, {t, 0, 5.5}]
```

```
Out[37]= {{x1 → InterpolatingFunction[{{0., 5.5}}, <>],
          y1 → InterpolatingFunction[{{0., 5.5}}, <>],
          x2 → InterpolatingFunction[{{0., 5.5}}, <>],
          y2 → InterpolatingFunction[{{0., 5.5}}, <>]}}
```

```
In[38]:= ParametricPlot[
          Evaluate[{{x1[t], y1[t]} /. twoorbits,
                    {x2[t], y2[t]} /. twoorbits}],
          {t, 0, 5.5}, AspectRatio -> Automatic];
```

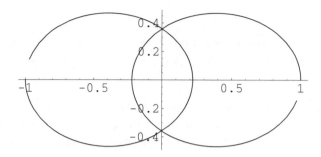

We chose the **t**-range so as to show not quite one complete orbit, making it easier to see how the masses are always located symmetrically with respect to their common center of gravity at the origin. The orbits are periodic, as one can see by increasing the **t**-range, and they appear to be ellipses again. The following picture shows the result if the initial velocities are changed to y1'[0] == 0.4 and y2'[0] == -0.2. The time interval is {t, 0, 10}.

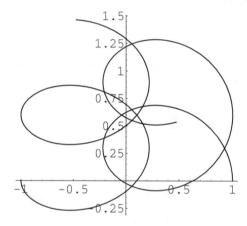

4.4.3 A boundary value problem

In general, *Mathematica* will not solve differential equations with conditions given at different points. Here we look at a typical problem of a hanging chain which is to be supported at height 10 feet at two points 10 feet apart. The equation for a hanging chain is

In[39]:= **chainEq[k_] = u''[x] - k** $\sqrt{1 + u'[x]^2}$ **== 0;**

Here **k** is a parameter depending directly on the weight and inversely on the tension. A good value for **k** is 0.35. If one tries to solve this directly using

```
(* NDSolve[
   {chainEq[0.35], u[0]==10, u[10]==10}, u, {x, 0, 10}] *)
```

then, on my machine anyway, *Mathematica* refuses even to try. Instead, we can try the "shooting method". This method tries to find a value for **u'[0]** so that the value of **u[10]** is 10. First set up the system of equations with an unknown derivative **s** at 0.

In[40]:= **eqns[s_, k_] := {chainEq[k], u[0] == 10, u'[0] == s}**

The generic solution to the initial value problem is given by

In[41]:= **g[s_, k_] := NDSolve[eqns[s, k], u, {x, 0, 10}]**

First, just guess a value for s.

In[42]:= **sol = g[-2, 0.35]**

Out[42]= $\{\{u \rightarrow \text{InterpolatingFunction}[\{\{0., 10.\}\}, <>]\}\}$

The interpolating function itself is given by

In[43]:= **(u /. sol)[[1]]**

Out[43]= $\text{InterpolatingFunction}[\{\{0., 10.\}\}, <>]$

Its value at 10 is

In[44]:= **(u /. sol)[[1]][10]**

Out[44]= 14.9619

This is clearly not the desired solution. The shooting method to solve the boundary value problem consists in varying **s** until the desired solution is found. This can be done using **FindRoot**, copying the preceding steps, and trying values of **s** between –1.5 and –11. This elegant solution is due to Allan Hayes.

In[45]:= **svalue =**
       FindRoot[(u /. Evaluate[g[s, 0.35][[1]]])[10] == 10,
          {s, -1.5, -11}]

Out[45]= $\{s \rightarrow -2.79041\}$

Using this value of **s**, we can find the new solution.

In[46]:= **newSol = (g[s, 0.35] /. svalue)**

Out[46]= {{u → InterpolatingFunction[{{0., 10.}}, <>]}}

Check that its value at 10 is correct.

In[47]:= (u /. newSol)[[1]][10]

Out[47]= 10.

If we make a picture, then it is clear that this is the kind of solution that we are looking for.

In[48]:= Plot[Evaluate[u[x] /. newSol], {x, 0, 10},
        PlotRange - <All,
        AspectRatio - <Automatic];

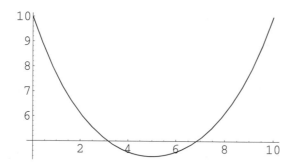

With a little bit of programming, this whole procedure can be automated.

4.5 Runge–Kutta Methods

See the Examples in Chapter 8, Section 4.4.

4.6 Series Solutions

One way to try to solve an ordinary differential equation is to assume that the dependent variable, **y**, is given by a power series with unknown coefficients in the independent variable, **x**. Substituting the power series in the differential equation leads to a collection of simultaneous algebraic equations for the coefficients. For instance, to solve $(dy/dx)^2 - y = x$, first construct a finite series approximation to **y** centered at 0 with unknown coefficients labeled **a[i]** for **i** between 0 and 6.

In[1]:=y[x_] := SeriesData[x, 0, Table[a[i], {i, 0, 6}]]

Substituting the series for **y** in the differential equation gives the following equation.

In[2]:= **seriesDiffEQ = D[y[x], x]^2 - y[x] == x**

Out[2]= $\left(-a[0] + a[1]^2\right) + (-a[1] + 4\,a[1]\,a[2])\,x +$
$\left(-a[2] + 4\,a[2]^2 + 6\,a[1]\,a[3]\right)\,x^2 +$
$(-a[3] + 12\,a[2]\,a[3] + 8\,a[1]\,a[4])\,x^3 +$
$\left(9\,a[3]^2 - a[4] + 16\,a[2]\,a[4] + 10\,a[1]\,a[5]\right)\,x^4 +$
$(24\,a[3]\,a[4] - a[5] + 20\,a[2]\,a[5] + 12\,a[1]\,a[6])\,x^5 + O[x]^6 == x$

Then use **LogicalExpand** to construct the equations given by setting equal the coefficients of powers of **x** on both sides of this equation.

In[3]:= **coefficientEQ = LogicalExpand[seriesDiffEQ]**

Out[3]= $-a[0] + a[1]^2 == 0\ \&\&\ -1 - a[1] + 4\,a[1]\,a[2] == 0\ \&\&$
$-a[2] + 4\,a[2]^2 + 6\,a[1]\,a[3] == 0\ \&\&$
$-a[3] + 12\,a[2]\,a[3] + 8\,a[1]\,a[4] == 0\ \&\&$
$9\,a[3]^2 - a[4] + 16\,a[2]\,a[4] + 10\,a[1]\,a[5] == 0\ \&\&$
$24\,a[3]\,a[4] - a[5] + 20\,a[2]\,a[5] + 12\,a[1]\,a[6] == 0$

These equations can be solved for a[1] through a[6] in terms of a[0], but the expressions are quite complicated, so we just assign a value to a[0] at the beginning.

In[4]:= **coefficientSol = Solve[{coefficientEQ, a[0] == 1},**
                    Table[a[i], {i, 0, 6}]]

Out[4]= $\{\{a[0] \to 1,\ a[6] \to 0,\ a[5] \to 0,\ a[4] \to 0,$
$a[3] \to 0,\ a[2] \to 0,\ a[1] \to -1\},\ \{a[0] \to 1,\ a[6] \to \dfrac{469}{11520},$
$a[5] \to -\dfrac{41}{960},\ a[4] \to \dfrac{5}{96},\ a[3] \to -\dfrac{1}{12},\ a[2] \to \dfrac{1}{2},\ a[1] \to 1\}\}$

Then substitute these two solutions into **y** to get the resulting series approximations.

In[5]:= **seriesSol = y[x]/.coefficientSol**

Out[5]= $\left\{1 - x + O[x]^7,\ 1 + x + \dfrac{x^2}{2} - \dfrac{x^3}{12} + \dfrac{5\,x^4}{96} - \dfrac{41\,x^5}{960} + \dfrac{469\,x^6}{11520} + O[x]^7\right\}$

These are still *Mathematica* series and have to be converted to normal expressions in order to be plotted.

In[6]:= **?Normal**

> Normal[expr] converts expr to a normal
>   expression, from a variety of special forms.

In[7]:= **Plot[Evaluate[Normal[seriesSol]], {x, 0, 3}];**

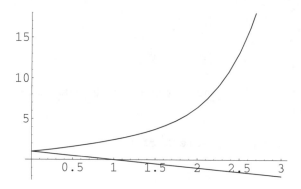

Finally, we can check that the differential equation is satisfied by both solutions up to order 6.

In[8]:= **seriesDiffEQ /. coefficientSol**

Out[8]= $\{x + O[x]^6 == x, x + O[x]^6 == x\}$

In fact, converted into normal expressions, the solutions satisfy the differential equation exactly.

In[9]:= **Normal[%]**

Out[9]= {True, True}

In[10]:= **Clear[y]**

4.7 Laplace Transforms

4.7.1 The Laplace transform package

Non-homogeneous, linear equations are frequently solved by Laplace transform techniques. In order to use Laplace transforms, the appropriate package has to be loaded, which takes a while.

In[11]:= **Needs["Calculus`LaplaceTransform`"]**

To learn how to use it, see the excellent discussion in the Help Browser, or for a brief account, use ?.

In[12]:= **?LaplaceTransform**

> LaplaceTransform[expr, t, s] gives a function of s,
>   which is the Laplace transform of expr, a function of
>   t, t >= 0. It is defined by LaplaceTransform[expr,
>   t, s] = Integrate[Exp[-s t] expr, {t, 0, Infinity}].

Here are a number of standard examples.

In[13]:= **{LaplaceTransform[1, t, s],**
        LaplaceTransform[t, t, s],
        LaplaceTransform[E^(a t), t, s],
        LaplaceTransform[t^n, t, s],
        LaplaceTransform[Cos[w t], t, s],
        LaplaceTransform[Cosh[w t], t, s]}

Out[13]= $\left\{ \dfrac{1}{s}, \dfrac{1}{s^2}, \dfrac{1}{-a+s}, s^{-1-n} \text{Gamma}[1+n], \dfrac{s}{s^2+w^2}, \dfrac{s}{s^2-w^2} \right\}$

The following relationships are the reason why Laplace transforms can be used to solve differential equations.

In[14]:= **{LaplaceTransform[y'[t], t, s],**
        LaplaceTransform[y''[t], t, s]}

Out[14]= $\{ s \text{LaplaceTransform}[y[t], t, s] - y[0],$
        $s^2 \text{LaplaceTransform}[y[t], t, s] - s y[0] - y'[0] \}$

4.7.2 A single differential equation

The simplest use of Laplace transforms is for linear differential equations with constant coefficients whose right-hand sides consist of terms whose Laplace transforms are known. Here is an example.

In[15]:= **ltDiffEq1 = y''[t] - 3 y'[t] + 2 y[t] == 4 t + E^{3 t};**

The Laplace transform turns this differential equation into an algebraic equation for the Laplace transform of **y**.

In[16]:= **ltAlgEq1 = LaplaceTransform[ltDiffEq1, t, s]**

Out[16]= 2 LaplaceTransform[y[t], t, s] + s² LaplaceTransform[y[t], t, s] −

$\quad$ 3 (s LaplaceTransform[y[t], t, s] − y[0]) −

$$s\,y[0] - y'[0] == \frac{1}{-3+s} + \frac{4}{s^2}$$

Next, solve this algebraic equation for **LaplaceTransform[y[t], t, s]**.

In[17]:=**algSolution1 =**
$\quad$**Solve[ltAlgEq1, LaplaceTransform[y[t], t, s]]**

Out[17]= $\left\{\left\{\text{LaplaceTransform}[y[t], t, s] \rightarrow \right.\right.$

$$\left.\left.-\frac{-\frac{1}{-3+s} - \frac{4}{s^2} + 3\,y[0] - s\,y[0] - y'[0]}{2 - 3\,s + s^2}\right\}\right\}$$

Finally, we want the inverse Laplace transform of this substitution.

In[18]:= **?InverseLaplaceTransform**

```
InverseLaplaceTransform[expr, s, t]
  gives a function of t, t >= 0, which is  the
  inverse Laplace transform of expr, a function of s.
```

We have to apply the inverse Laplace transform to both parts of the algebraic solution to find the value of **y[t]**. This is done by the **Map** function that will be explained in Chapter 6.

In[19]:=**diffSolution1 =**
$\quad$**Map[InverseLaplaceTransform[#, s, t]&,**
$\quad$**algSolution1, {3}]**

Out[19]= $\left\{\left\{y[t] \rightarrow \frac{E^t}{2} - E^{2\,t} + \frac{E^{3\,t}}{2} + 4\left(\frac{3}{4} - E^t + \frac{E^{2\,t}}{4} + \frac{t}{2}\right) + \right.\right.$

$$\left.\left.(-E^t + 2\,E^{2\,t})\,y[0] - (-E^t + E^{2\,t})\,(3\,y[0] - y'[0])\right\}\right\}$$

Finally, we check that this is actually a solution of the original differential equation. The solution here involves **y[t]**, whereas we would rather have **y** as a pure function. We have to make the conversion ourselves.

In[20]:=**diffSol1 =**
$\quad$**{y -> Evaluate[Evaluate[y[t]/.diffSolution1[[1]]/.t -> #]&]}**

Out[20]= $\left\{y \rightarrow \left(\frac{E^{\#1}}{2} - E^{2\,\#1} + \frac{E^{3\,\#1}}{2} + 4\left(\frac{3}{4} - E^{\#1} + \frac{E^{2\,\#1}}{4} + \frac{\#1}{2}\right) + \right.\right.$

$$\left.\left.(-E^{\#1} + 2\,E^{2\,\#1})\,y[0] - (-E^{\#1} + E^{2\,\#1})\,(3\,y[0] - y'[0])\,\&\right)\right\}$$

In[21]:=`ltDiffEq1 /. diffSol1 //Simplify`

Out[21]= `True`

4.7.3 Non-constant coefficients

Certain differential equations with non-constant coefficients can also be solved by Laplace transform techniques. Consider the following example.

In[22]:=`ltDiffEq2 = t y''[t] - t y'[t] - t == 0;`

In[23]:=`ltDiffDiffEq2 = LaplaceTransform[ltDiffEq2, t, s]`

Out[23]= $-\dfrac{1}{s^2}$ + LaplaceTransform[y[t], t, s] -

2 s LaplaceTransform[y[t], t, s] + y[0] +

s LaplaceTransform$^{(0,0,1)}$[y[t], t, s] -

s^2 LaplaceTransform$^{(0,0,1)}$[y[t], t, s] == 0

The term here of the form `LaplaceTransform`$^{(0,0,1)}$`[y[t], t, s]` is a form of the derivative. Its actual input form is as follows:

In[24]:=`Derivative[0, 0, 1][LaplaceTransform][y[t], t, s]`

Out[24]= LaplaceTransform$^{(0,0,1)}$[y[t], t, s]

So, this time the result involves both the Laplace transform of **y** and its derivative; i.e., we get a first order differential equation for the Laplace transform of **y**, rather than an algebraic equation. Unfortunately, **DSolve** is unable to deal with this equation directly, so we have to replace the Laplace transform by a generic function **g[s]** and its derivative by **g'[s]**.

In[25]:=`newDiffEq =`
`    ltDiffDiffEq2 //.`
`      LaplaceTransform[y[t], t, s] -> g[s] //.`
`      Derivative[0, 0, 1][LaplaceTransform][y[t], t, s] -> g'[s]`

Out[25]= $-\dfrac{1}{s^2}$ + g[s] - 2 s g[s] + y[0] + s g'[s] - s^2 g'[s] == 0

Now we can solve this equation for **g[s]** and then replace **g[s]** by the Laplace transform of **y** again.

In[26]:=`diffSolution2 = DSolve[newDiffEq, g[s], s]`

Out[26]= $\left\{\left\{g[s] \rightarrow E^{-Log\,[-1+s]-Log\,[s]} \, C[1] + \dfrac{E^{-Log\,[-1+s]-Log\,[s]} \, (1 + s^2 \, y[0])}{s}\right\}\right\}$

Unfortunately, the E to the Log term is not simplified, so we have to do that ourselves.

In[27]:=`algSolution2 = diffSolution2//.`
            `{E^(a_ + b_) :> E^a E^b,`
            `E^(-Log[x_])  :> 1/x,`
            `g[s] -> LaplaceTransform[y[t], t, s]}`

Out[27]= $\left\{\left\{\text{LaplaceTransform}[y[t], t, s] \rightarrow \dfrac{C[1]}{(-1+s)\,s} + \dfrac{1 + s^2\,y[0]}{(-1+s)\,s^2}\right\}\right\}$

Finally, apply the inverse Laplace transform to this.

In[28]:=`answer =`
        `Map[InverseLaplaceTransform[#, s, t]&, algSolution2, {3}]`

Out[28]= $\{\{y[t] \rightarrow -1 - t + (-1 + E^t)\,C[1] + E^t\,(1 + y[0])\}\}$

Again, find **y** as a pure function.

In[29]:=`diffSol2 =`
        `{y -> Evaluate[Evaluate[y[t]/.answer[[1]]]/.t -> #]&}`

Out[29]= $\{y \rightarrow (-1 + (-1 + E^{\#1})\,C[1] - \#1 + E^{\#1}\,(1 + y[0])\,\&)\}$

The check then proceeds without difficulty.

In[30]:=`ltDiffEq2/.diffSol2//Simplify`

Out[30]= True

4.7.4 A system of two differential equations

The real power of the Laplace transform comes in using it for systems of linear ordinary differential equations with constant coefficients. In this example, **y1[t]** and **y2[t]** are two functions of **t** which are related by a pair of second order differential equations.

In[31]:=`ltDiffSystem =`
        `{y1''[t] == k (y2[t] - 2 y1[t]),`
        `y2''[t] == k (y1[t] - 2 y2[t])};`

At present, *Mathematica* is unable to solve this system using **DSolve**. So, instead we apply the Laplace transform to this system of differential equations.

In[32]:=`ltAlgSystem = LaplaceTransform[ltDiffSystem, t, s]`

Out[32]= $\{s^2\,$ LaplaceTransform$[y1[t], t, s] - s\,y1[0] - y1'[0] ==$
        $k\,(-2\,$ LaplaceTransform$[y1[t], t, s] +$
            LaplaceTransform$[y2[t], t, s]),$
        $s^2\,$ LaplaceTransform$[y2[t], t, s] - s\,y2[0] - y2'[0] ==$
        $k\,($ LaplaceTransform$[y1[t], t, s] -$
            $2\,$ LaplaceTransform$[y2[t], t, s])\}$

The procedure is the same as with a single equation. First solve this system of algebraic equations for `LaplaceTransform[y1[t], t, s]` and `LaplaceTransform[y2[t], t, s]`.

In[33]:=`algSystemSolution =`
        `Solve[ltAlgSystem,`
            `{LaplaceTransform[y1[t], t, s],`
            `LaplaceTransform[y2[t], t, s]}]`

Out[33]= $\{\{$ LaplaceTransform$[y1[t], t, s] \rightarrow$
        $-(-2\,k\,s\,y1[0] - s^3\,y1[0] - k\,s\,y2[0] - 2\,k\,y1'[0] - s^2\,y1'[0] -$
            $k\,y2'[0])\,/\,(3\,k^2 + 4\,k\,s^2 + s^4),$ LaplaceTransform$[y2[t], t, s] \rightarrow$
        $-(-k\,(-s\,y1[0] - y1'[0]) - (2\,k + s^2)\,(-s\,y2[0] - y2'[0]))\,/$
            $\left(k^2 - (2\,k + s^2)^2\right)\}\}$

Then apply the inverse Laplace transform to these solutions. We suppress printing the solution since it is rather long.

In[34]:=`diffSystemSolution =`
        `Map[InverseLaplaceTransform[#, s, t]&,`
            `algSystemSolution, {3}];`

If we choose initial conditions carefully, this simplifies considerably.

In[35]:= `initialConditions =`
        $\{$`y1[0]` $\leftarrow 1,$ `y2[0]` $\leftarrow 1,$
        `y1'[0]` $\leftarrow \sqrt{3}\,\sqrt{k},$ `y2'[0]` $\leftarrow \sqrt{3}\,\sqrt{k}\}$;

In[36]:=`initialSystemSolution =`
        `diffSystemSolution /. initialConditions`

Out[36]= $\Big\{\Big\{y1[t] \to \text{Cos}\Big[\sqrt{k}\ t\Big] + \text{Sin}\Big[\sqrt{3}\ \sqrt{k}\ t\Big],$

$\qquad y2[t] \to \text{Cos}\Big[\sqrt{k}\ t\Big] - \text{Sin}\Big[\sqrt{3}\ \sqrt{k}\ t\Big]\Big\}\Big\}$

As before, convert the solutions to pure functions.

In[37]:= `initSysSol =`
`    {y1 -> Evaluate[Evaluate[y1[t]/.`
`            initialSystemSolution[[1]]]/.t -> #]&],`
`     y2 -> Evaluate[Evaluate[y2[t]/.`
`            initialSystemSolution[[1]]]/.t -> #]&]}`

Out[37]= $\Big\{y1 \to \Big(\text{Cos}\Big[\sqrt{k}\ \#1\Big] + \text{Sin}\Big[\sqrt{3}\ \sqrt{k}\ \#1\Big]\ \&\Big),$

$\qquad y2 \to \Big(\text{Cos}\Big[\sqrt{k}\ \#1\Big] - \text{Sin}\Big[\sqrt{3}\ \sqrt{k}\ \#1\Big]\ \&\Big)\Big\}$

Check the result.

In[38]:= `ltDiffSystem /. initSysSol // Simplify`

Out[38]= `{True, True}`

Of course, we can make a picture of this solution, treating `y1[t]` and `y2[t]` as determining a parametric curve.

In[39]:= `ParametricPlot[`
`    Evaluate[{y1[t], y2[t]}/.initialSystemSolution/.k -> 2],`
`    {t, 0, 10}];`

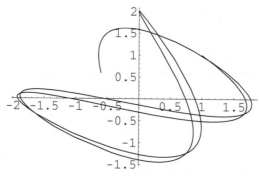

This is very curious behavior. Trying different plots, one sees that the curve starts at (1, 1) with **x** increasing and **y** decreasing. It follows the lower track around to near the point (0, 2) where there is an apparent singularity at **t** $\approx$ 4.5. The curve turns around and seems to go back through the point (1, 1) at **t** $\approx$ 8.95. Actually, there is no singularity and the curve misses (1, 1) the second time. For instance, near **t** = 4.5 we have the situation

```
In[40]:= ParametricPlot[
            Evaluate[{y1[t], y2[t]}/.
                    initialSystemSolution/.k -> 2],
            {t, 4.45, 4.6}];
```

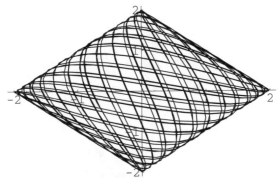

Thus, the curve is smooth as it goes past this value. Over a long period of time, the curve appears to fill out a region in space.

```
In[41]:= ParametricPlot[
            Evaluate[{y1[t], y2[t]}/.
                    initialSystemSolution/.k -> 2],
            {t, 0, 100}];
```

Looking carefully, one can see that there are other sharp bends in the curve, for instance, near $t \approx 37.5$. It is interesting to look at the plots for **t** from 0 to 500, or 0 to 1000, but we omit those here. Again, this whole procedure can be automated with a little bit of programming.

5 *Practice*

1.   {N[Pi], N[E], N[I], N[Degree],
            N[GoldenRatio], N[EulerGamma], N[Catalan]}

```
2.    N[Sin[60 Degree]]

3.    {Re[2 + 3 I], Im[2 + 3 I], Conjugate[2 + 3 I]}

4.    {Re[a + b I], Im[a + b I], Conjugate[a + b I]}

5.    Needs["Algebra`ReIm`"]

6.    a/: Im[a] = 0

7.    b/: Im[b] = 0

8.    {Re[a + b I], Im[a + b I], Conjugate[a + b I]}
```

9.   `Options[NumberForm]` (Try out various options.)

```
10.   BaseForm[1/3, 2]

11.   Table[{ToExpression[ToString[N[Pi, n]]],
            N[Cos[ToExpression[ ToString[N[Pi, n]]]], 11] },
            {n, 1, 5} ] // TableForm

12.   Simplify[Sin[x]^2 + 2 Cos[x]^2]

13.   FindRoot[Sin[x]/x == 0, {x, 2}]

14.   FindRoot[x Cos[x] == 1, {x, 10}]

15.   Random[Integer, {0, 10}]

16.   Table[Random[Real, {1, 2}], {20}]

17.   ColumnForm[NSolve[{2 x y + 3 x + 4 y == 5,
                      6 x^2 - 7 x - 8 y == 9}]]

18.   Eliminate[{x^2 + 2 a x + a^2 y == 0,
               y^2 - 2 b y + a b x == 0}, a]

19.   {ToRules[%]}

20.   Solve[{x^2 + 2 a x + a^2 y == 0,
            y^2 - 2 b y + a b x == 0}, {x, y}]

21.   LinearSolve[{{1, 4, 3}, {4, 2, 3}, {3, 3, 1}},
                 {1, 2, 3}]
```

```
22.   Solve[{x^2 + y^2 == 1, x^3 + y^3 == 2}, {x, y}]

23.   N[%]

24.   NSolve[{ x^2 + y^2 == 1, x^3 + y^3 == 2}, {x, y}]

25.   Rationalize[N[Pi], 0]

26.   Union[Table[Rationalize[N[Pi], (0.1)^n], {n, 20}]]

27.   RowReduce[Table[3 i - 2 j, {i, 3}, {j, 4}]]

28.   NDSolve[
          {y1''[t] == 2 (y2[t] - 2 y1[t]),
           y2''[t] == 2 (y1[t] - 2 y2[t]),
           y1[0]    == 1, y2[0] == 1,
           y1'[0]   ==  Sqrt[6],
           y2'[0]   == -Sqrt[6]},
          {y1, y2}, {t, 0, 10}]

29.   ParametricPlot[Evaluate[{y1[t], y2[t]}/.
                  newSolution],
              {t, 0, 10}]

30.   {$Version, $TimeUnit, $RecursionLimit}

31.   ??Solve   (try out some of the options)
```

32.   Try the two body problem with other unsymmetrical initial velocities.

```
33.   ?N*
```

6 Exercises

Give names to all of the expressions, equations, and solutions you use in the following problems. For instance, in problem 1, call the equation there **equation1** and the list of solutions **solution1**, etc.

1. Solve the equation and check the results.

$$x^4 + \frac{17 x^3}{14} - \frac{31 x^2}{7} + \frac{37 x}{14} - \frac{3}{7} = 0$$

2. Solve the equation

$$x^5 - \frac{x^2}{2740} - \frac{3}{9704700} = 0$$

with 10 digit accuracy; with $MachinePrecision and $MachinePrecision + 1 digit accuracy. Check your answers. (You may need to use the built-in function Chop.)

3. Solve the pair of equations

$$x^2 y + y = 2, \quad y - 4 x = 8$$

exactly for x and y. Suppress the answer but check the results.

4. Solve the three equations

$$a x + b y - z = 3 b,$$
$$x - 4 y - 5 c z = 0,$$
$$x + a y - b z = c$$

exactly for x, y, and z. Show the answer and a check of its correctness. Also solve for a, b, and c and check the answer.

5. Determine the values of **a** for which the equation has a solution

$$\sqrt{1 - x} + \sqrt{1 + x} = a$$

Consider both real and complex values of **a** and both real and complex solutions.

6. Use the built-in operation DSolve to solve the following differential equations. Check your solutions.

i) $y' = y \tan(x)$

ii) $y' - y \tan(x) = \sec(x)$

iii) $y' - 2 x y = 1$

iv) $x^2 y' + 3 x y = (\sin x) / x$

v) $y' = x^2 / ((x^3 + 1) y)$

vi) $y' = x y^2 + y^2 + x + 1$

vii) $y'' + x y' + y = 0$

viii) $x^2 y'' - 3 x y' + 4 y = 0$ (Euler's equation)

ix) $y'' - 5 y' + 6 y = 2 e^x$

7. Solve the following differential equations. Check your solutions.

i) $- x^2 y' + y^2 + 3 x y + x^2 = 0$

ii) $(x^2 e^y + \sin(x) + 2) y' + 2 x e^y + y \cos(x) = 0$

iii) $x^2 y' + x y (x y + 4) + 2 = 0$

iv) $x y' + a x y^2 + 2 y + b x = 0$

v) $y''[x] + 4 y'[x] + 4 y[x] = E^{\wedge}(-2 x)/x^{\wedge}2$

vi) $y'' + 2 y' - 3y = 0$ (make a picture)

vii) $y'' - 2 y' + y = 0$ (make a picture)

viii) $y'' + 5 y' = 0$ (make a picture)

8. Try to use **DSolve** to solve the system of differential equations

$x'(t) = 2 x(t) - x(t) y(t) - 2 x(t)^2$

$y'(t) = y(t) - (1/2) x(t) y(t) - y(t)^2$

$x(0) = 2$
$y(0) = 2.$

When that fails, solve it numerically for **t** between 0 and 10 and plot the solution.

9. Use Laplace transforms to solve the following differential equations.

i) $y'' - w^2 y = 0$

ii) $y'' - 4 y' + 4 y = t^2$

iii) $y'' - 5 y' + 4 y = e^{2t}$

iv) $y'' + 2 y' + 2 y = t$

v) $y'' + 2 y' + 2 y = e^{-t} \sin t$

vi) $y1' = - 3 y1 + 4 y2 + \cos t$
    $y2' = - 2 y1 + 3 y2 + t$

10. Define a function `pascalTriangleRow[n_]` which displays the nth row of Pascal's triangle. (Note: there is a built-in function `Binomial[m, n]`.) Use this to write another function `pascalTriangle[n]` which shows the first n rows of Pascal's triangle in triangular form.

11. Define a function `completeTheSquare[expr_]` that takes an expression of the form

$$a\,x^2 + b\,x + c$$

and writes it in the form $a\,(x + b\,/\,2\,a\,)^2 + c - b^2\,/\,4\,a^2$. You may find it necessary to define some auxilary functions to extract the coefficients from the expression.

12. i) Jacobian matrices (Look them up in your advanced calculus book): Define a function `jacobian[funlist_, varlist_]` which takes as arguments a list of functions and a list of variables. It calculates the Jacobian matrix of the functions with respect to the variables. (The (i, j) entry is the partial derivative of the ith function with respect to the jth variable.) Include `Simplify` in the definition of the function. Note that the length of a list is given by `Length[list]`.
ii) Calculate the Jacobian matrix for the pair of functions

$$u = x^2 + y^2 \qquad v = -\,2\,x\,y$$

with respect to x and y. Name this matrix `jak`. Note that `jak` is expressed in terms of the variables x and y.

iii) Solve for x and y as functions of u and v. There will be four complicated solutions.

iv) In particular, the third solution in part iii) gives x and y as functions of u and v. Use this to calculate the jacobian matrix of x and y with respect to u and v. Name this matrix `invjak`. Note that `invjak` is expressed in terms of the variables u and v.

v) Let `jak'` be `invjak` expressed in terms of x and y rather than u and v. That is, substitute the values of u and v in terms of x and y into `invjak` to get `jak'`.

vi) Show that `jak.jak' = IdentityMatrix[2]`.

13. (More Stoutemyer experiments.)

i) Is $e^{\pi\,\sqrt{163}}$ an integer? How precisely does it have to be calculated to determine the answer?

ii) Determine how *Mathematica* deals with $\infty-\infty$, ∞/∞, $0\;\infty$, 1^∞.

iii) Does *Mathematica* solve the equation Sqrt[x] = 1 − x correctly?

iv) Does *Mathematica* calculate the definite integral of $\frac{1}{x^2}$ from -3 to 2 correctly?

CHAPTER

4

Built-In Graphics and Sound

Pictures, pictures everywhere.

1 Plotting Commands and Optional Arguments

For many users, graphics commands are the most important feature of *Mathematica*. Either they want to know what some built-in or user-defined function looks like, or they have data from somewhere else that they want to plot. In either case, the basic plotting commands are very simple to use. We have already seen a number of examples using **Plot**, **Plot3D**, **ParametricPlot**, etc. The main thing to be learned is how to use the optional arguments to these functions. First get all of the possible commands that end in the term **Plot** or **Plot3D**. All such commands are given by **?*Plot** and **?*Plot3D**.

In[1]:= **?*Plot**

```
ContourPlot          MovieDensityPlot
DensityPlot          MovieParametricPlot
ListContourPlot      MoviePlot
ListDensityPlot      ParametricPlot
ListPlot             Plot
MovieContourPlot
```

In[2]:= **?*Plot3D**

```
ListPlot3D           ParametricPlot3D
MoviePlot3D          Plot3D
```

These are the built-in plotting commands that automatically produce a picture. Each of these plotting commands can take a number of optional arguments. The basic graphics options are found in the command **Graphics**, which will be discussed in Chapter 10.

In[3]:=**Options[Graphics]**

$$\Big\{\text{AspectRatio} \to \frac{1}{\text{GoldenRatio}} \text{, Axes} \to \text{False, AxesLabel} \to \text{None,}$$

AxesOrigin → Automatic, AxesStyle → Automatic,
Background → Automatic, ColorOutput → Automatic,
DefaultColor → Automatic, Epilog → {}, Frame → False,
FrameLabel → None, FrameStyle → Automatic, FrameTicks → Automatic,
GridLines → None, ImageSize → Automatic, PlotLabel → None,
PlotRange → Automatic, PlotRegion → Automatic,
Prolog → {}, RotateLabel → True, Ticks → Automatic,
DefaultFont :→ $DefaultFont, DisplayFunction :→ $DisplayFunction,
FormatType :→ $FormatType, TextStyle :→ $TextStyle}

Length[%]

25

The command **Plot** adds a few more options that can be found as follows.

In[5]:= **Complement[Options[Plot], Options[Graphics]]**

{Axes → Automatic, Compiled → True, MaxBend → 10.,
PlotDivision → 30., PlotPoints → 25, PlotStyle → Automatic}

The total of 30 entries here are in the form of substitutions that give the default values for the indicated optional arguments. For instance, **AspectRatio** is the ratio of the height to the width of the final plot. Possible values are any real number, or **Automatic**, which means that the distances on the two axes are the same. The possible values of **Axes** are **True** (meaning draw axes), **False** (meaning don't draw axes), **{Boolean, Boolean}** (where **Boolean** is **True** or **False**) means draw one but not both axes and **Automatic** (meaning the program will decide where to draw the axes). Fortunately, most of these various choices make sense, so one doesn't have to try to remember which are in effect for any given command. The option **GoldenRatio** for **AspectRatio** is a built-in constant.

In[6]:=**N[GoldenRatio]**

1.61803

It was Plato who claimed that the golden ratio was the ideal shape for a picture. The possible values for an optional argument are not always evident. There are certain standard values that frequently work.

Automatic	Use an optimal internal algorithm.
All	Include everything.
None	Do not include this.
True	Do this.
False	Don't do this.
\<number>	Use this number as the value.
\<list>	Use the entries in the list as the values.

The **Help Browser** contains extensive information about the optional arguments in various commands and their meanings. First look up the command in the browser and then type in the name of an optional argument in the Go To box to find out everything about it. In Chapter 12 we will see how to define functions with their own optional arguments. In that case, it will be up to us to decide what the possible values should be and what effect they should have.

2 Two-Dimensional Graphics

2.1 Plot

Plot has two arguments, the first being either a function or a list of functions and the second an iterator. The best way to understand the 30 possible options is to try out various combinations of them.

2.1.1 A simple plot

First, plot a single function of one variable, using no options. The default **AspectRatio** for **Plot** is **1/GoldenRatio** as shown earlier, so the plot region is always about 1.6 times as long as it is high. Note that the scales on the two axes are completely different.

In[7]:= **Plot[Sin[x], {x, 0, 2 Pi}];**

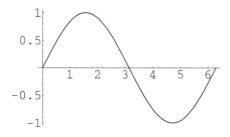

2.1.2 Several functions at once

Now plot several functions of one variable together just by making the first argument a list of functions, again with no options.

In[8]:=**Plot[{Sin[x], -Sin[x], Cos[x], -Cos[x]},**
        {x, 0, 2 Pi}];

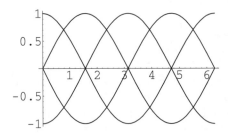

2.1.3 Square plots with labels

The iterator **{x, xmin, xmax}** specifies what range of values should be plotted. Optional arguments are added in a sequence in any order after the iterator. This way of using optional arguments in one of the strengths of *Mathematica*, since you are not forced to give options in a particular order or even know anything at all about options you are not using. As a first example, make the plotting region a square, put a frame around the plot, and add a label. Notice that the default value for **Frame** is **False** and the default value for **PlotLabel** is **None**.

In[9]:=**Plot[Sin[x], {x, 0, 2 Pi},**
          AspectRatio -> 1,
          Frame   -> True,
          PlotLabel   -> "A sin curve"];

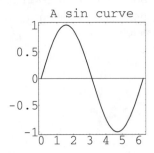

2.1.4 AspectRatio -> Automatic

Next make the **x** and **y** scales the same by making **AspectRatio** equal to **Automatic** and add a background shading. On a color screen, use **Hue** instead of **GrayLevel** to get a colored background. Also, add grid lines.

```
In[10]:= Plot[Sin[x], {x, 0, 2 Pi},
            AspectRatio -> Automatic,
            Background  -> GrayLevel[0.8],
            GridLines   -> Automatic];
```

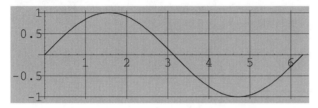

Thus, we see that **AspectRatio -> Automatic** makes the distances the same on both axes and **GridLines -> Automatic** draws grid lines where the program has inserted values on the axes. If you want to know what value was actually used for **AspectRatio**, you can find out as follows.

```
In[11]:= FullOptions[%, AspectRatio]
```

```
    0.318309
```

The actual value for **GridLines** is more complicated.

```
In[12]:= FullOptions[%%, GridLines]
```

```
    {{{0., {RGBColor[0., 0., 0.5], AbsoluteThickness[0.25]}},
      {1., {RGBColor[0., 0., 0.5], AbsoluteThickness[0.25]}},
      {2., {RGBColor[0., 0., 0.5], AbsoluteThickness[0.25]}},
      {3., {RGBColor[0., 0., 0.5], AbsoluteThickness[0.25]}},
      {4., {RGBColor[0., 0., 0.5], AbsoluteThickness[0.25]}},
      {5., {RGBColor[0., 0., 0.5], AbsoluteThickness[0.25]}},
      {6., {RGBColor[0., 0., 0.5], AbsoluteThickness[0.25]}}},
     {{-1., {RGBColor[0., 0., 0.5], AbsoluteThickness[0.25]}},
      {-0.5, {RGBColor[0., 0., 0.5], AbsoluteThickness[0.25]}},
      {0., {RGBColor[0., 0., 0.5], AbsoluteThickness[0.25]}},
      {0.5, {RGBColor[0., 0., 0.5], AbsoluteThickness[0.25]}},
      {1., {RGBColor[0., 0., 0.5], AbsoluteThickness[0.25]}}}}}
```

The first sublist contains the vertical lines and the second the horzontal ones. The lines are very thin and, on a color monitor, pale blue.

2.1.5 Properties of axes

Now, shift the origin of the axes, make them thicker using **AxesStyle**, and add labels to the axes.

```
In[13]:= Plot[Sin[x], {x, 0, 2 Pi},
            AxesOrigin -> {1, 0.5},
            AxesLabel  -> {"x-axis", "y-axis"},
            AxesStyle  -> Thickness[0.01]];
```

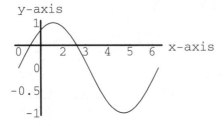

2.1.6 The smoothness of plots

This time, change the number of plot divisions to make the curve as jaggedy as possible and cut off the bottom half.

```
In[14]:= Plot[Sin[x], {x, 0, 6 Pi},
            PlotPoints    -> 7,
            PlotDivision  -> 1,
            MaxBend       -> 45,
            PlotRange     -> {0, 1}];
```

The way two-dimensional graphics works is to first find the values of the function at the default value of **PlotPoints**, which is the x-axis subdivided into 25 points. The program then looks at the angles between successive line segments, and if these angles are greater than the specified **MaxBend** in degrees, it adds more divisions until that is the maximum angle. We have chosen the minimum value for **PlotDivision**, the maximum value for **MaxBend**, and a choice for **PlotPoints** that gives a surprising result. For our last example of **Plot**, make a nice **Sin** curve by putting in labels along the x-axis at intervals of $\frac{\pi}{2}$, while allowing the y-axis intervals to be given automatically by the program. The **FontForm** graphics command gives us control over the appearance of the text. Whether this works or not on the screen is platform dependent, but it will always print correctly.

In[15]:= **Plot$\big[$Sin[x], {x, 0, 2 π},**

        Ticks ⟵**{{{0, "0"}, {$\frac{3.14}{2}$, "$\frac{\pi}{2}$"}, {3.14, "π"},**

              {$\frac{3}{2}$ 3.14, "$\frac{3\pi}{2}$"}, {2 3.14, "2π"}},

          Automatic},

      PlotLabel ⟵**FontForm["A better sin curve",**
{"Palatino-Bold", 12}]$\big]$;

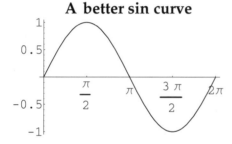

Two important options that we have not discussed are **Epilog** and **Prolog**. These allow one to add Graphics primitives to built-in graphics functions and will be treated in Chapter 10.

2.1.7 $DisplayFunction

The command **Show** will display several pictures at the same time on the same set of axes, as we saw in Chapter 1. There we preplotted the pictures before applying **Show**. This time, let's do it all at once. Now, the output of **Plot** is a graphics object as is indicated by the actual output –Graphics–, but the picture is a side effect that happens during the evaluation of a **Plot** command. This causes a problem in showing several plots, since any intermediate plots will be displayed also. Thus, the following gives three pictures.

In[16]:= **Show[Plot[Sin[x], {x, 0, 2Pi}],**
         Plot[Cos[x], {x, 0, 2Pi}]];

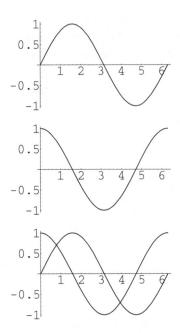

The first two are produced as side effects to evaluating the **Plot[Sin - - -]** and **Plot[-Cos - - -]** commands, while the third is the side effect of the final evaluation of **Show**. The cure to this is to turn off the display of the two intermediate pictures and then turn the picture for **Show** back on. This is done with the **DisplayFunction** option. Possible values are

DisplayFunction -> $DisplayFunction
(the default value which displays the drawing on the screen)

DisplayFunction -> Identity
(the graphics is calculated but no picture is displayed)

DisplayFunction -> Function[Display["file name", #]]
(send the PostScript code to the named file).

The following command does what we want. Note that it doesn't matter if the **Plot** commands are put in a list or not.

```
In[17]:= Show[{Plot[Sin[x], {x, 0, 2Pi},
            DisplayFunction -> Identity],
        Plot[Cos[x], {x, 0, 2Pi},
            DisplayFunction -> Identity]},
        DisplayFunction -> $DisplayFunction];
```

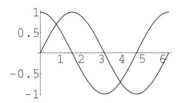

2.2 ListPlot

First check the options of **ListPlot**.

In[18]:=**Complement[Options[ListPlot], Options[Graphics]]**

  {Axes → Automatic, PlotJoined → False, PlotStyle → Automatic}

It has one new option, **PlotJoined**, but is missing three of the options of **Plot** concerning fineness of the plot. Basically, **ListPlot** just plots points. A list of single values is treated as the y-values for x-coordinates from 1 to the number of points.

In[19]:=**ListPlot[{3.2, 5.1, 1.4, 0.5, 4.4}];**

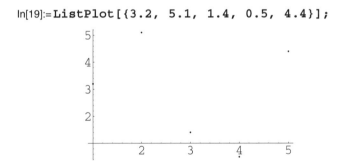

Note that the options for **Axes** and **AxesOrigin** are **Automatic** and the axes here do not go through the origin. Next, increase the size of the points so we can see them better, and adjust the shape of the plot region by specifying ranges for both for the x and y values. Make the **AspectRatio** 1 to get a more realistic picture. **PlotStyle** here is a catch-all argument that takes as its value either **Automatic** or a list of directions concerning properties of points or lines. We'll look at a number of possible values for it in what follows. It will also be discussed further in Chapter 10.

In[20]:=**ListPlot[{3.2, 5.1, 1.4, 0.5, 4.4},**
        PlotRange -> {{0, 6}, {0, 6}},
        AspectRatio -> 1,
        PlotStyle -> {PointSize[0.02]}];

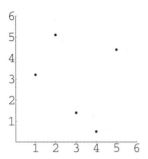

2.2.1 PlotJoined

Now try the new option, **PlotJoined**, which adds a line between successive points.

```
In[21]:= ListPlot[{3.2, 5.1, 1.4, 0.5, 4.4},
           PlotRange -> {{0, 6}, {0, 6}},
           AspectRatio -> 1,
           PlotJoined -> True];
```

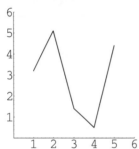

2.2.2 Fitting curves to data

Next, let's generate some data to illustrated with **ListPlot**.

```
In[22]:= data = Table[N[{x, Sin[x]}], {x, 0, 2 Pi, Pi/5}]
```

```
{{0, 0}, {0.628319, 0.587785}, {1.25664, 0.951057},
 {1.88496, 0.951057}, {2.51327, 0.587785}, {3.14159, 0},
 {3.76991, -0.587785}, {4.39823, -0.951057},
 {5.02655, -0.951057}, {5.65487, -0.587785}, {6.28319, 0}}
```

If the first argument of **ListPlot** is a list of pairs, then each pair is treated as the x and y coordinates of a point.

```
In[23]:= ListPlot[data, PlotStyle -> {PointSize[0.04]}];
```

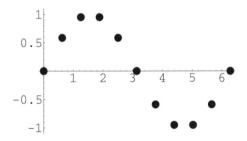

In many applications, the data to be plotted is in some other file and the main problem is to import the data into *Mathematica*. This can be more or less complicated depending on the form of the data. As a very simple example, we put data into a file named `"storage"` and then read it back into a **ListPlot** function.

In[24]:= **Put[OutputForm[data], "storage"]**

To see what is in the file, use the following command.

In[25]:= **! ! storage**

However, this form cannot be used within **ListPlot** to make a picture of this list. Instead, use the form

In[26]:= **ListPlot[Get["storage"], PlotStyle - ⟨{ PointSize[0.04]}];**

This gives exactly the same picture as before, so it is omitted.
         In the picture, these points look as if they could lie on a third degree curve, so we find the best cubic curve that approximates these points.

In[27]:= **fitCurve = Fit[data, {1, x, x^2, x^3}, x]**

$$-0.063509+1.70678\,x-0.805274x^2+0.0854422x^3$$

We can of course combine **Plot** and **ListPlot** in a **Show** command, so we can plot in the same picture both the data and the curve that tries to fit the data.

In[28]:= **Show[Plot[fitCurve, {x, -0.5, 6.7},**
            DisplayFunction -> Identity],
         ListPlot[data, PlotStyle -> {PointSize[0.04]},
            DisplayFunction -> Identity],
         DisplayFunction -> $DisplayFunction];

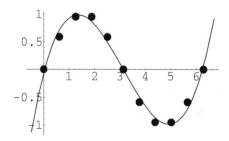

2.3 ParametricPlot

ParametricPlot plots parametric curves. The options for **ParametricPlot** are exactly the same as for **Plot**. As a first example, we plot a Lissajou figure with a frequency ratio of 2/3.

In[29]:= `ParametricPlot[{Sin[2 t], Sin[3 t]}, {t, 0, 2 Pi}];`

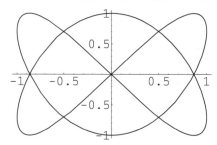

One can also give a list of parametric curves as the first argument. In the next picture, we have changed the color of the curves and made the first one dashed and the second one thick by using **PlotStyle**, whose value here is a list of Graphics primitives, one for each parametric curve. On a monochrome monitor, the curves appear to be shaded.

In[30]:= `ParametricPlot[{{Sin[2 t], Sin[3 t]}, {Sin[t], Sin[4 t]}},`
         `{t, 0, 2 Pi},`
         `PlotStyle ->`
             `{{Dashing[{0.05, 0.03}], RGBColor[1, 0, 0]},`
              `{Thickness[0.01],        RGBColor[0, 1, 0]}} ];`

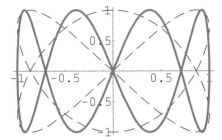

ParametricPlot can be combined with **Plot** or **ListPlot** in a **Show** command.

2.4 ContourPlot

Contour plots are two-dimensional pictures in which the curves are drawn along which a function of two variables takes on constant values. It is not clear whether contour plots are two-dimensional or three-dimensional plots, so we have put them in between. **ContourPlot** adds a number of options concerned with the rendering of the contours and changes some of the other options of **Plot**. The following list shows those options that are different from the options of **Plot**.

In[31]:= **Complement[Options[ContourPlot], Options[Plot]]**

```
{AspectRatio → 1, Axes → False,
 ColorFunction → Automatic, ContourLines → True,
 Contours → 10, ContourShading → True, ContourSmoothing → True,
 ContourStyle → Automatic, Frame → True, PlotPoints → 15}
```

In the three following pictures, the function **Sin[x] Cos[y]** is shown over a range encompassing two maxima and two minima. We use **ContourSmoothing -> Automatic** in these plots, although it then takes considerably longer for them to be drawn. It can also be **None**, which gives jaggedy pictures, or an integer, which specifies how often grid lines should be subdivided in estimating where contours cross the grid lines.

In[32]:= **ContourPlot[Sin[x] Cos[y],**
                {x, 0, 2 Pi}, {y, -Pi/2, 3 Pi/2},
                ContourSmoothing -> Automatic];

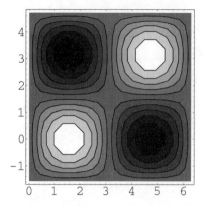

The picture is more interesting, at least on a color monitor, if it is colored using **ColorFunc-tion** by a function that depends on the values of the function. The total range of values is scaled for 0 (lowest) to 1 (highest) and the indicated pure function is applied to these values. **Hue** is a graphics primitive which is also discussed in Chapter 10.

```
In[33]:= ContourPlot[Sin[x] Cos[y], {x, 0, 2 Pi},
                {y, -Pi/2, 3 Pi/2},
                ContourSmoothing -> Automatic,
                ColorFunction -> (Hue[#/2]&)];
```

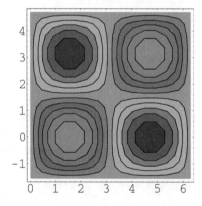

2.5 DensityPlot

DensityPlot uses shading instead of contours to indicate the values of a function of two variables. The picture improves dramatically with increased **PlotPoints**, but plotting time can become very long. Besides the usual options, **DensityPlot** adds an option **Mesh** whose default value is **True**. If **Mesh** is turned off, the picture may look much smoother. Adding color improves the picture. This time we use **RGBColor** rather than **Hue**. (See Chapter 10.)

```
In[34]:=DensityPlot[Sin[x] Cos[y],
            {x, 0, 2 Pi}, {y, -Pi/2, 3 Pi/2},
            PlotPoints -> 50,
            Mesh -> False,
            ColorFunction -> (RGBColor[1 - #, #, 0]&)];
```

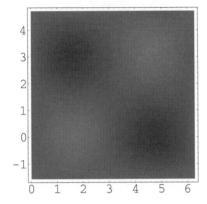

2.6 Two-Dimensional Graphics Commands in Packages

There are many other two-dimensional plotting commands to be found in the packages distributed with *Mathematica*. For full details, see the Technical Report "Guide to Standard *Mathematica* Packages" from Wolfram Research Inc. Most of these commands are located in the Graphics directory found in the directory **Mathematica 3.0 : Add Ons : Standard Packages**. One way to find out what commands are made available by loading the package **Graphics`Master`** is to go to that directory and open the file **Master.m** in *Mathematica*. The **Help Browser** contains information on some of these commands, but not all of them. These commands work just like the built-in graphics commands and they take the same kinds of optional arguments. They can be made available by loading the appropriate package if you know what it is, but it is a more convenient to use the Master file.

```
In[35]:=Needs["Graphics`Master`"]
```

What this does is to load a master file that gives all of the graphics commands the attribute **Stub**. This has the effect that whenever one of these commands is used or mentioned, then the appropriate package is automatically loaded. Here are a couple of examples.

```
In[36]:=??PolarPlot
```

```
    Graphics`Graphics`PolarPlot
    Attributes[PolarPlot] = {Stub}


    PolarPlot = ":Graphics:Master.m"
```

In[37]:= `PolarPlot[Sin[2 theta], {theta, 0, 2 Pi},`
               `PlotStyle - <{Thickness[0.01]},`
               `Ticks - <None,`
               `AxesLabel - <{"polar axis", None}];`

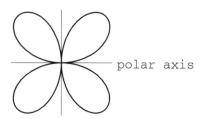

The following used to give the same picture in Version 2.1, but apparently because of a change in **Solve**, it now only finds half of the curve.

In[38]:= `ImplicitPlot[(x^2 + y^2)^(3/2) == 2 x y, {x, -1, 1},`
                `Frame        -> True,`
                `FrameStyle -> {Thickness[0.01]},`
                `FrameTicks -> None,`
                `Ticks        -> None];`

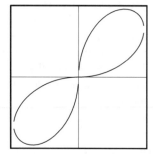

3 Three-Dimensional Graphics

3.1 Plot3D

Plot3D adds many new options, although some of them are also options for **DensityPlot**.

```
In[1]:=Complement[Options[Plot3D], Options[Graphics]]
```

{AmbientLight → GrayLevel[0], AspectRatio → Automatic,
 Axes → True, AxesEdge → Automatic, Boxed → True,
 BoxRatios → {1, 1, 0.4}, BoxStyle → Automatic,
 ClipFill → Automatic, ColorFunction → Automatic, Compiled → True,
 FaceGrids → None, HiddenSurface → True, Lighting → True,
 LightSources → {{{1., 0., 1.}, RGBColor[1, 0, 0]}, {{1., 1., 1.},
    RGBColor[0, 1, 0]}, {{0., 1., 1.}, RGBColor[0, 0, 1]}},
 Mesh → True, MeshStyle → Automatic, Plot3Matrix → Automatic,
 PlotPoints → 15, Shading → True,
 SphericalRegion → False, ViewCenter → Automatic,
 ViewPoint → {1.3, -2.4, 2.}, ViewVertical → {0., 0., 1.}}

First make a simple three-dimensional picture, labeling the axes to see where they are.

```
In[2]:= Plot3D[Cos[x y], {x, 0, Pi}, {y, 0, Pi},
          AxesLabel - <{"x-axis", "y-axis", "z-axis"},
          AspectRatio - <1];
```

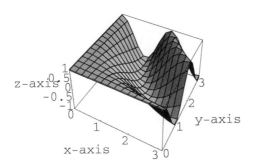

It is not possible to plot several surfaces by changing the first argument to a list of surfaces. Instead, one can change the surface shading by replacing the first argument with a pair consisting of the function and shading functions that also depends on x and y. This can be either a **GrayLevel**, a **Hue**, or an **RGBColor** specification.

```
In[3]:=Plot3D[{Cos[x y], GrayLevel[Abs[x - y]/(2 Pi)]},
          {x, -Pi, Pi}, {y, -Pi, Pi},
          Boxed -> False, Axes -> False, PlotPoints -> 25];
```

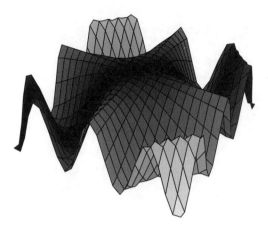

3.2 ParametricPlot3D

ParametricPlot3D plots parametric surfaces and parametric curves in three-dimensional space.

```
In[4]:= ParametricPlot3D[
          {Sin[t],Cos[t],Sin[t]^2}, {t, 0, 2 Pi},
            Axes -> False, BoxRatios -> {1, 1, 1}];
```

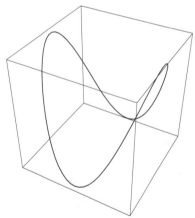

```
In[5]:= ParametricPlot3D[
          {r Cos[omega], r Sin[omega], omega/6},
          {r, 0, 1}, {omega, -Pi, 4 Pi},
          PlotPoints -> {8, Floor[N[16 Pi]]},
          Boxed -> False, Axes -> False];
```

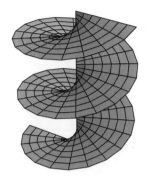

```
In[6]:=ParametricPlot3D[ {r Cos[t], r Sin[t], r^2 Cos[2 t]},
       {t, 0, 2 Pi}, {r, 0, 1},
       Axes -> False, BoxRatios -> {1, 1, 1},
       FaceGrids -> {{-1, 0, 0}, {0, 1, 0}, {0, 0, -1}},
       PlotPoints -> 40];
```

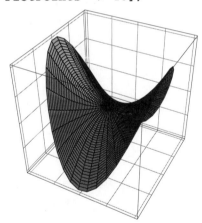

3.3 Three-Dimensional Graphics Commands in Packages

As with two-dimensional graphics, there are many 3-dimensional graphics commands in Packages. These are also made available when the **Graphics`Master`** Package is loaded. Here are three examples.

```
In[7]:= Needs["Graphics`Master`"];
```

In[8]:=`SphericalPlot3D[2 Sin[φ],`
`        {φ, 0, π, π/10},`
`        {Θ, 0, 3 π/2, π/10}];`

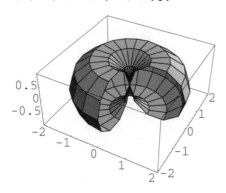

In[9]:=`ShadowPlot3D[Sin[x] Cos[y], {x, 0, 2π}, {y, 0, 2π}];`

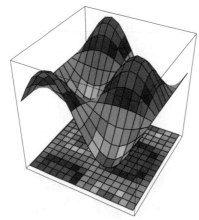

In[10]:=`SurfaceOfRevolution[3 Cos[x], {x, 0, (5/2) π},`
`                    ViewPoint->{1.965, -2.551, 1.040}];`

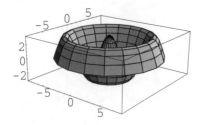

4 Animation

Animations are most conveniently made in a notebook front end environment by using a **Do** loop that evaluates a sequence of expressions, controlled by a simple iterator. We will make an animation of a vibrating plucked string. The initial position is on the interval from 0 to 2, but the function describing its position has to be extended to be an odd, periodic function of period 4. We do this by giving several rules for the function **shape[x]** controlled by the clauses that follow the **/;**'s. (Read **/;** as "provided"; see Chapter 7, Section 5.2.)

```
In[11]:= shape[x_]  :=  x/4              /;  0 <= x < 1;
         shape[x_]  :=  (2 - x)/4        /;  1 <= x < 2;
         shape[x_]  :=  -shape[-x]       /;  x < 0
         shape[x_]  :=  -shape[x - 2]    /;  2 <= x
```

A picture shows that this has the desired properties. The extra lines at y = 0.3 and y = - 0.3 are added to control the shape of the picture.

```
In[15]:= Plot[{0.3, -0.3, shape[x]}, {x, -2, 6},
             AspectRatio -> Automatic,
             Ticks -> None];
```

The position of the string as a function of time is given by the following function of two variables.

```
In[16]:= string[x_, t_]  :=  0.5 (shape[x - t] + shape[x + t])
```

Now we construct 11 pictures showing the positions of the string for time intervals between 0 and 2.

```
In[17]:= Do[Plot[Evaluate[{0.3, -0.3, string[x, 0.2 t]}],
               {x, 0, 2},
               Ticks -> None,
               PlotRange -> All],
           {t, 0, 10}]
```

The output is omitted since we can't actually show an animation in a book. In a notebook front end, one would select these 11 plots and animate them. Use the controls to slow down the animation and to make it cycle back and forth. By making say 40 plots rather than 10, one can get a much smoother action at the expense of a much longer plot time and much more memory to store the result. As an alternative to showing the animation, we can make a graphics array of the output showing all 11 (actually 12) pictures in one drawing.

```
In[18]:= Show[GraphicsArray[
        Table[Plot[
            Evaluate[{0.3, -0.3, string[x, 0.2 (4 i + j)]}],
                {x, 0, 2},
                Ticks -> None,
                PlotRange -> All,
                DisplayFunction -> Identity],
            {i, 0, 2}, {j, 0, 3}]
    ], DisplayFunction -> $DisplayFunction];
```

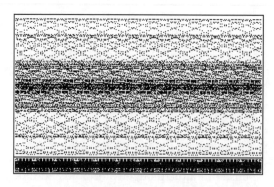

There's one extra plot to fit the 3 × 4 array.

5 Sound

Sounds are created in very much the same way as pictures. Specify a sound wave, for instance, as a sine wave of an appropriate frequency and "show" it by using **Play**. Here is a major triad.

```
In[19]:= Play[{Sin[440 2 Pi t], Sin[440 5/4 2 Pi t],
        Sin[440 3/2 2 Pi t]}, {t, 0, 2}];
```

See [Gray and Glynn 1] for many inventive uses of sound and combinations of sound and graphics. The packages **Miscellaneous`Audio`** and **Miscellaneous`Music`** contain many constants and operations that are useful in constructing functions to be used with **Play**.

6 *Practice*

1. ```
ListContourPlot[{{1, 1, 1, 0}, {2, 1, 2, 1},
                 {3, 2, 1, 0}, {1, 2, 3, 1}}];
```

2. ```
ListDensityPlot[{{1, 1, 1, 0}, {2, 1, 2, 1},
                 {3, 2, 1, 0}, {1, 2, 3, 1}}];
```

3. ```
ListDensityPlot[{{1, 1, 1, 0}, {2, 1, 2, 1},
                 {3, 2, 1, 0}, {1, 2, 3, 1}},
    ColorFunction -> (Hue[#/2, (1 - #/3), 1]&)];
```

4. ```
ListPlot3D[{{1, 1, 1, 0}, {2, 1, 2, 1},
            {3, 2, 1, 0}, {1, 2, 3, 1}},
           AxesLabel -> {x, y, z}];
```

5. ```
ListPlot3D[Table[i/j, {i, 10}, {j, 10}],
            Table[Hue[Random[], Random[], 1],
            {i, 9}, {j, 9}]];
```

6. (after loading **Graphics`Master`**)
```
PointParametricPlot3D[{u v, u + v, u - v},
                      {u, 0, 1}, {v, 0, 1}];
```


# 7 *Exercises*

1. Investigate the meaning of **Automatic** for other optional values.

2. Investigate the option **FaceGrids** for **Plot3D**.

3. Try out **PieChart** and **BarChart** in the graphics packages.

4. Look up in a differential equations book the function that describes a vibrating circular or square membrane (i.e., a drum) and make an animation of this.
```

II

Mastering
Mathematica *as a*
Programming Language

CHAPTER 5

The Mathematica Language

---

1 Everything Is an Expression

1.1 Types and Atoms

In programming languages, types are a device for dividing expresssions into different kinds of entities, mainly for the purpose of checking that certain expressions are correctly formed. For instance, if there is a type called "Integer" and a function whose argument is supposed to be an integer, then, providing the function has some way to know the type of the argument it is being given, there is the possibility of generating an error message when the function tries to evaluate a wrong kind of argument. This is very useful, especially in complicated programs. On the other hand, languages that demand that a type be declared for every entity before it can even be defined can become very cumbersome to use. *Mathematica* tries to have it both ways, and, in some sense, succeeds. In a certain sense, every entity in *Mathematica* has a type, but the type doesn't ever have to be declared and we generally don't have to use it unless we want to.

There are three fundamental types in *Mathematica*, called *symbols*, *numbers*, or *strings*. Expressions of these types are called *atoms*. An atom which is a symbol is any sequence of letters and integers (and possibly $) not starting with an integer. (Letters have ASCII codes from 160 to 255.) Thus, `a2Cd` is a symbol. An atom is a number if it is an integer or a real number. The other four types of numbers–rationals, Gaussian integers, Gaussian rationals, and complex numbers–are not atoms. Thus, 123 and 12.3 are atoms, but 3/4 and 2 + I are not. In these number types, any number of digits is allowed, any exponent is allowed, and any base from 2 to 36 is allowed. Finally, strings are sequences of any ASCII characters between double quotes, i.e., "A word". In the strict sense, there are no other built-in types. Everything else is just an expression.

1.2 Expressions

The heading of this section, "Everything Is an Expression", is to be understood in a very literal sense. Many of the things we have looked at in the first two chapters don't look like expressions as we are about to characterize them, but in fact they are. Expressions are going to be described recursively, and the recursion has to start somewhere. The place it starts is with atoms.

1.2.1 Syntax of expressions

An expression is either an atom or of the form

$$f[a_1, a_2, \ldots, a_n], \quad n \leq 0,$$

where $f, a_1, a_2, \ldots, a_n$ are expressions. Note that $f[\ ]$ is allowed (i.e., n can be 0), but $[a]$ is not. Compare this notation with the LISP notation, $(f \ a_1 \ a_2 \ \ldots \ a_n)$, where the intended meaning is that the first argument inside the parentheses determines a function of the other arguments. Note that there are no commas in the LISP notation, but there are commas in *Mathematica*. If you are familiar with LISP, you can often think of a *Mathematica* expression as a LISP expression with different bracketing. A typical *Mathematica* expression might look like $f[x, y[w1, w2], z]$ where all of the symbols here are atoms. However, we will frequently prefer to write it in a "pretty printed" form with all the atoms written out as complete words rather than being abbreviated as single letters.

```
function[
    argument1,
    argument2[
        subargument1,
        subargument2
    ],
    argument3
]
```

Here, each new argument level is indented and closing brackets are written directly under the first letter of the name of the function they are closing. Sometimes we won't be so strict and allow a modified form that is just as legible, in which we line up the arguments at a given level, except at the bottom level if there is room for them on one line.

```
function[   argument1,
         argument2[subargument1, subargument2],
         argument3]
```

If the expression is given in the form $\text{exp} = \text{f}[\text{a}_1, \text{a}_2, \ldots, \text{a}_n]$, then f is the *head* of the expression; i.e, $\text{Head[exp]} = \text{f}$. The entries a_1, a_2, $\ldots$, a_n, are called the *elements* or *arguments* of exp, and the *length* of exp is n. The ith argument can be accessed by the command exp[[i]]. Expressions are characterized recursively, which means that the head and arguments can be atoms or other expressions; e.g.,

$$\text{h[k[m, n]][a, b[b}_1\text{, b}_2\text{, b}_3\text{[b}_{11}\text{, b}_{22}\text{]], c[c}_1\text{], d, e[e}_1\text{[e}_2\text{]]]}$$

is a perfectly good expression. Writing this out in modified pretty printed form, it looks like:

```
headFunction[kFunction[mArgument, nArgument]
][  aArgument,
    bFunction[
        b1Argument,
        b2Argument,
        b3Function[b11Argument, b22Argument]
    ],
    cFunction[cArgument],
    dArgument,
    eFunction[e1Function[e2Argument]]
]
```

There is only one way to parse this as an expression. Its head is the expression h[k[m, n]], its first argument is a, its second argument is $\text{b[b}_1\text{, b}_2\text{, b}_3\text{[b}_{11}\text{, b}_{22}\text{]]}$, its third is $\text{c[c}_1\text{]}$, its fourth is d, and its fifth is $\text{e[e}_1\text{[e}_2\text{]]}$.

1.2.2 Meaning of expressions

There are several ways to think about expressions that help in understanding how to use them. Some of these are suggested by the following table.

Interpretation	Example
Function[argument];	Sin[x]
Command[argument];	$\text{Expand[(x + y)}^{10}\text{]}$
Operator[operands];	Plus[x, y]
Type[parts];	List[a, b, c, d]

The differences between **Function**, **Command**, and **Operator** as descriptions of heads are purely psychological. We might think of something as a function if it takes numbers as arguments and produces numbers as values. If there are several numerical arguments all on the same level, then we might regard the head as an operator, even when it is used with symbolic arguments. On the other hand, something that takes expressions as arguments and rewrites them in different forms, or carries out some complicated procedure, might be regarded as a command. But what do we mean by a type with parts?

This kind of ambiguity is very helpful, since a single semantic model for the behavior of *Mathematica* expressions is not forced on the user. For instance, on the one hand, **List** just holds its arguments together and doesn't do anything to them. On the other hand, it takes a number of different entities and produces something new out of them, namely, the list containing them, so it can be looked at as a function. Similarly, we usually think of **Sin** as a function, but when it is applied to an integer in *Mathematica*, nothing happens, so it is just holding its argument and producing an entity of type **Sin**.

1.2.3 Forms of expressions

You would certainly be justified in being skeptical about this description of *Mathematica*, since many of the things we have used don't resemble expressions in this sense at all. What the description really applies to is *Mathematica*'s own internal represention of expressions. However, this internal description can also be used for entering expressions as inputs, which will be very important later on. It turns out that everything has a head, even atoms. The internal form can be accessed by the command **FullForm**. Here are many examples, each presented as list consisting of the input form of some expression, its **Head**, and its **FullForm**. The **FullForm** of atoms does not include the **Head**, but for everything else it does, so we will omit calculating the **Head** separately for non-atoms.

In[1]:= `{abc, Head[abc], FullForm[abc]}`

Out[1]= {abc, Symbol, abc}

In[2]:= `{27, Head[27], FullForm[27]}`

Out[2]= {27, Integer, 27}

In[3]:= `{27.35, Head[27.35], FullForm[27.35]}`

Out[3]= {27.35, Real, 27.35}

In[4]:= `{"A word", Head["A word"], FullForm["A word"]}`

Out[4]= {A word, String, "A word"}

Notice that the output form of a string does not include the quotation marks. Here is what happens with rational and complex numbers.

In[5]:= $\{ \frac{3}{4},$ **Head**$\left[\frac{3}{4} \right],$ **FullForm**$\left[\frac{3}{4} \right] \}$

Out[5]= $\{ \frac{3}{4},$ Rational, Rational[3, 4] $\}$

In[6]:= **{3 + 5 I, Head[3 + 5 I], FullForm[3 + 5 I]}**

Out[6]= $\{3 + 5\,I,$ Complex, Complex[3, 5] $\}$

In particular, the **FullForm** of a rational number or a complex number includes the head Rational or Complex, so these are compound expressions rather than atoms. Now consider some more complicated expressions.

In[7]:= **{x + y + z, FullForm[x + y + z]}**

Out[7]= $\{x + y + z,$ Plus[x, y, z] $\}$

This shows that + is just the infix form of the head **Plus**. Furthermore, **Plus** can take any number of arguments, not just two. One can of course use **Plus** instead of the + sign in an input.

In[8]:= **Plus[2, 3, 4, 5]**

Out[8]= 14

In order to see this form using numbers, we have to prevent *Mathematica* from automatically evaluating a sum of numbers. This is done by wrapping the operation **Hold** around the sum.

In[9]:= **{3 + 4, FullForm[Hold[3 + 4]]}**

Out[9]= $\{7,$ Hold[Plus[3, 4]] $\}$

Multiplication is just like addition, using the head **Times**. Subtraction is not a separate operation internally, but is replaced by **Plus** and **Times**.

In[10]:= **{x - y, FullForm[x - y]}**

Out[10]= $\{x - y,$ Plus[x, Times[-1, y]] $\}$

Exponentiation is a separate operation, but division is not.

In[11]:= $\left\{ x^n,\ \texttt{FullForm}[x^n],\ \dfrac{x}{y},\ \texttt{FullForm}\left[\dfrac{x}{y}\right]\right\}$

Out[11]= $\left\{ x^n,\ \texttt{Power}[x, n],\ \dfrac{x}{y},\ \texttt{Times}[x, \texttt{Power}[y, -1]]\right\}$

Curly brackets are just a "circumfix" form for the head **List**.

In[12]:= `{{x, y, z}, FullForm[{x, y, z}]}`

Out[12]= `{{x, y, z}, List[x, y, z]}`

The dot "." in the dot product of vectors is the infix form of **Dot**.

In[13]:= `{{a, b}.{2, 3}, FullForm[Hold[{a, b}.{1, 2}]]}`

Out[13]= `{2 a + 3 b, Hold[Dot[List[a, b], List[1, 2]]]}`

The arrow used in substitutions is the infix form of the head **Rule** and the **/.** symbol used in applying substitutions is the infix form of the head **ReplaceAll**.

In[14]:= `{FullForm[Hold[x ➝y]], FullForm[Hold[x /. y ➝z]]}`

Out[14]= `{Hold[Rule[x, y]], Hold[ReplaceAll[x, Rule[y, z]]]}`

The two kinds of equals signs used in making assignments and function definitions are the infix forms of the heads **Set** and **SetDelayed**, respectively.

In[15]:= `{FullForm[Hold[x = y]], FullForm[Hold[f := y]]}`

Out[15]= `{Hold[Set[x, y]], Hold[SetDelayed[f, y]]}`

The symbols **f** and **x** now have values, which is a nuisance later, so we clear them here.

In[16]:= `Clear[x, f]`

There are three symbolic forms of function application: **f@a** and **a//f**, both of which mean exactly the same as **f[a]**, and an additional form **a ~f~ b** for functions of two variables, which means the same as **f[a, b]**. Since these are themselves expressions in the proper form, that is what is returned in these cases by **FullForm**.

In[17]:= `{f@a, FullForm[f@a], a//f, FullForm[Hold[a//f]]}`

Out[17]= `{f[a], f[a], f[a], Hold[f[a]]}`

In[18]:= **{a ~f~ b, FullForm[a ~f~ b]}**

Out[18]= {f[a, b], f[a, b]}

Double square brackets are the postfix notation for part extraction, here denoted by the head **Part**.

In[19]:= **FullForm[x[[i]]]**

    Out[19]//FullForm=
    Part[x, i]

Here are the **FullForm**s of some of the various size comparisons for numbers.

In[20]:= **{FullForm[a <= b], FullForm[a == b], FullForm[a >= b]}**

Out[20]= {LessEqual[a, b], Equal[a, b], GreaterEqual[a, b]}

Various notions of infinity all have a **FullForm** using DirectedInfinity.

In[21]:= **{FullForm[Infinity], FullForm[-Infinity],
    FullForm[ComplexInfinity], FullForm[I Infinity]}**

Out[21]= {DirectedInfinity[1], DirectedInfinity[-1],
    DirectedInfinity[], DirectedInfinity[Complex[0, 1]]}

Finally, some of the more mysterious symbols also correspond to reasonable heads.

In[22]:= **{FullForm[Hold[%]], FullForm[Hold[%%]], FullForm[Hold[%5]]}**

Out[22]= {Hold[Out[]], Hold[Out[-2]], Hold[Out[5]]}

In[23]:= **{FullForm[_], FullForm[x_], FullForm[x_Integer]}**

Out[23]= {Blank[], Pattern[x, Blank[]], Pattern[x, Blank[Integer]]}

In[24]:= **{FullForm[#], FullForm[#&]}**

Out[24]= {Slot[1], Function[Slot[1]]}

1.3 Cells, Notebooks, etc.

In Version 3.0, the slogan "Everything Is an Expression" extends from the kernel to the front end, since everything in the front end is also an expression.

1.3.1 Cells

A part of the front end enclosed by a single bracket on the right is called a *cell*. Input goes in Input cells and output appears in Output cells. There are text cells, section cells, title cells, etc. What one normally sees on the screen is the displayed form of such cells. However, if you select a cell and then choose **Show Expression** in the **Format** menu, then you will see something completely different. An expression with head **Cell** takes two arguments. The first is a description of what you normally see, while the second says what kind of a cell it is. In addition, there can be many optional arguments. For instance, the cell holding the name of this subsubsection looks like

```
Cell[TextData[{
          CounterBox["Section"],
           ".",
          CounterBox["Subsection"],
           ".",
          CounterBox["Subsubsection"],
           " Cells"
        }], "Subsubsection", CellTags->{"Cell"}]
```

This means that the actual description of this cell is a **Cell** expression whose first argument is a **TextData** expression consisting of three **CounterBox**es (giving the numbers for the subsubsection) separated by ".",'s, followed by the string " Cells". The second argument shows that the type is "Subsubsection". The third optional argument, CellTags ->{"Cell"}, was added by me, simply by typing it in the **Show Expression** form of this cell. (One can also use the **Add/Remove Cell Tags...** item in the **Find** menu.) Cell tags were used to generate the index of this book.

1.3.2 Notebooks

An entire notebook is just an expression with head Notebook. It takes one argument which is a list of cells. There can also be optional arguments, but at the time of writing, these are not documented anywhere. There is a whole little language for handling notebooks. For instance, the command

```
In[25]:= Table[NotebookPut[Notebook[
          {Cell[StringJoin["Section[", ToString[i], "]"], "Section"],
           Cell[StringJoin["Something[", ToString[i], "]"],
             "Text"]}]], {i, 3}]
```

```
Out[25]= {- NotebookObject -, - NotebookObject -, - NotebookObject -}
```

as a side effect, puts up three notebooks (piled on top of each other) consisting of a Section cell labeled **Section[i]** and a Text cell containing **"Something[i]"**, for **i** from 1 to 3. The corresponding NotebookObjects in the kernel can be read back into the current session using the following command.

```
In[26]:= NotebookGet[%[[2]]]
```

```
Out[26]= Notebook[
          {Cell[CellGroupData[{Cell[TextData[Section[2]], Section],
              Cell[TextData[Something[2]], Text]}, Open]]},
          FrontEndVersion → Macintosh 3.0, ScreenRectangle →
           {{0., 832.}, {0., 604.}}, WindowSize → {520., 509.},
          WindowMargins → {{156., Automatic}, {Automatic, 38.}},
          MacintoshSystemPageSetup → 00<0001804P000000]
            P2:?oQon82n@960dL5:0?10080001804P000000]P2:001
          0000I00000400`<300000BL?00400@
            0000000000000006P801T1T000000@0000
          00000000004000000000000000000000]
```

Here we see that the actual description of the notebooks we produced is more complicated than our input. The **CellGroupData** item means that the cells are grouped together in one enclosing bracket and this outer cell, by default, is **Open**. The optional arguments show some of the complexity of actually putting up a window.

1.3.3. Palettes and Buttons

The default configuration of *Mathematica* when it is first opened includes a palette consisting of buttons that can be clicked to produce effects. Such buttons are simple to produce. The following command produces one.

In[27]:= **ButtonBox["Expand[■]", Active   ↩True,**
        ButtonStyle   ↩CopyEvaluateCell"] // DisplayForm

Out[27]//DisplayForm=

The first argument is entered by typing **"Expand["\"[SelectionPlaceholder]]"**
omitting the quotation marks around the backslash. This produces the funny little square. The
optional argument **Active ↩True** means that the button actually does something and the
ButtonStyle argument determines what that is. To see how this button works, create some
input, but don't evaluate it. Then select the input and click on the button.

In[28]:= $(1 + x)^4$

In[29]:= **Expand[$(1 + x)^4$]**

Out[29]= $1 + 4x + 6x^2 + 4x^3 + x^4$

The result is to wrap **Expand** around the selection, put this in a new cell, and evaluate that
cell.

Buttons can be assembled in palettes by using the **Create Table/Matrix/Palette...** item in the
Input menu and defining actions for each button. Then use the **Generate Palette from Selection** item in the **File** menu to produce a new palette.

This should be enough to convince you that, internally at least, everything *is* an expression.

1.4 Analysis of Expressions

1.4.1 Types revisited

Some heads of expressions cause a computation to be performed involving the arguments of
the expression. Others, such as **List**, don't do anything except hold their arguments together
as a single entity. It is certainly a reasonable point of view to regard **List** as a type in the
sense of type theory for programming languages. But then why not regard any head as a type,
as suggested in the section at the beginning of the chapter? Then every expression has a type,
but we don't have to do anything special about declaring types. (This is not what is usually
understood in type theory, where complicated expressions are supposed to have types that
are derived somehow from the types of their constituents.) However, if we decide to think this
way, then, as will be seen in Chapter 7, *Mathematica* supports this idea by allowing type
checking for any head.

1.4.2 Parts of expressions

Parts of expressions are described by a numbering scheme which can be used either forwards or backwards.

```
In[30]:= exp = f[a1, a2, a3, a4];
```

```
In[31]:= {exp[[0]], exp[[1]], exp[[2]], exp[[3]], exp[[4]]}
```

```
Out[31]= {f, a1, a2, a3, a4}
```

Negative numbers inside double square brackets count from the right-hand end of the expression.

```
In[32]:= {exp[[-4]], exp[[-3]], exp[[-2]], exp[[-1]]}
```

```
Out[32]= {a1, a2, a3, a4}
```

1.4.3 Tree structure of expressions

If we ask for the **FullForm** of a more complicated expression, then the result is again a *Mathematica* expression built up from the **FullForms** of the parts. For instance:

```
In[33]:= exp1 = x³ + (1 + z)²;
```

```
In[34]:= FullForm[exp1]
```

```
Out[34]//FullForm=
Plus[Power[x, 3], Power[Plus[1, z], 2]]
```

This kind of format is derived by forcing all operators to be given in prefix form. It is rather like "forward Polish notation" with explicit bracketing. Expressions can be displayed in another format, which is sometimes more informative, using the command **TreeForm**.

```
In[35]:= TreeForm[exp1]
```

```
Out[35]//TreeForm=
Plus[ |              ,  |                      ]
          Power[x, 3]   Power[ |          , 2]
                                Plus[1, z]
```

This is intended as a representation of the tree

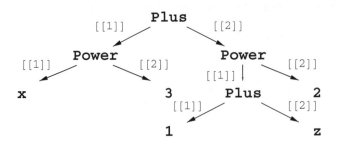

In the drawing we have labeled the edges with the part extraction command that leads to each particular argument. From this tree form of an expression, one can see how to access any part of the expression by a multiple part extraction. Thus, the expression corresponding to the subtree starting at any node can be displayed by giving the path of edges from the root **Plus** to that node as a sequence of numbers inside double square brackets.

In[36]:= **exp1[[1]]**

Out[36]= x^3

In[37]:= **exp1[[1]][[2]]**

Out[37]= 3

Instead of first extracting the first argument of **exp1** and then extracting the second argument of the result, there is the following abbreviated form.

In[38]:= **exp1[[1, 2]]**

Out[38]= 3

Try finding some other subexpression.

In[39]:= **{exp1[[2, 1]], exp1[[2, 1, 2]], exp1[[2, 1, 0]]}**

Out[39]= $\{1 + z, z, \text{Plus}\}$

Note: The initial 0 is not given, since **exp1[[0, 1]]** would mean the first part of the head of **exp1**, which doesn't exist here. However, it does in the following example.

In[40]:= **(f[z][x])[[0, 1]]**

Out[40]= z

A *partspec* is a positive or negative number, n or -n, or a sequence of such numbers (n1, n2, . . .) describing the position of an argument in an expression. It is what goes inside `[[]]`.

If the expression is larger, then it is hard to display its tree form. There are two facilities for examining such expressions, `Short` and `Shallow`.

```
In[41]:=bigexp :=
        Sum[Product[Sum[x[i, j, k],
             {i, 1, 5}], {j, 1, 5}], {k, 1, 5}];
```

```
In[42]:=Short[bigexp, 4]
```

```
Out[42]//Short=
(x[1, 1, 1] + x[2, 1, 1] + x[3, 1, 1] + x[4, 1, 1] + x[5, 1, 1])
   (x[1, 2, 1] + x[2, 2, 1] + x[3, 2, 1] + x[4, 2, 1] + x[5, 2, 1]) (≪1  ≪
    (≪1  ≪(x[1, 5, 1] + x[2, 5, 1] + x[3, 5, 1] + x[4, 5, 1] + x[5, 5, 1]) +
   ≪1  ≪≪1  ≪(≪1  ≪≪3  ≪≪1  ≪ ≪1  ≪
```

`Short` shows us part of the complete detail of `bigexp`. (The 4 means show about four lines of the expression. <<n>> means n things are left out). We see that it is a sum of five terms, and the first term is a product of five terms, each of which is a sum of five terms of the form `x[i, j, k]`, three of which are omitted here.

```
In[43]:=Shallow[bigexp]
```

```
Out[43]//Shallow=
+ ≪5   ≪≪5   ≪≪5   ≪≪5   ≪≪5   ≪
+ ≪5   ≪≪5   ≪≪5   ≪≪5   ≪≪5   ≪
+ ≪5   ≪≪5   ≪≪5   ≪≪5   ≪≪5   ≪
+ ≪5   ≪≪5   ≪≪5   ≪≪5   ≪≪5   ≪
+ ≪5   ≪≪5   ≪≪5   ≪≪5   ≪≪5   ≪
```

`Shallow` just displays some of the top of the expression tree. Here we see that the output is a sum of five terms, each of which is a product of five terms, and each of these is again a sum of five terms. It is only the lowest level subtrees `x[i, j, k]` that are compressed. Together `Short` and `Shallow` give us a fairly good idea of what `bigexp` is like. Both `Short` and `Shallow` take optional arguments which allow a great deal of fine control over what is displayed. The following example is from [Wei].

$$In[44]:= \ \textbf{badexp = Together}\left[\textbf{Normal}\left[\textbf{Series}\left[\frac{1}{2 - \textbf{Sin[t - a]}}, \ \{\textbf{t, 0, 4}\}\right]\right]\right]$$

$$Out[44]= \ \left(384 + 192 \ t \ Cos[a] - 32 \ t^3 \ Cos[a] + 96 \ t^2 \ Cos[a]^2 - 32 \ t^4 \ Cos[a]^2 + \right.$$
$$48 \ t^3 \ Cos[a]^3 + 24 \ t^4 \ Cos[a]^4 + 768 \ Sin[a] + 96 \ t^2 \ Sin[a] -$$

$$8\, t^4\, \mathrm{Sin}[a] + 288\, t\, \mathrm{Cos}[a]\, \mathrm{Sin}[a] + 48\, t^3\, \mathrm{Cos}[a]\, \mathrm{Sin}[a] +$$
$$96\, t^2\, \mathrm{Cos}[a]^2\, \mathrm{Sin}[a] + 40\, t^4\, \mathrm{Cos}[a]^2\, \mathrm{Sin}[a] + 24\, t^3\, \mathrm{Cos}[a]^3\, \mathrm{Sin}[a] +$$
$$576\, \mathrm{Sin}[a]^2 + 144\, t^2\, \mathrm{Sin}[a]^2 + 12\, t^4\, \mathrm{Sin}[a]^2 + 144\, t\, \mathrm{Cos}[a]\, \mathrm{Sin}[a]^2 +$$
$$72\, t^3\, \mathrm{Cos}[a]\, \mathrm{Sin}[a]^2 + 24\, t^2\, \mathrm{Cos}[a]^2\, \mathrm{Sin}[a]^2 +$$
$$28\, t^4\, \mathrm{Cos}[a]^2\, \mathrm{Sin}[a]^2 + 192\, \mathrm{Sin}[a]^3 + 72\, t^2\, \mathrm{Sin}[a]^3 +$$
$$18\, t^4\, \mathrm{Sin}[a]^3 + 24\, t\, \mathrm{Cos}[a]\, \mathrm{Sin}[a]^3 + 20\, t^3\, \mathrm{Cos}[a]\, \mathrm{Sin}[a]^3 +$$
$$24\, \mathrm{Sin}[a]^4 + 12\, t^2\, \mathrm{Sin}[a]^4 + 5\, t^4\, \mathrm{Sin}[a]^4 \Big) \Big/ \Big(24\, (2 + \mathrm{Sin}[a])^5\Big)$$

In[45]:= **Short [badexp]**

Out[45]//Short=

$$\frac{384 + \ll\!31\quad\ll\!5\, t^4\, \mathrm{Sin}[a]^4}{24\, (2 + \mathrm{Sin}[a])^5}$$

In[46]:= **Shallow [badexp]**

Out[46]//Shallow=

$$\frac{1}{(+\ \ll\!2\quad\succ\!\!\!\prec}^5$$

$$\Big(\frac{1}{24}\ (384 + \mathrm{Times}[\ll\!3\quad\succ\!+ \mathrm{Times}[\ll\!3\quad\succ\!+ \mathrm{Times}[\ll\!3\quad\succ\!+ \mathrm{Times}[\ll\!3\quad\succ\!+$$
$$\mathrm{Times}[\ll\!3\quad\succ\!+ \mathrm{Times}[\ll\!3\quad\succ\!+ \mathrm{Times}[\ll\!2\quad\succ\!+$$
$$\mathrm{Times}[\ll\!3\quad\succ\!+ \mathrm{Times}[\ll\!3\quad\succ\!+ \ll\!20\quad\succ\!\!\!\prec$$

Short gives us some vague idea of the form of the expression, but **Shallow** has lost all the important detail. However, the following gives a fair idea of what the function actually is.

In[47]:= **Short [badexp, 4]**

Out[47]//Short=

$$\Big(384 + 192\, t\, \mathrm{Cos}[a] - 32\, t^3\, \mathrm{Cos}[a] + \ll\!26\quad\ll$$
$$24\, \mathrm{Sin}[a]^4 + 12\, t^2\, \mathrm{Sin}[a]^4 + 5\, t^4\, \mathrm{Sin}[a]^4 \Big) \Big/ \Big(24\, (2 + \mathrm{Sin}[a])^5\Big)$$

1.4.4 Levels of expressions and levelspecs

Recall **exp1 = x^3 + (1 + z)^2** from above. There is another way to describe the parts of an expression. Each part occurs at some specific level. The top of the tree is at level 0, the next row is level 1, etc. Levels are described by levelspecs, which are numbers **n** or **-n**, single numbers in curly brackets **{n}** or **{-n}**, pairs in curly brackets **{n1, n2}** or **Infinity**. To see the parts at exactly level 2, use

In[48]:=**Level[exp1, {2}]**

Out[48]= $\{x, 3, 1 + z, 2\}$

Actually, this gives the subtrees, written as expressions, whose root is exactly at level 2. (By a subtree, we mean some node together with everything below it. The node is the root of the subtree.) To see the parts (i.e., subtrees) at level 2 and higher, omit the curly brackets:

In[49]:=**Level[exp1, 2]**

Out[49]= $\{x, 3, x^3, 1 + z, 2, (1 + z)^2\}$

A levelspec of the form **{n1, n2}** gives the subtrees whose root is between **n1** and **n2**. For instance,

In[50]:=**Level[f0[f1[f2[f3[f4[f5]]]]], {2, 4}]**

Out[50]= $\{f4[f5], f3[f4[f5]], f2[f3[f4[f5]]]\}$

The depth of an expression is the maximum number of nodes along a path from the root to a leaf in the expression.

In[51]:=**Depth[exp1]**

Out[51]= 4

There are also negative levels which use negative numbers to count from the bottom up. What is actually counted is the depth of a subexpression. Thus, **{-1}** gives all subexpressions whose depth is exactly 1 (i.e., the leaves), whereas **-1** (without the curly brackets) gives all proper subexpressions of depth at least 1 (i.e., all proper subexpressions).

In[52]:=**{Level[exp1, {-1}], Level[exp1, -1]}**

Out[52]= $\{\{x, 3, 1, z, 2\}, \{x, 3, x^3, 1, z, 1 + z, 2, (1 + z)^2\}\}$

The level specification **{-2}** gives all proper subtrees of depth 2, whereas **-2** gives all proper subtrees of depth at least 2.

In[53]:=**{Level[exp1, {-2}], Level[exp1, -2]}**

Out[53]= $\{\{x^3, 1 + z\}, \{x^3, 1 + z, (1 + z)^2\}\}$

The levelspec **Infinity** also gives all proper subexpressions.

In[54]:= **Level[exp1, Infinity]**

Out[54]= $\{x, 3, x^3, 1, z, 1 + z, 2, (1 + z)^2\}$

1.4.5 Manipulating arguments of expressions

Start with a general expression with six arguments.

In[55]:= **generalExp = fun[a, b, c, d, e, f];**

There are a number of operations that change the arguments in some way. In **Drop** and **Take**, the description of which arguments are affected is given by a *sequencespec*. A single number **n** means the first n arguments. A single number **-n** means the last n arguments. A number with curly brackets **{n}** or **{-n}** means exactly the **n**th argument from the left or right. A pair of numbers in curly brackets means a range of arguments. In **Delete**, **Insert**, and **Replace-Part**, the second or third argument is a partspec, so a single number **n** or **-n** refers to the nth argument counted from the left or right, and a list refers to the part specification of a specific subtree. A list of lists refers to several such subtrees. Often these operations seem to make more sense if they are used just for lists, but in fact they work with arbitrary heads. The last two operations, as a side effect, change the value of **generalExp**. We have numbered the expressions for ease in seeing which operation has which effect, and we repeat **generalExp** at the beginning and end to show how it has been changed.

In[56]:= **{{1,   generalExp},**
         {2,   Drop[generalExp, 3]},
         {3,   Take[generalExp, 3]},
         {4,   Take[generalExp, {2, 4}]},
         {5,   Delete[generalExp, -3]},
         {6,   Insert[generalExp, hello, 3]},
         {7,   ReplacePart[generalExp, hello, 3]},
         {8,   ReplacePart[generalExp, hello, {{2}, {-2}}]},
         {9,   Join[generalExp, Reverse[generalExp]]},
         {10, RotateRight[generalExp]},
         {11, RotateLeft[generalExp]},
         {12, First[generalExp]},
         {13, Rest[generalExp]},
         {14, Reverse[generalExp]},
         {15, Partition[generalExp, 2]},
         {16, Prepend[generalExp, yesterday]},
         {17, Append[generalExp, tomorrow]},
         {18, PrependTo[generalExp, yesterday]},
         {19, AppendTo[generalExp, tomorrow]},
         {20, generalExp}}//TableForm

```
Out[56]//TableForm=
1          fun[a, b, c, d, e, f]
2          fun[d, e, f]
3          fun[a, b, c]
4          fun[b, c, d]
5          fun[a, b, c, e, f]
6          fun[a, b, hello, c, d, e, f]
7          fun[a, b, hello, d, e, f]
8          fun[a, hello, c, d, hello, f]
9          fun[a, b, c, d, e, f, f, e, d, c, b, a]
10         fun[f, a, b, c, d, e]
11         fun[b, c, d, e, f, a]
12         a
13         fun[b, c, d, e, f]
14         fun[f, e, d, c, b, a]
15         fun[fun[a, b], fun[c, d], fun[e, f]]
16         fun[yesterday, a, b, c, d, e, f]
17         fun[a, b, c, d, e, f, tomorrow]
18         fun[yesterday, a, b, c, d, e, f]
19         fun[yesterday, a, b, c, d, e, f, tomorrow]
20         fun[yesterday, a, b, c, d, e, f, tomorrow]
```

There are other ways to operate on the arguments of expressions that take account of *what* they are rather than *where* they are. These will be discussed in detail in Chapter 7, but here is an example. **Positive** is a predicate on numbers which is True just for numbers greater than 0. **Select** with second argument **Positive** drops all arguments that are not positive.

```
In[57]:= Select[f[-3, 3, -2, 2, -1, 1, 0], Positive]
```

```
Out[57]= f[3, 2, 1]
```

These operations are all immensely useful, as will be seen in the later chapters. Here we just call your attention to their existence since they will be needed in the Exercises.

2 Lists, Arrays, Intervals, and Sets

Some programming languages have special types for arrays, matrices, lists, etc. In *Mathematica* all of these concepts are represented by lists. Since lists are such an important aspect of the language, there are many special features for dealing with them.

2.1 *Listability*

When many built-in operations are applied to a list, they automatically apply themselves to the entries in the list. Such an operation is called **Listable**. Start with a simple list.

In[1]:= **list = {2, 3, 4};**

The only sensible meaning for **Sin** of this list is the list of values of **Sin** applied to the entries in the list.

In[2]:= **Sin[list]**

Out[2]= {Sin[2], Sin[3], Sin[4]}

This happens by itself without our doing anything about it. In other words, **Sin** commutes with (or distributes over) **List**. Certain functions have the attribute of being **Listable** which is shown by the operation **Attributes**. For example,

In[3]:= **Attributes[Sin]**

Out[3]= {Listable, NumericFunction, Protected}

Many other operations have this property. (See Chapter 11, Section 3, for a list of all of them.) For instance, arithmetic operations, etc., automatically propagate down lists.

In[4]:= **newlist = x^{list} - 1**

Out[4]= $\{-1 + x^2, -1 + x^3, -1 + x^4\}$

Thus, for instance, to show the squares of the entries in a list in expanded form, just expand the list raised to the power 2.

In[5]:= **Expand[newlist^2]**

Out[5]= $\{1 - 2 x^2 + x^4, 1 - 2 x^3 + x^6, 1 - 2 x^4 + x^8\}$

To find the derivatives of these functions at the point 3, use the facts that differentiation is **Listable** in its first argument as is substitution.

In[6]:= **D[%, x] /. x $\to$ 3**

Out[6]= {96, 1404, 17280}

Since everything here is `Listable`, this could have all been done in one step.

In[7]:= $D\left[\text{Expand}[x^{\{2, 3, 4\}} - 1]^2, x\right] /. x \to 3$

Out[7]= $\{96, 1404, 17280\}$

Listability actually means more than just that single operations automatically map themselves down lists. Consider what happens if several lists are multiplied.

In[8]:= `{2, 3, 4} {a, b, c} {x, y, z}`

Out[8]= $\{2\,a\,x, 3\,b\,y, 4\,c\,z\}$

Thus, if `Times` is given several lists of the same length, then it forms the list given by multiplying corresponding entries. Any operation that is `Listable` behaves the same way.

In[9]:= $\{x, y, z\}^{\{2, 3, 4\}}$

Out[9]= $\{x^2, y^3, z^4\}$

This way of handling lists is characteristic of *Mathematica*, and we shall make frequent use of it.

2.2 Construction of Lists – Table and Range

We have already seen how to create a list using the `Table` command. For instance:

In[10]:= $\text{Table}\left[x^i + 2\,i, \{i, 1, 5\}\right]$

Out[10]= $\{2 + x, 4 + x^2, 6 + x^3, 8 + x^4, 10 + x^5\}$

The second argument in `Table`, and in a number of other commands like `Integrate` for definite integrals, `Plot`, etc., is called an *iterator*. It comes in several forms.

`{i, imin, imax, step}`	This gives `i` values from `imin` to `imax` in steps of size `step`.
`{i, imin, imax}`	This gives `i` values from `imin` to `imax` in steps of size 1.
`{i, imax}`	This gives `i` values from 1 to `imax` in steps of size 1.
`{imax}`	This repeats something that doesn't depend on an index `imax` times.

Here are examples of all four kinds of iterators.

In[11]:= $\texttt{Table}\left[\texttt{i, } \left\{\texttt{i, 3, 6, } \frac{1}{2}\right\}\right]$

Out[11]= $\left\{3, \frac{7}{2}, 4, \frac{9}{2}, 5, \frac{11}{2}, 6\right\}$

In[12]:= $\texttt{Table[i, \{i, 3, 6\}]}$

Out[12]= $\{3, 4, 5, 6\}$

In[13]:= $\texttt{Table[i, \{i, 6\}]}$

Out[13]= $\{1, 2, 3, 4, 5, 6\}$

In[14]:= $\texttt{Table[Random[Integer, \{0, 12\}], \{12\}]}$

Out[14]= $\{9, 5, 6, 6, 0, 11, 2, 3, 5, 9, 9, 5\}$

There is another way to create lists without having some variable take on successive values. This is done as follows:

In[15]:= $\texttt{Range[10]}$

Out[15]= $\{1, 2, 3, 4, 5, 6, 7, 8, 9, 10\}$

In[16]:= $\texttt{Range[-3, 8]}$

Out[16]= $\{-3, -2, -1, 0, 1, 2, 3, 4, 5, 6, 7, 8\}$

Range can also take a third argument specifying the step size.

In[17]:= $\texttt{Range}\left[\texttt{-3, 4, } \frac{1}{2}\right]$

Out[17]= $\left\{-3, -\frac{5}{2}, -2, -\frac{3}{2}, -1, -\frac{1}{2}, 0, \frac{1}{2}, 1, \frac{3}{2}, 2, \frac{5}{2}, 3, \frac{7}{2}, 4\right\}$

The arguments to **Range** are like the first three kinds of iterators without the variable **i**. Once a list of index values has been constructed by the **Range** operation, then other lists can be created using listability by replacing the variable with the appropriate value of **Range**. Thus, in the **Table** constructed at the beginning of this section, replace **i** by **Range[5]** to get exactly the same output.

In[18]:= $\texttt{x}^{\texttt{Range[5]}} \texttt{ + 2 Range[5]}$

Out[18]= $\{2 + x, \ 4 + x^2, \ 6 + x^3, \ 8 + x^4, \ 10 + x^5\}$

Another way to operate on ranges using the **Map** function will be discussed in Chapter 7. What about multidimensional lists? Consider the Hilbert matrix of size 3, which is given by the following **Table** construction.

In[19]:= **Table$\left[\dfrac{1}{i + j - 1}, \ \{i, \ 3\}, \ \{j, \ 3\}\right]$**

Out[19]= $\left\{\left\{1, \ \dfrac{1}{2}, \ \dfrac{1}{3}\right\}, \ \left\{\dfrac{1}{2}, \ \dfrac{1}{3}, \ \dfrac{1}{4}\right\}, \ \left\{\dfrac{1}{3}, \ \dfrac{1}{4}, \ \dfrac{1}{5}\right\}\right\}$

This can be constructed using the operation **Outer**, which applies its first argument (a function of two or more variables) to all choices of entries from each of two lists. Thus, for instance,

In[20]:= **Outer[Plus, Range[3], Range[3]]**

Out[20]= $\{\{2, \ 3, \ 4\}, \ \{3, \ 4, \ 5\}, \ \{4, \ 5, \ 6\}\}$

So the matrix we want is given by the following construction.

In[21]:= $\dfrac{1}{\textbf{Outer[Plus, Range[3], Range[3]] - 1}}$

Out[21]= $\left\{\left\{1, \ \dfrac{1}{2}, \ \dfrac{1}{3}\right\}, \ \left\{\dfrac{1}{2}, \ \dfrac{1}{3}, \ \dfrac{1}{4}\right\}, \ \left\{\dfrac{1}{3}, \ \dfrac{1}{4}, \ \dfrac{1}{5}\right\}\right\}$

In general, **Outer** takes a (pure) function as its first argument and any number of lists (or expressions with the same head) as the rest of its arguments. If there are n lists, then the function must accept n arguments. **Outer** then constructs the multidimensional list of the function applied to all combinations of one argument from each list. **Outer** will be discussed further later. If the arguments to **Outer** are themselves multidimensional lists, then the behavior of **Outer** is more complicated. (See Section 3.3 below.)

Also note that **Range** is **Listable**, so we get the following unexpected result.

In[22]:= **Range$\left[\{1, \ 1\}, \ \{3, \ 5\}, \ \left\{\dfrac{1}{3}, \ \dfrac{1}{2}\right\}\right]$**

Out[22]= $\left\{\left\{1, \ \dfrac{4}{3}, \ \dfrac{5}{3}, \ 2, \ \dfrac{7}{3}, \ \dfrac{8}{3}, \ 3\right\}, \ \left\{1, \ \dfrac{3}{2}, \ 2, \ \dfrac{5}{2}, \ 3, \ \dfrac{7}{2}, \ 4, \ \dfrac{9}{2}, \ 5\right\}\right\}$

which is the same as

In[23]:= $\left\{\textbf{Range}\left[1, \ 3, \ \dfrac{1}{3}\right], \ \textbf{Range}\left[1, \ 5, \ \dfrac{1}{2}\right]\right\}$;

2.3 Arrays

Symbolic arrays of given sizes can be constructed by the command **Array**. For instance, we can make an array of indexed values of **aa**'s.

In[24]:=**Array[aa, {3, 4}]**

Out[24]= {{aa[1, 1], aa[1, 2], aa[1, 3], aa[1, 4]},
          {aa[2, 1], aa[2, 2], aa[2, 3], aa[2, 4]},
          {aa[3, 1], aa[3, 2], aa[3, 3], aa[3, 4]}}

If desired, we can format the **aa**'s with subscripts and superscripts.

In[25]:=**Format[aa[i_, j_]] = Subscripted[aa[i, j], {1}, {2}];**

Let's give a name to the formatted form to use later.

In[26]:=**aaArray = Array[aa, {3, 4}]//TableForm**

Out[26]//TableForm=

aa_1^1	aa_1^2	aa_1^3	aa_1^4
aa_2^1	aa_2^2	aa_2^3	aa_2^4
aa_3^1	aa_3^2	aa_3^3	aa_3^4

Values can be assigned to the entries if one wishes to make **aaArray** into an array of numbers.

In[27]:=**aa[i_, j_] := i + j**

In[28]:=**aaArray**

Out[28]//TableForm=

2	3	4	5
3	4	5	6
4	5	6	7

2.4 Flatten

Flatten removes all inner brackets in lists.

In[29]:=**list = {{{{a, b}, {c, d}},{{e, f}, {g, h}}},
            {{{i, j}, {k, l}},{{m, n}, {o, p}}}};**

In[30]:=`Flatten[list]`

Out[30]= {a, b, c, d, e, f, g, h, i, j, k, l, m, n, o, p}

Flatten can also take a second argument which is a level specification.

In[31]:=`Flatten[list, 1]`

Out[31]= {{{a, b}, {c, d}}, {{e, f}, {g, h}},
         {{i, j}, {k, l}}, {{m, n}, {o, p}}}

Often it takes a good deal of experimentation to discover the appropriate level specification for a desired outcome. Actually, all that **Flatten** requires is that all heads be the same, so it works for arbitrary heads instead of just **List**.

In[32]:=`Flatten[f[f[a, b], f[c, d]]]`

Out[32]= f[a, b, c, d]

FlattenAt flattens parts of expressions just at specific locations given by a position list. Study the following example carefully to understand what **FlattenAt** actually does.

In[33]:=`FlattenAt[list, {{1, 1}, {2}}]`

Out[33]= {{{a, b}, {c, d}, {{e, f}, {g, h}}},
         {{i, j}, {k, l}}, {{m, n}, {o, p}}}

2.5 *Real Intervals*

There is a facility **Interval** similar to **Range** for dealing with intervals of real numbers. It arises as the value of certain functions; e.g.,

In[34]:=`real = Limit[Cos[x], x -> Infinity]`

Out[34]= Interval[{-1, 1}]

One can carry out limited calculations with real intervals similar to the calculations with **Range**. For instance:

In[35]:= `N[2`^{real}`] + 2 real`

Out[35]= Interval[{-1.5, 4.}]

Much of *Mathematica*'s extended precision arithmetic is done using intervals.

2.6 Set Operations

Some operations on lists treat them as though they were sets. Sets can be though of as lists where elements are not repeated and whose order doesn't matter. We start with a long list with repeated entries.

In[36]:= **longlist = Table[Random[Integer, {0, 9}], {20}]**

Out[36]= {6, 2, 9, 0, 2, 1, 7, 9, 4, 4, 1, 5, 8, 5, 3, 7, 9, 3, 0, 3}

Union of a single list turns it into a set by removing duplicate elements and ordering the result. **Union** of several lists first joins them together and then turns the result into a set. **Complement** starts with its first entry, deletes the elements of the remaining entries, and then turns what remains into a set. Finally, **Intersection** forms the set theoretic intersection of its arguments.

In[37]:= **{Union[longlist],**
       Union[longlist, {5, 4, 3}, {12, 14, 16}],
       Complement[longlist, {1, 2}, {5}, {7, 8}],
       Intersection[{a, b, c, d, c}, {b, c, d, c, e},
                      {c, d, e, c, f}, {f, g, d, a, c}]}//MatrixForm

Out[37]//MatrixForm=

$$\begin{pmatrix} \{0, 1, 2, 3, 4, 5, 6, 7, 8, 9\} \\ \{0, 1, 2, 3, 4, 5, 6, 7, 8, 9, 12, 14, 16\} \\ \{0, 3, 4, 6, 9\} \\ \{c, d\} \end{pmatrix}$$

3 Other Aspects

There are a number of other aspects of the *Mathematica* language that require consideration. For two of them, the virtual operating system and the string language, we limit ourselves to some brief comments and examples. The third, programming facilities, will be the subject of the next three chapters.

3.1 The Virtual Operating System

For a detailed description of the virtual operating system, see The *Mathematica* Book, Chapter 2, Section 10. Some aspects, such as reading and writing external files, will be discussed as part of the examples treated in Chapter 8 here. Other aspects, those enabling one to manipulate files and run external programs from within *Mathematica* are what really constitute the virtual operating system. The effect of such commands is, of course, system dependent. Thus, the command **Run["date"]** produces the date in a Unix environment and 0 on a Macintosh. On the other hand, the commands **Directory[]**, which gives the current directory, and **FileNames[]**, which lists the files in the current directory, seem to work anywhere.

In[1]:=**Directory[]**

Out[1]= Russell:*Mathematica* 3.0 Files

In[2]:=**FileNames[]**

Out[2]= {AddOns, Configuration, Documentation, Executables,
        Installer Log File, *Mathematica* 3.0, numbers1, Registration,
        sessionHistory, storage, SystemFiles, SystemThings}

In addition, there are a number of commands starting with **$** that either give information about the current machine and its operating system, or concern how *Mathematica* interacts with it. You can see all of them by typing **Names["$*"]**.

For instance, **$Path** tells *Mathematica* where to look for files to be loaded using **Needs**.

In[3]:=**$Path**

Out[3]= {Russell:System Folder:Preferences:*Mathematica*:3.0:Kernel,
        Russell:System Folder:Preferences:*Mathematica*:
          3.0:AddOns:Autoload, Russell:System Folder:
          Preferences:*Mathematica*:3.0:AddOns:Applications, :,
        Russell:*Mathematica* 3.0 Files:AddOns:StandardPackages,
        Russell:*Mathematica* 3.0 Files:AddOns:StandardPackages:StartUp,
        Russell:*Mathematica* 3.0 Files:AddOns:Autoload,
        Russell:*Mathematica* 3.0 Files:AddOns:Applications,
        Russell:*Mathematica* 3.0 Files:AddOns:ExtraPackages,
        Russell:*Mathematica* 3.0 Files:SystemFiles:Graphics:Packages}

Here are some others.

```
In[4]:= {$OperatingSystem, $Packages, $SessionID}
```

```
Out[4]= {MacOS, {Global`, System`}, 20063583463014506288}
```

3.2 The String Language

Strings form a special data type in *Mathematica* which in some sense reflects within itself the whole *Mathematica* language. The two commands **ToString** and **ToExpression** take expressions to strings and vice versa. Similarly, **ToCharacterCode** and **FromCharacterCode** convert between strings and ASCII code. Furthermore, there is a collection of operations that mimic those for manipulating expressions. Start with some string.

```
In[5]:= generalString = "The quick brown fox ";
```

Here is a sample of things that can be done with a string.

```
In[6]:= {{1, StringDrop[generalString, {5, 10}]},
         {2, StringTake[generalString, {11, 15}]},
         {3, StringInsert[generalString, " stupid", 4]},
         {4, StringReplace[generalString,
                         {"b" -> "g", "x" -> "e"}]},
         {5, StringJoin[generalString,
                      StringReverse[generalString]]}}//
      TableForm
```

```
Out[6]//TableForm=
   1       The brown fox
   2       brown
   3       The stupid quick brown fox
   4       The quick grown foe
   5       The quick brown fox  xof nworb kciuq ehT
```

We will make frequent use of **StringJoin**, in its infix form **<>**, in some of the later programs. The main uses I can think of for these operations are to massage data that has been imported from or will be exported to an external program and to produce output dependent text for graphics.

3.3 Programming Facilities

There are many facilities in *Mathematica* for dealing with various styles of programming. These are so important that the next three chapters will be devoted to them, in the following order: functional programming, rule based programming, and procedural programming.

4 Practice

```
1.    ??Set
2.    ??SetDelayed
3.    {FullForm[x && z], FullForm[x || z], FullForm[!x]}
4.    {FullForm[a < b], FullForm[a > b]}
5.    FullForm[f']
6.    FullForm[f'''']
7.    Drop[Range[10], 3]
8.    Take[Range[10], 3]
9.    Delete[Range[10], 3]
10.   Insert[Range[10], hello, 3]
11.   ReplacePart[Range[10], hello, 3]
12.   Reverse[Range[10]]
13.   Partition[Range[10], 4, 2]
14.   Select[Range[-3, 10], Negative]
15.   StringLength[ToString[Range[10]]]
16.   StringDrop[ToString[Range[10]], 3]
17.   StringTake[ToString[Range[10]], 3]
18.   StringInsert[ToString[Range[10]], ", hello", 3]
19.   StringReplace[ToString[Range[10]],
          Table[ToString[i] -> ToString[11 - i], {i, 10}]]
20.   StringReverse[ToString[Range[10]]]
21.   Distribute[g[f[a, b], f[1, 2]], f, g, ff, gg]
22.   Distribute[{{x^2 + y^2, 2 x y}, {x, y}},
                 List, List, List, D]
23.   ??!
24.   Names["$*"]
25.   {{a, b, c}, {d, e, f}, {g, h, i}}^2
26.   {{a, b, c}, {d, e, f}, {g, h, i}}^0
27.   Permutations[{1, 2, 3}]
28.   Range[{1, 1, 1, 1, 1}, {5, 4, 3, 2, 1}]//TableForm
29.   Limit[Tan[x], x -> Infinity]
30.   N[ArcTan[%]]
```

5 Exercises

1. In problem 13 of the exercises in Chapter 3, the third solution of the transformation

$$u = x^2 + y^2$$
$$v = -2 x y$$

for (x, y) in terms of (u, v) was used to construct **invjak**, which was then expressed in terms of x and y to get the matrix **jak'**. It satisfies **jak.jak' = Id**. This time:

i) Modify your definition of the Jacobian function using the notions introduced in this chapter.

ii) Find **invjak(n)**, $1 \le n \le 4$ for each of the four solutions of x and y in terms of u and v. Keep **invjak(n)** as an expression in u and v.

iii) For each of the four solutions for (x, y) in terms of (u, v), express the original matrix **jak** in terms of u and v instead of x and y, giving four Jacobians **jak(n)**, $1 \le n \le 4$ in terms of u and v.

iv) For each n show that **jak(n).invjak(n) = Id**. Your final output should be a list of four 2 by 2 identity matrices.

2. Find the greatest common divisor of the *n*th row of Pascal's triangle, omitting the 1's. To do this:

i) Define a modified function **pascal(n)**, from problem 11 of the exercises in Chapter 3, which gives the entries of this row without the 1s.

ii) Then define a function **gcd(n)** which gives the greatest common divisor of the entries in **pascal(n)**. Note that there is a built-in function **GCD**.

iii) Make a table of the first 20 values of **gcd(n)** and conjecture the value of **gcd(p)** for **p** a prime number.

iv) Use *Mathematica* to check that your conjecture is correct for the first 50 primes. Note that there is a built-in function **Prime[n]**.

v) Use your head to prove your conjecture for all primes. You may assume that binomial coefficients are integers.

vi) Guess the values of **gcd(n)** for **n** a power of a prime, and for **n** a number with at least two different prime factors. (You might want to extend your table to **n** = 50, or even to **n** = 100, to get more evidence for your guess.)

3. Which rows of Pascal's triangle have all odd entries? For which rows of Pascal's triangle are all entries, except for the initial and final 1's, even? Note that there are built-in predicates **EvenQ** and **OddQ**. Check your conjectures for the first 2^{10} rows.

4. Define **f[m, r] = b[m + r - 1, r]**, where **b[m, n]** is the binomial coefficient function. This rotates the usual Pascal's triangle by 45 degrees. Make a table showing these values in an upper-left triangular form corresponding to the usual table up to size 10. Pascal's Corollary 4 asserts that in this table, each entry is equal to the sum of all the entries to the north west of it plus 1. Verify this for a number of small values of **m** and **r**.

5. Implement the Gram–Schmidt method for orthogonalizing vectors with respect to the dot product. It should take as input a list of n-dimensional vectors over the reals and output a list of n-dimensional orthonormal vectors. You may assume that the original list is linearly independent. It is sufficient to do this for $n = 3$. Check your algorithm on a list of three three-dimensional vectors with random real components. (Think about the case of four three-dimensional vectors with random real components.)

CHAPTER 6

Functional Programming

Pascal is for building pyramids–imposing, breathtaking, static structures built by armies pushing heavy blocks into place. Lisp is for building organisms–imposing, breathtaking, dynamic structures built by squads fitting fluctuating myriads of simpler organisms into place.
[Abelson]

We, of course, intend to replace "Pascal" by "C" and "Lisp" by "*Mathematica*."

1 Some Functional Aspects of Mathematica

What is a functional programming language? Basically, it's a language in which functions can be defined and applied to arguments. It is important that the arguments of a function can themselves be other functions applied to other arguments, etc. There is an abstract, theoretical functional programming language called the lambda calculus (to be discussed in detail in Chapter 11, Section 7) which has exactly three operations: function definition, function application, and substitution–which is essentially a rewrite rule for function application. There are only a few pure functional languages (e.g., Haskel and Miranda). Most so-called functional languages, such as Scheme or ML, are impure in the sense that they have many other features to make programming more convenient. *Mathematica* belongs to the category of languages that have a pure functional component, but also include many other features. It has the advantage over Scheme and ML of having many built-in mathematical functions to start with that can be combined with each other to create new functions. Later in this chapter we'll look at exactly what constitutes functional programming in *Mathematica*.

Another way to look at the question is to change it to: what properties characterize functional programming languages? Some possible answers to this are:

i) Higher order functions. This property means that functions can be arguments and values of other functions. We'll give examples of this in *Mathematica* later. Mathematically, for instance, composition of functions is represented by the equation $(f \circ g)(x) = f(g(x))$. The operation on the left-hand side, $(f \circ g)$, is a higher order function. It takes two functions, f and g, as its arguments and returns another function, their composition, as its value. Another example is the special case of twice, where twice is defined by the equation

$$(twice(f))(x) = f(f(x));$$

i.e., twice(f) = f ∘ f.

ii) *Referential transparency.* A programming language is *referentially transparent* if the value of an expression depends only on the values of its subexpressions. For instance, the value of m + n should depend only on the values of m and n (as well as the value of +). Thinking mathematically, it is hard to imagine this failing. What else could it depend on? Well, it might depend on the order in which m and n are evaluated because evaluating one of them could, as a side effect, change the result of evaluating the other. (Look in Chapter 8, Section 2.1, to see how this can happen in *Mathematica*.) One way to make the notion of "depending only on the values of subexpressions" more precise is to define it to mean that any subexpression (by which we mean a subtree of the tree form of the expression) can be replaced by any other expression with the same value. As a practical matter, it comes to the same thing–no side effects; i.e., assignment statements like a = 5 are forbidden.

iii) *Functions have no memory.* Every time a function is evaluated (with, of course, the same values for the arguments) it returns the same value. Thus, a history of successive evaluations of a function would be very dull, consisting just of the same value over and over again. We'll see in Chapter 9 that this is one of the differences between applying a function to a value and sending a message to an object.

iv) *Lazy evaluation.* A language uses *lazy evaluation* if arguments to functions are evaluated only when they are needed. In LISP, functions with this property are called *special forms.* In *Mathematica*, they are operations that hold some or all of their arguments. This is an important property because it allows computations to proceed even if some of their arguments are not well defined. For instance, the definition

```
bad[x_] := If[x == 0, good, 1/x]
```

works perfectly well at x = 0, even though 1/0 is undefined.

Mathematica shares features with several other programming languages, such as C, Pascal, Lisp, or APL, but it has extended and modified these features as well as adding its own original constructs. In this chapter we examine those features that qualify it as a functional programming language.

2 *Defining Functions*

2.1 *Pure Functions*

The function **Map** is a feature of virtually every functional language. Normally, it takes two arguments (a function and a list), and it applies the function to every entry in the list, producing a list as the output. For instance:

In[1]:= **Map[Sin, {a, b, c}]**

Out[1]= {Sin[a], Sin[b], Sin[c]}

When using **Map** and similar operations (see Section 3 for more details), it is convenient not to have to name the function being mapped over a list or other expression. For instance, suppose we have two functions **f** and **g** and we want to form the sum **f[x] + g[x]** for all the entries in a list. One way would be to define a new function **h** to be the sum of **f** and **g**, and then map it down the list.

In[1]:=**h[x_] := f[x] + g[x]**

In[2]:=**Map[h, {a, b, c}]**

Out[2]= {f[a] + g[a], f[b] + g[b], f[c] + g[c]}

Instead, without defining a separate function **h**, the form **Function[{x}, f[x] + g[x]]** can be used.

In[3]:=**Map[Function[{x}, f[x] + g[x]], {a, b, c}]**

Out[3]= {f[a] + g[a], f[b] + g[b], f[c] + g[c]}

The expression **Function[{x}, body]** is a "pure function" with one bound variable **x**. This notation is essentially the lambda notation of the lambda calculus, discussed in Chapter 11, Section 7, where the form λx . body is written for the same thing. The operation **Function[{x}, body]** or λx . body is a canonical name for the function that does to any argument whatever **body** describes as being done to **x**. Such a function is applied to a value in the same way that built-in functions are. For instance:

In[4]:= **Function[{x}, x^2][5]**

Out[4]= 25

One can of course give a name to such an expression and then use it as a function:

In[5]:= **t = Function[{x}, Expand[(1 + x)^3]];**

In[6]:= **t[a]**

Out[6]= $1 + 3\,a + 3\,a^2 + a^3$

The form **Function[{x}, body]** is the same as **Function[body]** or, in postfix notation, **body&**, where **x** in **body** is replaced by **#** (whose **FullForm** as we have seen is **Slot[1]**.) This is a much more convenient form to use in **Map** and, as will be seen, in many other places. For instance, our original example now becomes

In[7]:= **Map[(f[#] + g[#])&, {a, b, c}]**

Out[7]= $\{f[a] + g[a],\ f[b] + g[b],\ f[c] + g[c]\}$

The three expressions

    i) **Function[{x}, f[x] + g[x]]**

    ii) **Function[f[#] + g[#]]**

    iii) **(f[#] + g[#])&**

all produce the same value when applied to an argument. The third is clearly the most concise form.

    Consider a simpler example, the squaring function, written in pure function form as **#^{2} &**. It is applied to an argument using square brackets, as usual.

In[8]:= **#^{2} &[5]**

Out[8]= 25

To map it down a list, use **Map**, or **/@**.

In[9]:= **Map[#^{2} &, {a, b, c}]**

Out[9]= $\{a^2,\ b^2,\ c^2\}$

In[10]:= **#^{2} &/@{a, b, c}**

Out[10]= $\{a^2,\ b^2,\ c^2\}$

A combination of symbols like #² & /@ can be hard to read. It is somewhat improved by extra spaces in the form #^2& /@ {a, b, c}, but in general we will avoid such combinations (although I am personally very fond of them) unless they force themselves on us.

For functions of several arguments, the slots are numbered.

In[11]:= $\dfrac{\texttt{m[\#1, \#2]}}{\texttt{n[\#1, \#2]}}$ &[a, b]

Out[11]= $\dfrac{\texttt{m[a, b]}}{\texttt{n[a, b]}}$

One can also operate on pure functions and get pure functions as the output. For instance, the derivative of a function **f** can be written as **f'**.

In[12]:= **Sin'**

Out[12]= Cos[#1] &

Notice that **Sin** with no argument is a pure function and the output of **Sin'** is written explicitly as a pure function with a # and an &, although presumably, a simple **Cos** would have been sufficient. The following also work:

In[13]:= {#² & ', #⁷ & ' ' ' ' '}

Out[13]= $\left\{2\ \#1\ \&,\ 7\left(6\left(5\left(4\left(3\ \#1^{2}\right)\right)\right)\right)\ \&\right\}$

Pure functions written in this form with # and & are a distinctive and very attractive feature of the *Mathematica* programming language.

2.2 Four Kinds of Function Definition

Functions are such an important feature of *Mathematica* that they are represented in (at least) four different ways.

2.2.1 Expressions

An expression like **x^2 + b x + c** contains a symbol **x** that is intended to be interpreted (by us) as a variable; that means, we are intended to interpret the whole expression as describing a function of the variable **x**. However, it is only we who know this; there is no way for *Mathematica* to know it unless we somehow tell *Mathematica* what the variable is. Thus, in commands like

```
Solve[ x^2 + b x + c == 0, x],
Plot[Sin[Cos[x] + Tan[x]], {x, 0, Pi}],
Sum[ i³, {i, 0, 10}]
Product[(x + i), {i, 1, 4}], etc.,
```

the first argument is such an expression and the second argument includes a description of the appropriate variable, either by just naming it or by including it as the first argument of an iterator.

Furthermore, if an expression is intended to be regarded as describing a function of some variable, then there should be some way to substitute an actual value for the variable. This is where **ReplaceAll** or, in infix notation, **/.**, comes in. Thus, if **expr** is some expression involving **x**, then its value at **a** is given by **expr/.x** ←a. (See Chapter 7, Section 2.2, for a through discussion of **/.** and →This, of course, may cause some further simplification to be carried out. For instance:

In[14]:= **2 x + 5 /. x** ←2

Out[14]= 9

2.2.2 Named pure functions

In a function definition such as **f[x_] := x²**, the thing being defined is **f**, which should be thought of as the "function in itself". The **x** in this definition is a dummy variable which is not really there at all. (In Chapter 7, Section 3.2, we'll see that **x_** is a pattern.) If we insist on thinking of **x** as the variable in the definition of **f**, then it is a bound variable in the sense of logic. The **f** defined here is in fact a name for a pure function and can be used wherever pure functions are appropriate, just like the names of built-in functions. For example:

In[15]:= **f[x_] := x²;**

In[16]:=**Map[f, {a, b, c}]**

Out[16]= {a², b², c²}

Thus, defining a function using **SetDelayed** (i.e., **:=**) gives a name to a pure function. The definitions

```
square[x_] := x²,
square = Function[{x}, x²] and
square = #²&
```

are essentially equivalent. However, there are subtle differences described in Chapter 11, Section 7.

It is not possible, for instance, to plot the function **square** by using the command **Plot[square, {x, 0, 10}]** or to integrate it by the command **Integrate[square, {x, 0, 10}]** because it is nowhere indicated how **x** is involved with **square**. In order to plot or integrate **f** we have to turn it into an expression involving a variable, whose name of course doesn't matter, which is described in the second argument to **Plot**. Thus, **Plot[square[x], {x, 0, 10}]** and **Plot[square[y], {y, 0, 10}]** both give the same picture. It is necessary to be aware of those commands that require expressions in this sense with variable names together with some other information about those variables.

2.2.3 Nameless pure functions with bound variables

Expressions like **Function[{x}, x^2]** define functions using a syntax that is essentially the same as the lambda calculus. Such a definition involves a bound variable, in this case **x**, whose name clearly doesn't matter; i.e., **Function[{y}, y^2]** describes the same function. The function itself has no other name attached to it, so it is a nameless pure function. Functions of more than one variable are allowed, for example

In[17]:=**Function[{x, y}, x + y][2, 3]**

Out[17]= 5

but they need to be given the proper number of values as arguments (two here). This is to be distinguished from the following "curried" version, which is a function of one variable returning as value another function of one variable.

In[18]:=**Function[{x}, Function[{y}, x + y]][2]**

Out[18]= Function[{y$}, 2 + y$]

In[19]:=**%[3]**

Out[19]= 5

2.2.4 Anonymous functions

The point of the syntax using **#** and **&** is that it is possible to construct a nameless pure function with no variables, bound or otherwise, i.e., an *anonymous* function. Anonymous pure functions are functions named by a canonical variable-free name, i.e., a **#** - **&** expression. When there is more than one slot, enough arguments have to be given to fill all slots. "Currying" as above is not possible with anonymous functions, except by rather grotesque contortions.

In[20]:=**Evaluate[(#1 + #2)&[2]]&/. #2 -> #1**

Out[20]= 2 + #1 &

In[21]:=%[3]

Out[21]= 5

2.2.5 Conversion between forms of functions

Each of these notions is appropriate in its own place. It is possible to convert from one to the other.

    i) Starting with an expression exp = x^2+ 2x, we can get

       a) A named pure function f[x_] := exp

       b) A pure function with bound variable Function[{x}, exp]

       c) An anonymous function Evaluate[exp/.x ←#]&. (The Evaluate is required here because & holds its argument.)

    ii) Starting with the named pure function f[x_] := x^2 + 2x, we can get the following:

       a) An expression f[x]

       b) The actual pure function itself f

       c) A pure function with bound variable

$$Function[\{x\}, Evaluate[f[x]]$$

       d) An anonymous function Evaluate[f[#]]&

    iii) Starting with a pure function with a bound variable, f = Function[{x}, x^2 + 2x], we can get the following:

       a) An expression f[x]

       b) The named pure function f

       c) An anonymous function, Evaluate[f[#]]&

    iv) Starting with an anonymous function ($\#^2$ + 2#)&, we can get the following:

       a) An expression ($\#^2$ + 2#)&[x]

       b) A named pure function f = ($\#^2$ + 2#)&

       c) A pure function with bound variable

$$Function[\{x\}, Evaluate[(\#^2 + 2\#)\&[x]]]$$

For instance, the anonymous function $(\#^2 + 2\#)\&$ can be plotted using iv) a) above:

```
In[22]:= Plot[(#² + 2 #) &[x], {x, 0, 10}];
```

2.3 The fundamental dictum of Mathematica programming

The purpose of all of these operations based on **Map** is to make it possible to treat lists as a whole. For instance, one really poor way to square the entries in a list is as follows:

```
In[23]:= list = {a, b, c};
```

```
In[24]:= squares = Table[list[[i]]², {i, Length[list]}]
```

Out[24]= $\{a^2, b^2, c^2\}$

What this does is to tear apart the original list by extracting its parts one at a time, carrying out the squaring operation on each part, and then reassembling the new parts into a new list. Another way, which is only marginally better, is to use the typical Lisp style of recursive programming.

```
In[25]:= sqEls[{}] := {};
        sqEls[list_] := Join[{First[list]²}, sqEls[Rest[list]]]
```

```
In[26]:= sqEls[{a, b, c}]
```

Out[26]= $\{a^2, b^2, c^2\}$

Here, we have ripped out the first entry and squared it. Note that the procedure also has to be initialized by saying what it does to the empty list. This kind of disassembly of lists is forbidden in *Mathematica*-style functional programming. In a generalized form, here is the fundamental dictum of *Mathematica* programming.

> **Treat mathematical structures as wholes.**
> **Never tear them apart and rebuild them again.**

Mathematica allows one to bypass both of the preceding strategies and think of a mathematical structure as a whole in the Platonic sense of mathematics. Mathematical structures are complete entities, even though they may be made up of parts or elements, and in many cases one can carry out constructions on these entities directly without having to be concerned with their parts.

3 Applying Functions to Values

We have already discussed and used the *Mathematica* facilities for defining functions and applying them to arguments. For instance, $f[x\_] := x^2$ defines the squaring function and $f[2]$ or $f@2$ or $2//f$ applies it to the value 2. But there is more to practical functional programming than this. There are a number of built-in operations that take arbitrary functions as arguments and do something with them. For instance, there are several built-in commands that take a function as first argument and an arbitrary expression as second and apply the function to various parts of the expression. The first of these, **Map**, as we saw in Section 2.1, normally has two arguments, a function and a list, and it applies the function to every entry in the list, producing a list as the output. As usual in *Mathematica*, the list argument can be replaced by an arbitrary expression and levelspecs can be used to specify the level of the expression at which the function is applied.

3.1 Map and Its Relations

3.1.1 Map

The operation **Map[function, expr]** or in infix form **function /@ expr** applies **function** to each subexpression at level 1 in **expr**. The operation **Map** applied to a list, therefore, just applies the function to each entry in the list; e.g.,

In[1]:=**Map[Sin, {a, b, c}]**

Out[1]= {Sin[a], Sin[b], Sin[c]}

Furthermore, *Mathematica* can map a function down the arguments of an arbitrary expression, not just a list. For instance, continuing with **exp1** as in the preceding chapter,

In[2]:= **exp1 = $x^3 + (1 + z)^2$;**

In[3]:=**Map[Sin, exp1]**

Out[3]= $\text{Sin}[x^3] + \text{Sin}[(1 + z)^2]$

Be sure you understand the output here. Even more, **Map** takes a third argument, with a new effect. **Map[function, expr, levelspec]** applies **function** to the parts of **expr** described by **levelspec**. For instance, we can apply **Sin** to all of the leaves (described by the levelspec {-1}), or to all of the subexpressions whose depth is at least 2 (described by the levelspec -2).

In[4]:=**Map[Sin, exp1, {-1}]**

Out[4]= $\text{Sin}[x]^{\text{Sin}[3]} + (\text{Sin}[1] + \text{Sin}[z])^{\text{Sin}[2]}$

In[5]:=**Map[Sin, exp1, -2]**

Out[5]= $\text{Sin}[x^3] + \text{Sin}[\text{Sin}[1 + z]^2]$

This is a very powerful facility and is one of our main tools in manipulating expressions. The next sections describe some variations on **Map**.

3.1.2 MapAll

MapAll[function, expr] or **function //@ expr** applies **function** to every proper subexpression of **expr**. It is the recursive form of **function /@ expr**; i.e., it applies **function** to all of the arguments, then to all of the arguments of the arguments, etc.

In[6]:=**MapAll[Sin, exp1]**

Out[6]= $\text{Sin}\left[\text{Sin}[\text{Sin}[x]^{\text{Sin}[3]}] + \text{Sin}[\text{Sin}[\text{Sin}[1] + \text{Sin}[z]]^{\text{Sin}[2]}]\right]$

This operation is almost the same as using **Map** with the levelspec **Infinity** or **-1**.

In[7]:=**Map[Sin, exp1, Infinity]**

Out[7]= $\text{Sin}[\text{Sin}[x]^{\text{Sin}[3]}] + \text{Sin}[\text{Sin}[\text{Sin}[1] + \text{Sin}[z]]^{\text{Sin}[2]}]$

The difference is that the levelspec **Infinity** does not include the whole expression itself.

3.1.3 MapAt

MapAt[function, expr, positionlist] applies function to the parts of expr described by the list of partspecs in positionlist. Here a partspec, as usual, is described by the *list* of edges in the tree form of the expression from the root to the given subtree. We'll use this to apply Sin just to the variables x and z in exp1 by giving a list of two partspecs.

In[8]:=MapAt[Sin, exp1, {{1,1}, {2,1,2}}]

Out[8]= $\text{Sin}[x]^3 + (1 + \text{Sin}[z])^2$

See also MapThread and MapIndexed in The *Mathematica* Book.

3.1.4 Apply

The various versions of Map act on the arguments of a function. Apply acts only on the head of an expression. What Apply[head, expr] or head @@ expr does is to replace the head of expr by head (and possibly carries out a subsequent simplification or computation.) For instance:

In[9]:=Apply[Plus, {2, 3, 4}]

Out[9]= 9

Here, the head of {2, 3, 4} is List, which does nothing to its arguments. When List is replaced by Plus, then something happens, since Plus calculates the sum of its arguments. Here is a slightly trickier example.

In[10]:=Apply[Plus, 2 3 x]

Out[10]= 6 + x

This time the head of 2 3 x is Times, which multiplies out as much of the expression as it can, yielding Times[6, x]. When this head is replaced by Plus, we get the indicated result.

With a third argument, Apply[head, expr, levelspec] replaces heads in the parts of expr described by levelspec by head.

In[11]:= Apply[Sin, {{f[a], f[b]}, {f[c], f[d]}}, {2}]

Out[11]= {{Sin[a], Sin[b]}, {Sin[c], Sin[d]}}

Apply is frequently used if one wants to first prepare a number of ingredients and then apply some operation to them. The ingredients can be held in a list until they are ready and then the head of the list is changed to the appropriate operation by **Apply**.

3.1.5 Through

Once it is brought to your attention, **Map** seems like an obvious operation that ought to be available in any programming language. But it has an asymmetrical aspect: one function is applied to a list of values. What about applying a list of functions to a single value? That can be done too, using the operation **Through**. For instance:

In[12]:= **Through[{Sin, Cos, Tan}[a]]**

Out[12]= {Sin[a], Cos[a], Tan[a]}

If the functions in the list are themselves listable, then they can be applied to a list of values.

In[13]:= **Through[{Sin, Cos, Tan}[{a, b, c}]]**

Out[13]= {{Sin[a], Sin[b], Sin[c]},
        {Cos[a], Cos[b], Cos[c]}, {Tan[a], Tan[b], Tan[c]}}

3.2 Thread, Inner and Outer

3.2.1 Thread

The built-in operation **Thread** does the same thing for an arbitrary head that the attribute of listability does for heads with this property.

In[14]:= **Thread[ff[{2, 3, 4}, {a, b, c}, {x, y, z}]]**

Out[14]= {ff[2, a, x], ff[3, b, y], ff[4, c, z]}

Actually, **Thread** works with either lists of the same length or individual arguments.

In[15]:= **Thread[ff[{2, 3, 4}, {a, b, c}, 2, x]]**

Out[15]= {ff[2, a, 2, x], ff[3, b, 2, x], ff[4, c, 2, x]}

Furthermore, for **Thread** the arguments don't even have to be lists; they can also have an arbitrary head, which is then included as a second argument to **Thread**. (Note that **Listable** works only with lists.)

In[16]:=`Thread[ff[hh[2, 3, 4], hh[a, b, c], hh[x, y, z]], hh]`

Out[16]= $hh[ff[2, a, x], ff[3, b, y], ff[4, c, z]]$

Thread can be used, for instance, to construct lists used in substitutions. Consider the following fragment of *Mathematica* code.

In[17]:= `var = {x, y, z};`
        `point = {1, 2, 3};`
        `expr = x`^{2}` y + 2 y z - 4 x z`^{3}`;`

Thread can be used to produce a list of substitutions.

In[18]:=`Thread[Rule[var, point]]`

Out[18]= $\{x \rightarrow 1, y \rightarrow 2, z \rightarrow 3\}$

This can then be used to substitute the point into the expression.

In[19]:=`expr /. Thread[Rule[var, point]]`

Out[19]= -94

3.2.2 Outer

Outer was discussed briefly earlier. It basically does for functions of several variables what **Map** does for functions of one variable. **Outer[function, lists ...]** takes any number of lists as second through nth arguments and outputs the function applied to all choices of arguments from the lists arranged in a nested form to make this suitable for tensor computations. Thus, **Outer** of a function with two simple lists produces a matrix. Interestingly, this works with heads other than **List**. Note that the numbers of arguments in the lists do not have to be the same.

In[20]:=`{Outer[Times, {a, b, c}, {1, 2}],`
        `Outer[Times, set[a, b, c], set[1, 2, 3]]}//MatrixForm`

Out[20]//MatrixForm=

$$\begin{pmatrix} \{\{a, 2\,a\}, \{b, 2\,b\}, \{c, 2\,c\}\} \\ set[set[a, 2\,a, 3\,a], set[b, 2\,b, 3\,b], set[c, 2\,c, 3\,c]] \end{pmatrix}$$

Outer with three simple lists produces what could be regarded as a list of three matrices.

In[21]:=`Outer[Times, {a, b, c}, {1, 2, 3}, {u, v, w}]`

Out[21]= {{{a u, a v, a w}, {2 a u, 2 a v, 2 a w}, {3 a u, 3 a v, 3 a w}},
        {{b u, b v, b w}, {2 b u, 2 b v, 2 b w}, {3 b u, 3 b v, 3 b w}},
        {{c u, c v, c w}, {2 c u, 2 c v, 2 c w}, {3 c u, 3 c v, 3 c w}}}

Outer with two matrices as arguments produces something of depth 4.

In[22]:=**tensor = Outer[Times, {{a, b}, {c, d}},**
                  {{1, 2, 3}, {4, 5, 6}, {7, 8, 9}}]

Out[22]= {{{{a, 2 a, 3 a}, {4 a, 5 a, 6 a}, {7 a, 8 a, 9 a}},
        {{b, 2 b, 3 b}, {4 b, 5 b, 6 b}, {7 b, 8 b, 9 b}}},
       {{{c, 2 c, 3 c}, {4 c, 5 c, 6 c}, {7 c, 8 c, 9 c}},
        {{d, 2 d, 3 d}, {4 d, 5 d, 6 d}, {7 d, 8 d, 9 d}}}}

Transpose can be used to rearrange this in many ways.

In[23]:=**Transpose[tensor, {1, 4, 2, 3}]**

Out[23]= {{{{a, b}, {2 a, 2 b}, {3 a, 3 b}}, {{4 a, 4 b}, {5 a, 5 b}, {6 a, 6 b}},
        {{7 a, 7 b}, {8 a, 8 b}, {9 a, 9 b}}},
       {{{c, d}, {2 c, 2 d}, {3 c, 3 d}}, {{4 c, 4 d}, {5 c, 5 d}, {6 c, 6 d}},
        {{7 c, 7 d}, {8 c, 8 d}, {9 c, 9 d}}}}

See also **Distribute** and **Through**.

3.2.3 Inner

There are two more complicated ways of applying operations to arguments consisting of lists.
Inner is a generalization of **Dot**. The first example shows how to write **Dot** in terms of
Inner, while the fourth example here shows that, in fact, **Inner[f, list1, list2, g]** is
the same as **Apply[g, Thread[f[list1, list2]]]**. (See the next chapter for **Apply**.)
The fifth example shows that second and third arguments don't have to have head **List**. All
that matters is that the heads be the same and that they have the same number of arguments,
which are then extracted to be used by **f** and **g**.

In[24]:= **{Inner[Times, {a, b, c}, {1, 2, 3}, Plus],**
      Inner[Plus, {a, b, c}, {1, 2, 3}, Times],
      Inner[f, {a, b, c}, {1, 2, 3}, g],
      Apply[g, Thread[f[{a, b, c}, {1, 2, 3}]]],
      Inner[f, hello[a, b, c], hello[1, 2, 3], g]}//MatrixForm

Out[24]//MatrixForm=

$$
\begin{pmatrix}
a + 2\,b + 3\,c \\
(1 + a)\ (2 + b)\ (3 + c) \\
g[f[a, 1], f[b, 2], f[c, 3]] \\
g[f[a, 1], f[b, 2], f[c, 3]] \\
g[f[a, 1], f[b, 2], f[c, 3]]
\end{pmatrix}
$$

3.2.4 Distribute

In particular, **Distribute** can sometimes be used to do the same things as **Outer**, without going inside deeper list structures. The syntax is somewhat different.

In[25]:=**Distribute[{{a, b}, {c, d}}, List]**

Out[25]= {{a, c}, {a, d}, {b, c}, {b, d}}

In[26]:=**Distribute[{{{a, b}, {c, d}}, {{e, f}, {g, h}}}, List]**

Out[26]= {{{a, b}, {e, f}}, {{a, b}, {g, h}},
            {{c, d}, {e, f}}, {{c, d}, {g, h}}}

However, this fails to do the expected thing if the head of the first argument evaluates its arguments.

In[27]:=**Distribute[Plus[{a, b}, {c, d}], List]**

Out[27]= {{a + c, b + d}}

3.3 Nest and Fold

3.3.1 Nest, NestList, and FixedPoint

There are two more pairs of useful operations that fit the discussion here. They are common ingredients of functional programming languages. The first is **Nest** and its related operations. **Nest[function, x, n]** applies **function** to **x** and repeats it n times; e.g., **Nest[f, x, 3]** returns f[f[f[x]]], while **NestList[function, x, n]** makes a list of these repeated operations a total of n + 1 times (since it starts from 0). For instance:

In[28]:=**Nest[(# 2)&, a, 3]**

Out[28]= 8 a

In[29]:=**NestList[f, a, 3]**

Out[29]= {a, f[a], f[f[a]], f[f[f[a]]]}

In[30]:=**NestList[(# 2)&, a, 3]**

Out[30]= {a, 2 a, 4 a, 8 a}

The following works because, as we have seen, **D** is **Listable** in its first argument.

In[31]:=**NestList[D[#, y]&, r[y] == Sin[y] Cos[y], 4]//TableForm**

Out[31]//TableForm=
$$r[y] == Cos[y] Sin[y]$$
$$r'[y] == Cos[y]^2 - Sin[y]^2$$
$$r''[y] == -4 Cos[y] Sin[y]$$
$$r^{(3)}[y] == -4 Cos[y]^2 + 4 Sin[y]^2$$
$$r^{(4)}[y] == 16 Cos[y] Sin[y]$$

Here is an example producing a simple continued fraction.

In[32]:=**Nest[(1/(1 + #))&, x, 3]**

Out[32]= $\dfrac{1}{1 + \dfrac{1}{1 + \frac{1}{1+x}}}$

An operation that is closely related to **Nest** is **FixedPoint**, which nests its operation until there is no change. For instance, everyone is familiar with what happens if the **Cos** key on a pocket calculator is pushed repeatedly. In principle, **FixedPoint** is what happens if it is pushed forever.

In[33]:= **{Nest[Cos, 0.5, 6],**
     Nest[Cos, 0.5, 12],
     FixedPoint[Cos, 0.5]}

Out[33]= {0.719165, 0.737236, 0.739085}

Actually, **FixedPoint** stops after machine accuracy is achieved. Look up the options to **FixedPoint** to see how to change this. There is also an operation **FixedPointList**. The Practice section gives some examples. See Chapter 11, Section 6, for more serious uses of **FixedPoint**.

3.3.2 Fold and FoldList

The second pair of functions is **Fold** and **FoldList**, which do something similar to **Nest**, but for functions of two variables.

$$\text{Fold[f, seed, } \{a_1, \ldots, a_n\}]$$

takes a function **f** of two variables, a starting **seed** value, and a list of subsequent values and returns

$$f[f[\ldots f[\text{seed}, a_1], a_2], \ldots], a_n]$$

For instance:

In[34]:= **Fold[f, a, {b, c, d}]**

Out[34]= f[f[f[a, b], c], d]

Similarly, **FoldList** produces a list of the successive values of this procedure.

In[35]:= **FoldList[f, a, {b, c, d}]**

Out[35]= {a, f[a, b], f[f[a, b], c], f[f[f[a, b], c], d]}

Here are a couple of examples.

In[36]:= **FoldList[Plus, 0, {a, b, c}]**

Out[36]= {0, a, a + b, a + b + c}

In[37]:= **FoldList[Power, 2, {2, 3, 4, 5}]**

Out[37]= {2, 4, 64, 16777216, 1329227995784915872903807060280344576}

It is usually easy to see when it is appropriate to use the function **Nest**; namely, there is some top level operation that is to be repeated a number of times. (This is sometimes called "tail" recursion, although in *Mathematica* it might be better to call it "head" recursion.) However, it is not so easy to see when it is appropriate to use **Fold**. What happens is that, at each repetition of the operation, new information is fed in from the list of values in the third argument. That is, instead of the third argument being a number saying how many times the operation is to be performed, it is a list of values to be used in repeating the operation. In the rest of the book we shall see a number of non-trivial uses of **Fold**, each of which is a triumph of human ingenuity.

3.4 *Substitution*

Function application in a functional programming language usually means substitution of a value for a variable. Thus, we expect that defining the squaring function by `f[x_] := x^2` and then applying `f` to 2 should be the same as evaluating the substitution `x^2 /. x -> 2`. Of course it is, but as will be seen in Chapter 11, Section 4, this form of substitution sometimes doesn't work correctly. Furthermore, as will be discussed in the next chapter, this form of substitution is really an application of a local rewrite rule and should not be thought of as a substitution at all. However, there is another operation in *Mathematica* that exactly implements the idea of substitution in functional programming languages; namely, `With`. For instance,

In[38]:= `With[{x = 2}, x`^{2}`]`

Out[38]= 4

In many functional languages, this would be written in the form **let** $x = 2$ **in** x^2. Instead of giving the value where the function is to be applied in the first argument of `With`, one can also specify the function in this position.

In[39]:= `With[{square = #`^{2}` &}, square[2]]`

Out[39]= 4

In the theory of functional programming languages, based on the lambda calculus, an expression of the form **let** $x = 2$ **in** x^2, as above, is synonymous with applying the pure function $\lambda x.x^2$ to the value 2. Since applying pure functions to values is the only thing that is done in functional programming, such programs consist mainly of **let** expressions. This style of programming can be adopted in *Mathematica* and often has attractive results. For instance, consider the following method to calculate improper integrals with possible singularities at the end points.

In[40]:= `improperIntegrate[expr_, {x_, a_, b_}] :=`
     `With[{integral = Integrate[expr, x]},`
       `Limit[integral, x -> b, Direction -> 1] -`
       `Limit[integral, x -> a, Direction -> -1]]`

This works nicely on typical examples.

In[41]:= `improperIntegrate`$\left[\dfrac{1}{(2x-1)^{\frac{2}{3}}}, \left\{x, \dfrac{1}{2}, 2\right\}\right]$

Out[41]= $\dfrac{3 \; 3^{1/3}}{2}$

4 Functional Programs

Functional programs are nested sequences of "button pushes"; i.e., they are single expressions made up solely from built-in commands and built-in constants. Sometimes these are called "one-liners". It is possible to do many intricate operations just using such one-liners. There is a column devoted just to them in the *Mathematica* Journal. The basic rule for a "strict" one-liner, as it will be called here, is that the only ingredients allowed on the right-hand side are built-in operations and constants or argument names that occur in the left-hand side. This rules out nearly all **Table** constructions, since they require a (bound) variable in the iterator argument which does not occur in the left-hand side. It also rules out use of **Integrate** and **Solve** unless they are parts of function definitions which include the variable specification on the left-hand side. It rules out expressions of the form **Function[{x}, body]**, since that also includes the bound variable **x**, but anonymous pure functions do exactly the same things without introducing any bound variables, and this is what makes it possible to construct strict one-liners. Non-strict, or ordinary, one-liners have no such restrictions. The only condition for them is that, in theory at least, they should be written on one, possibly very long, line.

Writing such functional programs is an important part of *Mathematica* programming. Later on we will relax the rule that only built-in functions and constants are allowed on the right-hand side, and allow arbitrary user defined functions and constants as well, so that functions will be built up iteratively from the built-in base to yield more and more complicated constructions. Even when we consider other styles of programming–rewrite rule programming and procedural programming–the basic ingredients will still be such one-liners. What we are promoting here is a functional style of programming which is in stark contrast to the usual style of Pascal or C programs. In many cases, it is more efficient and easier to read than such programs. It certainly is much more in accord with mathematical ways of thinking about algorithms.

4.1 Simple Examples of Functional Programs

Some of the things we did interactively or in more than one step in the first three chapters can be put together to make simple functional programs. For instance, the interactive sequence of operations

```
Integrate[ x/(1 - x^3), x]
D[%, x]
Simplify[%]
```

can be put together into a single nested operation.

In[1]:= **Simplify**$\left[\,\textbf{D}\left[\,\textbf{Integrate}\left[\frac{\textbf{x}}{\textbf{1}-\textbf{x}^3}\,,\ \textbf{x}\right],\ \textbf{x}\right]\right]$

Out[1]= $\dfrac{x}{1-x^3}$

This violates the rules for a strict one-liner since it involves the bound variable **x**, but otherwise it is just a nested sequence of built-in commands. Here are the results of the same process applied to a number of other interactive constructions from Chapters 1 and 3. Note that we have replaced **Table** constructions by mapping a pure function down an index list constructed by **Range** or by using listability.

In[2]:= **ListPlot[N[Log[Map[#!&, Range[20]]]]];**

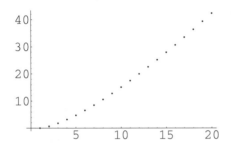

In[3]:= **Fit[N[Log[Map[#!&, Range[20]]]], {1, x, x^2}, x]**

Out[3]= $-2.02963 + 1.17902\,x + 0.0531166\,x^2$

Consider the construction **N[Log[Map[#!&, Range[20]]]]** here. There are, in fact, at least three ways to construct this list of numbers. First, just build a table as we did in Chapter 1.

In[4]:= **Table[N[Log[n]], {n, 20}]**

Out[4]= {0, 0.693147, 1.09861, 1.38629, 1.60944, 1.79176, 1.94591,
        2.07944, 2.19722, 2.30259, 2.3979, 2.48491, 2.56495, 2.63906,
        2.70805, 2.77259, 2.83321, 2.89037, 2.94444, 2.99573}

Second, map a pure function (the factorial function) down the list of desired numbers, constructed by the **Range** operation.

In[5]:= **N[Log[Map[#!&, Range[20]]]];**

Third, use the fact that **!**, **Log**, and **N** are **Listable** to get the result from a very brief command.

In[6]:=`N[Log[Range[20]!]];`

In the next three examples, we have to use **Map** because **ToString** and **Rationalize** are not **Listable**.

In[7]:=`Map[{ToExpression[ToString[N[Pi, #]]],`
       `        N[Sin[ToExpression[ToString[N[Pi, #]]]], 11] }&,`
       `        Range[5]] // TableForm`

Out[7]//TableForm=

3.	0.14112000806
3.1	0.041580662433
3.14	0.0015926529165
3.142	-0.00040734639894
3.1416	$-7.3464102068 \times 10^{-6}$

In[8]:= `Map[Rationalize[N[$\pi$], 0.1$^{\#}$] &, Range[10]]`

$$\text{Out[8]}= \left\{ \frac{22}{7}, \ \frac{22}{7}, \ \frac{355}{113}, \ \frac{355}{113}, \ \frac{355}{113}, \right.$$
$$\left. \frac{355}{113}, \ \frac{104348}{33215}, \ \frac{104348}{33215}, \ \frac{104348}{33215}, \ \frac{312689}{99532} \right\}$$

In[9]:= `Union[Map[Rationalize[N[$\pi$], 0.1$^{\#}$] &, Range[20]]]`

$$\text{Out[9]}= \left\{ \frac{80143857}{25510582}, \ \frac{245850922}{78256779}, \ \frac{5419351}{1725033}, \right.$$
$$\left. \frac{1146408}{364913}, \ \frac{312689}{99532}, \ \frac{104348}{33215}, \ \frac{355}{113}, \ \frac{22}{7} \right\}$$

This way of using *Mathematica* constitutes functional programming in *Mathematica*. It views the basic entities of *Mathematica* as functions (possibly of many variables) and the basic operation as composition of functions, or rather, the iterated application of functions to arguments. One-liners either can just carry out some specific calculation or can be used as definitions of functions whose arguments can then be given values to do something interesting.

4.2 Developing a Functional Program

The large stock of built-in functions make it possible to solve rather intricate problems in a straightforward way. For instance, in Section 3.1.3 above, starting with the expression

In[10]:= `exp1 = x$^3$ + (1 + z)$^2$;`

we applied **Sin** to the variables **x** and **z** using **MapAt**.

In[11]:=**MapAt[Sin, exp1, {{1,1}, {2,1,2}}]**

Out[11]= $Sin[x]^3 + (1 + Sin[z])^2$

You may have wondered how one would know what partspecs to give without carefully analysing the expression. But *Mathematica* will do this for you by itself, using **Position**.

In[12]:=**Position[exp1, x]**

Out[12]= $\{\{1, 1\}\}$

Thus, the following one-liner does it all.

In[13]:=**MapAt[Sin,
           exp1,
           Join[Position[exp1, x], Position[exp1, z]]]**

Out[13]= $Sin[x]^3 + (1 + Sin[z])^2$

However, this violates the strict rule by referring to the variables **x** and **z**. The idea here can be developed farther, by having *Mathematica* do more of the work. Perhaps you feel that *Mathematica* should also find the variables without our having to tell it what they are and that would turn this into a strict one-liner. In this kind of expression, the variables are particular leaves in the tree form of the expression. We would get all the leaves by the following:

In[14]:=**Level[exp1, {-1}]**

Out[14]= $\{x, 3, 1, z, 2\}$

We want to select **x** and **z** from this list. The predicate **Not[NumberQ[#]]&** is True just for them.

In[15]:=**Select[Level[exp1, {-1}], Not[NumberQ[#]]&]**

Out[15]= $\{x, z\}$

Next we need the position in **exp1** of each entry of this list, which we find by mapping the pure function **Position[exp1, #]&** down this list.

In[16]:=**Map[Position[exp1, #]&, %]**

Out[16]= $\{\{\{1, 1\}\}, \{\{2, 1, 2\}\}\}$

This has too many brackets, but **Flatten** with a levelspec will get rid of them.

In[17]:=**Flatten[%, 1]**

Out[17]= {{1, 1}, {2, 1, 2}}

Now we can put this all together, using the mouse to replace each % by its construction in the previous line, which yields a one-liner function definition that will take any expression **expr** (instead of **exp1**) and apply some given function **fun** (instead of **Sin**) just to the variables in it.

In[18]:=**mapVarsOnly[fun_, expr_] :=**
        MapAt[fun, expr,
            Flatten[
                Map[Position[expr, #]&,
                    Select[Level[expr, {-1}],
                        Not[NumberQ[#]]&]], 1]]

This is a true one-liner since the only ingredients on the right-hand side are built-in functions and constants, together with **expr** and **fun**, which occur on the left-hand side. Try this out with **Sin** and **exp1** to check that it works as expected.

In[19]:=**mapVarsOnly[Sin, exp1]**

Out[19]= $\text{Sin}[x]^3 + (1 + \text{Sin}[z])^2$

Now, try some other examples.

In[20]:= **mapVarsOnly[√# &, exp1]**

Out[20]= $x^{3/2} + \left(1 + \sqrt{z}\right)^2$

In[21]:= **mapVarsOnly[ArcTan, $\dfrac{(x - y^2 + 3)^w}{\sqrt{u^3 + 3\,v}}$]**

Out[21]= $\dfrac{(3 + \text{ArcTan}[x] - \text{ArcTan}[y]^2)^{\text{ArcTan}[w]}}{\sqrt{\text{ArcTan}[u]^3 + 3\,\text{ArcTan}[v]}}$

As an example of a one-liner, this is OK, but in fact it would fail on an expression that has what we would regard as a symbolic constant, e.g., **a**, because it would treat that as a variable, too. In the exercises we ask you to fix the definition so that **mapVarsOnly** only treats letters between **p** and **z** as variables.

4.3 Frequencies

List manipulations are important in functional programming. Suppose we want to write a function that takes a list as its argument and returns a list of the number of times each entry occurs in the original list. Start with a list to use as an example.

```
In[22]:= list =
        {a, d, s, f, d, a, s, a, d, f, d, f, g, d, a, f, g};
```

Union will give us the "set" of distinct entries in this list, written in canonical *Mathematica* order.

```
In[23]:= Union[list]
```

```
Out[23]= {a, d, f, g, s}
```

Our problem is to determine how many times each entry here occurs in **list**. There is a built-in function that will do that for a single entry.

```
In[24]:= Count[list, f]
```

```
Out[24]= 4
```

We don't just want the number 4, but we want it associated with the symbol **f** so we know what it means. Thus, we want the pair {f, 4} as output. This is easily constructed by a pure function.

```
In[25]:= {#, Count[list, #]}&[f]
```

```
Out[25]= {f, 4}
```

So all we have to do is put these together correctly in order to design a function that gives each distinct element and the number of times it occurs.

```
In[26]:= frequencies[list_] :=
        Map[{#, Count[list, #]}&, Union[list] ]
```

Try this out on our example.

```
In[27]:= frequencies[list]
```

```
Out[27]= {{a, 4}, {d, 5}, {f, 4}, {g, 2}, {s, 2}}
```

This is read as saying that **a** occurs **4** times, **d 5** times, etc. We will use **frequencies** later in constructing our own **BarChart** graphics function.

4.4 Newton's Method

4.4.1 One variable

Newton's method is a procedure for finding a zero of a function. There is of course the built-in function **FindRoot**, but we want to construct our own version to see how it works. Given an expression representing a function of **x**, e.g., **expr = x^2 - 3**, and some starting value **x0**, then the procedure calculates a sequence of values given by the iterative formula

$$x_0 = x0;$$
$$x_{n+1} = x - \frac{expr}{D[expr, x]} \text{ /. } x \rightarrow x_n$$

This is repeated until the results change by less than some specified error. We will use **Fixed-Point** and let *Mathematica* decide when to stop. What we have to do is to take the right-hand side of the iterative formula and turn it into a pure function, which is then repeatedly applied to **x0**. It is clearest to do this as a separate function called **oneNewtonStep** as follows:

In[28]:= **oneNewtonStep[expr_, {x_, x0_}] :=**
$$x - \frac{expr}{D[expr, x]} \text{ /. } x \rightarrow N[x0]$$

To force the computation to be done numerically rather than exactly, we use **N[x0]** instead of **x0** as the last argument here. The value of **oneNewtonStep** at stage n is what is to be used as the starting point for stage n + 1. That means we want to consider it as a pure function of the initial point **x0**. We can either use **Nest** some given number of times, or let *Mathematica* decide how often to iterate this procedure by using **FixedPoint**.

In[29]:=**newton[expr_, {x_, x0_}] :=**
      FixedPoint[oneNewtonStep[expr, {x, #}]&, N[x0]]

Here are a couple of examples.

In[30]:=**newton[x^2 - 3, {x, 1.0}]**

Out[30]= 1.73205

In[31]:=`newton[x - Cos[x], {x, 0.5}]`

Out[31]= `0.739085`

4.4.2 Several variables

Essentially the same formula works for n functions of n variables. Newton's method then finds values of all the variables so that all of the functions are zero. We just imagine that **x** means an n-dimensional vector and **expr** means n functions of n variables. The derivative **D** becomes the Jacobian matrix, and division means multiplication in the sense of the **Dot** product by the inverse. The formula becomes

$$x_{n+1} = (x - Inverse[jacobian[expr, x]] . expr[x] /. x \to x_n$$

Recall that the Jacobian is given by the operation

In[32]:=`jacobian[exprs_, vars_] := N[Outer[D, exprs, vars]]`

Our intention now is that **exprs** is to be a list of expressions and **vars** a list of variables. The initial value will be a list **vars0** of values. We have to change the notation slightly so that **newton** won't become confused about being given one function or a list of functions. (For a better way to handle this, see the next chapter.)

In[33]:= `oneNewtonStep[exprs_, vars_, vars0_] :=`
        `(vars - Inverse[jacobian[exprs, vars]] . exprs) /.`
        `Thread[vars -> N[vars0]]`

In the single-variable case, we just said $x \to x_0$, but now we need a list of rules. **Thread** does exactly the right thing; e.g.,

In[34]:= `Thread[{x, y, z} -> {1, 2, 3}]`

Out[34]= $\{x \to 1, y \to 2, z \to 3\}$

The final function is almost the same as the one variable case.

In[35]:=`newton[exprs_, vars_, vars0_] :=`
        `FixedPoint[oneNewtonStep[exprs, vars, #]&, N[vars0]]`

Here is a simple example.

In[36]:= `exprs1 = {x^2 + y^2 - 13, x^3 - y^3 - 19};`

In[37]:= `newton[exprs1, {x, y}, {2, 1}]`

Out[37]= $\{3., 2.\}$

In the exercises, you are asked to restructure this program so that the answer is a list of substitutions.

5 *Practice*

1. `MapThread[Rule, {x, y, z}, {1, 2, 3}]`

2. `MapIndexed[Nest[Sin, #1, Sequence@@#2]&, {a, b, c}]`

3. `#^2 &[anything]`

4. `# &[anything]`

5. `1&[anything]`

6. `something&[anything]`

7. `polys = Table[1 - x^n, {n, 10}];`

8. `Factor /@ polys // ColumnForm`

9. `Expand[#^2] & /@ polys // ColumnForm`

10. `FixedPointList[Cos, .5]`

11. `FixedPointList[Cos, .5,`
    `SameTest ⭲(Abs[#1 - #2] < 10^{-6} &)]`

The following are taken from the One-Liners column of the *Mathematica* Journal, Vol. 1, 1991. Try to understand what they do and how they work. In some cases, minor or major changes have been made to comply with the canon that one-liners should not introduce any extraneous variables on their right-hand sides.

```
12. rootPlot[poly_, z_] :=
        ListPlot[{Re[z], Im[z]}/. Solve[N[poly == 0], z],
                 Prolog -> PointSize[0.04]]
```

```
13. poly[n_, z_] :=
        z^n - Plus@@(Power[z, #] & /@ Range[0, n - 1])
```

```
14. rootPlot[poly[20, z], z]
```

15. $\text{newtonRoot}[f\_, x0\_] := \text{FixedPoint}\left[\left(\# - \dfrac{f[\#]}{f'[\#]}\right) \&, x0\right]$

```
16. newtonRoot[(# - Cos[#])&, 0.5]
```

17. $\text{ListDensityPlot}\left[\right.$

$\quad\quad \text{Outer}\left[\text{If}\left[\text{IntegerQ}\left[\sqrt{\#1^2 + \#2^2}\right., 0, 1\right] \&,\right.$

$\quad\quad\quad\quad \text{Range}[50], \text{Range}[50]\left]\right]$

```
18. phasePlot[f_, {x_, xmin_, xmax_}] :=
        ParametricPlot[
            Evaluate[{f, D[f, x]}], {x, xmin, xmax}]
```

19. `phasePlot[Sin[x^2], {x, 0, 2 π}]`

```
20. reverseInteger[n_] :=
        Dot[Power[10, #]& /@ Range[0, Floor[N[Log[10, n]]]],
            IntegerDigits[n] ]
```

```
21. reverseInteger[123456789]
```

6 Exercises

Observe the fundamental dictum of *Mathematica* programming in working these exercises.

1. Solve Exercise 13 in Chapter 3 about Jacobians again in a functional style. Hint: Figure out how to combine Thread and Dot.

2.   i) Implement your own version of Newton's method to find a zero of a differentiable function near a given starting value. The basic function should be of the form

```
newton[expr, {x, x0, n}]
```

where **expr** is some expression involving some independent variable **x**, **x0** is the starting value of **x**, and **n** is the number of times the operation in Newton's method is to be iterated. Define another function **newton[expr, {x, x0}]** which continues iterating until there is no change. Then there should be two extra functions,

```
newtonList[expr, {x, x0, n}] and
```

```
newtonList[expr, {x, x0}, opt]
```

which produce a list of successive approximations to the final value. The optional argument **opt** should allow a test to determine when the iteration should stop. See **Nest**, **NestList**, **FixedPoint**, and **FixedPointList**.

ii) Define a function

```
newtonPicture[expr, {x, xmin, xmax}, {x0, n}]
```

which makes a plot of the function defined by the expression for values between **xmin** and **xmax**, together with a line illustrating the first **n** successive approximations starting from **x0**. The line should show the successive tangents to the curve at each approximation point. Test your routine with the example:

```
newtonPicture[Cos[x^3], {x, 0.8, 1.5}, {.8788, 6}]
```

iii) Adapt your functions so they work for **n** functions of **n** variables.

iv) Restructure these operations so the output is a substitution.

v) Try some test examples and check your results.

3. Define a function **continuedFraction[list]** which takes a list as its only argument and returns the continued fraction whose numerators are given by the entries in the list in the given order. Thus, **continuedFraction[{a, b, c, d}]** returns

$$a / (1 + b / (1 + c /(1 + d)))$$

displayed in a nice form. Hint: try **Fold**.

4. What does the function

```
power1[x_, n_, base_] :=
    Fold[(#1^2 #2)&, 1, x^IntegerDigits[n, base]]
```

calculate when **base** is 2? when **base** is 3?

5. i) In Exercise 5 of Chapter 5, the Gram–Schmidt algorithm was implemented for orthogonalizing ordinary vectors with respect to the usual dot product. Generalize this procedure so that it works for vectors from an arbitrary vector space with respect to an arbitrary inner product called **innerProduct[v, w]**. The new procedure should have two arguments, the first being a list of vectors and the second being the inner product. Continue assuming that the given list of vectors is linearly independent. (There is a very nice way to do this using **Fold**.) Include a separate normalization function that also uses **innerProduct[v, w]**. Also include a procedure to check that a given list of vectors is orthonormal with respect to **innerProduct[v, w]**. The standard case should be recovered by setting **innerProduct** to be **Dot**.

ii) The matrix

8	3	0	0
3	2	1	2
0	1	2	2
0	2	2	14

is positive definite and symmetric and hence determines an inner product for 4-dimensional vectors. Orthogonalize and normalize the four standard unit vectors in 4-space using this inner product. Check the result.

iii) Apply the Gram–Schmidt algorithm to orthogonalize the list of functions {**1, x, x^2, x^3, x^4**} with respect to the inner product given by

$$\text{legendre (f, g)} = \int_{-1}^{1} f(x) g(x) dx$$

Check the result.

iv) Normalize the result of part iii). This does not give the first five terms in the usual sequence of Legendre polynomials. Why not? Fix things so that you get the first five Legendre polynomials. Make a plot of the first five Legendre polynomials.

6. Modify the definition of `mapVarsOnly` so that it only treats letters between p and z as variables. Hint: look up the operations `ToString`, `ToCharacterCode`, `Less`, and `Greater`.

7. i) The function `Fold` is sometimes called foldright because it "folds" in its arguments from the right. Define a function `foldleft` so that `foldleft[f, {a, b, c}, d]` gives the output `f[a, f[b, f[c, d]]]`.

ii) Write your own function `composeList` that works just like the built-in operation with the same name, using `FoldList`. Conversely, write your own function `foldList` that works just like the built-in operation with the same name, using `ComposeList`.

8. Somewhere in the first 1000 digits in the decimal expansion of π, there is a sequence of six successive 9's. Use `IntegerDigits`, `Partition`, and `Position` to find where this occurs. Avoid displaying large intermediate results. What other digits also occur more than twice in succession in this partial decimal expansion? (Based on a problem from [Blachman1].)

9. i) Write your own functions `map` and `through` that work just like the built-in operations with the same names, using `Outer`, `Flatten` , `#`, `&`, and `@`. (That is, if pure functions can be written, then `Map` and `Through` are special cases of `Outer`, suitably flattened.

ii) Generalize this to construct an operation that applies a list of arbitrary functions (not necessarily listable ones) to a list of values.

CHAPTER 7

Rule Based Programming

What we need in the future are systems which support both algorithmic coding of the basic mathematics and a rule-driven interface for the user to direct the semantic flow of the calculations in as flexible a manner as possible. [Hearn]

1 Introduction

The basic ingredient in a *Mathematica* program is the one-liner. If built-in operations are the words in the *Mathematica* language, then one-liners are the sentences. We now want to turn our attention to paragraphs. There are essentially three ways to construct larger and more complicated programs.

i) Remain in the functional programming paradigm and construct sequences of one-liners, each of which uses some of the functions defined in the previous one-liners. This is the way that LISP works and is the main *modus operandi* of all functional programming languages. Such constructions are essentially sequential. Actually, treelike is a better description. The final function constructed in terms of earlier functions can always be expanded into a very complicated one-liner, so, in this paradigm, paragraphs are just very long sentences.

ii) Defining a function by an expression of the form `f[x_] := body` is just a special case of a rewrite rule of the form `f[pattern] := body`. Rule based programming exploits this observation by giving many rules for the same function name `f`, depending on the form of the pattern of its arguments; i.e., these rewrite rules are conditional rules where the conditions can be given by general *Mathematica* expressions. Each rewrite rule itself is a one-liner. Constructions of this form are essentially parallel, consisting of many trees, so paragraphs look like forests. This is the topic of this chapter. As we will see, the additional facility of *local* rewrite rules is a special feature in *Mathematica* which has surprising uses.

iii) Use *Mathematica* as a block structured language with the usual control structures of an imperative language. This is the topic of the next chapter.

Functional programming languages evaluate their expressions by using essentially just one kind of rewrite rule embodying substitution. (See the discussion in Chapter 11, Section 4.) But they usually do not allow users to add their own rewrite rules. It seems that up to the appearance of Reduce, general programming languages did not incorporate generic procedures for adding such rules. *Mathematica* contains very powerful facilities for adding rewrite rules. Of course, such systems of rewrite rules have been extensively studied and used in special purpose languages intended for dealing with equationally defined data types. An equational data type (or theory) is described by giving a number of operations together with equations satisfied by various combinations of these operations. If the equations are directed from left to right, then they can be regarded as rewrite rules.

For a very simple example of this kind of a calculation using ordinary mathematical notation, consider the recursive definition of addition in terms of 0 and succ; i.e., $0 + m = m$ and $succ(n) + m = succ(n + m)$. Turn these into rewrite rules by directing them from left to right.

$$0 + m \Rightarrow m$$
$$succ(n) + m \Rightarrow succ(n + m)$$

Then we would like to prove that $2 + 2 = 4$, i.e., that the equation

$$succ(succ(0)) + succ(succ(0)) = succ(succ(succ(succ(0))))$$

holds in the system. (Alternatively, we could say that we just want to evaluate $2 + 2$.) This can be done by using the rewrite rules to turn the left-hand side into the right-hand side.

$succ(succ(0)) + succ(succ(0))$	
$\Rightarrow succ(succ(0) + succ(succ(0)))$	by the second rule
$\Rightarrow succ(succ(0 + succ(succ(0))))$	by the second rule
$\Rightarrow succ(succ(succ(succ(0))))$	by the first rule.

This says several interesting, useful, and perhaps liberating things about the equation $2 + 2 = 4$.

i) It shows that $2 + 2$ rewrites to 4.

ii) The (operational) meaning of $2 + 2$ is 4.

iii) The normal form of $2 + 2$ is 4.

The last is the best. It means that the calculation of $2 + 2$ is done by reducing $2 + 2$ to normal form. (A normal form is an expression to which no further rewrite rules apply.) This is the way in which rewrite rule systems do calculations.

2 Rewrite Rules in Mathematica

Instead of viewing the expressions $x = a$ and $f[y\_] := y^2$ as assigning the value **a** to **x** and defining the squaring function, respectively, we can regard them as establishing rewrite rules. That means we think that they mean the following:

   i) Whenever **x** occurs, rewrite it as **a**.

   ii) Whenever **f[anything]** occurs, rewrite it as **anything**^{2}.

Mathematica supports this interpretation of these expressions in two different forms: as global rules and as local rules. We discuss each form in turn.

2.1 Global Rules

Global rules are rules that are applied whenever the appropriate left-hand side is encountered. The basic operation of the built-in functions in *Mathematica* can be thought of as either applying operations that are hard-coded in C or applying built-in global rewrite rules. Whenever **Plus[2, 3, 4]** is encountered, *Mathematica* rewrites it as **9**, etc. There are two kinds of user defined global rewrite rules, those using **=** and those using **:=**. The distinction between the two lies in *when* the right-hand side is evaluated. Furthermore, for each kind of rule there are two forms depending on *where* the rule is stored, indicated by **=**, **^=**, **:=**, and **^:=**.

2.1.1 = rules

Up to now, we have viewed rules using **=** as assignment statements, in analogy with traditional imperative programming languages. Thinking of them instead as rewrite rules, the characteristic property of rules using **=** is that they evaluate the right-hand side immediately and all subsequent occurances of the left-hand side are replaced by the evaluated right-hand side. For instance:

In[1]:=**x = a;**

In[2]:=**x + 5**

Out[2]= 5 + a

In[3]:= **Clear[x]**

In traditional programming languages, the left-hand side of an assignment statement is required to be a simple identifier (i.e., a symbol). Here, the left-hand side can be arbitrarily complicated. For instance:

In[4]:=**magic[7 + z[5, two]] = Expand[(1 + y)^4]**

Out[4]= $1 + 4\,y + 6\,y^2 + 4\,y^3 + y^4$

Note that the output of an = expression is the evaluated form of the right-hand side. The left-hand side should be regarded as a *pattern* such that whenever something is found that matches that pattern, then it is replaced by the evaluated right-hand side. For instance:

In[5]:= **(magic[z[1 + 4, two] + 2 + 5]+ 5)^2 + magic[6 + z[5, two]]**

Out[5]= $\left(6 + 4\,y + 6\,y^2 + 4\,y^3 + y^4\right)^2 + \text{magic}[6 + z[5, two]]$

In this evaluation, the pattern **magic[z[1 + 4, two] + 2 + 5]** simplifies to **magic[7 + z[5, two]]**, which is replaced by $1 + 4\,y + 6\,y^2 + 4\,y^3 + y^4$. Hence the first part of the input, which includes + 5, simplifies to the first term in the output. The second term does not match any pattern involving **magic** and so is left in unevaluated form. This rule is stored with **magic**, which is shown by the following output.

In[6]:= **??magic**

```
Global`magic
magic[7 + z[5, two]] = 1 + 4*y + 6*y^2 + 4*y^3 + y^4
```

Again we see the evaluated form on the right.

Now, there are some problems associated with left-hand sides that are not symbols. For instance, suppose we try to make the following rule.

In[7]:=**a + c = d**

```
Set::write : Tag Plus in a + c is Protected.
```

Out[7]= d

We get an error message saying that **Plus** is **Protected**. Let us check this. **Protected** is an attribute of functions.

In[8]:=**Attributes[Plus]**

Out[8]= {Flat, Listable, NumericFunction,
          OneIdentity, Orderless, Protected}

We already know what **Listable** means. What **Protected** means is that one cannot make up new rules for **Plus**. There is a (possibly gigantic) table of rules for each built-in operation and we are not allowed to add new rules for them. That makes a certain amount of sense since every time *Mathematica* encounters **Plus**, it searches through its rules for **Plus** to see if something applies, using the first rule it finds. If we add a new rule for **Plus**, then that rule would have to be looked at every subsequent time that we wanted to do an addition. Actually, in Version 3.0, **Plus** and **Times** are distinguished from other built-in functions, which normally use user defined rules first, by using built-in rules first. Thus, built-in rules involving numbers, for instance, are always used first, but when a rule of the form **a + c = d** is given, *Mathematica* interprets that as a rule of the form **Plus[a, c] = d**. Rules have to be stored somewhere, and the default place is with the rules for the head of the left-hand side. Of course, maybe we really want to make a rule for **Plus**, in which case we can unprotect **Plus**, make the rule, and then reprotect it.

In[9]:=**Unprotect[Plus]**

Out[9]= {Plus}

In[10]:=**a + c = d**

Out[10]= d

In[11]:=**Protect[Plus]**

Out[11]= {Plus}

Now whenever *Mathematica* sees **a + c**, it rewrites it as **d**.

In[12]:=**a + c + m**

Out[12]= d + m

However, there is a much less drastic way to achieve the same goal. Definitions that attach a value to the head of the left-hand side are called *down values* of the head. There are also *up values* which try to associate the rule with the leftmost unprotected argument of the left-hand side. They are written

In[13]:=**m + n ^= p**

Out[13]= p

Note the caret ^ before the = sign. This rule is associated with the symbol **m**.

In[14]:= **? ?m**

> Global`m
> m /: m + n = p

The m/: followed by the rule means that this rule has been deliberately stored with **m**. One can, in fact, use this form of the syntax instead of the sign ^= directly.

In[15]:= **q/: q + r = s**

Out[15]= s

We already know that = is the infix form of **Set**. What is the real name of ^=?

In[16]:= **FullForm[Hold[q + r ^= s]]**

> Out[16]//FullForm=
> Hold[UpSet[Plus[q, r], s]]

Thus, the symbol ^= is the infix form of **UpSet**. A given symbol can have both up and down values. A down value is one that arise by default when the symbol is the head of the left-hand side. Let's give **q** a down value as well as its up value.

In[17]:= **q[x_] := 27 x^3**

Then looking at **q** shows both kinds of values. The first is an up rule, indicated by the **q/:**, and the second is a down rule.

In[18]:= **? ?q**

> Global`q
> q /: q + r = s
> q[x_] := 27 * x^3

Finally, we can access the up values and down values individually.

In[19]:= **{UpValues[q], DownValues[q]}**

Out[19]= {{HoldPattern[q + r] :→ s}, {HoldPattern[q[x_]] :→ 27 x^3}}

We'll explain later why the left-hand side of these substitutions is wrapped in **HoldPattern** and the substitution is written :→ rather than ->. Note that **HoldPattern** replaces the form **Literal** from earlier versions of the program.

Now let us try to naively define the squaring function using an = rule.

In[20]:= **g [x] = x^2**

Out[20]= x^2

This works all right for the symbol **x** but not for anything else.

In[21]:= **{g[x], g[y], g[2]}**

Out[21]= $\{x^2, g[y], g[2]\}$

This is where the special symbol _ comes in. The form **x_** means a *pattern named* **x**.

In[22]:= **FullForm[x_]**

Out[22]//FullForm=
Pattern[x, Blank[]]

An underscore _ in a pattern matches anything, so it is a kind of "wild card". If it appears on the left-hand side of an "=" rule with a name, like **x**, then the left-hand side is rewritten as the right-hand side with **x** replaced by the anything. Use this to redefine **g**.

In[23]:= **g [x_] = x^2**

Out[23]= x^2

In[24]:= **{g[3], g[x], g[y], g[z + w]}**

Out[24]= $\{9, x^2, y^2, (w + z)^2\}$

Thus, we can use = rules to define functions.

2.1.2 := rules

Rules using **:=** are characterized by the property that they do not evaluate the right-hand side immediately, but instead leave it unevaluated until the function is actually used. They can be used with simple left-hand sides or with left-hand sides containing patterns. For instance, here are two rules that differ only by using = or **:=**.

In[25]:= **a1   = Expand[(1 + x)^2];
     a2 := Expand[(1 + x)^2];**

If these are evaluated, they give the same result.

In[26]:= **{a1, a2}**

Out[26]= $\{1 + 2 x + x^2, 1 + 2 x + x^2\}$

If we now give **x** a value, then **a1** and **a2** will use that value in different ways.

In[27]:= **x = z + w;**

In[28]:= **{a1, a2}**

Out[28]= $\{1 + 2 (w + z) + (w + z)^2, 1 + 2 w + w^2 + 2 z + 2 w z + z^2\}$

In[29]:= **Clear[x, a1, a2]**

In **a1**, the **Expand** was evaluated when the rule was entered and is no longer present. Hence the **(w+z)** terms are not expanded further. On the other hand, if the left-hand side of a **:=** rule contains a pattern, then on a subsequent occurrence of the left-hand side with actual arguments, the formal arguments (or names of patterns) on the right-hand side are replaced by their values from the left-hand side, and then the right-hand side is evaluated. Thus, each time the left-hand side of such a rule matches something, it is replaced by a new evaluation of the right-hand side. To see the difference, we again set up two rules, differing only by **=** or **:=**.

In[30]:= **ff[u_] = Expand[u²];**
        gg[u_] := Expand[u²];

Now, try out these two definitions on the same value.

In[31]:= **{ff[1 + y], gg[1 + y]}**

Out[31]= $\{(1 + y)^2, 1 + 2 y + y^2\}$

The right-hand side of the rule for **ff** is evaluated immediately when it is entered. Since there is nothing to expand, it just evaluates to u^2. The rule for **gg**, on the other hand, retains the whole expression **Expand[u²]**. When the two functions are subsequently used, **ff[1 + y]** is just replaced by $(1 + y)^2$, while **gg[1 + y]** is replaced by **Expand[(1 + y)²]**, which evaluates to **1 + 2y + y²**. The internal representation of such a definition has the following form.

In[32]:= **FullForm[Hold[h[x_] := p]]**

    Out[32]//FullForm=
    Hold[SetDelayed[h[Pattern[x, Blank[]]], p]]

Thus, the symbol `:=` is is the infix form of **SetDelayed**. We can also check what *Mathematica* knows about **ff** and **gg**.

In[33]:= **??ff**

```
Global`ff
ff[u_] = u^2
```

In[34]:= **??gg**

```
Global`gg
gg[u_] := Expand[u^2]
```

This makes dramatically clear the distinction between evaluation when the rule is given and evaluation when the rule is used.

2.1.3 The order of rules

If several rules are given for the same operation, then *Mathematica* puts them in order of increasing generality so more specific rules are listed first. When *Mathematica* uses these rules it starts at the beginning and uses the first one that applies. If *Mathematica* is unable to decide which of two rules is more general, then it stores them in the order in which they were entered. For instance,

In[35]:= **foo[a_, 2] := bar;**
     foo[2, b_] := barbar;

In[37]:= **foo[2, 2]**

Out[37]= bar

DownValues will give us the order in which these are stored.

In[38]:= **DownValues[foo]**

Out[38]= {HoldPattern[foo[a_, 2]] :→ bar, HoldPattern[foo[2, b_]] :→ barbar}

If we don't like this order, then it can be changed (and even have more rules added to it) by reassigning some new value to **DownValues[foo]**. For instance:

In[39]:= **DownValues[foo] = Reverse[DownValues[foo]]**

Out[39]= {HoldPattern[foo[2, b_]] :→ barbar, HoldPattern[foo[a_, 2]] :→ bar}

Then we get the other result for **foo[2, 2]**.

In[40]:= **foo[2, 2]**

Out[40]= barbar

This example illustrates a well-known problem with rewrite rules. If more than one rule applies to a particular expression, then which one should be used first? It would be nice if the order of application of rules didn't make any difference. Such systems of rewrite rules are called *confluent* or Church–Rosser (after a famous theorem about the lambda calculus). Since *Mathematica* does use a definite order, we often make use of that knowledge in setting up systems of rewrite rules which are not confluent when, with a bit more care, they could be written in a confluent form.

2.2 *Local Rules*

Local rewrite rules are rules that are applied only to a single expression. The basic syntactical ingredient of a local rewrite rule is an arrow,   ←Such a rule is applied to an expression using the operation **/.**.

2.2.1 ←rules

Local rules using an arrow have already been encountered in checking the solution of an equation.

In[41]:= **equation = x^2 – 5 x + 6 == 0;**
         solution = Solve[equation, x]

Out[41]= $\{\{x \to 2\}, \{x \to 3\}\}$

In[42]:= **equation /. solution**

Out[42]= {True, True}

The output of **Solve** is a list of lists of local rules. Thus, $x \to 2$ is a local rule which is the analogue of the global rule x = 2. The rule is applied to an expression by using **/.**, so the expression **equation /. solution** means "use the rewrite rule $x \to 2$ just in **equation**". The result of this is the expression 2^2 – 5*2 + 6 == 0, which simplifies to 0 == 0, which is then evaluated as True. The usual form of the right-hand side of **/.** is a list of local rules for some of the symbols that appear on the left-hand side. For example:

In[43]:= **x y z /. {x ->2, y ->3}**

Out[43]= 6 z

If the right-hand side is a list of lists of local rules, then **/.** behaves as though it were **Listable** above the bottom level, so it moves inside the first layer of brackets in this case and returns a list of results.

In[44]:= **x y z /. {{x ->2, y ->3}, {y ->4, z ->5}}**

Out[44]= {6 z, 20 x}

Actually, **/.** is rather clever about decoding the list structure of the second argument.

In[45]:= **x y z /. {{{x ->2, y ->3}, {y ->4, z ->5}},**
**   {{x ->6, y ->7}, {y ->8, z ->9}},**
**   {x ->10, z ->11}}**

Out[45]= {{6 z, 20 x}, {42 z, 72 x}, 110 y}

Local rules with " ->"share with "=" rules the property that they evaluate their right-hand sides immediately.

2.2.2 :> rules

The local analogue of a "**:=**" rule is a "**:>**" rule; i.e., a local rule that evaluates its right-hand side only when it is used, or as the computer scientists say, only when it is called. We can make an experiment similar to the one we made with "**=**" and "**:=**".

In[46]:= **fff[1 + u] /. fff[v_] ->Expand[(3 + v)^2]**

Out[46]= $9 + 6 (1 + u) + (1 + u)^2$

In[47]:= **ggg[1 + u] /. ggg[v_] :> Expand[(3 + v)^2]**

Out[47]= $16 + 8 u + u^2$

This time, the left-hand sides of the local rules involve patterns rather than just symbols. The difference is that the local rule for **fff[v_]** replaces it by the evaluation of **Expand[(3 + v)^2]**, which equals $9 + 6v + v^2$. Hence, when this is used with **v** equal to **1 + u**, we get the result $9 + 6 (1 + u) + (1 + u)^2$. On the other hand, the local rule for **ggg[v_]** replaces it by the unevaluated **Expand[(3 + v)^2]** which, when used with **v** equal to **1 + u**, gives **Expand[(3 + (1 + u))^2]**. This is then simplified to $16 + 8u + u^2$. We can check how *Mathematica* represents these expressions internally.

In[48]:= **FullForm[Hold[m /. n ←p]]**

   Out[48]//FullForm=
   Hold[ReplaceAll[m, Rule[n, p]]]

In[49]:= **FullForm[Hold[m /. n :> p]]**

   Out[49]//FullForm=
   Hold[ReplaceAll[m, RuleDelayed[n, p]]]

Thus, **/.** is the infix form of ReplaceAll, the arrow ←is the infix form of Rule, and the arrow **:>** is the infix form of RuleDelayed (corresponding to Set and SetDelayed for = and **:=**). I read the symbol **/.** as "where". It can be regarded as the postfix form of the construction "Let n = p in m" in functional programming languages, at least when used with **:>**.

2.2.3 Application of rules using /. and //.

There is another form of **/.** given by **//.** which applies a local rule repeatedly until there is no further change in the expression. Note that this is the normal mode for application of global rules; they are always applied wherever possible. Internally, **//.** is represented by

In[50]:= **FullForm[Hold[m //. n -> p]]**

   Out[50]//FullForm=
   Hold[ReplaceRepeated[m, Rule[n, p]]]

In[51]:= **FullForm[Hold[m //. n :> p]]**

   Out[51]//FullForm=
   Hold[ReplaceRepeated[m, RuleDelayed[n, p]]]

Thus, **//.** is the infix form of ReplaceRepeated. I read the symbol **//.** as "where rec". It is the postfix form of the construction "Let rec n = p in m" in functional languages. Here is an example of the difference between **/.** and **//.** These examples use a list of rules rather than just a single rule. When a list of rules is applied to a single expression, then each rule for each symbol is tried from the left until a match is found. In the following example, the right-hand side of the **/.** expression consists of a list of two rules for the same symbol, **fac**. This list is searched from the left until a pattern is found that matches the left-hand side of the **/.** expression. In the first case, as soon as a match is found, the evaluation is finished. In the second case, the rules are tried repeatedly from the left on the output of the previous evaluation until no matches are found.

In[52]:= `fac[5] /. {fac[1] -> 1, fac[n_] -> n fac[n - 1]}`

Out[52]= `5 fac[4]`

In[53]:= `fac[5] //. {fac[1] -> 1, fac[n_] -> n fac[n - 1]}`

Out[53]= `120`

In the first case, the left-hand side of the rule `fac[1] -> 1` doesn't match anything in `fac[5]`, but `fac[n_] -> n fac[n - 1]` does with n_ equal to 5, so the output is `5 fac[4]`. In the second case, the left-hand side of the rule `fac[n_] -> n fac[n - 1]` continues to match a part of the existing expression until one arrives at `120 fac[1]`. Then the left-hand side of the rule `fac[1] -> 1` matches leading to `120*1`, which simplifies to `120` where neither rule matches, so the output is `120`.

If such rules are given globally, as in Chapter 2, Section 2.2.3, then the order in which they are given doesn't matter since *Mathematica* will put the more specific rule, `fac[1] = 1`, first. However, in a list of local rules, applied with `//.`, we are completely responsible for the ordering. Thus, the following gave the "wrong" answer in previous versions. It now gives the "right" answer, apparently because of a bug.

In[54]:= `fac[5] //. {fac[n_] -> n fac[n - 1], fac[0] -> 1}`

Out[54]= `0`

2.2.4 Named lists of rules

Lists of rules can also be named to be used wherever desired.

In[55]:= `facrules = {fac[1] -> 1, fac[n_] -> n fac[n - 1]};`

In[56]:= `{fac[7] /. facrules, fac[7] //. facrules}`

Out[56]= `{7 fac[6], 5040}`

Look at the packages `Trigonometry.m` and `LaplaceTransform.m` to see large examples of named lists of delayed rules.

2.2.5 Simultaneous substitution

If several local rules are given for different symbols, then each rule is tried once on each part of the expresion. This makes it appear that these rules are applied simultaneously. For instance,

In[57]:= `{x, y, z}/. {x -> y, y -> z, z -> w}`

Out[57]= `{y, z, w}`

If the substitutions are carried out sequentially, then the results are quite different.

```
In[58]:= {x, y, z} /. {x →y} /. {y →z} /. {z →w}
```

```
Out[58]= {w, w, w}
```

This is the same result as carrying out the original substitution recursively. In particular, this means that variables can be interchanged without introducing an intermediate temporary variable.

```
In[59]:= {x, y} /. {x →y, y →x}
```

```
Out[59]= {y, x}
```

2.3 Summary of Transformation Rules

□	Global rules stored with the head	Global rules stored with an argument	Local rules	Application of rules
Evaluate rhs	=	^=	←	/. = where
Delay rhs	:=	^:=	: <	//. = where rec

3 Pattern Matching

3.1 Patterns

A basic ingredient of rule based programming is pattern matching. There is no problem with a rule like **x = a** where the only thing that has to be matched is **x**. But in more complicated circumstances, like the rule for **magic** above, there is something to be done to discover that some expression involving **magic** matches the appropriate pattern. Even more so, there is something to be done for expressions involving _ , perhaps in several locations. Rules using _ are not just simple rules; they are rule schemes having the effect that anything of a given form is rewritten in some other specified form. One can think of underscores as "wild cards" that match anything, except that **x_** does not mean "**x** followed by a wild card"; it means a pattern named **x**. Thus, the full form of **symbol_** is **Pattern[symbol, Blank[]]**. The symbol here is called the "name" of the pattern. Note that **symbol** must be a symbol in the *Mathematica* sense of the term.

A compound pattern is an expression with zero or more of these simple patterns as subexpressions. One can consider a compound pattern as a template for an expression. An expression **expr** matches a compound pattern **patt** containing simple patterns $p_1, \ldots p_n$, if there are subterms $t_1, \ldots t_n$, of **expr** such that **patt**, with $p_1, \ldots p_n$, replaced by $t_1, \ldots t_n$, is the same as **expr**. (Note that some of the t_i's can themselves be patterns.) In all probability, what *Mathematica* actually does is equivalent to working in the reverse order by replacing subterms of **expr** by **Pattern** to see if the expression **patt** can be derived in this way. There are various techniques with names like *resolution* and *narrowing* for complicated pattern matching, but Wolfram Research, Inc., has not revealed exactly how *Mathematica* does it.

3.2 Underscore Rules

3.2.1 Rules with _

The symbol _ by itself, without any symbol on the left, can be used to describe a pattern. (Recall that the **FullForm** of _ is Blank[]). For instance, the expression _^- matches anything of the form x^y, where **x** and **y** are any expressions. However, there is no way to use the things that match the _'s on the right-hand side. Here is an example and three instances of it.

In[1]:= **f1 [_^-, _] := p**

In[2]:= **{f1 [a^a, a], f1 [a^b, c], f1 [magic^{clown}, what]}**

Out[2]= {p, p, p}

3.2.2 Rules with x_

A pattern of the form **x_** is matched by any expression and then **x** is bound to the expression for purposes of evaluating the right-hand side. Thus, if a definition is given in the form **f [x_] := x^2**, then the result of **f [a]** is the same as evaluating x^2 /. x ←a. If there are two instances of **x_** on the left-hand side of a rule, then they must be filled with the same expression. Here are a pair of examples with three instances of each.

In[3]:= **f2 [x_^{y-}, z_] := p[x q[y, z]]**

In[4]:= **{f2 [a^a, a], f2 [a^b, c], f2 [magic^{clown}, what]}**

Out[4]= {p[a q[a, a]], p[a q[b, c]], p[magic q[clown, what]]}

In[5]:= **f3 [x_$^{y-}$, x_] := p[x q[y]]**

In[6]:= **{f3 [a^a, a], f3 [a^b, a], f3 [magic^{clown}, what]}**

Out[6]= {p[a q[a]], p[a q[b]], f3 [magic^{clown}, what]}

3.2.3 Rules with x_Head

A pattern of the form **x_Head** is matched by any expression whose head is **Head**. Here is an example with three instances.

In[7]:= **f4 [x_$^{y\_Integer}$, z_] := p[x q[y, z]]**

In[8]:= **{f4 [a^a, a], f4 [a^b, c], f4 [magic^2, what]}**

Out[8]= {f4 [a^a, a], f4 [a^b, c], p[magic q[2, what]]}

The only expression here that matches the pattern is **f [magic^2, what]**. The head can be anything; e.g.,

In[9]:= **f5 [x_$^{y\_foo}$, z_] := p[x q[y, z]]**

In[10]:= **{f5 [a^a, a], f5 [a^b, c], f5 [magic$^{foo [b]}$, what]}**

Out[10]= {f5 [a^a, a], f5 [a^b, c], p[magic q[foo [b], what]]}

The internal forms of **_**, **x_**, and **x_Head** are

In[11]:= **{FullForm[_], FullForm[x_], FullForm[x_head]}**

Out[11]= {Blank[], Pattern[x, Blank[]], Pattern[x, Blank[head]]}

Thus **x_Head** is a restricted form of a wild card that only can be filled by expressions whose head is **Head**. As we have seen, one way to think about heads is as types; i.e., the type of an expression is its head. Then a pattern of the form **x_Head** only applies to entities of type **Head**. We will exploit this point of view later. In all of the examples above, an expression that doesn't match the left-hand side is returned in unevaluated form. But note that something is always returned as the output. The program does not crash or report an error. (In a certain sense, the normal thing is for an expression to be returned without change. Only in "special" circumstances is it rewritten in a different form.)

3.2.4 Double and triple underscores

If we give a rule for an expression involving two separate underscores, then we are constructing a function of two variables. It only works when it is then given exactly two arguments.

In[12]:= **f6[x_, y_] := x + y**

In[13]:= **{f6[a], f6[a, b], f6[a, b, c]}**

Out[13]= $\{f6[a], a + b, f6[a, b, c]\}$

For the point of view of ordinary mathematics, this is the only thing that makes sense. A function depends on some specified number of arguments. However, from the point of view of rewrite rules, all that matters is the pattern on the left-hand side, and we as well as the computer are able to distinguish the pattern consisting of "one or more arguments", or "zero or more arguments". *Mathematica* has a provision for using such patterns. Besides rule schemes using a single underscore _, there are rule schemes using a double or triple underscore. A double underscore __ , is matched by one or more expressions, separated by commas, while a triple underscore ___ , is matched by zero or more arguments. The form **x__** means a sequence of one or more expressions, named **x**, and **x__Head** means a sequence of one or more expressions, named **x**, all of whose heads are **Head**. Similarly, the form **x___** means zero or more expressions, named **x**, and **x___Head** means zero or more expressions, named **x**, all of whose heads are **Head**. Here is an example of a function whose output is the square of the number of arguments it has been given. In the first case, it accepts one or more arguments, and in the second, it accepts zero or more arguments.

In[14]:= **f7[x__] := Length[{x}]^2**

In[15]:= **f8[x___] := Length[{x}]^2**

In[16]:= **{f7[], f7[a], f7[a, b], f7[a, b, c]}**

Out[16]= $\{f7[], 1, 4, 9\}$

In[17]:= {f8[], f8[a], f8[a, b], f8[a, b, c]}

Out[17]= {0, 1, 4, 9}

Note that some of the built-in functions allow zero or more arguments. For example,

In[18]:= {Plus[], Plus[3], Plus[3, 5], Plus[3, 5, 7]}

Out[18]= {0, 3, 8, 15}

In[19]:= {Times[], Times[3], Times[3, 5], Times[3, 5, 7]}

Out[19]= {1, 3, 15, 105}

The case of zero arguments for these built-in operations produces the unit for the operation. Note: It is hard to think of a way to use **x__** or **x___** in a way that does not either turn them into a list by using **{x}** on the right-hand side or pass them to some built-in function that knows what to do with a variable number of arguments.

3.2.5 Optional arguments

Default values and double or triple underscores are important techniques in giving optional arguments to functions, in the sense that any number of arguments can be given to such a function. However, there is another sense in which a specific argument can be optional. Let us go back to a modification of our first example of a pattern above.

In[20]:= f9[x_$^{y\_}$, z_] := p[x y z]

In[21]:= {f9[a^b, c], f9[a, c]}

Out[21]= {p[a b c], f9[a, c]}

We might think that **f[a, c]** should match the pattern with an understood exponent 1, which would mean that it should be rewritten as p[a c], but of course *Mathematica* can't guess that this is what we intend. However, there is a provision to take care of such default values that are meant to be inserted in a pattern if they are missing. When a pattern is intended to have a default value, **v**, this is indicated by writting **_:v**. So, the effect we wanted to achieve is given by the form

In[22]:= f10[x_$^{y\_:1}$, z_] := p[x y z]

In[23]:= {f10[a^b, c], f10[a, c]}

Out[23]= {p[a b c], p[a c]}

In this case, the default value 1 for the exponent is the natural and obvious choice, and *Mathematica* knows this. It has standard built-in default values for a number of such positions. The notation `_.` tells *Mathematica* to use the built-in default value. Note the almost invisible period after the underscore. Thus, the effect we wanted at the beginning is given by a tiny modification of the original form.

```
In[24]:= f11[x_^y_., z_] := p[x y z]

In[25]:= {f11[a^b, c], f11[a, c]}

Out[25]= {p[a b c], p[a c]}
```

Here is another example involving an optional second argument, whose default value is the pure function **Tan**.

```
In[26]:= apply[argument_, function_:Tan] := function[argument]

In[27]:= {apply[3], apply[3.1], apply[3.1, Cos], apply[3, #^2 &]}

Out[27]= {Tan[3], -0.0416167, -0.999135, 9}
```

There is still another way that optional arguments occur in *Mathematica*. Some functions, such as **Plot**, can take optional named arguments, such as **AspectRatio -> 1**. By incorporating such functions into our own definitions, we can also use such optional named arguments. (We will see in Chapter 11 how to write our own functions with optional named arguments.) Consider an example of a plotting function.

```
In[28]:= plotWithSin[function_, var_] :=
            Plot[{Sin[var], function[var]}, {var, 0, 2π}]

In[29]:= plotWithSin[Cos[2 #]&, x];
```

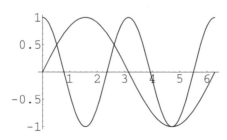

We would like to be able to use the optional arguments for **Plot** in our **plotWithSin** command. The way to do this is to add a triple underscore pattern to its form with the name **opts** and then just pass **opts** to **Plot**.

```
In[30]:=plotWithSinOpts[function_, var_, opts___] :=
        Plot[{Sin[var], function[var]}, {var, 0, 2π}, opts]
```

```
In[31]:=plotWithSinOpts[E^(-#/2)&, x, AspectRatio -> 1];
```

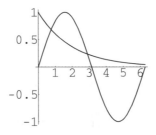

3.2.6 Names for compound patterns

In an expression of the form $f[x\_^{n\_Integer}, z\_]$, the patterns are named with the symbols x, n, z, but there is no name for the whole compound pattern $x\_^{n\_Integer}$. There is a way to give names to such compound patterns so that they can be referred to directly on the right-hand side. This is used frequently in some of the packages that are shipped with *Mathematica*. The syntax consists of a name, followed by a colon, followed by the compound pattern. (Don't confuse **_:symbol** with **name:pattern**.) One often sees this compound pattern written with an explicit head, but that is not necessary. Here is an example with two different rules, where the output depends on whether the exponent is an integer or a real number.

```
In[32]:= f12[expr : x_^n_Integer, z_] := z ExpandAll[expr];
         f12[expr : x_^n_Real, z_] := z expr
```

$$In[33]:= \left\{ f12\left[(1 - x^2)^3, 2 \right], f12\left[(1 - x^2)^{-3.2}, 2 \right] \right\}$$

$$Out[33]= \left\{ 2 (1 - 3 x^2 + 3 x^4 - x^6), \frac{2}{(1 - x^2)^{3.2}} \right\}$$

3.2.7 Repeated patterns

An interesting kind of pattern that is not covered by the preceding devices is a list of arbitrary length, all of whose entries match some specified pattern. The pattern {entries___Integer} is matched only by a (possible empty) list of integers, but if we want them to all be of the form $x\_^{n\_Integer}$, then some new description of the pattern is necessary. This is given by **..** and **...** in the following examples. As usual, **..** means one or more repetitions. The *Mathematica* Book says that **...** means zero or more repetitions, but this doesn't seem to work.

In[34]:= `f13 [list : { (_`$^{-\text{Integer}}$`) ..}] := Apply[Plus, list]`

In[35]:= `{f13[{a`^{2}`, b`^{3}`, c`^{4}`}], f13[{a`^{x}`, b`^{x}`, c`^{x}`}]}`

Out[35]= $\{a^2 + b^3 + c^4,\ f13[\{a^x,\ b^x,\ c^x\}]\}$

In[36]:= `f14[list:{{_, _}...}] := Map[Apply[Plus, #]&, list]`

In[37]:= `{f14[{}], f14[{{1, 2}, {3, 4}, {5, 6}}]}`

Out[37]= $\{\{\},\ \{3,\ 7,\ 11\}\}$

See the package **Statistics`DataManipulation`** for a number of examples.

4 Using Patterns in Rules

Patterns play an important role in both global and local rules.

4.1 Patterns in Global Rules

4.1.1 Logarithms

In Chapter 1, rules were given for defining a logarithm-like function.

In[1]:= `log[a_ b_] := log[a] + log[b];`
        `log[a_`$^{b_-}$`] := b log[a];`

These rules cover some unexpected cases because global rules are applied as often as possible.

In[2]:= `{log[a b c d], log[a b^3 c], log[a/b], log[Sqrt[b]]}//TableForm`

Out[2]//TableForm=
$\log[a] + \log[b] + \log[c] + \log[d]$
$\log[a] + 3 \log[b] + \log[c]$
$\log[a] - \log[b]$
$\frac{\log[b]}{2}$

4.1.2 Differentiation

Here is another simple program using rewrite rules to try to differentiate polynomials in a single variable.

In[3]:= $\texttt{diffr[x\_}^{\texttt{n\_.}}\texttt{, x\_] := n x}^{\texttt{n-1}}\texttt{;}$
       $\texttt{diffr[a\_ + b\_, x\_] := diffr[a, x] + diffr[b, x];}$

Notice the default value for **n** in the first rule. Try it out on some typical values.

In[4]:= $\left\{\texttt{diffr[x}^{\texttt{3}}\texttt{, x], diffr[y, y],}\right.$
       $\left.\texttt{diffr}\left[\texttt{w}^{\frac{1}{3}}\texttt{, w}\right]\texttt{, diffr[r}^{\texttt{3.1}}\texttt{, r], diffr[x}^{\texttt{2}}\texttt{ + x}^{\texttt{3}}\texttt{, x]}\right\}$

Out[4]= $\left\{3\,x^2,\ 1,\ \dfrac{1}{3\,w^{2/3}},\ 3.1\,r^{2.1},\ 2\,x + 3\,x^2\right\}$

But notice that **diffr** doesn't know what to do with a constant times **x**, or just a constant for that matter, and we have no obvious way as yet to teach it what to do.

In[5]:= $\texttt{\{diffr[a, x], diffr[a x, x]\}}$

Out[5]= $\{\text{diffr}[a, x], \text{diffr}[a\,x, x]\}$

We could try the following: First, give a rule for products.

In[6]:= $\texttt{diffr[a\_ b\_, x\_] := a diffr[b, x] + diffr[a, x] b}$

Using this for **a x** gives

In[7]:= $\texttt{diffr[a x, x]}$

Out[7]= $a + x\,\text{diffr}[a, x]$

It doesn't know that **a** is supposed to be a constant, so we have to tell it that explicitly.

In[8]:= $\texttt{diffr[a, x] = 0;}$

Then it gives the "correct" answer.

In[9]:= $\texttt{diffr[a x, x]}$

Out[9]= a

However, this is not very satisfactory. We would like some general way to say that **a** is not a function of **x**. This will be dealt with in the section below on restricting pattern matching with predicates.

4.2 *Patterns in Local Rules*

An important way to use patterns is on the left-hand sides of local rules. The first example is just to change the appearance of a matrix. Use **Array** to make a matrix with indexed entries and then use a local rule to display the indexes as subscripts.

```
In[10]:=MatrixForm[
        Array[a, {2, 5}]/. a[i_, j_] :> Subscripted[a[i, j]] ]
```

Out[10]//MatrixForm=

$$\begin{pmatrix} a_{1,1} & a_{1,2} & a_{1,3} & a_{1,4} & a_{1,5} \\ a_{2,1} & a_{2,2} & a_{2,3} & a_{2,4} & a_{2,5} \end{pmatrix}$$

Our next example shows that there can be a rule which depends only on the length of a list. Whenever the list here tries to grow longer than three entries, the first four entries are multiplied together pairwise to decrease the length of the list by two.

```
In[11]:=Table[
        Range[n]//.{x_, y_, z_, w_, u___} -> {x y, z w, u},
           {n, 1, 10}]
```

Out[11]= {{1}, {1, 2}, {1, 2, 3}, {2, 12}, {2, 12, 5}, {24, 30},
          {24, 30, 7}, {720, 56}, {720, 56, 9}, {40320, 90}}

The next example was the 1991 *Mathematica* programming competition question. The problem is to write a function called **runEncode** which detects repeated adjacent entries in a list. The output is a list of pairs which encodes the entries and how often they are repeated. (Note: this is not the same as **frequencies**, discussed in the preceding chapter.) Here is one of the best procedures.

```
In[12]:=runEncode[list_List] :=
        Map[{#, 1}&, list] //.
           {u___, {v_, r_}, {v_, s_}, w___} -> {u, {v, r + s}, w}
```

And here is a random list of **a**'s and **b**'s to try it on.

In[15]:=`newlist = Map[{a, b}[[Random[Integer, {1, 2}] ]]&, Range[20]]`

Out[15]= {a, a, a, a, a, a, b, a, a, b, a, a, a, b, b, a, b, b, b, b}

In[16]:=`runEncode[newlist]`

Out[16]= {{a, 6}, {b, 1}, {a, 2}, {b, 1}, {a, 3}, {b, 2}, {a, 1}, {b, 4}}

5 Restricting Pattern Matching with Predicates

So far, all of the rules we have considered have been "context free" rewrite rules. Whenever the pattern is matched, the rewriting is carried out. There can be a restriction on the head of the matching expression included in the pattern. However, there are also conditional rewrite rules which are only applied when some condition is satisfied. First of all, we have to discuss predicates in *Mathematica*, since the conditions will always be expressed in terms of them.

A predicate is a function that returns the value `True` or `False`. Predicates can be thought of as another way to construct types. In this view, types are subsets of the (infinite) universe of all *Mathematica* expressions. A predicate **P** corresponds to the type, or set of all expressions, **expr**, such that **P[expr]** evaluates to `True`. (There is actually a version of set theory proposed by Von Neumann in the 1920s that defined sets to be exactly such predicates on a preexisting universe of elements.) Thus, we have at least two ways to think about types in *Mathematica*, as heads or as predicates.

5.1 Examples of Predicates

First of all, there are predicates defined only on numbers.

In[1]:=`{1 == 2, 1 < 2, 1 <= 2, 1 >= 2, 1 > 2}`

Out[1]= {False, True, True, False, False}

In[2]:=`{Positive[3], Positive[-3], Negative[3], Negative[-3]}`

Out[2]= {True, False, False, True}

If these are used for symbols, the results are unevaluated.

In[3]:=`{a == b, a < b, a <= b, a >= b, a > b, a == a, a <= a}`

Out[3]= {a == b, a < b, a ≤ b, a ≥ b, a > b, True, a ≤ a}

In[4]:= **{Positive[a], Positive[-a], Negative[a], Negative[-a]}**

Out[4]= {Positive[a], Positive[-a], Negative[a], Negative[-a]}

However, there is a predicate defined for all expressions that is similar to ==.

In[5]:= **{expr === expr, a === b}**

Out[5]= {True, False}

In[6]:= **FullForm[Hold[a === b]]**

Out[6]//FullForm=
Hold[SameQ[a, b]]

Thus, === is the infix form of **SameQ**, which returns True if the left- and right-hand sides are syntactically identical, and False otherwise. (Recall that == is the infix form of **Equal**.) All built-in predicates defined for all expressions end with **Q**. It's easy to display all of them.

In[7]:= **??*Q**

ArgumentCountQ	AtomQ
DigitQ	EllipticNomeQ
EvenQ	ExactNumberQ
FreeQ	HypergeometricPFQ
InexactNumberQ	IntegerQ
IntervalMemberQ	InverseEllipticNomeQ
LegendreQ	LetterQ
LinkConnectedQ	LinkReadyQ
ListQ	LowerCaseQ
MachineNumberQ	MatchLocalNameQ
MatchQ	MatrixQ
MemberQ	NameQ
NumberQ	NumericQ
OddQ	OptionQ
OrderedQ	PartitionsQ
PolynomialQ	PrimeQ
SameQ	StringMatchQ
StringQ	SyntaxQ
TrueQ	UnsameQ
UpperCaseQ	ValueQ
VectorQArgumentCountQ	

Try some of the obvious ones.

In[8]:= $\left\{ \texttt{NumberQ[5.3]}, \texttt{NumberQ}\left[\dfrac{3}{5}\right], \right.$

$\left. \texttt{NumberQ[3 + 5 i]}, \texttt{NumberQ[yesterday]} \right\}$

Out[8]= {True, True, True, False}

In[9]:= $\left\{ \texttt{IntegerQ[27]}, \texttt{IntegerQ[5.3]}, \texttt{IntegerQ}\left[\dfrac{3}{5}\right] \right\}$

Out[9]= {True, False, False}

In[10]:= {EvenQ[4], OddQ[4], PrimeQ[31]}

Out[10]= {True, False, True}

In[11]:= { PolynomialQ[2 x^3 + 3 y, {x, y}],
        PolynomialQ[a x^3 + by, {x, y}],
        PolynomialQ[a x + b],
        PolynomialQ[Sin[x + 1], {x}]}

Out[11]= {True, True, True, False}

The second argument to **PolynomialQ** is the list of variables such that the first argument is a polynomial in them. If it is missing, then the single argument must be a polynomial in all of its leaves.

In[12]:= {VectorQ[{a, b, c}], VectorQ[a], VectorQ[{a}]}

Out[12]= {True, False, True}

OrderedQ asks if the entries in a list are ordered according to the canonical built-in ordering which is defined for any two expressions.

In[13]:= OrderedQ[{3, 5, a, w}]

Out[13]= True

In[14]:= {AtomQ[a], AtomQ[Sin[a]], AtomQ[5]}

Out[14]= {True, False, True}

ValueQ asks if its argument has a current value.

In[15]:= **ValueQ[f]**

Out[15]= False

The predicates **MemberQ** and **FreeQ** are less obvious in the way they work.

In[16]:= {MemberQ[{x, y, z}, x], MemberQ[{x, y, z}, s],
        MemberQ[{x, x^n}, n], MemberQ[{x, x^n}, n, Infinity],
        MemberQ[{x^2, y^2}, x_], MemberQ[Plus[x, y, z], x],
        MemberQ[(x + y) z, x + y]}

Out[16]= {True, False, False, True, True, True, True}

To determine what is going on here, look up the help entry for **MemberQ**.

In[17]:= **?MemberQ**

> MemberQ[list, form] returns True if an element of list
>   matches form, and False otherwise. MemberQ[list, form,
>   levelspec] tests all parts of list specified by levelspec.

This is not completely clear. First of all, "list" doesn't have to be a list, but then "element" means "occurs at level one". As the fourth example above shows, the way to find out if something occurs at some other level than 1 is to add a levelspec (here **Infinity**). The "form" in the second argument can be a symbol, or a pattern, or a possible subexpression. All of these, of course, are patterns, but some of them are very specific patterns that are matched by just one thing. The opposite, in an appropriate sense, of **MemberQ** is **FreeQ**. It also can take a levelspec as third argument.

In[18]:= {FreeQ[x y z, x], FreeQ[x y z, s], FreeQ[{x, x^n}, n],
        FreeQ[{x, x^n}, n, {1}], FreeQ[{x^2, y^2}, x_],
        FreeQ[Plus[x, y, z], x], FreeQ[(x + y) z, x + y]}

Out[18]= {False, True, False, True, False, False, False}

Finally, there is a predicate that tells if a given expression matches a pattern, in this case, the pattern of having its head be **Complex**.

In[19]:= **MatchQ[5 + 3I, x_Complex]**

Out[19]= True

5.2 Using Predicates

Predicates are used to control pattern matching. However, the position of the predicate in an expression can make it appear that predicates are being used in different ways. Predicates are applied using /;, which is the infix form of `Condition`.

In[20]:= `FullForm[Hold[m /; n]]`

> Out[20]//FullForm=
> `Hold[Condition[m, n]]`

5.2.1 Restricting rule application

If the predicate is placed at the end of a global rule definition, then it appears to be used to restrict the application of the rule. For instance, define a multiplicative function as follows:

In[21]:= `h[a_ b_] := a h[b] /; FreeQ[a, x]`

In[22]:= `h[2 (1 + x) x^2] + h[a b x]`

Out[22]= $a\,b\,h[x] + 2\,h[x^2\,(1+x)]$

I read /; as "provided" rather than `Condition`. Rules given this way can be considered to be conditional rewrite rules, in contrast to previous rules which are unconditional, i.e., which are applied whenever something matches their pattern. Using **rule /; Predicate** restricts the rule to those situations in which the predicate evaluates to **True**; i.e., to those expressions belonging to the type given by the predicate. An unrestricted rule is the same as a conditional rule where the predicate always equals **True**.

5.2.2 Differentiation revisited

Predicates can be used to extend our definition of differentiation to deal with arbitrary polynomials in a very natural way by adding a single conditional rule.

In[23]:= `diffr[x_^n_., x_] := n x^(n-1);`
`       diffr[a_ + b_, x_] := diffr[a, x] + diffr[b, x];`
`       diffr[a_ b_, x] := a diffr[b, x] + diffr[a, x] b;`
`       diffr[a_, x_] := 0 /; FreeQ[a, x];`

Now, constants and products are handled properly.

In[24]:= {diffr[a x, x], diffr[3 x² + 5 x + 2, x]}

Out[24]= {a, 5 + 6 x}

In[25]:= {diffr[a x, x], diffr[(3 x² + 2) (5 - 7 x³), x]}

Out[25]= {a, -21 x² (2 + 3 x²) + 6 x (5 - 7 x³)}

5.2.3 Restricting simple patterns–factorial functions

The other place to put a predicate is immediately after the pattern being affected. For instance, our simple construction of a factorial function uses two rules.

In[26]:= factorial[1] = 1; factorial[n_] := n factorial[n - 1]
This works perfectly well if we give it positive integers as arguments.

In[27]:= factorial[3]

Out[27]= 6

However, if we give it some other kind of argument, then it fails badly.

In[28]:= {factorial[today], factorial[-3]}

$RecursionLimit::"reclim": "Recursion depth of \!\(256\) exceeded."

A very large output is omitted. What happens, of course, in these cases is that the value 1 is never encountered as an argument, so the function keeps calling itself recursively until the built-in recursion limit is reached. Note also that these rules are not confluent. When they do work correctly, the result depends crucially on always trying to use the first rule before the second one. We could correct this bad behavior by using a conditional rule and, incidentally, make the system confluent.

In[29]:= factorial1[1] = 1;
    factorial1[n_] := n factorial1[n - 1] /; n > 1

In[31]:= {factorial1[5], factorial1[-3], factorial1[today]}

Out[31]= {120, factorial1[-3], factorial1[today]}

There is another way to express this using the observation that the condition only involves one argument on the left-hand side of the rule. One can use the form _?Predicate , which restricts the pattern to something for which the predicate evaluates to True. To keep things confluent, we start the system at 0.

```
In[32]:= factorial2[0] = 1;
         factorial2[n_?Positive] := n factorial2[n - 1]
```

```
In[34]:= {factorial2[5], factorial2[-3], factorial2[today]}
```

```
Out[34]= {120, factorial2[-3], factorial2[today]}
```

Notice that in the form **_?Predicate**, it is required that **Predicate** be a pure function. Another version of the syntax in Version 2.0 and higher for this is

```
In[35]:= factorial3[0] = 1;
         factorial3[n_/; Positive[n]] := n factorial3[n - 1]
```

```
In[37]:= {factorial3[5], factorial3[-3], factorial3[today]}
```

```
Out[37]= {120, factorial3[-3], factorial3[today]}
```

Note the distinction in form. In **n_?Positive** , **Positive** is a pure function, while in the form using **/;** , the condition is the value of the predicate for the name of the pattern. In either case, **Positive** or **Positive[n]** is a positive test in the sense that the pattern is matched and the rule applied only if the test succeeds. These two forms are equivalent, but *Mathematica* 's internal representation of them is different. (See the Practice section.)

Now let's try **factorial3** on a real number and see what happens.

```
In[38]:= factorial3[5.3]
```

```
Out[38]= 67.4607 factorial3[-0.7]
```

What happens is that 5.3, 4.3, 3.3, 2.3, 1.3, and 0.3 are all **Positive**, so the rule is applied until the value -0.7 is reached, in which case no pattern is matched, so **factorial3[-0.7]** is returned in unevaluated form. Notice that there is no error message, because no error has been committed. It is not an error for a rule not to match.

Of course, the real problem is that we only intend **factorial** to apply to integers. But this additional restriction can easily be added by mixing together the two type systems–the one based on heads and the other based on predicates. The pattern **n_Integer?Positive** is only matched by something whose head is **Integer** and which in addition is positive.

```
In[39]:= factorial4[0] = 1;
         factorial4[n_Integer?Positive] := n factorial4[n - 1]
```

```
In[41]:= {factorial4[5], factorial4[-3],
          factorial4[today], factorial4[5.3]}
```

```
Out[41]= {120, factorial4[-3], factorial4[today], factorial4[5.3]}
```

This also has an alternative form in Version 2.0 and higher. We revert to starting at 1.

```
In[42]:=factorial5[1] = 1;
        factorial5[n_Integer/;n > 1] := n factorial5[n-1]

In[44]:={factorial5[5], factorial5[-3],
        factorial5[today], factorial5[5.3]}

Out[44]= {120, factorial5[-3], factorial5[today], factorial5[5.3]}
```

We also check in the Practice section below that the internal representations of these two restrictions are different. Of course the predicate that appears after **/;** or **?** can also be a user defined expression.

```
In[45]:=p[x_Integer?(# > 3&)]:= x + 1

In[46]:= {p[1], p[2], p[3], p[4], p[5]}

Out[46]= {p[1], p[2], p[3], 5, 6}
```

Here is the same thing in Version 2.0 and higher. Note the difference in syntax.

```
In[47]:=pp[x_Integer /; x > 3] := x + 1

In[48]:= {pp[1], pp[2], pp[3], pp[4], pp[5]}

Out[48]= {pp[1], pp[2], pp[3], 5, 6}
```

5.2.4 Restricting compound patterns

The form **?predicate** can only be used after single slots, but the form **/; predicate** can be used after any pattern, simple or compound. For instance,

```
In[49]:= mm[x_, n_] /; OddQ[n + x] := x^n;
         mm[x_, n_] /; EvenQ[n + x] := x^-n;

In[50]:= {mm[2, 3], mm[3, 3], mm[3, 4], mm[4, 4]}
```

$$Out[50]= \left\{8, \frac{1}{27}, 81, \frac{1}{256}\right\}$$

The *Mathematica* Book suggests that it is better to place the predicate as close to the pattern being affected as possible. However, the pattern has to be a complete expression, so in the following example, the list brackets are essential.

In[51]:= `nn[{x_, n_} /; OddQ[n + x]] := x`^{n}`;`
        `nn[{x_, n_} /; EvenQ[n + x]] := x`$^{-n}$`;`

In[52]:= `{nn[{2, 3}], nn[{3, 3}], nn[{3, 4}], nn[{4, 4}]}`

Out[52]= $\left\{ 8, \dfrac{1}{27}, 81, \dfrac{1}{256} \right\}$

Named compound patterns can be treated the same way.

In[53]:= `nnn[expr : x_`$^{n\_Integer}$` /; MemberQ[x, n, Infinity], z_] :=`
        `ExpandAll[z expr]`

In[54]:= $\left\{ \text{nnn}\left[(1 - x^2)^2, \ w + z \right], \ \text{nnn}\left[(1 - x^2)^3, \ w + z \right] \right\}$

Out[54]= $\left\{ w - 2\,w\,x^2 + w\,x^4 + z - 2\,x^2\,z + x^4\,z, \ \text{nnn}\left[(1 - x^2)^3, \ w + z \right] \right\}$

5.2.5 Manipulating lists

Predicates also play an important role in manipulating lists. We have already made frequent use of the **Select** operation. Recall a simple example.

In[55]:= `Select[Range[-3, 3], Positive]`

Out[55]= $\{ 1, 2, 3 \}$

There is a similar operation called **Cases** whose second argument is a pattern rather than a predicate.

In[56]:= `Cases[{a + b, a b, a`^{b}`, a - b, x`^{x}`}, _`$^-$`]`

Out[56]= $\{ a^b, x^x \}$

The pattern in the second argument can be restricted by a predicate, in either of the two usual forms.

In[57]:= `Cases[Range[-3, 3], _?Positive]`

Out[57]= $\{ 1, 2, 3 \}$

In[58]:= `Cases[Range[-3, 3], x_ /; x > 0]`

Out[58]= $\{ 1, 2, 3 \}$

The opposite of **Cases** is **DeleteCases**, which drops all entries not matching some pattern.

In[59]:= **DeleteCases[Range[10]^2, x_ /; OddQ[x]]**

Out[59]= {4, 16, 36, 64, 100}

There is a related operation called **Position** whose second argument is also a pattern. It gives the parts list for all arguments that match the pattern.

In[60]:= **Position[Range[-3, 3], x_ /; x > 0]**

Out[60]= {{5}, {6}, {7}}

Strangely, there is no operation doing the same thing as **Position** but using a predicate rather than a pattern. But, as this example demonstrates, that is no restriction, since the pattern can be that of an expression that satisfies some predicate. Note that **Cases, Delete-Cases**, and **Position** can all take a third argument which is a levelspec. There is also another form of **Cases** in which some operation is applied to the entries that are selected.

In[61]:= **Cases[Range[-3, 3], (x_ /; x > 0) :> Sqrt[x]]**

Out[61]= $\left\{1, \sqrt{2}, \sqrt{3}\right\}$

Finally, the operation **Scan** applies some pure function to each element in a list, starting at the left, just like **Map**, except that no output is returned. If the operation has some side effect, then that will be carried out. For instance:

In[62]:= **Scan[Print, Range[3]]**

```
1
2
3
```

Frequently, **Scan** is used to find the first entry satisfying some property. In order to see the result it is necessary to break out of the scanning procedure when this happens. For instance,

In[63]:= **Scan[If[# ≮ 4, Return[#]] &, Range[-3, 3]^2]**

Out[63]= 9

In fact, all of these operations work for expressions with arbitrary heads, not just for lists.

6 Examples of Restricted Rewrite Rules

6.1 Global Rules

6.1.1 Subsets of a set

This example appeared in [Simon 1]. Given a finite set (presented as a list) and an integer **k**, it finds all **k**-element subsets of the set. Note that the arguments are protected everywhere so that the function only applies to a pair of inputs consisting of a list and an integer.

```
In[1]:= kSubsets[list_List, 0] := {{ }};
    kSubsets[list_List, 1] := Partition[list, 1];
    kSubsets[list_List, k_Integer?Positive] :=
            {list} /; (k == Length[list]);
    kSubsets[list_List, k_Integer?Positive] :=
            Join[(Prepend[#, First[list]])& /@
                kSubsets[Rest[list], k - 1],
                kSubsets[Rest[list], k] ] ;
```

The rules correspond directly to the usual proof that the number of **k**-element subsets of an **n**-element set is given by the binomial coefficient (**n**, **k**). Thus, the set of 0-element subsets consists of just the empty set. The set of 1-element subsets consists of the singleton subsets. If **k** = **n**, then there is just one **k**-element subset, namely, the set itself. Finally, in general, the **k**-element subsets consist of the **k**-element subsets of the set given by dropping the first element together with the first element added to the (**k** − 1)-element subsets of the same set.

```
In[5]:= kSubsets[{1, 2, 3, 4, 5, 6}, 3]
```

```
Out[5]= {{1, 2, 3}, {1, 2, 4}, {1, 2, 5}, {1, 2, 6}, {1, 3, 4},
        {1, 3, 5}, {1, 3, 6}, {1, 4, 5}, {1, 4, 6}, {1, 5, 6},
        {2, 3, 4}, {2, 3, 5}, {2, 3, 6}, {2, 4, 5}, {2, 4, 6},
        {2, 5, 6}, {3, 4, 5}, {3, 4, 6}, {3, 5, 6}, {4, 5, 6}}
```

To get all subsets from this version, we have to join together the lists of **k**-element subsets for all **k** up to the size of the set.

```
In[6]:= subsets[list_List] :=
        Join@@Table[kSubsets[list, k], {k, 0, Length[list]}]
```

```
In[7]:= subsets[{1, 2, 3, 4}]
```

```
Out[7]= {{}, {1}, {2}, {3}, {4}, {1, 2}, {1, 3}, {1, 4}, {2, 3}, {2, 4},
        {3, 4}, {1, 2, 3}, {1, 2, 4}, {1, 3, 4}, {2, 3, 4}, {1, 2, 3, 4}}
```

Another way to calculate all subsets of a set, via a functional strict one-liner, was found by I. Vardi [Vardi], based on the distributive law. Observe first how **Distribute** works on three factors.

In[8]:=**Distribute[(1 + a) (1 + b) (1 + c)]**

Out[8]= 1 + a + b + a b + c + a c + b c + a b c

This result is clearly related to the set of all subsets of {a, b, c}. The plus sign has to be replaced by a comma and the multiplication has to be replaced by **List** somehow. *Mathematica* has a more general form of **Distribute** in which one can specify that **f** is to be distributed over **g**. (Actually, the final **f** in this expression is unnecessary.)

In[9]:=**Distribute[f[g[x, y], g[x, y]], g, f]**

Out[9]= g[f[x, x], f[x, y], f[y, x], f[y, y]]

So here is a first step in getting all subsets of {a, b, c}. We distribute **List** over **List** as follows:

In[10]:=**trial =**
      Distribute[List[{{}, {a}}, {{}, {b}}, {{}, {c}}], List]

Out[10]= {{{}, {}, {}}, {{}, {}, {c}}, {{}, {b}, {}}, {{}, {b}, {c}},
        {{a}, {}, {}}, {{a}, {}, {c}}, {{a}, {b}, {}}, {{a}, {b}, {c}}}

One way to turn this into the list that we want is to **Flatten** each of the inner lists.

In[11]:=**Map[Flatten, trial]**

Out[11]= {{}, {c}, {b}, {b, c}, {a}, {a, c}, {a, b}, {a, b, c}}

Another way is to change the head of each argument of this list to **Union**.

In[12]:=**Map[Apply[Union, #]&, trial]**

Out[12]= {{}, {c}, {b}, {b, c}, {a}, {a, c}, {a, b}, {a, b, c}}

So, all we have to do is construct the strange list of pairs consisting of the empty set together with a singleton set from the original set. This is also easy to do.

In[13]:=**Map[({{}, {#}})&, {a, b, c}]**

Out[13]= {{{}, {a}}, {{}, {b}}, {{}, {c}}}

Thus, the desired one-liner can be written in two forms. We also **Sort** the result to get things in their usual order.

```
In[14]:= subsets1[list_List] :=
        Sort[Map[Flatten,
                Distribute[ Map[({{}, {#}})&, list], List]]]
```

```
In[15]:= subsets2[list_List] :=
        Sort[Map[Apply[Union, #]&,
            Distribute[ Map[({{}, {#}})&, list], List]]]
```

```
In[16]:= {subsets1[{a, b, c}], subsets2[{a, b, c}]}
```

```
Out[16]= {{{}, {a}, {b}, {c}, {a, b}, {a, c}, {b, c}, {a, b, c}},
          {{}, {a}, {b}, {c}, {a, b}, {a, c}, {b, c}, {a, b, c}}}
```

Actually, Vardi's version is somewhat different. Note that the output is not sorted.

```
In[17]:= subsetsFunctional[list_List] :=
        Distribute[{{}, {#}}& /@ list, List, List, List, Union]
```

See The *Mathematica* Book or the online documentation for this form of **Distribute**.

```
In[18]:= subsetsFunctional[{a, b, c}]
```

```
Out[18]= {{}, {c}, {b}, {b, c}, {a}, {a, c}, {a, b}, {a, b, c}}
```

6.1.2 Laplace transforms

As a more complicated example of a rule based program, consider a simple version of the Laplace transform. Here is a list of rules that will calculate the Laplace transform for many simple functions.

```
                    laplace[function, t, s]
```

means the Laplace transform of the function **function** which depends on the variable **t**, expressed as a function of the variable **s**. If the function is a constant **c**, then its Laplace transform is **c/s**, giving us the first rule.

```
In[19]:= laplace[c_, t_, s_] := c / s /; FreeQ[c, t]
```

The Laplace transform is a linear function of its first argument. This is expressed by two rules.

```
In[20]:= laplace[a_ + b_, t_, s_] :=
          laplace[a, t, s] + laplace[b, t, s]
```

```
In[21]:= laplace[c_ a_, t_, s_] :=
          c laplace[a, t, s] /; FreeQ[c,t]
```

If the function is of the form t^n for an integer n, then the Laplace transform has a simple form.

$$In[22]:= \text{laplace}[t\_^{n\_\cdot}, t\_, s\_] := \frac{n!}{s^{n+1}} \text{/; (FreeQ}[n, t] \&\& n \geq 0)$$

If the function is a product where one factor is of the form t^n, then the Laplace transform is somewhat more complicated.

```
In[23]:= laplace[a_ t_^n_., t_, s_] :=
          (-1)^n D[laplace[a, t, s], {s, n}] /; (FreeQ[n, t] && n ≥ 0)
```

The Laplace transform of a function divided by t can sometimes be calculated.

$$In[24]:= \text{laplace}\left[\frac{a\_}{t\_}, t\_, s\_\right] := \int_s^\infty \text{laplace}[a, t, v] \, dv$$

Finally, a function involving E to an exponent which is linear in t can be reduced to a simpler form.

```
In[25]:= laplace[a_. e^{b_.+c_. t_}, t_, s_] :=
          laplace[a e^b, t, s-c] /; FreeQ[{b, c}, t]
```

Note that these rules are mutually recursive. Let us try a few examples.

```
In[26]:= laplace[c t^2, t, s]
```

$$Out[26]= \frac{2c}{s^3}$$

```
In[27]:= laplace[(t^3 + t^4) t^2, t, s]
```

$$Out[27]= \frac{720}{s^7} + \frac{120}{s^6}$$

In[28]:= **laplace[t^2 e^{2+3t}, t, s]**

Out[28]= $\dfrac{2\,E^2}{(-3+s)^3}$

In[29]:= **laplace$\left[\dfrac{e^{2+3t}}{t},\ t,\ s\right]$**

Integrate::idiv : Integral of $\dfrac{E^2}{-3+v}$ does not converge on $\{s,\ \infty\}$.

Out[29]= $E^2 \displaystyle\int_s^\infty \dfrac{1}{-3+v}\,dv$

See the packages **LaplaceTransform.m** and **Trigonometry.m** for programs making extensive use of lists of rules with intricate patterns and conditions.

6.2 Local Rules

Patterns can be used on the left-hand sides of local rules, so restrictions on them using predicates can appear in this position also.

6.2.1 maxima

This example was the 1992 *Mathematica* programming competition question. The problem is to write a function called **maxima** that starts with a list of numbers and constructs the sublist of the numbers bigger than all previous ones from the given list. For instance, **maxima[{4, 7, 5, 2, 7, 9, 1}]** should return {4, 7, 9}. The winning entry uses a pattern with a condition in a local rule.

In[30]:=**maxima[list_List] :=**
              list //. {a___, x_, y_, b___} /; y <= x -> {a, x, b}

In[31]:=**maxima[{4, 7, 5, 2, 7, 9, 1}]**

Out[31]= {4, 7, 9}

6.2.2 complexSort

Complex numbers are sorted in *Mathematica* first by increasing real part and then by increasing imaginary part. This example, adapted from one on the network, shows how to sort complex numbers so that conjugate numbers are placed next to each other.

```
In[32]:=complexSort[cplxs_List] :=
      Flatten[
        Sort[cplxs]//.
          ({a___, z_, b___, zbar_, c___} /;
            Length[{a, b, c}] > 0 && z == Conjugate[zbar]) :>
            {a, If[Im[z] < Im[zbar], {z, zbar}, {zbar, z}],
             b, c}]
```

```
In[33]:=cplxs[n_] := Outer[Plus, Range[-n, n], I Range[-n, n]]
```

```
In[34]:=Flatten[Sort[cplxs[1]]]
```

Out[34]= $\{-1 - I, -1, -1 + I, -I, 0, I, 1 - I, 1, 1 + I\}$

```
In[35]:=complexSort[cplxs[1]]
```

Out[35]= $\{-1 - I, -1 + I, -1, -I, I, 0, 1 - I, 1 + I, 1\}$

6.2.3 intervalUnion

This next example comes from John Lee, University of Washington, in response to discussions on the network about a program to compute the union of a set of possibly overlapping intervals.

```
In[36]:=intervalUnion[listOfIntervals_List] :=
       Sort[listOfIntervals] //.
         {a___, {b_, c_}, {d_, e_}, f___} :>
            {a, {b, Max[c, e]}, f} /; d <= c
```

```
In[37]:=intervalUnion[{{1, 2}, {3, 4}, {1.5, 3.5}}]
```

Out[37]= $\{\{1, 4\}\}$

```
In[38]:=intervalUnion[{{1, 2}, {3, 4}, {3, 5}}]
```

Out[38]= $\{\{1, 2\}, \{3, 5\}\}$

In Version 3.0, there is a built-in function with the same name. It works for sequences of things with the explicit head **Interval**.

```
In[39]:= IntervalUnion[Interval[{1, 2}],
       Interval[{3, 4}], Interval[{3.6, 5}]]
```

Out[39]= Interval$[\{1, 2\}, \{3, 5\}]$

6.2.4 Discussion

In each of these examples, a list is rewritten in a non-trivial way by describing how a typical pattern in the original list is to be rewritten in the new list. Arbitrary locations in the list are accessed by using patterns of the form **a___** involving zero or more arguments, and conditions are placed on whether the rewriting should take place by following the left-hand side by a **/;** clause. This is a powerful technique which is a valuable tool in *Mathematica* programming.

6.3 Dynamic Programming and $RecursionLimit

Recall the final version of the factorial function.

```
In[40]:= factorial[1] = 1;
         factorial[n_Integer /; n > 1] := n factorial[n - 1]
```

Can one actually use this definition to calculate **factorial[n]** for large values of **n**? It turns out that there is a specific limit that cannot be exceeded.

```
In[42]:= factorial[255]
```

The output is omitted because it is not interesting. What is interesting is the next output.

```
In[43]:= factorial[256]
```

> $RecursionLimit::"reclim": "Recursion depth of \!\(256\) exceeded."

```
Out[43]= 4289088876714213270595411358406163125788907601397428099298278255·
         1886347262765737946887201456802257042251879426711682921530785·
         8417346848237661144644248713012839818666281684393221337603813·
         9728009398443398576057165385103876332322573235459366305041643·
         1628514094903868358907270851252615093042476595340691287405351·
         6408779729738493517332856369069643102617378404109430350601805·
         4157604675097371855455086348413143080313183121751142047209570·
         2123079680000000000000000000000000000000000000000000000000000·
         0000000000 factorial[2]
```

As we have programmed it, **factorial** is a recursive function. In order to calculate **factorial[n]**, it first has to calculate **factorial[n - 1]**, etc., so it builds up a sequence of unevaluated terms until it finally gets to **factorial[1]**, which has an explicit value, so then all the other terms can be evaluated. Precisely, it builds a nested sequence of values as shown in the following computation.

In[44]:= **Trace[factorial[3]]**

Out[44]= {factorial[3], {3 > 1, True}, 3 factorial[3 - 1],
        {{3 - 1, 2}, factorial[2], {2 > 1, True}, 2 factorial[2 - 1],
         {{2 - 1, 1}, factorial[1], 1}, 2 1, 2}, 3 2, 6}

Mathematica has a built-in limit, called **$RecursionLimit**, which by default is set to 256 so that it will not carry out more than 256 such steps. In our calculation of **factorial[256]** it got as far as the last step, **factorial[2]**, but then it had to give up. In earlier versions, one way to proceed was to release the hold by using **ReleaseHold[%]** immediately after the calculation. It would then proceed for a maximum of 256 more steps. In Version 3.0, this result is no longer explicitly held, and so far we have been unable to discover how to tell the calculation to continue. However, once we are certain that we are not in an infinite loop, we can set **$RecursionLimit** higher to calculate larger values.

In[45]:= **$RecursionLimit = 2000;**

Try timing successive multiples of 200 to see how long these computations take.

In[46]:= **Table[{200 n, Timing[factorial[200 n];][[1]]}, {n, 1, 6}]**

Out[46]= {{200, 0.05 Second}, {400, 0.233333 Second},
        {600, 0.483333 Second}, {800, 0.75 Second},
        {1000, 0.75 Second}, {1200, 0.9 Second}}

Thus, the time to calculate **factorial[200 n]** is approximately linear in **n**. Let us check what *Mathematica* knows about **factorial**.

In[47]:= **??factorial**

    Global`factorial
    factorial[1] = 1

    factorial[n_Integer /;
     n > 1] := n * factorial[n - 1]

It knows just the rules that we gave it. There is another way to write the program for **factorial** so that *Mathematica* will remember the values that it has already calculated and hence not have to recalculate them each time it goes through such a recursive procedure. This is called dynamic programming. The syntax is very simple.

```
In[48]:= factorialDyn[1] = 1;
    factorialDyn[n_Integer /; n > 1] :=
        factorialDyn[n] = n factorialDyn[n - 1]
```

If we calculate **factorialDyn[n]**, then *Mathematica* will have calculated and remembered all smaller values.

```
In[50]:= factorialDyn[6]
```

```
Out[50]= 720
```

```
In[51]:= ??factorialDyn
```

```
    Global`factorialDyn
    factorialDyn[1] = 1
    factorialDyn[2] = 2
    factorialDyn[3] = 6
    factorialDyn[4] = 24
    factorialDyn[5] = 120
    factorialDyn[6] = 720
    factorialDyn[n_Integer /; n > 1] :=
            factorialDyn[n] = n * factorialDyn[n - 1]
```

If we want to calculate a higher value, then the recursion will only have to go down to the value 6 instead of 1. We can use this to calculate large values without increasing **$Recursion-Limit** as much as before.

```
In[52]:= Table[{200 n, Timing[factorialDyn[200 n];][[1]]}, {n, 1, 10}]
```

```
Out[52]= {{200, 0.216667 Second}, {400, 0.266667 Second},
        {600, 0.266667 Second}, {800, 0.616667 Second},
        {1000, 0.5 Second}, {1200, 0.9 Second},
        {1400, 1.06667 Second}, {1600, 1.26667 Second},
        {1800, 1.93333 Second}, {2000, 1.75 Second}}
```

At each step in the table the recursion only has to go back to the previous step. We won't ask *Mathematica* what it knows about **factorialDyn** now because that would cause it to display 1200 rules, which is more than we want to look at. The timing for each step appears to be almost constant, or only growing slowly until the values get to 1800. But apparently the total time to get to 2000, which is the sum of all of the preceding times, is now significantly longer that the time for the single computation. In the exercises, we will treat an example where dynamic programming has a more significant effect, making possible calculations that were simply not possible without it. (However, in this case, special methods work even better.)

7 *Practice*

1. `FullForm[x ^= y]`

2. `FullForm[x ^:= y]`

3. `??UpValues`

4. `??DownValues`

5. `??Global`*`

6. `???`

7. `FullForm[x__Head]`

8. `FullForm[x___Head]`

9. `FullForm[x_:v]`

10. `FullForm[x:v]`

11. `FullForm[n_ /; n > 0]`

12. `FullForm[n_Integer?Positive]`

13. `FullForm[n_Integer /; n > 1]`

14. `FullForm[gg[fun:Power[x_, n_Integer]]]`

15. `subsets11[list_List] :=`
    `  Sort[Flatten/@`
    `    Distribute[{{}, {#}}&/@list, List]]`

16. `subsets22[list_List] :=`
    `  Sort[Union@@#&/@`
    `    Distribute[{{}, {#}}&/@list, List]]`

8 *Exercises*

1. Find all values of the form $n = m/3$ between -10 and 10 such that *Mathematica* can evaluate the following integral:

$$\int \frac{\left(1 - \frac{1}{u}\right)^{\frac{4}{3}}}{u^n} \, d u$$

Hint: make a table and use **Select** and **FreeQ**.

2. In Exercise 5 of Chapters 6 and 7, the Gram–Schmidt procedure was developed. The procedure there only works if the given list of vectors is linearly independent. Make several changes in the procedure so it still works even if the given list of vectors is linearly dependent.

i) Restrict the functions so they only work for arguments of the proper kinds.

ii) Include a separate rule to deal with the projection of a vector on a zero vector.

iii) The resulting list of orthogonal vectors may then contain a zero vector. Add a new operation, **nozeros,** to remove such zero vectors. Note that the notion of a zero vector depends on the vector space under consideration.

iv) Test your procedure on a long list of random three-dimensional vectors with real entries.

v) Test your procedure using the Legendre inner product and various polynomials including the powers of x up to x^4.

3. i) Write a function **type** of one variable such that **type** takes the value 0 for integer arguments, the value $1/2$ for rational arguments, the value 1 for real numbers, the value 2 for complex numbers, and the value ∞ for anything else.

ii) Change the definition of **type** so that it takes the value 10 for "algebraic expressions". An algebraic expression is one which is built up recursively from symbols (i.e., variables) and numbers (integers, rationals, reals, and complexes) by using addition, subtraction, multiplication, division, and exponentiation. (Hint: use pattern matching recursively to define a predicate **algexpQ** which takes the value True just for algebraic expressions. For instance, one such rule is: **algexpQ[u_ + v_] := algexpQ[u] && algexpQ[v].**)

iii) Test your predicate **algexpQ** on the following inputs.

```
x^2 + (y + 2)^3
x^2 + (Sin[y] + 2)^3
(5 x y)^(z + w)
Sqrt[5 x y]^(z + w)
x^(x^(x^(x^x)))
(y + w)^(x + 2)
(x + 2 I) (3 + y I)^(5 + 4I)
(2x + y) + I (z w + u)
Tan[x^2 + y^2]
```

iv) Test your type function on the following inputs.

```
{anything, 24, 3/7, 3.64, (5 + 3 I), -(x + y z)^(z - 3 w),
(x + 2 I) (3 + y I)^(5 + 4I), Sin[anything] + 4}
```

4. i) Extend the definition of **diffr** further so that it differentiates restricted algebraic expressions correctly, where algebraic expressions are as above, but restricted means that the only kinds of exponents that are allowed are numbers and symbols.

ii) Extend the definition of **diffr** further so that it differentiates "calculus expressions" correctly. Here "calculus expressions" are expressions which are built up recursively from symbols, numbers, trigonometric functions, the exponential function, and the logarithm function by using addition, subtraction, multiplication, division, and restricted exponentiation, where restricted exponentiation now means that either the base or the exponent is a constant (i.e., a number or a symbol).

iii) Extend the definition of **diffr** further to higher order and mixed derivatives.

5. The function **maxima** described in the Examples section above can also be implemented by a strict one-liner functional program. A one-liner using **FoldList, Infinity, Max, Rest,** and **Union** was the most efficient function found in the contest. Write this function and do a **Timing** comparison with the pattern matching version.

6. This is an exercise in calculating the Fibonacci numbers by different methods. Part of the exercise is to attempt to estimate the complexity of the various methods.

i) The recursive definition:

```
fibr[1] = 1;
fibr[2] = 1;
fibr[n_] := fibr[n - 1] + fibr[n - 2]
```

ii) Dynamic programming:

```
fibd[1] = 1;
fibd[2] = 1;
fibd[n_] := fibd[n] = fibd[n - 1] + fibd[n - 2]
```

iii) Iteration:

```
fibi[n_] :=
    Module[{an1 = 1, an2 = 1},
        Do[{an1, an2} = {an1 + an2, an1}, {i, 3, n}];
        an1
    ]
```

iv) Formula for the nth number:

```
e1  = (1 + Sqrt[5]) / 2;
e2  = (1 - Sqrt[5]) / 2;
b1  = (5 + Sqrt[5]) / 10;
b2  = (5 - Sqrt[5]) / 10;
fibf[n_] := Expand[b1 e1^(n - 1) + b2 e2^(n - 1).
```

v) The **fibf** version can be speeded up by replacing **Sqrt[5]** by a suitable numerical approximation which depends on **n**. Try to do this if you see how. Call this version **fibfn[n]**.

vi) The powers of the matrix {{1, 1}, {1, 0}} are related to the Fibonacci numbers. Use this to give yet another way to calculate them called **fibm**.

Suggestions for analyzing the algorithms: In each case, experiment to find appropriate maximal sizes for **n**. Then make a table of values and timings up to the appropriate size. Make a plot of these values to see what the timings look like. Try to fit your timing data to an appropriate curve and use that to find out how long it would take to calculate the millionth Fibonacci number. In the last five cases, you will probably want to use input data of the form 2^n, rather than **n**. You might want to combine all of the plots into a single plot showing the relations between the methods.

CHAPTER 8

Procedural Programming

1 Introduction

In this chapter, we turn to the third alternative mentioned in the previous chapter: Use *Mathematica* as a block structured language with the usual control structures of an imperative language. These block structures are constructed by the command **Module** in *Mathematica*, although the term **Block** was used in Version 1.x. The language of while-programs is an abstract version of such a language. It consists of exactly four kinds of commands:

> **assignment** commands
> **if_then_else_** commands
> **composition** commands
> **while_do_** commands

These commands work in quite a different way than the operations in a functional or rewrite rule language, both of which deal with expressions and their reduction to normal form. An imperative language deals with states of a computer. To explain this concept, suppose there is a fixed finite set of variables $\{x_1, \ldots, x_K\}$, where K is the number of memory locations in some computer; e.g., $K = 2^{32}$. We will, in fact, think of x_j as the name of a specific memory location, the jth one. Suppose further that each memory location can hold a value, which could be a number or a bit, or some other choice for values. Let V be the set of values and consider V^K, the Cartesian product of V with itself K times. An element of V^K is a K-tuple of values, $v = (v_1, \ldots, v_K)$. We can regard the jth component of such a K-tuple as the contents of the jth memory location and call v a *state* of the computer. Thus, a state is some assignment of a value to each memory location, and V^K is the set of all states of the computer. The action of a command is to change the state by changing the values at some of the memory locations; i.e., commands produce mappings from V^K to itself. We have to explain exactly what mapping corresponds to each kind of command.

1.1   The Language of While-Programs

More formally, the language of while-programs with values in the set $V = N$ of natural numbers consists of the following structures:

1.1.1 Arithmetic terms

These consist of the constant **0**, variables x_j, and terms **succ**(A), **pred**(A), **plus**(A, A'), **minus**(A, A'), and **times**(A, A') whenever A and A' are arithmetic terms. Arithmetic terms are thought of as functions from N^K to N; e.g., the value of the arithmetic term **succ**(**plus**($x_1 + x_2$)) on a K-tuple is the successor of the sum of the first two entries.

1.1.2. Predicates or Boolean terms

These consist of the constants **tt** and **ff** and terms (A **==** A'), (A < A'), (A > A'), (B **and** B'), (B **or** B'), (B **implies** B'), and **not** (B) whenever A and A' are arithmetic terms and B and B' are predicates. Predicates are thought of as functions from N^K to the set Bool = {True, False}; e.g., the value of the predicate

$$\mathbf{minus}(\mathbf{times}(x_1, x_1), \mathbf{times}(x_2, x_2)) == \mathbf{times}(\mathbf{plus}(x_1, x_2), \mathbf{minus}(x_1, x_2))$$

is True.

1.1.3 Commands

i) An **assignment** command is one of the form $x = A$, where x is a variable and A is an arithmetic term. For instance, if n_j is stored at memory location j, then the assignment command $x_j = 5$ denotes the mapping from N^K to itself that changes the value of n_j to 5 and leaves all other values unchanged. If A contains variables, they are given the values they have in the current state.

ii) A **composed** command is one of the form **begin** $C_1; \ldots; C_n$ **end**, where $C_1, \ldots, C_n$ are command terms. The interpretation of a composition command is just the composition (in the sense of functions) of the interpretations of the C_i's as mappings from N^K to itself.

iii) A **conditional** command is one of the form **if** B **then** C **else** C', where B is a predicate and C and C' are command terms. The interpretation of a conditional command as a mapping from N^K to itself is the interpretation of C (resp., C') in the current state if the value of B in the current state is **tt** (resp., **ff**).

iv) A **loop** command is one of the form **while** B **do** C, where B is a predicate and C is a command term. The interpretation of a loop command is more complicated. Let n be the present state. If the interpretation of B in state n is **ff**, then the command leaves the state unchanged. If it is **tt**, then the command C is executed, leading to a new state. B is evaluated again in this new state. If the result is now **ff**, then the new state is the result of the command. Otherwise, C is executed again. This continues until B evaluates to **ff**, in which case the state at that point is the result. If B never evaluates to **ff**, then the loop continues forever. In this case, one says that the command *diverges*.

Here is a simple example of a while-program to calculate y ! .

$$
\begin{array}{l}
\textbf{begin} \\
x = 0; \\
z = \text{succ}(0); \\
\textbf{if } y == 0 \textbf{ then } z = \text{succ}(0) \textbf{ else} \\
\qquad \textbf{while } \textbf{not}(x == y) \textbf{ do} \\
\qquad\qquad \textbf{begin} \quad x = \text{succ}(x); \\
\qquad\qquad\qquad\qquad z = \text{times}(z, x) \\
\qquad\qquad \textbf{end} \\
\textbf{end}
\end{array}
$$

To describe the interpretation of this program, suppose there are just three memory locations where x, y, and z are stored; i.e., $K = 3$. Let $\{x_0, y_0, z_0\}$ be the initial state before the program is run. After the first "initialization" steps, the state is $\{0, y_0, 1\}$. In the **if** statement, if y_0 is **0**, then the state, which is $\{0, 0, 1\}$, is returned as the result of the program. If y_0 is not **0**, then the **while** loop is entered. The condition **not**$(x == y)$ is clearly true so the **do** part is executed. The two assignment statements here change the state to $\{1, y_0, 1 * 1\}$. The predicate is checked again, and if y_0 is not equal to **1**, then the **do** part is executed again yielding the state $\{2, y_0, 1 * 1 * 2\}$. This continues until the first and second components of the state are y_0 and the third component is y_0 !. In either case the result of executing the program is that the x-location now has the value y_0 and the z-location now has the value y_0 !; i.e., the final state is $\{y_0, y_0, y_0 !\}$. Thus, the third memory location now stores the value y_0 !.

Mathematica of course has many more arithmetic terms and predicates than those described above. It also has operations that implement the imperative commands exactly. There are several forms of conditional and loop commands. However, instead of **begin–end** forms for programs, *Mathematica* uses blocks which are called **Module**s, although they were called **Block**s in Version 1.x. **Block**s still exist and are sometimes useful. Understanding the difference between **Module**s and **Block**s will turn out to be instructive. In *Mathematica*, the state is represented by values assigned to global variables. This kind of state is often called a *store* in impure functional languages, mainly to try to avoid the bad connotations of states in functional programs. In the context of a functional programming language, anything other than reducing an expression to normal form is regarded as a *side effect*. In particular, if there is a concept of state in the language, then changing the state is a side effect. In this sense, imperative languages work solely by side effects.

2 Basic Operations

2.1 Assignments and Composition

Assignment commands in *Mathematica* are mimicked by expressions of the form **x = a**; i.e., expressions with head **Set**. The composition or sequencing of commands is indicated by semicolons. The output of each operation serves as the input to the next. The output of the composed command is the output of the last command in the sequence. If we assume at the beginning that we just have assignment commands and arithmetic operations, then we can build up a composed command as follows:

In[1]:=**x = 1; x = x + 1; x = x + 1; x = x + 1**

Out[1]= 4

Notice that the output is 4, which is the output of the last command. In our machine metaphor, what is now stored in the **x** location is this value, as shown by querying the machine.

In[2]:=**x**

Out[2]= 4

As a side remark, recall that everything in *Mathematica* is an expression, so composed expressions must also be expressions. We check that this is true.

In[3]:=**FullForm[Hold[x = x + 1; x = x + 1]]**

    Out[3]//FullForm=
    Hold[CompoundExpression[Set[x, Plus[x, 1]], Set[x, Plus[x, 1]]]]

Thus, **;** is just the infix form of **CompoundExpression** in the same way that **+** is the infix form of **Plus**.

Assignment commands and composed commands are incompatible with functional programming constructs. For instance, the following two commands show that addition in not commutative.

In[4]:=**y = 6; Plus[(y = y + 1); 5, y]**

Out[4]= 12

In[5]:=**y = 6; Plus[y, (y = y + 1); 5]**

Out[5]= 11

Mathematica evaluates the arguments in **Plus** from left to right. So in the first case, inside the **Plus**, **y** is set to **7** when the first argument is evaluated. This happens as a "side effect" to the value of the first argument, which is **5**. When the second argument is evaluated, it finds that **y** is **7**, so the result is **12**. In the second case, when the first argument to **Plus** is evaluated, **y** is still **6**. When the second argument is evaluated, **y** is set to **7**, but that has no effect on the value of the first argument and also no effect on the value of the second argument, which is still **5**, so the result is **11**. The problem of adding assignment statements to functional languages in such a way as to control unfortunate effects like this is currently a research topic in computer science. What has happened in the first version is that the evaluation of the first argument affects a variable that is used in the evaluation of the second argument. Needless to say, fixing things so that this doesn't happen would add considerable complexity to the language. This particular example would be avoided if *Mathematica* evaluated its arguments in parallel; i.e., the state should be frozen until all arguments are evaluated, and then updated as necessary. (The problem with this solution is if two different arguments change the state in different ways, then what should the final state be?)

2.2 *Conditional Operations*

Conditional operations are used for branching; that is, depending on some condition, the program should continue following one path or another, but not both. The simplest conditional operation is the **if_then_else_** operation. In *Mathematica* everything is an expression, so this is represented by an expression with head **If** and three arguments: **If[test, then, else]**, where **test** is some predicate and **then** and **else** are any other two expressions. Besides this, there are two other related expressions,

```
Which[test₁, value₁, test₂, value₂, . . . ]
Switch[expr, form₁, value₁, form₂, value₂, . . . ]
```

which are explained below.

2.2.1 If

If[test, then, else] is just like the (**if_then_else_**) operation in Pascal. If **test** evaluates to **True**, then **then** is evaluated and if **test** evaluates to **False**, then **else** is evaluated. If **test** is not a Boolean expression (i.e., does not evaluate to **True** or **False**) then the **If** expression is not evaluated at all. There are two other forms for this expression:

If[test, then, else, unknown]

which returns the value of **unknown** if **test** does not evaluate to **True** or **False**. Finally

$$\texttt{If[test, then]}$$

returns **then** if **test** evaluates to True and Null if **test** evaluates to False. For instance:

In[6]:= {If [5 > 2, 1, 2],   If [5 < 2, 1, 2],
    If [a == b, 1, 2, 3], If [5 < 2, 1]}

Out[6]= {1, 2, 3, Null}

If we ask about the attributes of **If**, we find:

In[7]:=**Attributes[If]**

Out[7]= {HoldRest, Protected}

Thus, **If** holds its second and third arguments. This is important for an expression in which one of the arguments might diverge, but which are never evaluated in that case. For instance, define

In[8]:= **bad[x_]** := If$\left[$x == 0, 0, $\dfrac{1}{x}\right]$

Then **bad[0]** is perfectly well behaved, even though $\frac{1}{0}$ normally leads to an error message in *Mathematica*.

In[9]:= {bad[0], bad[1], bad[2]}

Out[9]= $\left\{0,\ 1,\ \dfrac{1}{2}\right\}$

A function definition of the form

$$\texttt{f[x\_]} := \texttt{If[test, then, else, unknown]}$$

where **test, then, else,** and **unknown** involve **x,** divides the universe of *Mathematica* expressions into three disjoint subsets: those expressions **exp** for which **test /. x -> exp** evaluates to True, in which case **then /. x -> exp** is evaluated; those for which **test /. x -> exp** evaluates to False, in which case **else /. x -> exp** is evaluated; and those for which **test /. x -> exp** evaluates to neither True nor False, in which case **unknown /. x -> exp** is evaluated.

One can use this in interesting ways in *Mathematica*. For instance, the resulting function can be plotted.

In[10]:= **f [x_]** := If [x < 0, x^2, -x^2]

In[11]:=**Plot[f[x], {x, -2, 2}];**

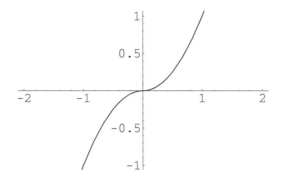

The same effect, of course, can be obtained by conditional rewrite rules:

In[12]:= **g[x_] := x² /; x ◁0;**
g[x_] := -x² /; x ≤ 0;

In[13]:=**Plot[g[x], {x, -2, 2}];**

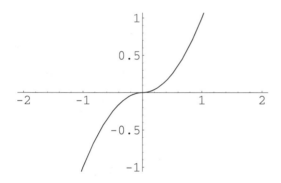

Interestingly, **f** can be differentiated but not **g**.

In[14]:= **{D[f[z], z], D[g[z], z]}**

Out[14]= **{If[z > 0, 2 z, -(2 z)], g'[z]}**

It is possible in *Mathematica* to get unintended results by using an expression that is only a predicate for numbers in a situation where more general inputs can arise. For instance, define an operation that depends on the head of an expression.

In[15]:= **heads1[exp_] := If[Head[exp] == Plus, exp², exp³]**

In[16]:= **{heads1[a + b], heads1[a b]}**

Out[16]= $\{(a+b)^2,$ If[Times == Plus, $(a\,b)^2,$ $(a\,b)^3]\}$

This works fine for expressions whose head is **Plus**, but for anything else, **Head[exp] == Plus** is unevaluated so the whole expression is returned. Presumably this is unintended, but it can be cured by using **===** instead of **==**.

In[17]:= **heads2[exp_]** := **If[Head[exp] === Plus, exp^2, exp^3]**

In[18]:= **{heads2[a + b], heads2[a b]}**

Out[18]= $\{(a+b)^2,$ $a^3\,b^3\}$

2.2.2 Which

Which[test$_1$, expr$_1$, test$_2$, expr$_2$, . . .] is just like the COND operation in LISP. It has an even number of arguments. Each odd-numbered argument expects a predicate. If **test$_i$** is the first predicate to evaluate to **True**, then **expr$_i$** is evaluated. For instance:

In[19]:= **Which[4 < 1, 1, 4 < 2, 2, 4 < 3, 3, 4 < 4, 4, 4 < 5, 5]**

Out[19]= 5

If no predicate evaluates to **True**, then the output is **Null**; i.e., there is no output.

In[20]:= **Which[4 < 1, 1, 4 < 2, 2, 4 < 3, 3, 4 < 4, 4]**

A function definition of the form

$$f[x\_] := \text{Which}[test_1, expr_1, test_2, expr_2, . . .]$$

where **test$_i$**, and **expr$_i$** involve **x**, for $1 \le i \le n$, divides the universe of *Mathematica* expressions into $n + 1$ disjoint subsets, where the **i**th subset consists of those expressions **exp** for which **test$_i$ /. x -> exp** is the first test which evaluates to **True**, in which case **expr$_i$ /. x -> exp** is evaluated. The $(n + 1)$th subset consists of those expressions for which no test evaluates to **True**, in which case the result is **Null**. Here is an example where there are three tests with their corresponding expressions.

In[21]:= **heads3[exp_]** :=
          Which[Head[exp] === Plus,   exp^2,
                 Head[exp] === Times,   exp^3,
                 Head[exp] === Power,   exp^4]

In[22]:= { heads3 [a + b], heads3 [a b], heads3 [a^b], heads3 [a && b] }

Out[22]= { (a + b)^2, a^3 b^3, a$^{4\,b}$, Null }

This is essentially the same as a list of conditional rewrite rules, except for the behavior on terms that fail to satisfy any of the conditions. For example.,

In[23]:= **heads4 [exp_] := exp^2 /; Head[exp] === Plus;**
heads4 [exp_] := exp^3 /; Head[exp] === Times;
heads4 [exp_] := exp^4 /; Head[exp] === Power;

In[24]:= { **heads4 [a + b], heads4 [a b], heads4 [a^b], heads4 [a && b] }**

Out[24]= { (a + b)^2, a^3 b^3, a$^{4\,b}$, heads4 [a && b] }

There is a possible difference in that the rewrite rules for heads may be reordered by *Mathematica*, which could change the output. Except for this possibility, a **Which** command with a final predicate **True** is the same as a list of conditional rewrite rules in which the last rule is unconditional. The *Mathematica* Book suggests that rules are more appropriate for *Mathematica* style programming.

2.2.3 Switch

Switch makes explicit use of *Mathematica* pattern matching. It is not really like anything else in other languages.

$$\texttt{Switch[expr, form}_1\texttt{, value}_1\texttt{, form}_2\texttt{, value}_2\texttt{,...]}$$

tries to match **expr** to one of the forms. It returns the value of the first one that it matches. Thus we can write:

In[25]:= **heads5 [exp_] := Switch[Head[exp],**
Plus, exp^2,
Times, exp^3,
Power, exp^4]

In[26]:= { **heads5 [a + b], heads5 [a b], heads5 [a^b], heads5 [a && b] }**

Out[26]= { (a + b)^2, a^3 b^3, a$^{4\,b}$, Switch[And,
Plus, (a && b)^2,
Times, (a && b)^3,
Power, (a && b)^4] }

As one sees from this example, **Switch** returns the entire **Switch** expression unevaluated if the expression fails to match any of the forms, so it is a good idea to include a final pair whose form is _, with some neutral value, e.g., **Null**.

```
In[27]:= heads6[exp_] := Switch[Head[exp],
                    Plus,   exp^2,
                    Times,  exp^3,
                    Power,  exp^4,
                    _,      Null]
```

```
In[28]:= {heads6[a + b],  heads6[a b],  heads6[a^b],  heads6[a && b]}
```

$$Out[28]= \{(a + b)^2, \ a^3 \, b^3, \ a^{4\,b}, \ Null\}$$

The forms don't have to be constants. They can be general patterns, so here is yet another way to write our operation.

```
In[29]:= heads7[exp_] := Switch[exp,
                    _Plus,   exp^2,
                    _Times,  exp^3,
                    _Power,  exp^4,
                    _,       Null]
```

```
In[30]:= {heads7[a + b],  heads7[a b],  heads7[a^b],  heads7[a && b]}
```

$$Out[30]= \{(a + b)^2, \ a^3 \, b^3, \ a^{4\,b}, \ Null\}$$

2.3 Loops

In the language of while-programs, the command that repeats an operation until some condition is satisfied is the **while_do_** command. In *Mathematica*, there are three built-in looping constructions, allowing a great variety of programming styles.

2.3.1 Do loops

The simplest loop construct is the **Do** loop. It is a function of two arguments consisting of an expression and an iterator of the form **Do[expr, {i, imin, imax, istep}]**. Notice that the form is exactly the same as that of the operations **Table, Sum, Product, Integrate,** etc. However, in contrast to these operations, **Do** has no output. The reason is that **Do** loops are used only for their side effects–changing the state, printing something, generating graphics, etc. A **Do** command evaluates **expr** a total of $((\text{imax} - \text{imin}) / \text{istep}) + 1$ times with the values   imin, imin + istep, imin + 2 istep, . . . imax successively substituted for **i** in **expr**. As usual, the "iterator" has abbreviated forms:

```
{i, imin, imax} = {i, imin, imax, 1}
{i, imax} = {i, 1, imax}
```
{imax}, if **expr** does not depend on **i**.

Note that if **imin** > **imax** and **istep** is negative, then the loop goes backwards. Here is a simple example. In order to see something happen, **expr** is a **Print** statement here, which as a side effect prints the values of its argument.

In[31]:= **Do[Print[i^2], {i, 3, 5}]**

```
9
16
25
```

If **expr** assigns a value to **expr$_1$**, then **expr$_1$** has the value it is given by the last repetition of the loop. Since **Do** itself does not return any value, in order to see the result, one has to ask for it explicitly. A typical construction might start with an "initialization" statement for some identifier, followed by a **Do** loop which does something to the initialized identifier, followed by calling the identifier itself. For example:

In[32]:= **y = 1; Do[y = (y + i)^2, {i, 5}]; y**

Out[32]= 5408554896900

Funny things are allowed because the only actual restriction on an iterator is that $((\text{imax} - \text{imin}) / \text{istep})$ has to be a number. The "variable of iteration", **i**, can be any expression. Thus, the following is legitimate.

In[33]:= **z = 1; Do[z = (z f[w])^2, {f[w], 3.2 ra, 6 ra, ra}]; z**

Out[33]= $9.25132 \times 10^7 \, ra^{14}$

Except for the funny things, **Do** is very much like the **For** loop operation in Pascal. Related operations are **Nest** and **Fold** and **FixedPoint[f, expr]** (which have been discussed earlier). Note that **y** and **z** now have values that have to be cleared.

In[34]:= **Clear[y, z]**

2.3.2 While loops

While[test, expr] is just like the command "**while** test **do** expr" in the language of while-programs. The **While** expression in *Mathematica* begins by evaluating **test**. If **test** is True, then it evaluates **expr**. Usually, **expr** includes a clause changing some parameter in **test**. Then **test** is reevaluated with the new value of the parameter. If it still evaluates to True, then **expr** is evaluated again. This continues until **test** evaluates to False. No value is returned by the **While** operation, but if **expr** assigns a value to **expr1**, then **expr1** has the value it had just before **test** evaluated to False. The use of **While** loops is the same as that of **Do** loops; e.g.,

In[35]:= **x = 1; While[x < 10, x = x + 1; y = x^2]; y**

Out[35]= 100

The last time the condition **x < 10** is evaluated with result True is when **x** is 9. In that case, the expression sets **x** to 10 and then gives **y** the value $10^2 = 100$. Note that **x** and **y** again have values that have to be cleared.

In[36]:= **Clear[x, y]**

2.3.3 For loops

For[start, test, step, expr] is almost exactly the same as a **for** loop in the language C, except that in C the clauses are separated by semicolons instead of commas. (Note that C uses commas for compound statements, so the roles of commas and semicolons in C are exactly the opposites of their roles in *Mathematica*.) A **For** loop first evaluates **start** and then repeatedly evaluates **expr** and **step** until **test** fails. They are evaluated in this order: **start, test, expr, step,** and then **test** again. Usually **start** initializes some variable and **step** causes it to change in some way that **test** will use to cause the **For** loop to eventually stop. As with **Do** and **While** loops, the output of a **For** loop is Null, so in the example we use a **Print** statement in order to see something.

In[37]:=**For[i = 1, i < 4, i++, Print[i]]**

> 1
> 2
> 3

We have used C slang in writting **step**. Here i++ is shorthand for i = i + 1. One can also write i += 1 with the same effect. Similarly, i-- is short hand for i -= 1 or i = i - 1.

Notice that the last value printed is 3. We can check that **test** was evaluated one more time to make **test** fail by asking for the value of i.

In[38]:=**i**

Out[38]= 4

This result also points up the unfortunate fact that evaluating this **For** loop has had the unintended "side effect" of giving a value to i, which we probably didn't want. We have to clear it to check the internal forms of i++ and i--.

In[39]:= **Clear[i]**

In[40]:= **{FullForm[i++], FullForm[i--]}**

Out[40]= {Increment[i], Decrement[i]}

Here is a more complicated example showing that **start** can initialize several variables in a compound statement and that **expr** can of course also be a compound expression.

In[41]:= **For[i = 1; t = x,**
** i^2 < 10,**
**   i++,**
**   t = t^2 + i; Print[Expand[t]]]; Clear[i]**

> $1 + x^2$
> $3 + 2 x^2 + x^4$
> $12 + 12 x^2 + 10 x^4 + 4 x^6 + x^8$

Note that outputs in the form of **Print** statements, which we have been forced to resort to in order to see something from loop statements, are generally not very useful since they are not available for further processing.

3 Modules and Blocks

3.1 Modules

In the first example of a compound operation above, after we had finished the calculation, the variable **x** ended up having the value 4. If all we cared about was the computation, then it would be unfortunate to give a value to **x** which might interfere with later computations. The solution to this is a mechanism that allows variables to be used just for one calculation and then erases any values they might have acquired during that computation. **Module**s and **Block**s are mechanisms that create such "local variables". Here is an example.

In[1]:=**Module[{x}, x = 1; x = x + 1; x = x + 1; x = x + 1]**

Out[1]= 4

This is the same value as before, but when we are done, **x** doesn't have any value.

In[2]:=**ValueQ[x]**

Out[2]= False

Furthermore, if **x** is given a value before starting; e.g., **x = 17**, and then the **Module** is evaluated, then **x** still has its original value. Thus, the **x** inside the module is independent of the **x** outside.

A **Module** expression takes two arguments, the first being a list of local variables and the second being any expression (usually it is a compound expression). If desired, initial values for local variables can be given within the first argument. The value of a **Module** expression is the value of the second argument; thus, when it is a compound expression, the value is the value of the last component of the compound expression.

In[3]:=**Module[{x = 1}, x = x + 1; x = x + 1; x = x + 1]**

Out[3]= 4

The local variables are just named, and initializations are given as above; e.g., **x = 1**. The body of the **Module** is separated from the list of variables by a comma. It is a single, possibly compound expression. Note that the semicolons in it bind more tightly than the comma, in distinction to most ordinary natural languages.

Modules are usually not used when working interactively. It is only when it is time to put some procedure into a more final form that they come into play. There are three reasons to use **Module**s: preventing variable clash, efficiency, and clarity. Preventing variable clash means insulating the local variables from any other variables with the same names outside the **Module**. This is highly desirable since there is no way to know in advance what other variables with values may be around when a particular function is used. As to efficiency, local variables serve to hold values of computations that may be required at several points in some procedure, so that they only have to be calculated once. Finally, the third use is to give names to intermediate steps in a computation for purposes of clarity. Look at the examples later in this chapter and determine which local variables are being used for iteration in some way, and so just have to be protected from the outside, which ones are used to store information that is used more than once, and which ones are there just for clarity.

3.2 *Blocks versus Modules*

There are two possible ways in which local variables can be insulated from global ones. **Block**s are just like **Module**s except in the way that they handle name clashes. Consider the following expression written with a **Block** statement.

In[4]:= **sumOfPowersB[x_]** := **Block**$\left[\{i\}, \text{Sum}\left[x^i, \{i, 5\}\right]\right]$

Try it on two examples.

In[5]:= **{ sumOfPowersB[a], sumOfPowersB[i] }**

Out[5]= $\{a + a^2 + a^3 + a^4 + a^5, 3413\}$

Now write the same function using a **Module** statement and try the same two examples.

In[6]:= **sumOfPowersM[x_]** := **Module**$\left[\{i\}, \text{Sum}\left[x^i, \{i, 5\}\right]\right]$

In[7]:= **{ sumOfPowersM[a], sumOfPowersM[i] }**

Out[7]= $\{a + a^2 + a^3 + a^4 + a^5, i + i^2 + i^3 + i^4 + i^5\}$

The difference between these two outcomes is the difference between *dynamic scoping* and *static scoping* of local variables. It is explained very well in The *Mathematica* Book. In the case of **Block**s, local variables have unique values but not unique names. When we ask for **sumOf-PowersB[i]**, what happens is that we get Sum[i^i, {i, 1, 5}], which is a number. The trouble is that the **i** from outside the **Block** is the same as the **i** inside it, so the scope of the **i** inside expands dynamically to the outside of the **Block**. In the case of **Module**s, local variables have unique values *and* unique names so that such a name clash is essentially impossible. The way this is done is to use new names for the local variables in **Module** every time the **Module** is used. The name given to a local variable in a **Module** is not actually used. It is replaced by a distinct name that does not occur anywhere else. Normally these new names are completely hidden so one never knows exactly what they are, but sometimes they accidentally (or deliberately) get out of the **Module**, as in the following example.

In[8]:=`Table[Module[{j}, j], {5}]`

Out[8]= `{j$5, j$6, j$7, j$8, j$9}`

Thus, **j** is replaced by `j$n` where n is an increasing sequence of numbers. The actual numbers depend on everything that has gone before; specifically on all local variables in all **Modules** that have been used in the current session. The numbers start with 1 and increase by 1 every time a local variable is used in a **Module**.

In[9]:=`Table[Module[{r}, r], {5}]`

Out[9]= `{r$10, r$11, r$12, r$13, r$14}`

Local variables can also be seen in **Trace** commands.

In[10]:=`Trace[Module[{t}, t = 3]]`

Out[10]= `{Module[{t}, t = 3], {t$15 = 3, 3}, 3}`

As long as variable names of the form **symbol$n** are never used, there is no possibility of name conflict.

3.3 Modules versus With

As remarked above, one use of **Modules** is just to give a name to some computation which will be used several times in a further expression. If this computation is used to initialize the name in the first argument of the **Module** and nothing is assigned to it in the body of the **Module**, then the **Module** command can be replaced by a **With** expression. See Chapter 6, Section 3.4.

4 Examples

We start with some simple examples and then turn to some more complicated ones showing how to translate programs in Pascal and C into *Mathematica* programs. In each case the direct translation can be replaced by a much shorter and clearer *Mathematica* program written in a functional or rewrite rule style.

4.1 A Procedural Factorial Function

As the first example, we write the while-program for the factorial function given in the introduction to this chapter in *Mathematica*. Notice that very little is changed.

```
In[1]:= factorialProc[y_] :=
        Module[{x = 0, z = 1},
            If[y == 0,
                1,
                While[!(x == y),
                    x = x + 1;
                    z = z x]];
        z
    ]
```

The main purpose served by the **Module** structure here is to prevent global values being given to **x** and **z**. The program works without being put inside a **Module** but then it would have the unfortunate side effect of giving **x** the value of **y** and **z** the value **y**!. Try this version on a pair of values.

```
In[2]:= {Timing[factorialProc[252];], Timing[factorialProc[1000];]}
```

```
Out[2]= {{0.0333333 Second, Null}, {0.166667 Second, Null}}
```

The timing for 252 is approximately the same as for the recursive version in Chapter 7, Section 6.3. For larger values there is no need to reset **$RecursionLimit** since no recursion is involved in this form of the function. (Here, what is **factorialProc[yesterday]** ? Explain the result.)

4.2 Continued Fractions

Any real number has finite continued fraction approximations. These are given as follows:

```
In[3]:= continuedFractionApprox[x_Real, n_Integer? Positive] :=
    Module[{integerPart, fractionPart = x, result = {}},
        Do[integerPart = Floor[fractionPart];
            AppendTo[result, integerPart];
                                            1
            fractionPart = ─────────────────────────────, {n}];
                            fractionPart - integerPart
        result]
```

In[4]:=`continuedFractionApprox[ N[π], 10 ]`

Out[4]= `{3, 7, 15, 1, 292, 1, 1, 1, 2, 1}`

The following functional one-liner will display a symbolic continued fraction, given the list of coefficients. Note that this is different from the form in Exercise 3 of Chapter 6.

In[5]:= `continuedFract[list_List]  :=`

$$\text{Fold}\left[\left(\#2 + \frac{1}{\#1} \ \&\right),\right.$$

$$\left.\text{First}[\text{Reverse}[\text{list}]], \ \text{Rest}[\text{Reverse}[\text{list}]] \ \right]$$

In[6]:=`continuedFract[{a, b, c, d}]`

Out[6]= $a + \dfrac{1}{b + \frac{1}{c+\frac{1}{d}}}$

To see the continued fraction approximation to π, we have to turn numbers into strings to prevent *Mathematica* from evaluating the continued fraction. We use functional programming again.

In[7]:=`continuedFractionPi =`
   `continuedFract[`
     `Map[ToString, continuedFractionApprox[ N[π], 10 ]]]`

Out[7]= $3 + \dfrac{1}{15+\dfrac{1}{1+\frac{1}{1+\frac{1}{1+\frac{1}{1+\frac{1}{1}+2}}}+292}+7}$

Unfortunately, *Mathematica* insists on writing some of the sums in the wrong order. Finally, this can be evaluated by using a functional program to turn the strings back into expressions.

In[8]:=`MapAt[ToExpression, continuedFractionPi,`
         `Position[continuedFractionPi, _String]]`

Out[8]= $\dfrac{1146408}{364913}$

In[9]:= `{N[%, 20], N[π, 20]}`

Out[9]= `{3.1415926535914039785, 3.1415926535897932385}`

4.3 A Procedural Program for Simple Differentiation

In Chapter 7, Sections 4.1.2 and 5.2.2, we wrote rule based programs for simple differentiation. It is much harder to write a procedural program for this. The problem is that if we don't use the pattern matching facilities of *Mathematica*, then we have to recognize the input expression by analyzing its structure directly; i.e., we have to construct our own parser. This is most easily organized in a **Which** statement rather than nested **If** statements.

```
In[10]:= diffw[y_, x_] :=
        Which[y === x, 1,
              Length[y] === 2 && y[[0]] === Power && y[[1]] === x,
                n = y[[2]]; n x^(n-1),
              y === log[x], 1/x ]
```

Try this out on some examples.

```
In[11]:= {diffw[x^3, x], diffw[y, y],
         diffw[log[z], z], diffw[w^(1/3), w], diffw[r^3.1, r]}
```

$$\text{Out[11]= } \left\{3\,x^2,\ 1,\ \frac{1}{z},\ \frac{1}{3\,w^{2/3}},\ 3.1\,r^{2.1}\right\}$$

Of course, if we use **Switch**, then we get a noticeably simpler program, because pattern matching is used.

```
In[12]:= diffs[y_, x_] :=
        Switch[y, x,          1,
                  x^n_,       y[[2]] x^(y[[2]]-1),
                  log[x],     1/x ]
```

Try this out on some examples.

```
In[13]:= {diffs[x^3, x], diffs[y, y], diffs[log[z], z],
         diffs[w^(1/3), w], diffs[r^3.1, r]}
```

$$\text{Out[13]= } \left\{3\,x^2,\ 1,\ \frac{1}{z},\ \frac{1}{3\,w^{2/3}},\ 3.1\,r^{2.1}\right\}$$

4.4 Runge–Kutta methods

Runge–Kutta methods are a technique for solving systems of first order ordinary differential equations of the form

$$x_1' = f_1(x_1, \ldots x_n)$$
$$- - - - - - - - - -$$
$$x_n' = f_n(x_1, \ldots x_n).$$

Here, prime means differentiation with respect to some independent variable *t* which does not occur explicitly on the right-hand sides of the equations. The built-in operation **NDSolve** finds solutions for more general systems of equations. The program to implement the Runge–Kutta method for finding approximate numerical solutions of such systems is similar to the program for Newton's method in Chapter 7. Starting from some list of initial values, there is a one-step move in the direction of an approximate solution. This new location is the initial point for another one-step move, etc. The fourth-order Runge–Kutta method utilizes the following one-step operation.

```
In[14]:= oneRungeKuttaStep[exprs_, vars_, vars0_, dt_] :=
    Module[{ k1, k2, k3, k4 },
        k1 = dt N[ exprs /. Thread[vars -> vars0] ];
        k2 = dt N[ exprs /. Thread[vars -> vars0 + k1/2] ];
        k3 = dt N[ exprs /. Thread[vars -> vars0 + k2/2] ];
        k4 = dt N[ exprs /. Thread[vars -> vars0 + k3] ];
        vars0 + (k1 + 2 k2 + 2 k3 + k4)/6
    ]
```

Here **exprs** is the list of right-hand sides of the system of equations and **dt** is the step size. The purpose of the **Module** structure here is to protect the local variables **k1**, **k2**, **k3**, and **k4**. They in turn serve to store intermediate results. One could substitute their values in the last line, starting with **k4**, and then both instances of **k3**, etc., to derive a purely functional operation, but that would require **exprs** to be evaluated 10 times instead of 4. This one-step operation is like the one in Newton's method, but it doesn't make sense to then use **FixedPoint** in the final operation since in general the solution will not converge to a fixed value. Instead, we use **NestList** to calculate the list of successive positions of the system. Here **n** is the number of steps to be carried out.

```
In[15]:= rungeKutta[exprs_, vars_, vars0_List, dt_, n_] :=
    NestList[oneRungeKuttaStep[exprs, vars, #, N[dt]]&,
             N[vars0], n ]
```

Here are some examples. (See also the package **ProgrammingExamples`RungeKutta** and [Maeder].)

4.4.1 Van der Pol's equation

Van der Pol's equation arises from the second order differential equation $x'' + x = \epsilon (1 - x^2) x'$ by converting it to a linear system of the form $x' = \text{xdot}$, $\text{xdot}' = \epsilon (1 - x^2) \text{xdot} - x$. Finding numerical solutions of this equation was an important research goal during the Second World War. We treat it for the value $\epsilon = 1$.

```
In[16]:= system1 = {xdot, (1 - x^2) xdot - x};
```

It is known that there is a closed solution through the point $\{2, 0\}$ and all other solutions approach it. We find three trajectories of this system starting at the points $\{0, 0.6\}$, $\{0, 2.2\}$, and $\{0, 3.6\}$. Note that all solutions move clockwise.

```
In[17]:= Show[Table[
            ListPlot[rungeKutta[system1, {x, xdot},
                {0, i}, 0.1, 70],
                PlotJoined -> True, AspectRatio -> Automatic,
                DisplayFunction -> Identity],
            {i, 0.6, 3.8, 1.6}],
        DisplayFunction -> $DisplayFunction];
```

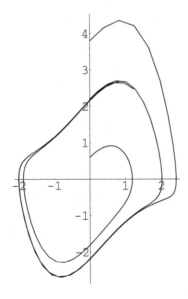

4.4.2 Gravitational attraction

We can compare the Runge–Kutta method with the built-in function **NDSolve**, using the example of two equal bodies under graviational attraction described in Chapter 3. We have to turn the system there into a system of first-order differential equations as usual. The four second-order equations become eight first-order equations.

In[18]:= **twoOrbitSystem =**

$$\left\{ \text{xdot1}, \quad \frac{-(x1 - x2)}{\left((x1 - x2)^2 + (y1 - y2)^2\right)^{\frac{3}{2}}}, \right.$$

$$\text{ydot1}, \quad \frac{-(y1 - y2)}{\left((x1 - x2)^2 + (y1 - y2)^2\right)^{\frac{3}{2}}},$$

$$\text{xdot2}, \quad \frac{-(x2 - x1)}{\left((x1 - x2)^2 + (y1 - y2)^2\right)^{\frac{3}{2}}},$$

$$\left. \text{ydot2}, \quad \frac{-(y2 - y1)}{\left((x1 - x2)^2 + (y1 - y2)^2\right)^{\frac{3}{2}}} \right\};$$

Eight function names and eight initial conditions are required to get a solution.

In[19]:= **twoOrbitSolution = rungeKutta[twoOrbitSystem,**
          {x1, xdot1, y1, ydot1, x2, xdot2, y2, ydot2},
          {1, 0, 0, 0.3, -1, 0, 0, -0.3}, 0.1, 60];

We just want to plot the two curves given by **{x1, y1}**, and **{x2, y2}**, so we have to extract their values from the list of eight values for each entry in the output of **NestList**.

In[20]:= **Show[{ListPlot[Map[{#[[1]], #[[3]]}&, twoOrbitSolution],**
          PlotJoined -> True, AspectRatio -> Automatic,
          PlotRange -> All, DisplayFunction -> Identity],
        ListPlot[Map[{#[[5]], #[[7]]}&, twoOrbitSolution],
          PlotJoined -> True, AspectRatio -> Automatic,
          PlotRange -> All, DisplayFunction ->Identity]},
        DisplayFunction -> $DisplayFunction];

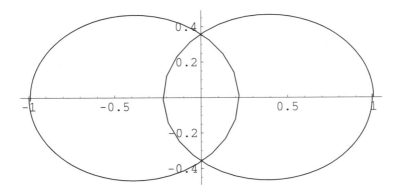

4.5 A Program from Oh! Pascal!

Oh! Pascal! by Doug Cooper and Michael Clancey, W. W. Norton and Co., 1982 [Cooper], is a standard book on Pascal programming. Here is the *Mathematica* translation of the program found on page 156 of that book. It concerns robbers who steal a number of gold bars. Secretly during the night, each one takes one-third for himself, each time leaving one left over, which he takes. In the morning, they divide what is left and find one still left over. The question is to find the number of bars. There are many solutions, so only the solutions less than or equal to 500 are given by this program. Here is the Pascal program:

```
program StolenGold(output)
var TrialNumber, DividedNumber: integer;
begin
    for TrialNumber:= 1 to 500
        do If (TrialNumber mod 3) = 1
            then begin
                DividedNumber:= 2*(TrialNumber div 3);
                If (DividedNumber mod 3) = 1
                then begin
                    DividedNumber:= 2*(DividedNumber div 3);
                    If (DividedNumber mod 3) = 1
                        then begin
                            DividedNumber:= 2*(DividedNumber div 3);
                            If (DividedNumber mod 3) = 1
                            then writeln(TrialNumber:3, 'is a solution.')
                                end
                    end
                end
end
```

This program can be recreated in *Mathematica*, almost word for word using a **Do** loop. The syntax of a **Do** loop is almost exactly the same as that of a **For** loop in Pascal, except the arguments are given in the reverse order. Functions like **If**, **Mod**, and **Quotient** are written in prefix form rather than infix or mixfix form as in Pascal. Finally, instead of writeln, we have to use **Print**. Perhaps the most noticable difference is that the **program** and **var** statements at the beginning of the Pascal program are replaced by the **Module** head and the local variable declarations in the **Module**.

```
In[21]:= Module[{trialNumber, dividedNumber},
     Do[
      If[Mod[trialNumber, 3] == 1,
         dividedNumber = 2 Quotient[trialNumber, 3];
         If[Mod[dividedNumber, 3] == 1,
           dividedNumber = 2 Quotient[dividedNumber, 3];
           If[Mod[dividedNumber, 3] == 1,
             dividedNumber = 2 Quotient[dividedNumber, 3];
             If[Mod[dividedNumber, 3] == 1,
               Print[trialNumber, " is a solution."]
        ]]]], {trialNumber, 1, 500}]]

79 is a solution.
160 is a solution.
241 is a solution.
322 is a solution.
403 is a solution.
484 is a solution.
```

Here is a strict one-liner doing the same thing. Note that it returns the values as an output list, available for further processing.

```
In[22]:= Select[Range[500],
     And @@ Map[(# == 1 &),
                Mod[NestList[2 Quotient[#, 3] &, #, 3], 3]] &]

Out[22]= {79, 160, 241, 322, 403, 484}
```

It is interesting that *Mathematica* is able to sort out the different #'s that occur in this function. The rightmost one is the one that gets filled by the entries from the list **Range[500]**. The next one to the left belongs to the pure function in the argument to **NestList**, while the leftmost one belongs to the predicate that is mapped down the resulting list. For instance, try the operations one at a time for the number 79:

In[23]:= **NestList[2 Quotient[#, 3] &, 79, 3]**

Out[23]= {79, 52, 34, 22}

Each of these numbers equals 1 modulo 3. (**Mod** is **Listable**.)

In[24]:= **Mod[NestList[2 Quotient[#, 3] &, 79, 3], 3]**

Out[24]= {1, 1, 1, 1}

Let *Mathematica* do the check that they are all 1's.

In[25]:= **Map[(# == 1 &), Mod[NestList[2 Quotient[#, 3] &, 79, 3], 3]]**

Out[25]= {True, True, True, True}

Get a single value True as the output by **And**ing together these values.

In[26]:= **And@@Map[(# == 1 &), Mod[NestList[2 Quotient[#, 3] &, 79, 3], 3]]**

Out[26]= True

Thus, the basic ingredient of our one-liner gives True for 79. Now try the numbers between 70 and 80.

In[27]:= **Map[Mod[NestList[2 Quotient[#, 3] &, #, 3], 3] &,**
        Range[70, 80]]

Out[27]= {{1, 1, 0, 2}, {2, 1, 0, 2}, {0, 0, 2, 2},
        {1, 0, 2, 2}, {2, 0, 2, 2}, {0, 2, 2, 2}, {1, 2, 2, 2},
        {2, 2, 2, 2}, {0, 1, 1, 1}, {1, 1, 1, 1}, {2, 1, 1, 1}}

In[28]:= **Map[And@@Map[(# == 1 &),**
        Mod[NestList[2 Quotient[#, 3] &, #, 3], 3]] &,
        Range[70, 80]]

Out[28]= {False, False, False, False, False,
        False, False, False, False, True, False}

So, 79 is the only number in this range for which the predicate returns the value True. Notice how **NestList** is used to help create the program itself. That is one aspect of what is meant by a higher order programming language. Of course, there is a much easier way to generate this series of numbers along with some bigger entries. We leave it to the reader to figure out why it works.

$$\text{In[29]:= } \mathbf{Table}\left[\frac{81\ (8\ k - 1) + 65}{8},\ \{\mathbf{k,\ 1,\ 10}\}\right]$$

Out[29]= {79, 160, 241, 322, 403, 484, 565, 646, 727, 808}

4.6 A Simple C Program

Our next example is a C program that prints out an interest table.

```
/*
 *  Generates a table showing interest accumulation. Allows
 * the user to input the interest rate, principal, and period.
 */
main()
{
        int     period,         /* length of period */
                year;           /* year of period   */
        float   irate           /* interest rate    */
                sum;            /* total amount     */

 printf ("Enter interest rate, principal, and period: ");
 if (scanf ("%f %f %d", &irate, &sum, &period) == 3)
       {
        printf ("Year\t  Total at %.2f%%\n\n", irate * 100.0);
        for (year = 0; year <= period;year++)
          {
            printf ("%5d\t   $ %10.2f\n", year, sum);
            sum += sum * irate;
          }
       }
    else
        printf ("Error in input. No table printed.\n");
}
```

This C program can be closely approximated in style and format by a *Mathematica* program.

```
In[30]:=Module[
        {irate, sum, per, year, scan},
        If[Length[
          scan =
            Input[
              "{interestRate?, principal?, period?} "]] == 3,
          {irate, sum, per} = scan;
          Print["Year,  Total at ", irate*100, "%"];
          For[year = 0, year <= per, year++,
              Print[PaddedForm[year, 2]," "        ", sum];
```

```
            sum += sum * irate],
        Print["Error in input. No table printed."]
    ]]
```

```
Year,   Total at 10.%
  0       10000
  1       11000.
  2       12100.
  3       13310.
  4       14641.
  5       16105.1
  6       17715.6
  7       19487.2
```

When this **Module** is evaluated, the first thing that happens is that the **Input** expression is evaluated. This asks the user for a list of three numbers. If a notebook interface is being used, then a dialogue box will appear asking for the input. Otherwise, a prompt will appear asking for it. The input given here was {0.1, 10000, 7}.}. Then the column headings are printed and a **For** loop is entered, printing out the values for each year, one at a time. If something other than a list of length three is entered, then an error message is printed. Notice how the **scan** statement is used inside the predicate **Length[scan = ...] == 3**, mirroring the way the scan statement is used in the C program. Two things are accomplished this way. There is a check if the input at least consists of three items, with an error message if it doesn't. The main purpose of setting the identifier **scan** to the list of three values is accomplished as a side effect. Every step in this program is a side effect.

Another way to handle the input statement, instead of typing the required information at the prompt, is to prepare a file containing the desired information using the form **Put[expr, "file"]**, or equivalently for a single expression, **expr >> file**.

In[31]:= **Put[{0.1, 10000, 7}, "test1"]**

Then, when the program asks for the information, respond with

In[32]:= **Get["test1"]**

Out[32]= {0.1, 10000, 7}

This could be written as a function taking a file name as its only argument.

In[33]:= **interestTable[file_String] :=**
        Module[
            {irate, sum, per, year, scan},
            If[Length[scan = Get[file]] == 3,
                {irate, sum, per} = scan;

```
            Print["Year,   Total at ", irate*100,  "%"];
            For[year = 0, year <= per, year++,
                Print[PaddedForm[year, 2]," ", sum];
                sum += sum * irate],
            Print["Error in input. No table printed."]
    ]]
```

In[34]:= `interestTable["test1"]`

```
  Year,    Total at 10.%
    0        10000
    1        11000.
    2        12100.
    3        13310.
    4        14641.
    5        16105.1
    6        17715.6
    7        19487.2
```

The most important differences between this program and the C program are that it is not necessary to declare a type for each local variable and that explicit directions don't have to be given for reading input and printing messages. Such a program would normally be written in *Mathematica* as a function whose arguments are **interestRate**, **principal**, and **period**. There is no need to include the **If** statement since if the wrong number of arguments are given, then *Mathematica* simply leaves the expression unevaluated.

In[35]:= `accumulation[interestRate_, principal_, period_] :=`

```
        Module[
            {sum = principal, year},
            Print["Year,   Total at ", interestRate*100,  "%"];
            For[year = 0, year <= period, year++,
                Print[PaddedForm[year, 2]," ", sum];
                sum += sum * interestRate]]
```

In[36]:= `accumulation[.10,  10000,  7];`

The **Print** statements are the same as before.

Here is a version even more in the spirit of *Mathematica* programming. Note that no local variables are required in the single **NestList** operation. The computation itself is simplified to the extent that most of the function consists of explicit directions for displaying the table and getting correct column headings as an output rather than a **Print** statement.

In[37]:= `accumulate1[interestRate_, principal_, period_] :=`

```
        PaddedForm[
```

```
        TableForm[
            NestList[{#[[1]] + 1, #[[2]] (1 + interestRate)}&,
                {0, principal}, period],
            TableHeadings ->
                {None,
                    {"    Year",
                     "Total at "<>ToString[100interestRate]<>"%"}},
            TableSpacing -> {0, 2}],
        5]
```

In[38]:=`accumulate1[0.1, 10000, 10]`

Out[38]//PaddedForm=

```
    Year     Total at 10.%
      0       10000
      1       11000.
      2       12100.
      3       13310.
      4       14641.
      5       16105.
      6       17716.
      7       19487.
      8       21436.
      9       23579.
     10       25937.
```

All numbers had to be padded to size five to get the numbers in the Year column to line up nicely, but not lose digits in the Total column. . (That's what the **PaddedForm** is about.) The column heading Total at 10.% had to be carefully constructed using the infix form **<>** of **StringJoin**. Notice that the final information is in output form and so is available for further processing. For example:

In[39]:=`%[[3]]`

Out[39]= `{2, 12100.}`

4.7 A C Program for a Histogram

4.7.1 The C program
Lastly, we consider a longer C program from the book *Programming in C*, by Lawrence H. Miller and Alexander E. Quilici, John Wiley & Sons, Inc. 1986 [Miller]. A histogram is a kind of a bar chart for displaying data. The data is separated into "buckets" of equal sizes according to the values of the data and then the number of data items in each bucket is plotted. The program below is divided into a main loop followed by the definitions of the two functions, fill_bkts and print_histo, which constitute the principal ingredients of the main loop. Thus, the main program is of the form

```
main()
{ - - -
     if  (fill_bkts - - )
     /*then*/
               print_histo
else
               error message
}
```

The procedure `fill_bkts` is a compound expression which first does all the work of putting
the data items into the correct buckets and then ends with a predicate asking if the variable
`inpress,` which is storing the read-in values, now has the value EOF ("EndOfFile"). Thus, the
predicate part of the if statement, as a side effect, does all of the real work of the command.
Assuming that all of the data has been read in, then the `print_histo` procedure is carried
out, which makes a picture of the histogram. Otherwise, an error message is printed.

```
/*
 *      Produce nice histogram from input values.
 */
#include <stdio.h>

#define MAXCOLS      50    /* columns available for markers */
#define MARKER       '*'   /* character marking columns */
#define MAXVAL       100   /* largest legal input value */
#define MINVAL       0     /* smallest legal input value */
#define NUMBKTS      11    /* number of buckets */
main()
 {
    int buckets[NUMBKTS],  /* buckets to place values in  */
        bktsize;           /* range bucket represents    */

    bktsize = (MAXVAL - MINVAL) / (NUMBKTS - 1);
    if (fill_bkts(buckets, bktsize))
       print_histo(buckets, bktsize);
    else
       printf("Illegal data value--no histogram printed\n");
 }
 /*
  * Read values, updating bucket counts, Returns nonzero
  * only if EOF was reached without error.
  */
 int fill_bkts(buckets, bktsize)
 int buckets[],         /* buckets to place values in */
     bktsize;           /* range of values in bucket */
  {
```

```
    int badcnt = 0,      /* count of out-or-range values */
        bkt,             /* next bucket to initialize */
        inpres,          /* result of reading input line */
        totcnt = 0,      /* count of values */
        value;           /* next input value */

    for (bkt = 0; bkt <= NUMBKTS; buckets[bkt++] = 0)
        ;                /* initialize bucket counts */
    while (inpres = scanf("%d", &value), inpres == 1)
    {
        if (value >= MINVAL && value <= MAXVAL)
            buckets[(value - MINVAL) / bktsize]++;
        else
            badcnt++;
        totcnt++;
    }
    if (!badcn)
        printf("All %d values in range\n", totcnt);
    else
        printf("Out of range %d, total %d\n", badcnt,totcnt);
    return inpres == EOF; /* did we get all the input? */
}
/*
 * Print a nice histogram, first computing a scaling factor
 */
print_histo(buckets, bktsize)
int buckets[],           /* buckets to place values in */
    bktsize;             /* range of vlues in bucket */
{
    int bottom,          /* first value in current bucket */
        bkt,             /* current bucket */
        markcnt,         /* number of marks written */
        most,            /* values in largest bucket */
        values;          /* number of values to write out */
    float scale;         /* scaling factor */
    /* compute scaling factor */

    for (bkt = most = 0; bkt < NUMBKTS; bkt++)
        if (most < buckets[bkt])
            most = buckets[bkt];
    scale = (most > MAXCOLS) ? (MAXCOLS / (float) most):1.0;
    /* print the histogram */

    putchar('\n');
    for (bkt=0, bottom=MINVAL; bkt<NUMBKTS; bottom+=bktsize,
```

```
    bkt++)
  {
     /* write range */

     printf("%3d-%3d |", bottom,
         (bkt=NUMBKTS - 1) ? MAXVAL : bottom + bktsize - 1);

     /* compute number of MARKERS to write, making sure
      * that at least one is written if there are any
      * values in the bucket
      */

     if (buckets[bkt] && !(values = buckets[bkt] * scale))
        values = 1;
     /* writes MARKERS and count of values */

     for (markcnt = 0; markcnt < MAXCOLS; markcnt++)
        putchar((markcnt < values) ? MARKER : ' ');
     if (buckets[bkt])
        printf(" (%d)", buckets[bkt]);
     putchar('\n');
  }
}
```

4.7.2 The direct *Mathematica* translation

First we give the direct translation, which attempts to be as close as possible in structure and spirit to the preceding C program. It is written as a function so that there is a reasonable way to use it. Also, the data is read from a file rather than from the keyboard as apparently is done in the C program.

```
In[1]:=histoGram[filename_String, {MINVAL_, MAXVAL_, NUMBKTS_}]:=
    Module[
      {   MARKER = "*",
          buckets,
          bktsize = (MAXVAL - MINVAL)/ (NUMBKTS - 1)
      },
      fillBkts[buckets_, bktsize_] :=
       Module[{bkt, snum, value, badcnt = 0, totcnt = 0},
         For[bkt = 0, bkt <= NUMBKTS, buckets[bkt++] = 0,Null];
           snum = OpenRead[filename];
           While[((value = Read[snum, Number];
                   Length[value] == 0) &&
                  (value =!= EndOfFile)),
             If[value >= MINVAL && value <= MAXVAL,
```

```
                buckets[Floor[(value - MINVAL) / bktsize] ]++,
                badcnt++];
             totcnt++];
        CloseRead[filename];
        If[!(badcnt > 0),
          Print["All ", totcnt," values in range\n"],
          Print["Out of range",badcnt,"total", totcnt,"\n"]];
        value === EndOfFile];
     printHisto[buckets_, bktsize_] :=
       Module[{bkt, markcnt, stars, bottom},
         For[bkt = 0; bottom = MINVAL, bkt < NUMBKTS,
           Print[PaddedForm[bottom, 3],
             " _",
             If[bkt == NUMBKTS - 1,
                PaddedForm[MAXVAL, 3],
                PaddedForm[bottom + bktsize - 1, 3]],
             " |",
             For[markcnt = 0; stars = {}, markcnt < 40,
                 markcnt++,
               If[markcnt < buckets[bkt+1],
                  AppendTo[stars, MARKER],
                  AppendTo[stars, " "]]];
             If[buckets[bkt+1] > 0,
                AppendTo[stars,
                  StringJoin[
                    "(", ToString[buckets[bkt+1]],")"] ]];
             StringJoin[stars]
           ];
           bottom += bktsize; bkt++]
         ];
    If[
       fillBkts[buckets, bktsize],
       printHisto[buckets, bktsize],
       Print["Illegal data value--no histogram printed\n"]
    ]
  ]
```

In order to use this program we construct a file, called **numbers1**, consisting of 200 random integers between 1 and 100.

```
In[2]:= OutputForm[
        TableForm[
          Table[Random[Integer, {1, 100}], {200}],
          TableSpacing -> {0, 2}
```

```
           ]
         ] >> numbers1

In[3]:=histoGram["numbers1", {0, 100, 11}]

      All 200 values in range

        0 -   9 |   ** ** ** ** ** ** ** ** ** ** ** **         (24)
       10 -  19 |   ** ** ** ** ** ** ** ** **                  (18)
       20 -  29 |   ** ** ** ** ** ** ** ** ** ** ** *          (21)
       30 -  39 |   ** ** ** ** ** ** ** **                     (16)
       40 -  49 |   ** ** ** ** ** ** ** ** ** ** *             (21)
       50 -  59 |   ** ** ** ** ** ** ** ** ** ** ** *          (23)
       60 -  69 |   ** ** ** ** ** ** ** **                     (16)
       70 -  79 |   ** ** ** ** ** ** ** ** **                  (18)
       80 -  89 |   ** ** ** ** ** ** ** ** ** ** ** **         (24)
       90 -  99 |   ** ** ** ** ** ** ** ** ** *                (19)
      100 - 100 |
```

This result, which comes from **Print** statements, is almost exactly the same as the output from the C program, which was our goal in this first translation.

4.7.3 Comments on the direct *Mathematica* translation

Now that the program is written in *Mathematica* rather than C, it is somewhat easier to follow the syntax. A number of things had to be changed in order to get a reasonable *Mathematica* program. In particular:

i) A **histoGram** is a function of two arguments, one of which is a list with three entries, using up three of the local variables in the C program: namely, MINVAL, MAXVAL, and NUMBKTS.

ii) MAXCOLS is omitted since the scaling factor computation is omitted, leaving only MARKER from the original local variables.

iii) The way in which the C program passes values around does not exactly match *Mathematica*'s functional style. Thus, **buckets** and **bktsize** are included in the top-level module so they have the same values everywhere.

iv) Inside the top-level **Module**, there are two other **Modules** as part of the definitions of the functions **fillBkts** and **printHisto**. Normally, these would be defined as separate functions outside the definition of **histoGram**, but there is no harm in putting them where they are.

v) In **printHisto**, **PaddedForm** is used several times to get the final **Print** statements lined up properly.

vi) The **fillBkts** operation reads in its data from a file rather than asking the user to type in numbers each time the program is run. The file is created using **>>**, which is the infix form of the command **Put**. We use the commands **OpenRead**, **Read**, and **Close-Read** to get the information from the file into the program. The **Put** construction created a file **numbers1** that we can examine as follows

In[4]:= **examine = OpenRead["numbers1"]**

Out[4]= InputStream[numbers1, 5]

Note that the name of the file must be a string. (The file name argument in the program is required to be a string.) The result of **OpenRead** is to open a stream communication with the file. We can then read from the stream using the command **Read**, which takes the name of the stream and the kind of data to be read as arguments. **Read** maintains a pointer to the last value read and on each use it returns the next value. Thus, the following **Table** returns the first 10 numbers in the file.

In[5]:= **Table[Read[examine, Number], {10}]**

Out[5]= {9, 41, 33, 98, 88, 43, 48, 85, 6, 80}

Repeated, it gives the next 10 values.

In[6]:= **Table[Read[examine, Number], {10}]**

Out[6]= {54, 86, 50, 44, 1, 29, 18, 15, 36, 81}

Finally, we close the stream using **Close**. It is always a good idea to close any open stream as soon as it is no longer needed.

In[7]:= **Close[examine]**

Out[7]= numbers1

A more elegant and symmetrical way to handle the construction of the file is to open a stream and write to it. The following **Write** construction is exactly opposite to the construction **Read**. Note that each **Write** statement writes one item, so we use a **Table** construction to write many items to the file. As with **Read**, we first open a stream to the file, write to it, and then close it.

```
In[8]:= sfile = OpenWrite["file"];
    Table[Write[sfile, Random[Integer, {0, 20}]], {40}];
    Close[sfile];
```

One can add numbers to the file as follows:

```
In[11]:= afile = OpenAppend["file"];
    Write[afile, 20]; Write[afile, 20];
    Close[afile];
```

To see the contents of the file, use:

```
In[14]:= !!file
```

We've omitted the output here since it is a long, single column of 42 numbers. To find out where this file is, use the following command, which returns the name of the current working directory. The output, of course, is machine dependent.

```
In[15]:= Directory[]
```

```
Out[15]= Russell:Mathematica 3.0 Files
```

4.7.4 A better *Mathematica* program

First of all, we agree with the general idea that the program has two main parts: the first part puts the data in the appropriate buckets and the second part makes a picture of the filled buckets. This second part will be implemented in Chapter 10 as an example of graphics programming. Here, we concentrate on putting the data in buckets. We assume, as does the C program, that the range of the data and the bucket size are given in advance (although it is easy to imagine a preprocessor that examines the data first and determines the actual range and an appropriate bucket size). Now, in Chapter 6, Section 4.3, we used the **Count** function to count how many times a given item occurs in a list. If we had the range divided into sublists of the size of each bucket, then we could just add up how many times each value in a given bucket occurs in the list of data. This is easy to arrange. Assume the range and bucket size are given in the form {**xmin**, **xmax**, **xstep**} so the values are between **xmin** and **xmax** and the bucket size is **xstep**. Then defining

```
(*buckets = Partition[Range[xmin, xmax], xstep]*)
```

would create the buckets as a list of lists. For example,

```
In[16]:= buckets = Partition[Range[0, 99], 10]
```

```
Out[16]= {{0, 1, 2, 3, 4, 5, 6, 7, 8, 9},
        {10, 11, 12, 13, 14, 15, 16, 17, 18, 19},
```

```
              {20, 21, 22, 23, 24, 25, 26, 27, 28, 29},
              {30, 31, 32, 33, 34, 35, 36, 37, 38, 39},
              {40, 41, 42, 43, 44, 45, 46, 47, 48, 49},
              {50, 51, 52, 53, 54, 55, 56, 57, 58, 59},
              {60, 61, 62, 63, 64, 65, 66, 67, 68, 69},
              {70, 71, 72, 73, 74, 75, 76, 77, 78, 79},
              {80, 81, 82, 83, 84, 85, 86, 87, 88, 89},
              {90, 91, 92, 93, 94, 95, 96, 97, 98, 99}}
```

gives us 10 non-overlapping sublists. In the C program, `buckets` is constructed one item at a time in a **For** loop, whereas here, obeying the fundamental dictum of functional programming, it is made by partitioning the existing list **Range[0, 99]**.

To test this, create a list of random integers between 1 and 100.

In[17]:=`data = Table[Random[Integer, {1, 100}], {500}];`

Then

In[18]:= `Map[Map[Count[data, #] &, #] &, buckets]`

Out[18]= `{{0, 2, 3, 4, 5, 3, 9, 6, 2, 2}, {4, 5, 6, 6, 5, 2, 3, 5, 2, 8},`
`{6, 5, 5, 3, 7, 7, 7, 6, 5, 6}, {3, 2, 6, 11, 8, 5, 6, 4, 7, 5},`
`{6, 3, 5, 3, 11, 4, 0, 2, 5, 6}, {5, 4, 8, 5, 5, 2, 7, 5, 7, 6},`
`{4, 7, 6, 5, 4, 5, 4, 4, 6, 9}, {3, 4, 4, 4, 2, 6, 8, 5, 3, 3},`
`{7, 3, 5, 6, 9, 3, 2, 4, 8, 5}, {10, 11, 5, 3, 5, 4, 3, 6, 4, 4}}`

tells how many times each bucket item occurs in the data, and

In[19]:= `Map[Plus @@ Map[Count[data, #] &, #] &, buckets]`

Out[19]= `{36, 46, 57, 57, 45, 54, 54, 42, 52, 55}`

adds up the items in each bucket. As with the **frequencies** command, these values should be combined with a description of the buckets. We choose to do this by giving the minimum and maximum values in each bucket; i.e.,

In[20]:= `Map[{Min[#], Max[#]} &, buckets]`

Out[20]= `{{0, 9}, {10, 19}, {20, 29}, {30, 39}, {40, 49},`
`{50, 59}, {60, 69}, {70, 79}, {80, 89}, {90, 99}}`

Finally, put this together with the values in each bucket.

```
In[21]:= Map[
         {{Min[#], Max[#]}, Plus@@Map[Count[data, #] &, #]} &, buckets]
```

```
Out[21]= {{{0, 9}, 36}, {{10, 19}, 46}, {{20, 29}, 57},
         {{30, 39}, 57}, {{40, 49}, 45}, {{50, 59}, 54},
         {{60, 69}, 54}, {{70, 79}, 42}, {{80, 89}, 52}, {{90, 99}, 55}}
```

Thus, the final program to calculate the values is a simple one-liner.

```
In[22]:= histogram[data_, {xmin_, xmax_, xstep_}] :=
         Map[{{Min[#], Max[#]}, Plus@@Map[Count[data, #]&, #]}&,
            Partition[Range[xmin, xmax], xstep]]
```

This corresponds to the fill_bkts part of the C program. Try this with data.

```
In[23]:= histo = histogram[data, {0, 99, 10}]
```

```
Out[23]= {{{0, 9}, 36}, {{10, 19}, 46}, {{20, 29}, 57},
         {{30, 39}, 57}, {{40, 49}, 45}, {{50, 59}, 54},
         {{60, 69}, 54}, {{70, 79}, 42}, {{80, 89}, 52}, {{90, 99}, 55}}
```

A different version of **histogram** can be based on the **BinCounts** function in the **DataManipulation** package.

```
In[24]:= histogram1[data_, {xmin_, xmax_, xstep_}] :=
         Module[
            {buckets  = Partition[Range[xmin, xmax], xstep],
             nbuckets = Ceiling[(xmax - xmin)/xstep],
             newdata  = Ceiling[(data - xmin)/xstep], i},
            Transpose[{
               Map[{Min[#], Max[#]}&, buckets],
               Table[Count[newdata, i], {i, nbuckets}]}]]
```

As an exercise, step through this program to see how it works. Then look up **BinCounts**.

In Chapter 10 we will show how to use *Mathematica* graphics primitives to construct a graphics object illustrating the output of **histogram** in order to see what the resulting output looks like. This will correspond to the print_histo part of the C program.

4.7.5 Comparison of the two *Mathematica* programs

The main difference between the C program, either in itself or as translated into *Mathematica*, and the better *Mathematica* program is the level on which data is treated. The C program only deals with individual items of data, while the *Mathematica* program deals directly with the data as a whole.

i) For instance, the array of empty buckets is created by a **For** loop, one bucket at a time, whereas the better *Mathematica* program creates the buckets by partitioning the already existing list given by the **Range** command. Next, in the fill_bkts part, the C program looks at each item of data in turn and increments the appropriate bucket. The *Mathematica* program, on the other hand, uses the technique of the **frequencies** function of Chapter 4 to run through the list of possible values in each bucket and add up the number of times that they occur in the data list, all by mapping appropriate constructions down lists.

ii) Similarly, in the print_histo part, for each bucket the C program calculates the lower and upper bounds of the bucket, prints them followed by a bar |, and then, one at a time prints a "*" for each item in the bucket, followed by individually calculated spaces " " to fill up each row. On the other hand, in the *Mathematica* graphics programs constructed in Chapter 10, a single construction will be applied to each pair in the output of **histogram** to build a graphics object which can then be displayed in various forms.

Good *Mathematica* style consists of dealing with mathematical objects as wholes, in accordance with the fundamental dictum of functional programming, never breaking them up into their constituent parts for later reconstruction in another form. It sometimes takes considerable thought to see how data in one form can be converted directly into data of another form, but that is one reason why *Mathematica* programming is interesting.

5 *Practice*

```
1.   Trace[y = 6; Plus[(y = y + 1); 5, y]]//TableForm

2.   Trace[y = 6; Plus[y, (y = y + 1); 5]]//TableForm

3.   Module[{t = 6, u = t}, u^2]

4.   Trace[Module[{t = 6, u = t}, u^2]]

5.   Table[Block[{r}, r], {10}]

6.   Trace[Block[{t}, t = 3]]

7.   ToCycle[perm_] :=
         Module[
             {a = {}, len = Length[perm], t, n, l, i},
             t = Table[True, {len}];
             For[i = 1, i <= len, i++,
                 If[t[[i]],
                     For[n = perm[[i]]; l = {},
                         t[[n]],
```

```
                           n = perm[[n]],
                           t[[n]] = False; AppendTo[l, n]]
                    AppendTo[a, l]
                    ]];
              Return[a]]
```

(See Chapter 12 for a functional version of this program, or write your own.)

```
    11. ToCycle[{3, 4, 15, 13, 2, 11, 7, 6,
              14, 9, 12, 1, 16, 5, 8, 10}]
```

6 Exercises

1. Many of the list operations in *Mathematica* are based on commands from the APL language. One that is not implemented is the function **deal** which is represented in APL by ?. Thus, L?R selects L integers at random from the population **Range[R]** without replacement.

   i) Write a more general *Mathematica* function **deal** so that **deal[list, n]** selects **n** entries at random from **list** without replacement.

   ii) A deck of cards consists of 52 cards divided into 4 suits called clubs, diamonds, hearts, and spades. Each suit consists of the cards 2, 3, 4, 5, 6, 7, 8, 9, 10, J, Q, K, 1. A bridge deal consists in giving 13 cards at random to each of 4 players. Define a *Mathematica* **deck** and a function **bridgeDeal[deck]** that generates and displays such a bridge deal.

2. Part of Problem 3 in Chapter 7 was to write a predicate **algexpQ** in pattern matching style. Write the same function in two different forms using:

   i) **Which**   ii) **Switch**.

3. Define a function **countTheCharacters[text_]** that takes a string **text** and turns it into a list of characters. It then returns a list whose entries are pairs with first entry a character in the list and second entry the relative frequency of the occurrences of the character in **text**, expressed as a percentage of the total number of characters in **text**. You may want to use the definition of **frequency** in Chapter 6, Section 4.3. Try to put the list in order of decreasing frequency.

4. i) Recreate the Pascal program "Stolen Gold"

      a) using a **For** loop in *Mathematica*,

      b) using a **While** loop in *Mathematica*.

   ii) Change the one-liner so it prints out the same results as the Pascal program. It should still be a strict one-liner.

5. Consider the two infinite sums with possible values

$$\textbf{1.}\ \sum_{n=1}^{\infty} \frac{a\,(2^n)}{2^n} = \frac{1}{99} \qquad \textbf{2.}\ \sum_{n=1}^{\infty} \frac{a\,(n)}{10^n} = \frac{10}{99}$$

Here, $a(n)$ is the number of odd digits in odd positions in the decimal expression for n. Thus, $a(901) = 2$, $a(1234) = 0$, $a(4321) = 2$, etc. Positions are counted from the right. At least one of the values is wrong and can be detected by a computation taking a reasonable length of time (i.e., < 10 seconds). Which one is it? [Borwein]

6. A perfect shuffle of a deck of $2n$ cards consists of dividing the deck in the middle into two decks of n cards each and then exactly interleaving the two decks. There are two ways to do this: either the first card of the first deck remains the first card, in which case the shuffle is called an out shuffle, or it becomes the second card, in which case the shuffle is called an in shuffle. Both shuffles determine a permutation of $2n$ cards. The two shuffles generate a subgroup of the group of all permutations of $2n$ cards by repeating and combining them. For instance, if $n = 3$, then there are six cards. Label them $\{1, 2, 3, 4, 5, 6\}$. An out shuffle produces the permutation $\{1, 4, 2, 5, 3, 6\}$ while an in shuffle produces the permutation $\{4, 1, 5, 2, 6, 3\}$.

i) Write functions **outShuffle** and **inShuffle** taking as argument a list of even length and permuting it by an out shuffle and an in shuffle.

ii) Since the group of all permutation is a finite group, both **outShuffle** and **inShuffle** have finite orders; i.e., there are integers **outOrder[n]** and **inOrder[n]** for each **n** such that if **outShuffle** is repeated **outOrder[n]** times and **inShuffle** is repeated **inOrder[n]** times, then the result is the identity permutation. Determine the orders of **outShuffle** and **inShuffle** for n between 1 and 50; i.e., for decks consisting of 2 to 100 cards, by finding experimentally how many times they have to be repeated to put the deck back into its original order. Note: For **n** = 26, i.e., for an ordinary deck of 52 cards, **outOrder[26]** = 8 and **inOrder[26]** = 52. Plot these values as a function of $2n$.

iii) It is a theorem that the order of **outShuffle** for a deck of $2n$ cards is the smallest k such that $2k = 1$ mod $2n - 1$, and the order of **inShuffle** is the same as the order of **outShuffle** for a deck consisting of 2 more cards. Write functions calculating these numbers and compare these numbers with the experimental results for n between 1 and 50.

iv) It is known that the group generated by **outShuffle** and **inShuffle** is isomorphic to the group of all symmetries of the n-dimensional generalization of the octahedron. (See [1] and [2] below.) For $n = 3$, it is the group of all symmetries of the usual octahedron. Using the values of the orders of **outShuffle[3]** and **inShuffle[3]**, show that there are symmetries of the required orders. Is there a nice graphical illustration of this result?

v) Generalize to the situation where a deck of $3n$ cards is divided into three equal parts which can then be shuffled perfectly in six different ways.

References:

[1] Diaconis, P., Graham, R. L., and Kantor, W. M., The mathematics of perfect shuffles, *Adv. Appl. Math.*, **4** (1983), 175–196.

[2] Medvedoff, S., and Morrison, K., Groups of perfect shuffles, *Mathematics Magazine*, **60** (1987), 3–14.

7. It is a non-trivial result in number theory that every positive integer can be written as the sum of four squares. (Zero is allowed as one of the summands.)

i) Write a program to find one such representation for each positive integer. Use it to find all integers between 1 and 1000 that are not sums of three squares.

ii) Write a program that finds all such representations for each positive integer.

iii) Not all integers can be written as the sum of four distinct non-zero integers. Find all integers between 1 and 1000 that don't have such a representation. (Warning: This takes 40 minutes on a SPARC workstation.)

8. There are various systematic methods for generating magic squares. Look up some of these methods and implement them in *Mathematica*.

CHAPTER 9

Object-Oriented Programming

1 Introduction

In the preceding three chapters we have discussed three distinct modes of computer programming–functional programming, rewrite rule programming, and imperative programming. All three have their appropriate roles and most *Mathematica* programmers use whatever style seems most appropriate to the thought being expressed, depending on the needs of the moment. The *Mathematica* Book [Wolfram] suggests that rule-based programming is the most appropriate. Others insist that only functional programming is acceptable, and presumably unregenerate C programmers will continue to write thinly disguised imperative programs. In discussing each style we have concentrated on producing operations that realize some definite mathematical or scientific goal.

But how do you proceed if you have more than one goal and if you want to produce software for others to use? Most large programs do many things, and the organization of the interactions between the pieces of a program can become a major task. There is a specific *Mathematica* facility–that of Packages, to be addressed in Chapter 11–that deals with one aspect of this problem. But in recent years a paradigm has emerged which has become increasingly popular in software engineering projects whose purpose is to create large programs–that of object-oriented programming (OOP). For instance, *Mathematica* is written in an object-oriented version of C. Also, essentially all graphical user interfaces are written in object-oriented languages.

It is possible to write programs in a pseudo object-oriented style in almost any higher order programming language, but certain languages like C++, Smalltalk, and Java are explicitly intended to be used only in this way. Although *Mathematica* does not provide any built-in support for object-oriented methods, a recent package by Roman Maeder in [Maeder 3], called `Classes.m`, implements a full-blown object-oriented extension to it. This package does not make it possible to do any calculations that couldn't be done before; it just makes it possible to completely rearrange the way in which they are carried out.

We have been subtly (and perhaps not so subtly) promoting the view that *Mathematica* is a heart a functional programming language. Such languages work by building up a myriad of smaller functions, each accomplishing one piece of a task, and then joining them together into one top level function which is applied to some data to produce a result. *Mathematica* adds to this the possibility of applying a given function to data in different forms with different outputs. It does this via the mechanism of pattern matching, using heads of expressions or predicates to restrict patterns. This facility is part of what is called *polymorphism*, which means exactly that the same operation works with data of different forms, usually resulting in similar outputs.

In the general situation, there will be many kinds of data and many operations. Some of the data will be acted on by more than one operation and some of the operations will act on more than one kind of data. It is this unexpected symmetry (or perhaps duality is a better term) between operations and data that led to the invention of object-oriented programming. Functional programming (or function-oriented programming) concentrates on the functions and their organization into hierarchies while object-oriented programming concentrates on the data and its organization into hierarchies.

There are enough subtleties involved in object-oriented programming to fill many books. Two that are very useful are [Budd] and [Meyer]. For a complete theoretical treatment, see [Abadi and Cardelli]. In this chapter we shall just explain the evolution and use of Maeder's implementation by means of some very simple examples. Section 2 is intended as motivation for the material in Section 3. In it we follow Maeder's discussion in [Maeder 2] of how to shift attention from the functions to the data. In Chapter 13, graph theory will be developed in a strictly object-oriented framework, in the hopes that a single comprehensive and comprehensible example is worth a hundred pages of philosophy.

2 *The Duality between Functions and Data*

The transition from functional programming to object-oriented progrmming is mediated by the notion of a dispatch table. For a thorough discussion in the context of LISP, see [Abelson]. Here we follow the treatment of [Maeder 2]. A standard example is given by points in the plane. Such points can be represented by Cartesian or polar coordinates and can be created in either form. Given a point in either representation, there are a number of things we would like to be able to calculate about it: i.e., its x-coordinate, its y-coordinate, its magnitude, and its polar angle. Furthermore, we would like to be able to make these calculations without worrying about which coordinate system is used to represent the point. Two somewhat different implementations of this idea will be given.

2.1 The First Implementation of Points in the Plane

As was discussed in Chapter 5, one meaning for the head of an expression is the *type* of the expression; e.g., the head of 2 is **Integer** and the head of {a, b, c} is **List**. In particular, we will use the heads **cartesian** and **polar** to identify Cartesian and polar coordinates, respectively, of points in the plane, thinking of these heads as representing two different types of points. Two functions are defined to create points of the given types just by wrapping the heads **cartesian** and **polar** around the values.

```
In[1]:=makeCartesian[{x_, y_}] := cartesian[x, y]
```

```
In[2]:=makePolar[{r_, theta_}] := polar[r, theta]
```

The four things we want to calculate–the *x*-coordinate, the *y*-coordinate, the magnitude, and the polar angle–now require two functions each, one for Cartesian points and one for polar points.

```
In[3]:=xCoordCartesian[cartesian[x_, y_]]      := x;
       yCoordCartesian[cartesian[x_, y_]]      := y;
       magnitudeCartesian[cartesian[x_, y_]]   := Sqrt[x^2 + y^2];
       polarAngleCartesian[cartesian[x_, y_]] := ArcTan[y/x];
```

```
In[7]:=xCoordPolar[polar[r_, theta_]]       := r Cos[theta];
       yCoordPolar[polar[r_, theta_]]       := r Sin[theta];
       magnitudePolar[polar[r_, theta_]]    := r;
       polarAnglePolar[polar[r_, theta_]]   := theta;
```

Consider the following table in which the rows represent the operations (given generic names) and the columns represent the types. The entries in the table are the actual functions that calculate the values for each type. Such a table is called a *dispatch table*. It *dispatches* the operations depending on the types of the arguments. (Cf. the *Mathematica* operation **Dispatch**.)

	cartesian	polar
xCoord	xCoordCartesian	xCoordPolar
yCoord	yCoordCartesian	yCoordPolar
magnitude	magnitudeCartesian	magnitudePolar
polarAngle	polarAngleCartesian	polarAnglePolar

We can construct functions that implement this table by using **Switch** to determine which concrete operation should be applied to arguments of each type. The four rows require four functions, each of which has to determine what to do with each type of argument. This is done by pattern matching using the head of the argument.

```
In[11]:=xCoord[point_] :=
        Switch[Head[point],
               cartesian, xCoordCartesian[point],
               polar,     xCoordPolar[point]]
```

```
In[12]:=yCoord[point_] :=
        Switch[Head[point],
               cartesian, yCoordCartesian[point],
               polar,     yCoordPolar[point]]
```

```
In[13]:=magnitude[point_] :=
        Switch[Head[point],
               cartesian, magnitudeCartesian[point],
               polar,     magnitudePolar[point]]
```

```
In[14]:=polarAngle[point_] :=
        Switch[Head[point],
               cartesian, polarAngleCartesian[point],
               polar,     polarAnglePolar[point]]
```

In each operation, the head of the argument is matched to the type to determine which operation should be applied.

Now, using these operations we can, for instance, add points irrespective of how they are represented.

```
In[15]:=add[point1_, point2_] :=
        makeCartesian[{xCoord[point1] + xCoord[point2],
                       yCoord[point1] + yCoord[point2]}]
```

As an example, construct a Cartesian and a polar point

$$In[16]:= \texttt{point1 = makeCartesian[\{2, 3\}];}$$
$$\texttt{point2 = makePolar}\left[\left\{2^{\frac{3}{2}}, \frac{\pi}{4}\right\}\right];$$

and then add them together.

```
In[17]:=add[point1, point2]
```

```
Out[17]= cartesian[4, 5]
```

In this organization, the information about each of the four basic functions is stored with the function itself as usual. For instance:

```
In[18]:= ?xCoord
```

```
Global`xCoord
xCoord[point_] := Switch[Head[point], cartesian,
  xCoordCartesian[point], polar, xCoordPolar[point]]
```

As we have seen in the answer to Exercise 8 of Chapter 8, such **Switch** statements are not the most efficient way to implement this kind of polymorphism. Parallel rewrite rules are better both stylistically and from the standpoint of efficiency. For instance, **xCoord** could be given by two rules:

```
xCoord[point_cartesian]  := xCoordCartesian[point];
xCoord[point_polar]      := xCoordPolar[point]
```

However, our ultimate goal is not to implement these operations but to explain the form of the argument to the operation **Class** described below, and that is best done using **Switch**.

2.2 The Second Implementation of Points in the Plane

A somewhat more intrinsic way to organize the same information is to group together the calculation of the x and y coordinates as a list and call it the Cartesian coordinates of a point. In the same way, the magnitude and polar angle are called the polar coordinates of a point. The calculations above can be grouped differently so that they represent translations between the two coordinate systems. This way, we only need two functions, one to turn polar points into Cartesian points, and the other to provide the opposite transformation. We keep the definitions of **makeCartesian** and **makePolar** from above. Here are the two required functions:

```
In[19]:=cartesianFromPolar[point_polar]  :=
        makeCartesian[{point[[1]] Cos[point[[2]]],
                       point[[1]] Sin[point[[2]]]}]
```

```
In[20]:=polarFromCartesian[point_cartesian]  :=
        makePolar[{Sqrt[point[[1]]^2 + point[[2]]^2],
                   ArcTan[point[[2]] / point[[1]] ]}]
```

For instance:

```
In[21]:=rr = makeCartesian[{3, 4}]
```

```
Out[21]= cartesian[3, 4]
```

```
In[22]:=pp = polarFromCartesian[rr]
```

Out[22]= $\text{polar}\left[5, \text{ArcTan}\left[\frac{4}{3}\right]\right]$

Now what we want to do is to extract the Cartesian and polar coordinates of a point independently of its type by functions to be called `cartesianCoords` and `polarCoords`. The corresponding dispatch table is somewhat simpler, and the entries look much simpler than our previous table. In particular, the diagonal entries just change the head of the point to `List`.

	cartesian	polar
cartesian Coords|	List@@point	List@@ cartesianFromPolar
polar Coords|	List@@ polarFromCartesian	List@@point

The *Mathematica* implementation of the rows of this table is similar to the first form.

```
In[23]:= cartesianCoords [point_] :=
         Switch [Head [point],
             cartesian, List@@point,
             polar,     List@@cartesianFromPolar [point]]
```

```
In[24]:= polarCoords [point_] :=
         Switch [Head [point],
             cartesian, List@@polarFromCartesian [point],
             polar,     List@@point]
```

The points **rr** and **pp** from above are really "the same" even though **rr** is a Cartesian point and **pp** is a polar point, in that they have the same Cartesian and polar coordinates.

```
In[25]:= {cartesianCoords [rr], cartesianCoords [pp]}
```

Out[25]= $\{\{3, 4\}, \{3, 4\}\}$

```
In[26]:= {polarCoords [rr], polarCoords [pp]}
```

Out[26]= $\left\{\left\{5, \text{ArcTan}\left[\frac{4}{3}\right]\right\}, \left\{5, \text{ArcTan}\left[\frac{4}{3}\right]\right\}\right\}$

Once we have these operations, we can implement others in terms of them; e.g., points can be translated by a vector and rotated about the origin.

```
In[27]:= translate [point_, vector_] :=
         makeCartesian [
             cartesianCoords [point] + vector]
```

```
In[28]:= rotate[point_, angle_] :=
        makePolar[
            {0, angle} + polarCoords[point]]
```

For instance:

```
In[29]:= translate[pp, {5, 5}]
```

Out[29]= cartesian[8, 9]

```
In[30]:= rotate[pp, π]
```

$$\text{Out[30]= } polar\left[5, \ \pi + ArcTan\left[\frac{4}{3}\right]\right]$$

2.3 *The Transition to OOP*

Instead of having operations that work with different kinds of data, the object-oriented paradigm designs data objects that respond to different kinds of messages. Furthermore, instead of applying functions to arguments, messages are sent to objects. Thus,

$$\frac{\text{messages}}{\text{functions}} = \frac{\text{objects}}{\text{data}}$$

Instead of functions knowing how to treat different kinds of arguments, the data itself knows how to process the messages. In this view, the data objects become the active participants, whereas the function messages are little more than passive names. In terms of the dispatch table, the columns play the main role rather than the rows. Nearly all object-oriented languages such as ObjectPascal, C++, and Java use a notation for sending messages to objects that doesn't look like applying a function to a value. A common form that we adopt here is

```
object . message
```

In order for this to work, we have to unprotect **Dot** and give our own rules for it.

```
In[31]:= Unprotect[Dot]; SetAttributes[Dot, HoldRest]; Protect[Dot]
```

At first it is hard to imagine how this can be achieved, but [Maeder 2] shows in a very simple way how it is done. In the following examples, in order to avoid confusion with the preceding operations, be sure to clear the previous definitions.

```
In[32]:= Clear[makeCartesian, makePolar, cartesianCoords, polarCoords]
```

Here is the new version of **makeCartesian** that creates an active object.

```
In[33]:=makeCartesian[{x_, y_}] :=
    Module[{cartesian},
      With[
        {dispatch =
          Function[{method},
            Switch[method,
                    cartesianCoords, {x, y},
                    polarCoords,
                      {Sqrt[x^2 + y^2], ArcTan[y/x]}]]},
        cartesian/:
          (cartesian.message_Symbol):= dispatch[message]/;
            MemberQ[{cartesianCoords, polarCoords}, message];
        cartesian]]
```

This operation creates Cartesian point *objects* (replacing the notion of a Cartesian point from above) from lists of two numbers. The intention is that the properties of a Cartesian point object will be the same as those of a Cartesian point. In particular, the data **cartesian[2, 3]** is replaced by the object that results from evaluating the operation **makeCartesian[{2, 3}]**, the function **cartesianCoords** is replaced by the message **cartesianCoords** (which is just a name), and the function **polarCoords** is replaced by the message **polarCoords**.

There are two ingredients in the definition of **makeCartesian**. Consider first the **With** expression:

```
With[{dispatch = Function[{method}, functionBody]},
     withBody]
```

In the **With** statement, a new variable **dispatch** is set equal to a pure function of yet another new variable **method**. The body, **functionBody**, of this pure function is like the dispatch table we had before, except that this time it is dispatching the method names instead of the data types. The function **dispatch** gets used in the body, **withBody**, of the **With** statement, which is essentially

```
cartesian.message_Symbol:= dispatch[message]
```

This tells the **cartesian** object how to respond to a message sent in the form **cartesian.message**; namely, use the dispatch table to match the variable **method** to **message** and output the appropriate result. This of course only works if **message** is either **cartesianCoords** or **polarCoords**. A pair like (**cartesianCoords**, {x, y}) is called a *method*. A method consists of two parts, the *methodName*, or *message* (e.g., **cartesianCoords**), and the *methodBody*, or *response* (e.g., {x, y}).

The information about how to respond to messages is stored with the local variable **cartesian** because the form

```
cartesian/: cartesian.message := dispatch[message]
```

is used. The clause following that form,

```
/; MemberQ[{cartesianCoords, polarCoords}, message]
```

restricts **message** to be one of the permissible messages. The final **cartesian** causes the output of the **Module** to be the local variable itself. For instance:

In[34]:= **pt = makeCartesian[{2, 3}]**

Out[34]= cartesian$1

Notice that **cartesian** is concatenated with **$n** since it is a local variable. Now we can try sending the messages **polarCoords** and **cartesianCoords**, as well as an illegal message **ff**, to **pt**.

In[35]:= **{pt . polarCoords, pt . cartesianCoords, pt . ff}**

Out[35]= $\left\{\left\{\sqrt{13}, \text{ArcTan}\left[\frac{3}{2}\right]\right\}, \{2, 3\}, \text{cartesian\$1 . ff}\right\}$

Thus, **pt** knows what its Cartesian and polar coordinates are, but it knows nothing about any other message, such as **ff**. The information about this object is stored with the name **cartesian$1**.

In[36]:= **?cartesian$1**

```
   Global`cartesian$1
   Attributes[cartesian$1] = {Temporary}

   cartesian$1 /: cartesian$1 . (message$_Symbol) :=
      Function[{method$}, Switch[method$, cartesianCoords, {2, 3},
         polarCoords, {Sqrt[2^2 + 3^2], ArcTan[3/2]}]][message$] /;
      MemberQ[{cartesianCoords, polarCoords}, message$]
```

What is stored with the data object **cartesian$6** is the information about how to respond to the messages **cartesianCoords** and **polarCoords** in terms of the parameter values, 2 and 3, used in defining it. Thus, we see in expanded form that sending the message **message&** to **cartesian$1** by evaluating **cartesian$1.message** applies the pure function **Function[{method}, functionBody]** to **message$**. When **message$** is substituted for **method$** in **functionbody**, the result is the **Switch** statement:

```
Switch[message$,
       cartesianCoords, {2, 3},
       polarCoords, {Sqrt[2^2 + 3^2], ArcTan[3/2]}]
```

provided, of course, that **message$** is either **cartesianCoords** or **polarCoords**. Thus, **message$** has to match one of the two patterns in the second and fourth arguments of the **Switch** statement, and hence, either the third or the fifth argument will be the output.

Polar objects are constructed analogously.

```
In[37]:=makePolar[{r_, theta_}] :=
     Module[{polar},
      With[
       {dispatch =
         Function[{method},
          Switch[method,
                 cartesianCoords,
                  {r Cos[theta], r Sin[theta]},
                 polarCoords, {r, theta}]]},
        polar/:
         polar.message_Symbol:= dispatch[message]/;
           MemberQ[{cartesianCoords, polarCoords}, message];
        polar
       ]]
```

For instance:

```
In[38]:=pt2 = makePolar[{2, Pi/3}]
```

Out[38]= polar$2

```
In[39]:= {pt2 . cartesianCoords, pt2 . polarCoords}
```

Out[39]= $\left\{\left\{1, \sqrt{3}\right\}, \left\{2, \frac{\pi}{3}\right\}\right\}$

Again, if we were just interested in these operations for themselves, rewrite rules would provide a much neater implementation. The columns of the dispatch table can equally well be given by such rules. For instance, **makeCartesian** could be written in the form:

```
In[40]:=makeCartesianRule[{x_, y_}] :=
        Module[{cartesian},
          cartesian/: cartesian.cartesianCoords = {x, y};
          cartesian/: cartesian.polarCoords =
                      {Sqrt[x^2 + y^2], ArcTan[y/x]};
          cartesian]
```

This works just as well as the more complex version creating a pure function. Thus,

```
In[41]:=ptr = makeCartesianRule[{3, 4}]
```

Out[41]= cartesian$3

```
In[42]:= {ptr . cartesianCoords, ptr . polarCoords}
```

Out[42]= $\left\{\{3, 4\}, \left\{5, \text{ArcTan}\left[\frac{4}{3}\right]\right\}\right\}$

However, this is not the way classes actually work. The operations **makeCartesian** and **makePolar** actually describe the *patterns* for Cartesian and polar objects. In object-oriented languages, such patterns are called *classes*. Thus, in summary, a class will be a pattern for a kind of object and an object will consist of some data bundled together with the information about how to respond to certain messages. Messages here are just names. Message passing looks like function application, but it actually consists of applying the object itself, in the guise of the pure function **dispatch**, to the message.

2.4 Messages with Parameters

So far we have achieved active data objects which contain within themselves all of the information needed to respond to messages. Note also that in principle there is no way to access the data in an object except through the messages that it recognizes. (In fact, of course, nothing is truly hidden in *Mathematica*.) Furthermore, these data objects are created as instances of general operations that play the role of classes.

However, that is not the whole story about object-oriented programming. Suppose, for instance, that we want to include **translate** as a message for Cartesian points whose effect is to translate the given point by some vector. There are two problems. First of all, **translate** will have to take a vector as a parameter, and second, it would have to be added to both **makeCartesian** and **makePolar** in order to work correctly for all points. We discuss these in turn.

The **translate** message takes a parameter–the vector of translation–whereas our other messages up to now don't require any additional input. The form of the methods and the way that messages are applied has to be changed to account for such possible parameters. The actual syntax of many object-oriented languages is

```
object . message[parameters]
```

In this form, a message without parameters has to be sent in the form

```
object . message[]
```

Here is the appropriate modification of the `makeCartesian` operation.

```
In[43]:= Clear[makeCartesian]

In[44]:=makeCartesian[{x_, y_}] :=
      Module[{cartesian},
        With[
         {dispatch =
           Function[{method},
            Switch[method,
                    cartesianCoords, {x, y}&,
                    polarCoords,
                      {Sqrt[x^2 + y^2], ArcTan[y/x]}&,
                    translate, makeCartesian[# + {x, y}]& ]
           ]},
         cartesian/:
           (cartesian.message_[args___]):=
              dispatch[message][args] /;
              MemberQ[{cartesianCoords, polarCoords,
                       translate}, message];
         cartesian
        ]]
```

The first change is that the `{x, y}` response to `cartesianCoords` in the previous version is replaced by the (constant) pure function `{x, y}&`, and similarly for the response to `polarCoords`. The response to the message `translate` is the pure function of one variable

```
makeCartesian[# + {x, y}]&
```

In fact, all of the possible outputs of the `Switch` expression have to be pure functions themselves. Furthermore, sending a message has to allow for the possibility of parameters and treat them correctly. This is accomplished by the new format for responding to messages.

```
cartesian.message[args___]:= dispatch[message][args]
```

This new format means that the response to a message must always be a pure function because it is going to be applied to a (possibly empty) sequence of arguments. This new form can be used just like the previous one, except that now a point knows how to translate itself by a given vector.

```
In[45]:=pt = makeCartesian[{3, 4}]
```

```
Out[45]= cartesian$4
```

```
In[46]:= pt . translate[{5, 5}]
```

```
Out[46]= cartesian$5
```

The result of a **translate** message is a new Cartesian point. To see what it is, we can ask for its Cartesian coordinates.

```
In[47]:= % . cartesianCoords[]
```

```
Out[47]= {8, 9}
```

Observe also that the presence of **makeCartesian** in the description of the response to **translate** makes the definition of **makeCartesian** recursive.

Yet again, messages with parameters could easily be added to a rewrite rule implementation. For example, redefine **makeCartesianRule** as follows:

```
In[48]:=makeCartesianRule[{x_, y_}] :=
        Module[{cartesian},
            cartesian/: cartesian.cartesianCoords[] = {x, y};
            cartesian/: cartesian.polarCoords[] =
                    {Sqrt[x^2 + y^2], ArcTan[y/x]};
            cartesian/: cartesian.translate[vector_] :=
                    makeCartesianRule[vector + {x, y}];
            cartesian]
```

For instance,

```
In[49]:=ptr = makeCartesianRule[{3, 4}]
```

```
Out[49]= cartesian$6
```

```
In[50]:= ptr . translate[{5, 5}] . cartesianCoords[]
```

```
Out[50]= {8, 9}
```

However, doing things this way wouldn't explain why the second component of a message, as discussed subsequently, has to be a pure function.

2.5 Inheritance

If we want **translate** to work for **polar** as well, then similar changes have to be made to the function **makePolar**. In this tiny example, that's harmless, but if we had many more kinds of objects to deal with, it might be very difficult to ensure that all of them were correctly updated when some new message was added. The solution to this problem is to organize objects into a hierarchy based on *inheritance*. To explain this notion, suppose that in addition to Cartesian points, we also want to have *colored* Cartesian points. Besides having a position, such points would also have a color, e.g., red, green, or blue. It would be very convenient if colored points could inherit all of their positional information from points and just add the color information themselves. We will say that **cartesian** is the superclass of **coloredCartesian** and, of course, that **coloredCartesian** is a subclass of **cartesian**. A class (or kind of object) has only one superclass, but it may have many subclasses. Thus, we would like to be able to write something like

```
makeColoredCartesian[{x_, y_}, colorname_] :=
    "the superclass is cartesian and the
     dispatch table has an additional
     pair <color, colorname&>. "
```

The **Classes** package of R. Maeder [Maeder 2] will provide a way to do this. What has to happen is that when the message **cartesianCoords** is sent to a colored point, **cpt**, then **cpt** has to recognize that it itself doesn't know how to deal with the message and so send it on to its superclass to see if the superclass can respond to it. Thus, the message **cartesianCord** sent to **cpt** will just be passed on to the class for Cartesian points which will return the answer **{x, y}**, whereas the message **color** sent to **cpt** will be answered by **cpt** itself.

There is an important proviso in inheritance. If the message **translate** is sent to **cpt**, then the result should again be a colored point, not just a point. That can't happen given the way our code is organized now, since the response to **translate** in the **makeCartesian** definition is of the form **makeCartesian[---]**, which produces a new Cartesian point, and there is no way to change that. Solving this problem requires a whole new mechanism for creating objects of a given kind. It is a generic method for creating objects, called **new**, and the appropriate form to make a Cartesian point will be

```
cartesian.new[{x, y}]
```

This still won't get us a colored point as the result of a **translate** message. One last ingredient is needed: a special variable name, **self**, that refers to the current object. Then the implementation of **cartesianPoint** can say as the response to a **translate** message:

```
Class[self].new[---]
```

meaning "make a new object just like yourself but with new parameters."

3 Object-Oriented Programming in Mathematica

3.1 Using Class

The key concepts in OOP are object, class, message, inheritance, **new**, **self**, and **super**. These are all implemented in Maeder's package **Classes.m** from [Maeder 3]. We won't attempt to explain how this package works. Suffice it to say that it is an ingenious combination of all of the facilities that are available in *Mathematica*, based on the ideas discussed above. The package is included in the diskette supplied with this book and will repay careful study. After some preliminary examples showing how to use inheritance, we will use the package to set up a small hierarchy of classes involving points. First, load the package.

In[1]:=**Needs["Classes`"]**

As far as the user is concerned, the main thing contained in this package is the command **Class**. What **Class** does is to create a pattern for constructing a particular kind of objects. It takes four arguments in the following form.

```
Class[ nameOfClass,
       nameOfSuperclass,
       listOfInstanceVariables,
       listOfMethods ]
```

The name of the class is whatever you want to call your class. The name of the superclass is the class one step up in the hierarchy of classes that you are constructing. If you don't have a hierarchy yet, then use **Object** as the superclass. The class **Object** is constructed in the package and serves as the absolute top of the class hierarchy. It actually implements certain standard methods to be explained later. The instance variables are variables like x and y in the function **makeCartesian**. They are required to be symbols, but when they are used, any expression can be substituted for them. Finally, a method is a pair of the form

```
{methodName, body&}
```

where **methodName** is a **Symbol** and **body&** is some (possibly compound) expression, which is a pure function, implementing the method. Thus, the list of methods looks like

```
{{methodName1, body1&},
   - - -
 {methodNamek, bodyk&}}
```

The **Switch** statements in the dispatch tables in the preceding section were deliberately organized to look just like this, without the parentheses. All classes have a method with the name **new** whose body creates a new object belonging to the class. An object is created by giving a command of the form

```
objectName =
    new[nameOfClass, "instantiate instance variables"]
```

For instance, we will construct a class **cartesianPoint** below by the command

```
Class[cartesianPoint, - - -]
```

This will write the **makeCartesian** definition from before in the background where we can't see it. To use this hidden definition, one uses the message **new** so that

```
new[cartesianPoint, 3, 4]
```

replaces **makeCartesian**[{3, 4}] from before.

   After an object has been created, then methods are invoked for the object by sending messages to it using the **Dot** notation described above. In order for this to work we have to again define our own rule for **Dot**, since the **Classes.m** package assumes that messages will be sent using a functional notation. Thus, our rule for **Dot** is the following:

```
In[2]:= Unprotect[Dot];
    SetAttributes[Dot, HoldRest]
    Dot[object_, method_[parameters___]] :=
      method[object, parameters]
    Protect[Dot];
```

Once this rule has been made, then messages are sent using the syntax

```
objectName.methodName["values of parameters"]
```

3.2 Examples

3.2.1 A bank account

The standard elementary example is a class representing bank accounts. A bank account has a *balance* value and money can be *deposited* and *withdrawn* from it. As a class it is implemented as follows:

```
In[5]:= Class[account, Object, {bal},
          {{new,        (super.new[]; bal = #)&},
```

```
          {balance,    bal&},
          {deposit,    (bal += #)&},
          {withdraw,   (bal -= #)&}}]
```

Out[5]= account

(If the output is not the name of the class, then there is a mistake somewhere.) The first method is one for **new**, which says how new instances (i.e., new objects) of the class **account** are made. Normally the first thing it does is to call **new** of the superclass (represented by the reserved word **super**). It is the responsibility of **new** to initialize the instance variables. In this case, **super.new[]** means **Object.new[]**, which doesn't do anything since there are no instance variables in **Object** to be initialized. The second expression in the compound statement, **(bal = #)**, sets the instance variable **bal** equal to the parameter value of **new**. Thus, an **account** object with initial balance of $1000 is created by the command

In[6]:= **ac = account . new[1000 dollars]**

Out[6]= -account-

The hyphens before and after account in the output conform to a general *Mathematica* format to indicate in an abstract way that the actual output is some generally uninformative, complicated expression that need not be examined further. (Cf. the output -Graphics- from a **Plot** command.) Note that **new** is a message sent to a class, in this case the class **account**. There are certain other messages that are also sent to classes (called factory methods because they are already provided by the program).

Normally messages are sent to objects. For instance, to check the balance of our account, use the message

In[7]:= **ac . balance[]**

Out[7]= 1000 dollars

and to withdraw $150 use the message

In[8]:= **ac . withdraw[150 dollars]**

Out[8]= 850 dollars

Note that the ouput is in fact the new balance, which should only have been returned by the message **balance**. This behavior will be changed in the next example.

In[9]:= **ac . balance[]**

Out[9]= 850 dollars

The left-hand side of the dot is the name of the object to which the message is sent. Observe that the body of the method with name **balance** is the constant pure function **bal&**, so it has no parameters, while the body of the method with name **withdraw** is **(bal += #)&**, which is a pure function of one variable, so **withdraw** requires a single parameter. Study this example carefully since everything else is an elaboration of it.

Notice that the two input lines **ac.balance[]** lead to different outputs even though they are identical in appearance. This illustrates an important difference between functional and object-oriented programs. As we remarked in the introduction to Chapter 6, functions have no memory; each time a function is invoked with the same arguments, it returns the same value. Sending a message to an object can be quite different since objects can have a memory. In particular, **ac** remembers that something has happened to it; namely, 150 dollars have been withdrawn, so the value of the message **balance** sent to **ac** changes accordingly. Objects can have a history, and this may be the most interesting thing to know about them.

3.2.2 An immutable balance bank account

In the preceding example, the value of the instance variable **bal** was changed by the action of making a withdrawal. Technically, this means that objects created by **Class** are *mutable*; i.e., the values of their instance variables can be changed in place. But recall that in our implementation of points in Section 2, the operation **translate** created a new point rather than changing the point on which it acted. Some object-oriented languages insist that a new object must be created with a new value by such an operation. This behavior can be imitated if we write the methods for **deposit** and **withdraw** in a different form so that only immutable objects can be created.

```
In[10]:=Class[immutableAccount, Object, {bal},
          {{new,      (super.new[]; bal = #)&},
           {balance,  bal&},
           {deposit,  immutableAccount.new[bal + #]&},
           {withdraw, immutableAccount.new[bal - #]&}}]
```

```
Out[10]= immutableAccount
```

In this form there are no assignment statements except in the method body for **new**. In particular, nothing changes the value of **bal**. Instead, the result of a **deposit** or **withdraw** message is to create a new object with the required new balance. For instance, create an immutable account and check its balance.

```
In[11]:=iac[1] = immutableAccount.new[1000 dollars]
```

```
Out[11]= -immutableAccount-
```

```
In[12]:= iac[1] . balance[]
```

Out[12]= 1000 dollars

Making a withdrawal will create a new **immutableAccount** object, which needs a name. We can make up a new name, or use the same one if we like, or we could name it with a time-stamp argument (using **Date[]**), etc. If we use the same name, then the behavior of these accounts will be almost identical to the behavior of the mutable accounts. We choose to number the objects here to keep track of new objects as they are created. So, make a series of 50 dollar withdrawals from this account.

In[13]:= **Do[iac[n + 1] = iac[n] . withdraw[50 dollars], {n, 10}]**

A withdrawal no longer returns the new balance, but just indicates a new object. To observe the balance, we *have* to use the **balance** method. We can, in fact, generate a report about the status of the bank account.

In[14]:= **Table[{StringForm["iac[``]", n], iac[n] . balance[]}, {n, 10}] //**
         TableForm

Out[14]//TableForm=

iac[1]	1000 dollars
iac[2]	950 dollars
iac[3]	900 dollars
iac[4]	850 dollars
iac[5]	800 dollars
iac[6]	750 dollars
iac[7]	700 dollars
iac[8]	650 dollars
iac[9]	600 dollars
iac[10]	550 dollars

3.2.3 Inheritance

So far, there has been no inheritance involved except for what is inherited from the class **Object**. This consists of factory methods (i.e., methods sent to classes), as well as a few methods that are available for all objects: namely, **ClassQ**, **SuperClass**, **InstanceVariables**, **Methods**, **Class**, **isa**, **delete**, and **NIM**. Here are examples (which should be self-explanatory) of each of these, except for **delete** and **NIM**, which will be treated later.

In[15]:= **account . ClassQ[]**

Out[15]= True

```
In[16]:= account . SuperClass[]
```

Out[16]= Object

```
In[17]:= account . InstanceVariables[]
```

Out[17]= {bal}

```
In[18]:= account . Methods[]
```

Out[18]= {balance, Class, delete, deposit, InstanceVariables,
        isa, Methods, new, NIM, SuperClass, withdraw}

```
In[19]:= ac . Class[]
```

Out[19]= account

```
In[20]:= ac . isa[account]
```

Out[20]= True

The list of methods of **account** includes all of the factory methods as well as the methods defined for **account**. This is essentially what inheritance means. Without our explicitly saying so, a class has available to it all of the methods of its superclass. Note that **Class** with a single argument returns the class to which the argument belongs, provided it is an object. Finally, **isa** is a predicate between objects and classes which is **True** providing the object "is a" member of the class.

3.2.4 Interest paying accounts

As an example of a subclass, we will create a new kind of a bank account that pays interest. It should inherit all the usual behavior of the class **account** and add one new method which changes the balance by adding an interest payment to it. This is easy to do.

```
In[21]:=Class[interestAccount, account, {},
        {{new, super.new[#]&},
         {payInterest, (bal += (# bal))&}}]
```

Out[21]= interestAccount

Notice that **interestAccount** has no instance variables of its own, but in the method for paying interest we can refer to the instance variable of the superclass, since it has been prepended to the (empty list) of instance variables of **interestAccount**. Also, the method for **new** just refers to **new[super, #]&** which takes one parameter since the superclass,

account, has one instance variable. (It need not always be true that the number of extra arguments to **new** is the same as the number of instance variables, but in these simple examples that will always be the case.)

In[22]:= **interestAccount . InstanceVariables[]**

Out[22]= {bal}

Let

In[23]:= **intAc = interestAccount . new[1000 dollars]**

Out[23]= -interestAccount-

Consider a sequence of interest payments and withdrawals.

In[24]:= **intAc . payInterest[0.03]**

Out[24]= 1030. dollars

In[25]:= **intAc . withdraw[350 dollars]**

Out[25]= 680. dollars

In[26]:= **intAc . payInterest[0.03]**

Out[26]= 700.4 dollars

Now try the same thing with the immutable version of accounts.

3.2.5 Immutable interest paying accounts

In[27]:= **Class[immutableInterestAccount, immutableAccount, {},**
          {{new, super.new[#]&},
           {payInterest,
               new[immutableInterestAccount,
                  ((1 + #) bal)]&}}]

Out[27]= immutableInterestAccount

Note that immutability requires that the **payInterest** method also creates a new object. Set up an account.

In[28]:= **imIntAc[1] = immutableInterestAccount . new[1000 dollars]**

Out[28]= -immutableInterestAccount-

Consider the same sequence of interest payments and withdrawals, checking the balance at each stage.

In[29]:= **imIntAc[2] = imIntAc[1] . payInterest[0.03]**

Out[29]= -immutableInterestAccount-

In[30]:= **imIntAc[2] . balance[]**

Out[30]= 1030. dollars

Make a withdrawal.

In[31]:= **imIntAc[3] = imIntAc[2] . withdraw[350 dollars]**

Out[31]= -immutableAccount-

In[32]:= **imIntAc[3] . balance[]**

Out[32]= 680. dollars

In[33]:= **imIntAc[4] = imIntAc[3] . payInterest[0.03]**

Out[33]= payInterest[-immutableAccount-, 0.03]

In[34]:= **imIntAc[4] . balance[]**

Out[34]= balance[payInterest[-immutableAccount-, 0.03]]

Whoops!! **imIntAc[3]** is only an object belonging to the class **immutableAccount**, not an **immutableInterestAccount** object, and it no longer pays interest; i.e., it does not respond to a **payInterest** message. The owner of the account will be very unhappy about this state of affairs. What has happened? The trouble lies in the way the methods in the superclass **immutableAccount** are written. The two methods for **deposit** and **withdraw** are as follows:

```
{deposit,  new[immutableAccount, bal + #]&},
{withdraw, new[immutableAccount, bal - #]&}
```

The problem is that they say to make a new **immutableAccount** object, and this is what is inherited by the class **immutableInterestAccount**. Note that this works perfectly well as far as immutable accounts are concerned. It is only when a subclass is defined that a problem turns up. The solution is to use the special variable **self** in place of **immutableAccount** here. Actually, what we need is **Class[self]**. So, we'll have to start over again and write a correct version of **immutableAccount**.

3.2.6 A better version of immutable accounts

```
In[35]:=Class[betterImmutableAccount, Object, {bal},
          {{new,       (super.new[]; bal = #)&},
          {balance,   bal&},
          {deposit,   self.Class[].new[bal + #]&},
          {withdraw,  self.Class[].new[bal - #]&}}]
```

Out[35]= betterImmutableAccount

Set up an account.

```
In[36]:= biac[1] = betterImmutableAccount . new[1000 dollars]
```

Out[36]= -betterImmutableAccount-

Check that it still works.

```
In[37]:= biac[2] = biac[1] . withdraw[200 dollars]
```

Out[37]= -betterImmutableAccount-

```
In[38]:= biac[2] . balance[]
```

Out[38]= 800 dollars

Define a **betterImmutableInterestAccount** class exactly as before, except that its superclass is **betterImmutableAccount** and the **payInterest** method is also implemented with **Class[self]** in case we want to have further subclasses.

```
In[39]:=Class[betterImmutableInterestAccount,
          betterImmutableAccount, {},
          {{new, super.new[#]&},
           {payInterest,
             new[Class[self], ((1 + #) bal)]&}}]
```

Out[39]= `betterImmutableInterestAccount`

Now everything works as it should.

In[40]:=**biIntAc[1] =
    betterImmutableInterestAccount.new[1000 dollars]**

Out[40]= `-betterImmutableInterestAccount-`

In[41]:=**biIntAc[2] = biIntAc[1].payInterest[0.03]**

Out[41]= `-betterImmutableInterestAccount-`

In[42]:=**biIntAc[2].balance[]**

Out[42]= `1030. dollars`

In[43]:=**biIntAc[3] = biIntAc[2].withdraw[350 dollars]**

Out[43]= `-betterImmutableInterestAccount-`

In[44]:=**biIntAc[3].balance[]**

Out[44]= `680. dollars`

In[45]:=**biIntAc[4] = biIntAc[3].payInterest[0.03]**

Out[45]= `-betterImmutableInterestAccount-`

In[46]:=**biIntAc[4].balance[]**

Out[46]= `700.4 dollars`

3.3 Discussion

The moral of this sequence of examples is that the basic underlying structure of object-oriented programming is very natural and elegant. However, methods of classes that are intended to have subclasses must be written very carefully to be sure that they do not fail in unexpected ways. The variables **self** and **super** are essential ingredients for doing this and it takes some practice to learn to use them correctly. One slightly confusing difference is that **self** refers to an object–the current object–while **super** refers to a class–the superclass of the class of the current object. So far, **self** has only occurred in the combination **Class[self]** but, as will be seen, it is even more important as a way for an object to refer to itself during a message execution.

4 The hierarchy of point classes

4.1 Cartesian and Polar Points

As a final exercise in OOP, we return to Cartesian and polar points and implement them as classes. In fact, we will construct a small hierarchy of points that looks as follows:

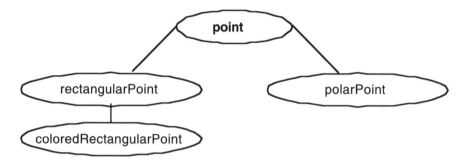

The classes **cartesianPoint** and **polarPoint** will be very similar to the constructions in Section 2. **Point** will be a new class that will contain the information about translating points and rotating them about the origin. Since both **cartesianPoint** and **polarPoint** are subclasses of **Point**, this information will be available to both of them. The class **coloredCar‐ tesianPoint** will add both a new instance variable and a new method. To begin with, **point** won't do anything, and will have to be redefined later.

In[1]:=**Needs["Classes`"]**

The cells defining various classes that follow have been made non-evaluatable since they are for illustrative purposes only. To try the examples, evaluate the cells in the implementation section at the end of this chapter.

```
Class[point, Object, {}, {{new, super.new[]&}}]
```

It is easy to define the classes for **cartesianPoint** and **polarPoint**.

```
Class[cartesianPoint, point, {x, y},
      {{new, (super.new[];(x = #1);(y = #2))&},
       {cartesianCoords, {x, y}&},
       {polarCoords, {Sqrt[x^2 + y^2], ArcTan[y/x]}& }}]
```

```
Class[polarPoint, point, {r, theta},
      {{new, (super.new[];(r = #1);(theta = #2))&},
       {cartesianCoords, {r Cos[theta], r Sin[theta]}& },
       {polarCoords, {r, theta}& }}]
```

For instance:

In[8]:=**pt = cartesianPoint.new[3, 4]**

Out[8]= -cartesianPoint-

In[9]:=**pt.polarCoords[]**

Out[9]= $\left\{5, \text{ArcTan}\left[\frac{4}{3}\right]\right\}$

Now we are ready to add methods **translate** and **rotate** to the class **point**. This definition replaces the one above. If the previous one has been evaluated, it is necessary to reevaluate the definitions of the classes **cartesianPoint** and **polarPoint**.after evaluating the new definition of **point**.

```
Class[point, Object, {},
   {{new, super.new[]&},
    {translate,
       cartesianPoint.new[
          Sequence@@(self.cartesianCoords[] + #)]&},
    {rotate,
       polarPoint.new[
          Sequence@@(self.polarCoords[] + {0, #})]&} }]
```

The methods here illustrate an important use of **self**. The class **point** is abstract; there are no objects belonging to this class. Every point is either a Cartesian or a polar point. Nevertheless, methods for **translate** and **rotate** can be implemented in the class **point** by using **self**. If a translate message is sent to a polar point **ppt**, then, when the method body

```
new[ cartesianPoint,
       Sequence@@ (cartesianCoords[self] + #)] &
```

is evaluated, **self** refers to **ppt** and so its response to **cartesianCoords** is used. Here are some sample computations.

In[10]:=**pt1 = pt.translate[5, 5]**

Out[10]= -cartesianPoint-

```
In[11]:=pt1.cartesianCoords[]
```

Out[11]= {8, 9}

$$\text{In[12]:= } \texttt{pt2 = pt1 . rotate}\left[\frac{\pi}{4}\right]$$

Out[12]= -polarPoint-

```
In[13]:=pt2.cartesianCoords[].N[]
```

Out[13]= {-0.707107, 12.0208}

4.2 Adding the Subclass coloredCartesianPoint: Overriding Methods

There is a problem in constructing the subclass **coloredCartesianPoint** of **cartesian-Point**. In the implementation of the class **point**, we made use of the observation that it is simple to **translate** a Cartesian point and equally simple to **rotate** a polar one, by explicitly creating a new Cartesian point or polar point in the appropriate place. For instance, if we **translate** a polar point, it will be turned into a Cartesian point. That's OK because all of our operations work for either kind of point and we don't want to worry about which representation is used for a particular point. But that means that in these messages, we cannot replace **new[cartesian, ---]** or **new[polar, ---]** by **new[Class[self], ---]** as we did with immutable interest accounts in Section 3.2.6. So how can we make these operations work for colored points, where we do want the output again to be colored?

It is possible for a class to override a method in its superclass. It does this just by including a method with the same name and a new body. The new method body doesn't necessarily have to have any relation to the body of the method in the superclass, and it is this new body that will be used when the method is invoked with an object belonging to the class. So, one solution to our problem is to just write new methods with the names **translate** and **rotate** for the class **coloredCartesianPoint**. However, another solution is to think a bit and realize that we would also like to have methods that will turn a Cartesian or polar point into a colored point and a method to forget the color of a colored point. If these are included in the appropriate classes, then we can write new methods in the class **coloredCartesianPoint** in a more elegant form. This involves changing the implementation of the class **point** by adding the following method.

```
{makeColored,
    coloredCartesianPoint.new[
        Sequence@@(self.cartesianCoords[]), #]&}
```

(See the version of the class **point** in the implementation section below. It has to be evaluated for the changes to take place.)

Now we can construct the class **coloredCartesianPoint**.

```
Class[coloredCartesianPoint, cartesianPoint, {colorname},
    {{new, (super.new[#1, #2]; (colorname = #3))&},
     {color, colorname&},
     {forgetColor,
        super.new[Sequence@@(self.cartesianCoords[])]&},
     {translate,
        self.forgetColor[].translate[#].
           self.color[].makeColored[]&},
     {rotate,
        self.forgetColor[].rotate[#].self.color[].makeColored[]&}
    }]
```

The methods for **translate** or **rotate** are implemented in terms of the methods in the top class **point**. Any changes made there will be propagated throughout the entire hierarchy of classes. Here is an example of translating a colored point.

In[14]:=**cpt = coloredCartesianPoint.new[3, 4, red]**

Out[14]= $-$coloredCartesianPoint$-$

In[15]:=**cpt1 = cpt.translate[{5, 5}]**

Out[15]= $-$coloredCartesianPoint$-$

In[16]:=**{cpt1.cartesianCoords[], cpt1.color[]}**

Out[16]= $\{\{8, 9\},$ red$\}$

In[17]:= **cpt2 = cpt1 . rotate$\left[\dfrac{\pi}{4}\right]$**

Out[17]= $-$coloredCartesianPoint$-$

In[18]:= **{cpt2 . cartesianCoords[], cpt2 . color[]}**

Out[18]= $\left\{\left\{\sqrt{145}\,\text{Cos}\left[\dfrac{\pi}{4} + \text{ArcTan}\left[\dfrac{9}{8}\right]\right], \sqrt{145}\,\text{Sin}\left[\dfrac{\pi}{4} + \text{ArcTan}\left[\dfrac{9}{8}\right]\right]\right\},\ \text{red}\right\}$

4.3 Isomorphism Testing

How can we decide if two points are the same? What does it mean for them to be the same? It turns out in object-oriented programming that this is not just a philosophical question. (See [Budd] for a thorough discussion.) If we examine what *Mathematica* thinks a point actually is, e.g.,

```
In[19]:= ??cpt1

    Global`cpt1
    cpt1 =
    Classes`Private`coloredCartesianPoint[colorname$19, x$19, y$19]
```

we see that it is a complicated expression involving some numbered local variables. Thus, any two points that we create are going to have different internal representations, and so they can never be identical in *Mathematica*'s sense of **Equal** or **SameQ**. Nevertheless, we would like to construct a message to send to a point asking if it is the same as another point. We arbitrarily decide that Cartesian or polar points are the same if they have the same Cartesian coordinates. Colored points are the same if they also have the same colors, and a colored point may or may not be the same as a non-colored point. This is easily done by adding the following methods to **point** and **coloredCartesianPoint**, respectively.

```
{isomorphicQ,
    cartesianCoords[self].SameQ[cartesianCoords[#]]&}


{isomorphicQ,
    (self.forgetColor[].isomorphicQ[
        #.forgetColor[]]  &&
    self.color[].SameQ[#.color[]])&}
```

These methods are included in the definitions in the Implementation section below. Note that **isomorphicQ** is a message that is sent to a point and has another point as a parameter. For instance, define several points

```
In[20]:= pt = cartesianPoint.new[3, 4];
    ppt = polarPoint.new[5, ArcTan[4/3]];
    cptred = coloredCartesianPoint.new[3, 4, red];
    cptgreen = coloredCartesianPoint.new[3, 4, green];
```

The Cartesian point and the polar point are the same.

```
In[24]:= pt.isomorphicQ[ppt]
```

Out[24]= True

The two colored points are different.

In[25]:=`cptred.isomorphicQ[cptgreen]`

Out[25]= `False`

If the Cartesian point and the first colored point are compared, then the result depends on who gets the message.

In[26]:=`pt.isomorphicQ[cptred]`

Out[26]= `True`

In[27]:=`cptred.isomorphicQ[pt]`

Out[27]= `False`

As always, the methods of the object receiving the message are used in responding to the message. It is slightly unsettling that isomorphism is not a symmetric relation, but that is often the case in object-oriented languages. Make sure that you understand why it happens here.

4.4 The Message NIM

One last concern is the message **NIM**, which is part of the class **Object** and hence available for all classes. It stands for Non-Implemented Method and is a catch-all method to allow for messages which have not been implemented in some class but are required to be implemented by all subclasses of the class. Its use is often a matter of housekeeping. For instance, the method **cartesianCoords** is required to be implemented in all subclasses of **point**, so we could add a method

```
{cartesianCoords,
    NIM[self, cartesianCoords]&}
```

to the class **point**, with a similar method for **polarCoords**. These would have no effect since they are in fact implemented in all subclasses. However, if we added a method

```
{color, NIM[self, color]&}
```

to the class **point** and then sent the message **color** to a polar point, for instance, then the result would be an error message saying that the method **color** was not implemented for polar points. We omit these methods here, but they are used in Chapter 13 on object-oriented graph theory.

5 *Exercises*

1. Add messages **dilate** and **affineTransform** to the class **point**. Here **dilate** takes one parameter, which is a real number, while **affineTransform** takes two parameters, a matrix and a vector.

2.   i) Consider only one representation for points, that of Cartesian coordinates, so there is no need for a separate class **point**. However, the class of Cartesian points (which might just as well be called simply **point** now) has two subclasses: colored points and directed points. A colored point has an extra attribute of color as before. A directed point is a point together with a unit vector specifying a direction. Make sure that when colored or directed points are translated or rotated the result is again colored or directed.

ii) Add a third subclass, that of rigid directed points in which the angle between the unit vector and the vector from the origin to the point is preserved by a translation or a rotation.

3. Implement three-dimensional points using representations via Cartesian, cylindrical, and spherical coordinates. Include methods for translation and dilation. What should be done about rotations?

6 *Implementation*

```
    Needs["Classes`"]

In[2]:= Unprotect[Dot];
    SetAttributes[Dot, HoldRest];
    Dot[object_, method_[parameters___]] :=
      method[object, parameters]
    Protect[Dot];

In[4]:=Class[point, Object, {},
      {{new, super.new[]&},
       {translate,
          cartesianPoint.new[
             Sequence@@(self.cartesianCoords[] + #)]&},
       {rotate,
          polarPoint.new[
             Sequence@@(self.polarCoords[] + {0, #})]&},
       {makeColored,
          coloredCartesianPoint.new[
```

```
            Sequence@@(self.cartesianCoords[]), #]&},
        {isomorphicQ,
            self.cartesianCoords[].SameQ[#.cartesianCoords[]]&}  }];

In[5]:= Class[cartesianPoint, point, {x, y},
            {{new, (super.new[]; (x = #1); (y = #2)) &},
             {cartesianCoords, {x, y} &},
             {polarCoords, {Sqrt[x^2 + y^2], ArcTan[y/x]} & }}];

In[6]:= Class[polarPoint, point, {r, theta},
            {{new, (super.new[];(r = #1);(theta = #2))&},
             {cartesianCoords, {r Cos[theta], r Sin[theta]}& },
             {polarCoords, {r, theta}& }}];

In[7]:= Class[coloredCartesianPoint, cartesianPoint, {colorname},
            {{new, (super.new[#1, #2]; (colorname = #3))&},
             {color, colorname&},
             {forgetColor,
                self.SuperClass[].new[
                    Sequence@@(self.cartesianCoords[])]&},
             {translate,
                self.forgetColor[].translate[#].makeColored[
                    self.color[]]&},
             {rotate,
                self.forgetColor[].rotate[#].makeColored[
                    self.color[]]&},
             {isomorphicQ,
                (self.forgetColor[].isomorphicQ[#.forgetColor[]]
                 && self.color[].SameQ[#.color[]])&}  }];
```

Graphics Programming

1 Introduction to Graphics Primitives

Producing complex and beautiful two- and three-dimensional graphics using the functions **Plot**, **Plot3D**, **ContourPlot**, **DensityPlot**, etc., is trivial in *Mathematica*. But *Mathematica* also includes a set of primitive graphics objects that you can use to build up complex pictures by employing all the facilities of the *Mathematica* programming language. For example, the following draws a point, a line, and a filled polygon:

```
In[1]:= Show[Graphics[
          {Point[{2, 0.5}],
           Line[{{0, 0}, {4, 4}}],
           Polygon[{{1, 1}, {1, 3}, {3, 3}, {3, 1}}]  } ]];
```

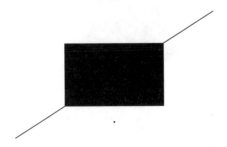

There are two kinds of graphics primitives: *geometric objects* and *graphics modifiers* or *directives* (to be called just modifiers here). Geometric objects are expressions with heads such as **Point**, **Line**, and **Polygon**. These heads are like **List** in that they don't process their arguments in any way; they just hold them together and indicate that they are particular kinds of objects. **Point** takes a single argument which is a "point". Both **Line** and **Polygon** take a single argument which is a list of "points", i.e., pairs interpreted as coordinates of points in the plane. **Line** will draw a line through the points in the same way as **ListPlot**, whereas **Polygon** draws the same line but then joins the last point to the first point and fills in the enclosed region. A *graphics object* is any expression with head **Graphics**. It takes one argument which is a list (of lists . . .) of graphics primitives, so, for instance, the expression

```
Graphics[{Point[{2, 0.5}], Line[{{0, 0}, {4, 4}}],
          Polygon[{{1, 1}, {1, 3},{3, 3}, {3, 1}}] }]
```

is a graphics object. A graphics object is displayed by using the command **Show**. **Show** is like **Print** for graphics; the actual picture is a side effect and the output is just the expression -Graphics-. Options can be specified either for **Graphics** or for **Show** as extra optional named arguments. It will make a difference where the optional arguments are placed when we consider **GraphicsArray**.

Graphics modifiers are graphics primitives that control various aspects of geometric objects. We can change the picture above by using the modifiers **PointSize**, **Thickness**, and **GrayLevel**. Each of these modifiers has to precede the geometric object it is intended to affect, and will affect everything (in the same sublist) that follows it.

```
In[2]:= Show[Graphics[
            {PointSize[0.05], Point[{2, 0.5}],
             Thickness[0.02], Line[{{0, 0}, {4, 4}}],
             GrayLevel[0.4],
             Polygon[{{1, 1}, {1, 3},{3, 3}, {3, 1}}] } ]];
```

If the package **Graphics`Colors`** is loaded, then color names can be used as modifiers, again preceding the geometric objects they modify. They can only be seen, of course, on a color monitor. For readability, we group each object with its modifiers separately.

```
In[3]:= Needs["Graphics`Master`"]
```

```
In[4]:= Show[Graphics[
            {{Red, PointSize[0.05], Point[{2, 0.5}]},
             {ForestGreen, Thickness[0.02], Line[{{0, 0}, {4, 4}}]},
```

```
{CornflowerBlue,
  Polygon[{{1, 1}, {1, 3}, {3, 3}, {3, 1}}]}} ]];
```

Here is a slightly more complicated design in which **Table** does some of the work.

```
In[5]:= Show[Graphics[
          {Table[{Hue[i/20], PointSize[i/250 + 1/50],
              Point[{Cos[Pi i/10], Sin[Pi i/10]}]},
            {i, 20}]}
        ], PlotRange -> {{-1.2, 1.2}, {-1.2, 1.2}},
          AspectRatio -> 1];
```

You can use the full power of *Mathematica* in producing the list of graphics primitives.

```
In[6]:= Show[Graphics[
          {Table[
            {Hue[i/20],
             PointSize[i/250 + 1/50],
             Point[{Cos[Pi i/10], Sin[Pi i/10]}],
             Polygon[{{0, 0},
               0.9 {Cos[Pi (2i-1)/20], Sin[Pi (2i-1)/20]},
               0.9 {Cos[Pi (2i+1)/20], Sin[Pi (2i+1)/20]}}] },
            {i, 20}] }
```

```
], PlotRange -> {{-1.2, 1.2}, {-1.2, 1.2}},
   AspectRatio -> 1];
```

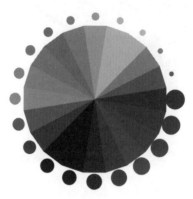

The following displays a bunch of random lines of varying thicknesses:

```
In[7]:= Show[Graphics[
    Table[{Thickness[0.005 + 0.0001 i],
          Line[{{Random[], Random[]},{Random[], Random[]}}]},
        {i, 30}] ]];
```

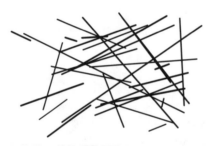

A slight variation shows a random filled polygon with 50 edges:

```
In[8]:= Show[Graphics[
    {GrayLevel[0.2],
     Polygon[Table[{Random[], Random[]}, {50}] ] } ]];
```

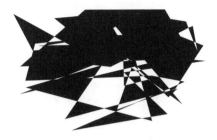

Note that a folded polygon is shaded using exclusive or; i.e., if a region is covered an even number of times then it appears white, while if it is covered an odd number of times then it is shaded.

2 Two-Dimensional Graphics Objects, Graphics Modifiers, and Options

2.1 Objects

Here is a list of the built-in two-dimensional graphics objects which can be displayed by a `Show[Graphics[---]]` command.

```
Point[{x0, y0}]
Line[{{x0, y0}, ...}]
Rectangle[{xmin, ymin}, {xmax, ymax}]
Polygon[{{x0, y0}, ...}]
Circle[{xcenter, ycenter}, radius],
    Circle[{xcenter, ycenter}, {semiaxis, semiaxis}]
    Circle[{xcenter, ycenter}, radius, {theta1, theta2}]
Disk[{xcenter, ycenter}, radius]
    Disk[{xcenter, ycenter}, {semiaxis, semiaxis}]
    Disk[{xcenter, ycenter}, radius, {theta1, theta2}]
Raster[numberArray]
    Raster[numberArray, rectangle]
RasterArray[modifierArray]
    RasterArray[modifierArray, rectangle]
Text[expr, {xcenter, ycenter}]
    Text[expr, {xcenter, ycenter}, {xoffset, yoffset}]
PostScript["string"]
```

2.1.1 Circle and disk

We have already discussed **Point**, **Line**, and **Polygon**. **Circle** and **Disk** can take other optional arguments giving ellipses and sectors of circles. Here are several examples.

```
In[9]:= Show[Graphics[
         {Table[Circle[{0, 0}, {1, 1 - i}], {i, 1, 0, -0.2}],
          Table[{GrayLevel[i], Disk[{2, 0}, {1, 1 - i}]},
               {i, 0, 0.8, 0.2}],
          Circle[{4, 0}, 1, {Pi/2, 3 Pi/2}],
          GrayLevel[0.5], Disk[{4, 0}, 1, {-Pi/2, Pi/2}]}
          ], AspectRatio -> Automatic];
```

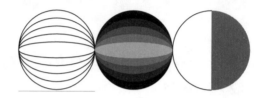

It is not possible to make a sector of an ellipse. A single such figure can be shown by using **AspectRatio**.

2.1.2 Raster and RasterArray

Raster and **RasterArray** produce rectangular arrays of gray or colored rectangles. The first argument of **Raster** has to be a matrix of values between 0 and 1, which are interpreted as gray levels. The optional second argument is the rectangle in which the array of gray levels should be drawn. The first argument of **RasterArray** is a matrix of modifiers–**GrayLevel**, **Hue**, or **RGBColor**.

```
In[10]:= Show[Graphics[
         {Raster[Table[Sin[x y],
                     {x, Pi/5, Pi, Pi/5}, {y, Pi/5, Pi, Pi/5}],
               {{0, 0}, {1, 1}}],
          RasterArray[Table[Hue[Sin[x y]],
                     {x, Pi/5, Pi, Pi/5}, {y, Pi/5, Pi, Pi/5}],
               {{1, 0}, {2, 1}}] }
          ], AspectRatio -> Automatic];
```

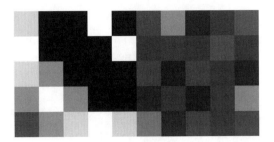

2.1.3 Text

Text can be included in a **Graphics** object using a **Text** object. The first argument of **Text** is an expression, which may or may not be a string, and the second argument describes the position of the text. The optional third argument describes how the text is offset from the center according to conventions described in The *Mathematica* Book. One can also choose a specific font for the text using the format illustrated below.

```
In[11]:= Show[Graphics[
        {Circle[{0, 0}, 1],
         Text[FontForm["Text in a circle", {"Chicago", 10}],
             {0, 0.5}],
         Text[center, {0, 0}],
         Text[right,  {0, 0}, {-3,  0}],
         Text[left,   {0, 0}, { 3,  0}],
         Text[above,  {0, 0}, { 0, -3}],
         Text[below,  {0, 0}, { 0,  3}]}
        ], AspectRatio -> 1];
```

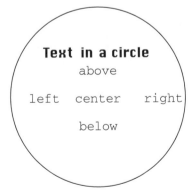

2.1.4 PostScript

Here is an example of a **PostScript** object. The single argument to **PostScript** is a string consisting of PostScript directions for making a drawing, written in the usual PostScript format.

```
In[12]:= Show[Graphics[
            {PostScript["
                0 0 moveto
                1 0 lineto
                1 1 lineto
                0 1 lineto
                closepath
                0.02 0.02 moveto
                0.98 0.98 lineto
                0.02 0.98 moveto
                0.98 0.02 lineto
                stroke"]}
            ], AspectRatio -> 1];
```

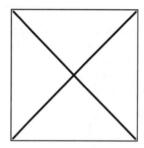

2.2 Modifiers

There are 10 modifiers that apply to two-dimensional graphics objects. In the following, **d** is any positive number, usually a small decimal.

```
PointSize[d]
AbsolutePointSize[d]
Thickness[d]
AbsoluteThickness[d]
Dashing[{d1, ..., }]
AbsoluteDashing[{d1, ..., }]
GrayLevel[r]                    0 ≤ r ≤ 1
Hue[r]                         0 ≤ r ≤ 1
    Hue[r, s, b]               0 ≤ r, s, b ≤ 1
RGBColor[r, g, b]              0 ≤ r, g, b ≤ 1
```

CMYKColor[c, m, y, b] $0 \leq c, m, y, b \leq 1$

2.2.1 PointSize and AbsolutePointSize, etc.

The difference between **PointSize** and **AbsolutePointSize** is that a **PointSize** dimension such as 0.01 means $1/100$ of the linear size of the displayed figure. If the figure is resized and made smaller, then the point will also be smaller. **AbsolutePointSize** dimensions are absolute lengths measured in units of printer's points, which are approximately $1/72$ of an inch.

In[14]:= **Show[Graphics[**
          {{PointSize[0.05], Point[{0, 0}]},
            {AbsolutePointSize[5], Point[{1, 0}]} }
      ], PlotRange -> {{-0.2, 1.2}, {-0.1, 0.1}}];

If the preceding graphic is resized, then the left-hand dot will change size while the right-hand one remains constant. The same comments apply to **Thickness** vs. **AbsoluteThickness** and **Dashing** vs. **AbsoluteDashing**.

2.2.2 Hue

Hue can take either one argument or three. In either case, the first argument refers to a color shade from the circumference of a color wheel, scaled between 0 and 1. The values 0 and 1 are both red and 0.5 is blue. Values smaller than 0.5 shade through green, yellow, and orange to red, while those larger that 0.5 shade through purple and violet to red. If **Hue** has a second and third argument, then the second is saturation and the third is brightness. The following pictures show the effect of varying both hue and saturation while keeping brightness equal to 1. The **Hue** scale starts at 0.1, whereas saturation starts at 0, where the colors are almost indistinguishable.

In[15]:= **Show[Graphics[{**
          Table[{Hue[i/10, j/10, 1],
              Rectangle[{i, j}, {i+1, j+1}]},
              {i, 1, 10}, {j, 0, 10}]}],
          AspectRatio -> 1];

2.2.3 RGBColor

RGBColor works differently. It takes three arguments, which are intensities of red, blue, and green color, respectively. The pure colors vary from black (= 1) to full intensity (= 1), as illustrated in the next drawing.

```
In[16]:= Show[Graphics[
           {Table[
              {PointSize[0.1], RGBColor[i, 0, 0],
               Point[{i, 0.2}]}, {i, 0, 1, 0.1}],
            Table[
              {PointSize[0.1], RGBColor[0, i, 0],
               Point[{i, 0.1}]}, {i, 0, 1, 0.1}],
            Table[
              {PointSize[0.1], RGBColor[0, 0, i],
               Point[{i, 0}]}, {i, 0, 1, 0.1}]}
           ], PlotRange -> {{-0.1, 1.1}, {-0.05, 0.25}}];
```

If the colors are combined, then the intensities add. Both **Hue** and **RGBColor** refer to transmitted light so adding colors in **RGBColor** is like adding colored lights. Here are the results of adding colors two at a time, keeping the total intensity equal to 1.

```
In[17]:= Show[Graphics[
            {Table[
              {PointSize[0.1], RGBColor[i, 1 - i, 0],
                Point[{i, 0.2}]}], {i, 0, 1, 0.1}],
            Table[
              {PointSize[0.1], RGBColor[0, i, 1 - i],
                Point[{i, 0.1}]}], {i, 0, 1, 0.1}],
            Table[
              {PointSize[0.1], RGBColor[1 - i, 0, i],
                Point[{i, 0}]}], {i, 0, 1, 0.1}]}
            ], PlotRange -> {{-0.1, 1.1}, {-0.05, 0.25}}];
```

To see the whole range of possible colors requires a three dimensional cube which one can peer into to any depth. We show three faces of such a cube, the green–red face, the blue–red face, and the blue–green face.

```
In[5]:= Show[GraphicsArray[
        {Graphics[
          {Table[{RGBColor[i, j, 0],
            Polygon[{{i, j}, {i, j+0.1},
                    {i+0.1, j+0.1}, {i+0.1, j}}]},
            {i, 0, 1, 0.1}, {j, 0, 1, 0.1}],
          Text["red 0 to 1", {0.5, -0.05}],
          Text["green 0 to 1 ", {-0.15, 0.5}, {0, 0}, {0, 1}]},
              AspectRatio -> Automatic,
              PlotRange->{{-.15, 1.1}, {-.15, 1.1}}],
        Graphics[
          {Table[{RGBColor[i, 0, j],
            Polygon[{{i, j}, {i, j+0.1},
                    {i+0.1, j+0.1}, {i+0.1, j}}]},
            {i, 0, 1, 0.1}, {j, 0, 1, 0.1}],
          Text["red 0 to 1", {0.5, -0.05}],
          Text["blue 0 to 1 ", {-0.15, 0.5}, {0, 0}, {0, 1}]},
              AspectRatio -> Automatic,
              PlotRange->{{-.15, 1.1}, {-.15, 1.1}}],
        Graphics[
          {Table[{RGBColor[0, i, j],
            Polygon[{{i, j}, {i, j+0.1},
```

```
                    {i+0.1, j+0.1}, {i+0.1, j}}]},
          {i, 0, 1, 0.1}, {j, 0, 1, 0.1}],
      Text["green 0 to 1", {0.5, -0.05}],
      Text["blue 0 to 1 ", {-0.15, 0.5}, {0, 0}, {0, 1}]},
          AspectRatio -> Automatic,
          PlotRange->{{-.15, 1.1}, {-.15, 1.1}}] }]];
```

The colors in the upper right-hand regions of these squares are yellow, magenta, and cyan. These appear in the next section on CMYK colors. One can look inside the cube by displaying the layers parallel to the blue–red face given by adding in green stepwise.

```
In[19]:= Show[GraphicsArray[
    {Table[Graphics[
      Table[{RGBColor[i, k, j],
       Polygon[{{i, j}, {i, j+0.2},
                {i+0.2, j+0.2}, {i+0.2, j}}]},
      {i, 0, 1, 0.2}, {j, 0, 1, 0.2}],
          AspectRatio -> Automatic,
          PlotLabel -> "green = "<>ToString[k]],
      {k, 0, 0.4, 0.2}],
    Table[Graphics[
      Table[{RGBColor[i, k, j],
       Polygon[{{i, j}, {i, j+0.2},
                {i+0.2, j+0.2}, {i+0.2, j}}]},
      {i, 0, 1, 0.1}, {j, 0, 1, 0.2}],
          AspectRatio -> Automatic,
          PlotLabel -> "green = "<>ToString[k]],
      {k, 0.6, 1.0, 0.2}] }]];
```

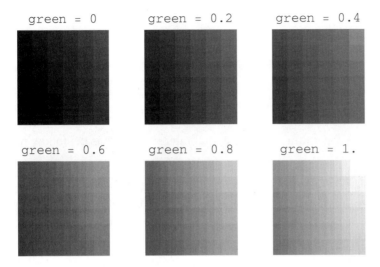

| green = 0 | green = 0.2 | green = 0.4 |
| green = 0.6 | green = 0.8 | green = 1. |

2.2.4 CMYKColor

CMYKColor is another scheme for specifying colors which is adapted to printing. The letters stand for cyan, magenta, yellow, and black and refer to specific printer's inks whose standards are carefully maintained. The three colors cyan, magenta, and yellow are essentially the complements of the colors red, green, and blue, and they work in the opposite way by removing colors (since they represent reflected colors) rather than adding them. Thus, for instance, a zero value represents white rather than black, as the following pure colors show.

```
In[20]:= Show[Graphics[
          {Table[
             {PointSize[0.1], CMYKColor[i, 0, 0, 0],
                Point[{i, 0.2}]}, {i, 0, 1, 0.1}],
           Table[
             {PointSize[0.1], CMYKColor[0, i, 0, 0],
                Point[{i, 0.1}]}, {i, 0, 1, 0.1}],
           Table[
             {PointSize[0.1], CMYKColor[0, 0, i, 0],
                Point[{i, 0}]}, {i, 0, 1, 0.1}]}
          ], PlotRange -> {{-0.1, 1.1}, {-0.05, 0.25}}];
```

Combining the colors two at a time, keeping a total intensity of 1, has the following effect.

```
In[21]:= Show[Graphics[
        {Table[
           {PointSize[0.1], CMYKColor[i, 1 - i, 0, 0],
              Point[{i, 0.2}]}, {i, 0, 1, 0.1}],
         Table[
           {PointSize[0.1], CMYKColor[0, i, 1 - i, 0],
              Point[{i, 0.1}]}, {i, 0, 1, 0.1}],
         Table[
           {PointSize[0.1], CMYKColor[1 - i, 0, i, 0],
              Point[{i, 0}]}, {i, 0, 1, 0.1}]}
        ], PlotRange -> {{-0.1, 1.1}, {-0.05, 0.25}}];
```

To see the whole range of possible colors would require a four dimensional cube this time which one could peer into to any depth. We ignore the effect of adding black, which decreases the intensity of the colors, and imagine a three-dimensional cube as before. We show three faces of such a cube, the magenta–cyan face, the yellow–cyan face, and the yellow–magenta face.

```
In[6]:= Show[GraphicsArray[
      {Graphics[
       {Table[{CMYKColor[i, j, 0, 0],
          Polygon[{{i, j}, {i, j+0.1},
                   {i+0.1, j+0.1}, {i+0.1, j}}]},
         {i, 0, 1, 0.1}, {j, 0, 1, 0.1}],
        Text["cyan 0 to 1", {0.5, -0.05}],
        Text["mag 0 to 1 ", {-0.15, 0.5}, {0, 0}, {0, 1}]},
           AspectRatio -> Automatic,
```

```
            PlotRange->{{-.15, 1.1}, {-.15, 1.1}}],
   Graphics[
   {Table[{CMYKColor[i, 0, j, 0],
     Polygon[{{i, j}, {i, j+0.1},
             {i+0.1, j+0.1}, {i+0.1, j}}]},
    {i, 0, 1, 0.1}, {j, 0, 1, 0.1}],
    Text["cyan 0 to 1", {0.5, -0.05}],
    Text["yel 0 to 1 ", {-0.15, 0.5}, {0, 0}, {0, 1}]},
        AspectRatio -> Automatic,
        PlotRange->{{-.15, 1.1}, {-.15, 1.1}}],
   Graphics[
   {Table[{CMYKColor[0, i, j, 0],
     Polygon[{{i, j}, {i, j+0.1},
             {i+0.1, j+0.1}, {i+0.1, j}}]},
    {i, 0, 1, 0.1}, {j, 0, 1, 0.1}],
    Text["mag 0 to 1", {0.5, -0.05}],
    Text["yel 0 to 1 ", {-0.15, 0.5},
             {0, 0}, {0, 1}]},
   AspectRatio -> Automatic,
        PlotRange->{{-.15, 1.1}, {-.15, 1.1}}] }]];
```

The colors in the upper right-hand regions of these squares are blue, green, and red, which appeared in the last section on RGB colors. One can look inside the cube by displaying the layers parallel to the yellow–cyan face given by adding in magenta stepwise.

```
In[23]:= Show[GraphicsArray[
    {Table[Graphics[
      Table[{CMYKColor[i, k, j, 0],
       Polygon[{{i, j}, {i, j+0.2},
               {i+0.2, j+0.2}, {i+0.2, j}}]},
      {i, 0, 1, 0.2}, {j, 0, 1, 0.2}],
          AspectRatio -> Automatic,
          PlotLabel -> "magenta = "<>ToString[k]],
     {k, 0, 0.4, 0.2}],
     Table[Graphics[
      Table[{CMYKColor[i, k, j, 0],
       Polygon[{{i, j}, {i, j+0.2},
               {i+0.2, j+0.2}, {i+0.2, j}}]},
      {i, 0, 1, 0.1}, {j, 0, 1, 0.2}],
```

```
      AspectRatio -> Automatic,
      PlotLabel -> "magenta = "<>ToString[k]],
{k, 0.6, 1.0, 0.2}] }]];
```

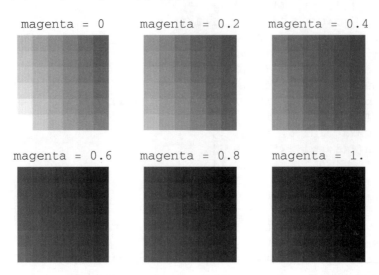

2.3 Options

The options available for **Graphics** are almost the same as those for **Plot** except for those options affecting the smoothness of a curve.

In[24]:=**Complement[Options[Graphics], Options[Plot]]**

Out[24]= {Axes → False}

Options can be given either inside the **Graphics** expression as last arguments or as last arguments to **Show**. **Show** does not have any specific options for itself, but uses anything that makes sense for the graphics objects it is displaying.

3 Combining Built-In Graphics with Graphics Primitives

There are two separate ways to combine built-in graphics with graphics primitives, depending on whether one is modifying a built-in graphics function by adding graphics primitives, or instead building up a graphics object by including elements produced by built-in graphics routines.

3.1 Modifying Built-in Graphics with Graphics Primitives

3.1.1 PlotStyle

We have already illustrated the use of **PlotStyle** to change the appearance of built-in graphics routines. The general format is **PlotStyle -> {{- - -}, ...}**, where each sublist applies to the curve in the corresponding position in the list of curves to be plotted. The entries in the sublist can be any graphics modifiers, e.g., **PlotStyle ->{-Thickness[0.02]}**.

3.1.2 Prolog and Epilog

Prolog and **Epilog** are options to all of the built-in graphics functions that allow one to add arbitrary graphic primitives to them. The difference between the two is that **Epilog** graphics are produced after the built-in graphics and hence print on top of them, while **Prolog** graphics are produced first so the built-in graphics print on top.

```
In[1]:= Plot[Sin[x], {x, 0, 2Pi},
     Epilog -> {
       {PointSize[0.05], Point[{Pi, -0.75}]},
       {Thickness[0.02], Line[{{0, -1}, {2Pi, 1}}]},
       {GrayLevel[0.4],
        Polygon[{{2, -0.5}, {4, -0.5}, {4, 0.5}, {2, 0.5}}],
        Text["Primitives on top", {Pi, 0.75}]} }];
```

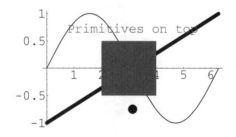

```
In[2]:= Plot[Sin[x], {x, 0, 2Pi},
     Prolog -> {
       {PointSize[0.05], Point[{Pi, -0.75}]},
       {Thickness[0.02], Line[{{0, -1}, {2Pi, 1}}]},
       {GrayLevel[0.4],
        Polygon[{{2, -0.5}, {4, -0.5}, {4, 0.5}, {2, 0.5}}],
        Text["Built-ins on top", {Pi, 0.75}]} }];
```

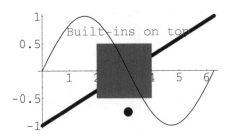

3.2 Adding Built-In Graphics to Graphics Objects

Show can be used to display several built-in graphics plots together just by giving several arguments to **Show**; i.e., **Show** can take any number of arguments which are graphics objects. In particular, **Show** will display both built-in graphics and graphics objects constructed from graphics primitives at the same time. Thus, the picture constructed in the previous section using **Epilog** can equally well be made as follows:

```
In[3]:= Show[Plot[Sin[x], {x, 0, 2 Pi}, DisplayFunction -> Identity],
    Graphics[
      {{PointSize[0.05], Point[{Pi, -1}]},
       {Thickness[0.02], Line[{{0, -1}, {2Pi, 1}}]},
       {GrayLevel[0.4],
        Polygon[{{2, -0.5}, {4, -0.5}, {4, 0.5}, {2, 0.5}}]}
      }], DisplayFunction -> $DisplayFunction];
```

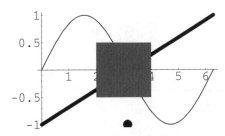

Here, things are displayed in the order in which they are given, so if the **Plot** and the **Graphics** were reversed, then the result would be the same as using **Prolog** instead of **Epilog**. **Show** will display any number of **Graphics** objects in any order. For instance, here is another way to add text to a **Plot**, just treating it as another graphics object.

```
In[4]:= Show[Plot[Sin[x], {x, 0, 2 Pi},
            DisplayFunction-> Identity],
        Graphics[{Text["sin and cos together", {Pi, 0.5}]}],
        Plot[Cos[x], {x, 0, 2 Pi},
```

```
        DisplayFunction -> Identity],
    DisplayFunction -> $DisplayFunction];
```

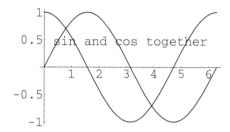

4 *Graphics Arrays and Graphics Rectangles*

4.1 *Graphics Arrays*

Show can actually take six kinds of arguments:

Graphics, GraphicsArray, Graphics3D,
SurfaceGraphics, ContourGraphics, and **DensityGraphics**

We have discussed the first above, and here we look at **GraphicsArray**. A **GraphicsArray**
object is a list or a matrix of graphics objects, where these can be any of the other five types
(since **GraphicsArray** is not a type of graphics). In order to have something to draw, recall
that the Fourier sine series for an odd, periodic function **f[x]** of period 2π is given by calculating the Fourier sine coefficients using the following formula.

In[5]:= **B[f_, n_, x_] :=** $\dfrac{2}{\pi} \displaystyle\int_0^\pi$ **f[x] Sin[n x] dx**

Then, the **n**th Fourier sine series approximation to **f[x]** is given by

In[6]:= **sinApprox[f_, n_, x_] :=** $\displaystyle\sum_{k=1}^{n}$ **B[f, k, x] Sin[k x]**

The step function, which is –1 between –π and 0 and +1 between 0 and π corresponds to the
constant function 1 between 0 and π made into an odd periodic function, so, for instance, its
fifth Fourier sine series approximation is

In[7]:= **sinApprox[1&, 5, x]**

Out[7]= $\dfrac{4 \, \text{Sin}[x]}{\pi} + \dfrac{4 \, \text{Sin}[3 \, x]}{3 \, \pi} + \dfrac{4 \, \text{Sin}[5 \, x]}{5 \, \pi}$

Note that the even approximations are the same as the preceding odd approximations. Define the step function by

In[8]:= `step[x_] := If[x > 0, 1, -1]`

The first six approximations to this square wave can be illustrated in a single plot.

```
In[9]:= Show[GraphicsArray[
    Table[
    Plot[Evaluate[{step[x],
                    sinApprox[1&, 2(3 i + j) + 1, x]}],
        {x, -Pi, Pi},
        DisplayFunction -> Identity,
        PlotStyle -> {Hue[1], Hue[0.7]},
        Axes -> False,
        PlotLabel -> "Approximation "<>ToString[3 i + j]],
    {i, 0, 1}, {j, 3}]
], DisplayFunction -> $DisplayFunction];
```

Approximation 1   Approximation 2   Approximation 3

Approximation 4   Approximation 5   Approximation 6

4.2 Graphics Rectangles

Instead of using **GraphicsArray** to display several drawing in the same picture, one can use the geometric object **Rectangle** with an optional third argument.

In[10]:= `??Rectangle`

Rectangle[{xmin, ymin}, {xmax, ymax}] is a two-dimensional
  graphics primitive that represents a filled rectangle, oriented
  parallel to the axes. Rectangle[{xmin, ymin}, {xmax, ymax},
  graphics] gives a rectangle filled with the specified graphics.
  Attributes[Rectangle] = {Protected}

```
In[11]:= Show[Graphics[
        Table[Rectangle[{i-0.5, i-0.5}, {i+0.5, i+0.5},
          Graphics[{
            Line[{{i-0.5, i+0.5}, {i+0.5, i-0.5}}],
            AbsolutePointSize[10 (i + 1)],
            Hue[i/4], Point[{i, i}]}],
          AspectRatio -> 1,
          DisplayFunction -> Identity]
        ], {i, 0, 3}]],
        DisplayFunction -> $DisplayFunction,
        AspectRatio -> 1];
```

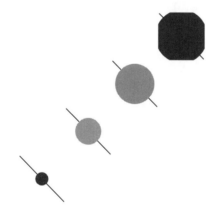

5 Examples of Two-Dimensional Graphics

5.1 Histogram plots

Recall from Chapter 8 the program **histogram**.

```
In[12]:= histogram[data_, {xmin_, xmax_, xstep_}] :=
        Map[{{Min[#], Max[#]}, Plus@@Map[Count[data, #]&, #]}&,
          Partition[Range[xmin, xmax], xstep]]
```

Here is a new set of data to use with it.

```
In[13]:=data = Table[Random[Integer, {1, 100}], {500}];
```

```
In[14]:=histo = histogram[data, {0, 99, 10}]
```

```
Out[14]= {{{0, 9}, 46}, {{10, 19}, 42}, {{20, 29}, 55},
        {{30, 39}, 57}, {{40, 49}, 55}, {{50, 59}, 46},
        {{60, 69}, 48}, {{70, 79}, 44}, {{80, 89}, 48}, {{90, 99}, 52}}
```

We will use *Mathematica* graphics primitives to construct a graphics object illustrating the output of **histogram** in order to see what it looks like. The output from the histogram function consists of pairs of the form {{a, b}, c}, where {a, b} is an interval on the x-axis giving the size of a bucket, and c is the number of items in the bucket. We want to plot this as a rectangle on the base {a, b} of height c, which in *Mathematica* is described by **Rectangle[{a,   0},   {b,   c}]**. Thus, we restructure the histogram, using the technique of local rewrite rules discussed at the end of Chapter 7, Section 6.2, to turn it into the appropriate form. Here is the first very simple version.

```
In[15]:=histoGraphics[histogram_] :=
       Graphics[
         histogram //.
           {{a_, b_}, c_?NumberQ} :> Rectangle[{a, 0}, {b, c}]]
```

Note that the condition on **c** is required to prevent infinite recursion. As usual, the *Mathematica* command **Show** displays this graphics object.

```
In[16]:=Show[histoGraphics[histo], Axes -> True];
```

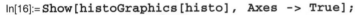

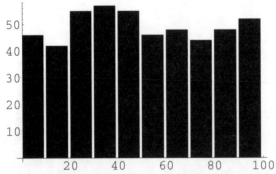

If all we care about is the final picture, then there is no reason to calculate it in two steps. We can just generate the desired list of rectangles directly.

```
In[17]:=histoGraphics1[data_, {xmin_, xmax_, xstep_}] :=
       Graphics[
         Map[Rectangle[{Min[#], 0},
           {Max[#], Plus@@Map[Count[data, #]&, #]}]&,
           Partition[Range[xmin, xmax], xstep]]]
```

```
In[18]:= Show[histoGraphics1[data, {0, 99, 10}]];
```

The picture is omitted since it is exactly the same as the previous picture.

　　Here is a somewhat more complicated graphics object that shades the buckets differently depending on their contents and includes a count of the number of items in each bucket at the top of each column. Since we need the height of each bar several times, it is necessary to first generate the histogram and then process it.

```
In[19]:= histoGraphicsCount[data_, {xmin_, xmax_, xstep_}] :=
      Module[
        {histo = histogram[data, {xmin, xmax, xstep}],
         maxnum},
        maxnum = Max[Map[#[[2]]&, histo]];
        Graphics[
          {histo //. {{a_, b_}, c_?NumberQ} :>
               {Hue[N[c/maxnum]], Rectangle[{a, 0}, {b, c}]},
           histo //.{{a_, b_}, c_} :>
               Text[c, N[{(a + b)/2, 1.05 c + 1}]]}] ]
```

```
In[20]:= Show[histoGraphicsCount[data, {0, 99, 10}], Axes -> True];
```

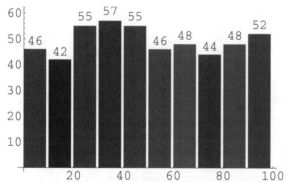

Finally, this program can also take its data from a file by embedding it in a larger routine which includes both **histogram** and **histoGraphicsCount**. Note that in the following program we use **ReadList**, rather than **Read**, because it reads in the entire contents of **numbs** as a list, which is exactly what we want as the argument to **histogram**. In order to compare this final program with the original C program, everything is written out in detail rather than using the previous definitions.

```
In[21]:= histoGraphicsFile[file_String, {xmin_, xmax_, xstep_}]:=
        Module[
          {numbs = ReadList[file], histo, maxnum},
          histo =
            Map[{{Min[#], Max[#]},
                 Plus@@Map[Count[numbs, #]&, #]}&,
```

```
            Partition[Range[xmin, xmax], xstep]];
     maxnum = Max[Map[#[[2]]&, histo]];
     Show[
      Graphics[
       {histo //. {{a_, b_}, c_?NumberQ} :>
           {Hue[N[c/maxnum]], Rectangle[{a, 0}, {b, c}]},
        histo //.{{a_, b_}, c_} :>
           Text[c, N[{(a + b)/2, 1.05 c + 1}]]}],
      Axes -> True]]
```

We reconstruct the file **numbers1** from Chapter 8, Section 4.7.

```
In[22]:=OutputForm[
     TableForm[
        Table[Random[Integer, {1, 100}], {200}],
        TableSpacing -> {0, 2}
     ]
   ] >> numbers1
```

```
In[23]:=histoGraphicsFile["numbers1", {0, 109, 10}];
```

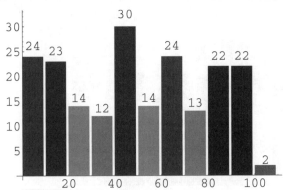

5.2 A simple bar chart

This example is related to the histogram example, except that this time we have data for given values and we want to plot the data by showing bars rather than by using something like **ListPlot**. We will use the frequencies command discussed in Chapter 6, Section 4.3, to generate the values to be plotted from a list of data.

```
In[24]:=frequencies[list_List] :=
        Map[{#, Count[list, #]}&, Union[list]]
```

Here is some sample data and the value of frequencies for it.

```
In[25]:=frequencies[data = {1, 2, 3, 4, 3, 2, 6, 5, 3, 7,
                            6, 9, 8, 5, 7, 6, 3, 5, 4}]
```

```
Out[25]= {{1, 1}, {2, 2}, {3, 4}, {4, 2},
         {5, 3}, {6, 3}, {7, 2}, {8, 1}, {9, 1}}
```

We would like to convert the output of frequencies into a rectangles by using local rewrite rules similar to the ones used in **histoGraphics**, of the form:

```
frequencies[data]//.{a_, b_} :>
                        Rectangle[{a, 0}, {a + 1, b}]
```

Interestingly, there does not seem to be any way to do this that doesn't lead to infinite recursion. So instead, we have to process the output of **frequencies** functionally. The shading is constructed as part of the same functional process.

```
In[26]:=barChart[data_] :=
        Module[{freq = frequencies[data],maxnum},
          maxnum = Max[Map[#[[2]]&, freq]];
          Show[Graphics[
            {Map[{Hue[N[#[[2]]/(maxnum)]],
                Rectangle[{#[[1]], 0},
                          {#[[1]] + 1, #[[2]]}]}&, freq]}],
          Axes -> True,
          AxesOrigin -> {Min[Map[#[[1]]&, freq]] - 1, 0} ] ]
```

Here is the plot for our simple data above.

```
In[27]:=barChart[data];
```

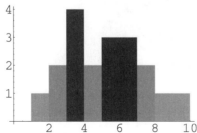

To get some more interesting data to plot, we use one of the statistical distributions in the packages.

In[28]:=**Needs["Statistics`ContinuousDistributions`"]**

We make a table of 50 scores selected at random from a normal distribution with mean 75 and standard deviation 10.

In[29]:=**scores =**
        Table[Floor[Random[NormalDistribution[75, 10]]], {50}];

In[30]:=**barChart[scores];**

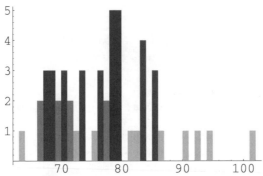

This doesn't look very much like a normal distribution, so we try again with 1000 scores, narrowing the standard deviation a bit.

In[31]:=**scores1 =**
        Table[Floor[Random[NormalDistribution[75, 9]]], {1000}];

In[32]:=**barChart[scores1];**

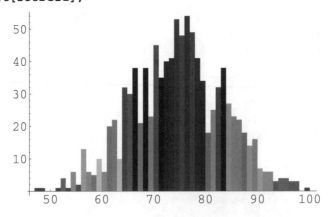

6 *Three-Dimensional Graphics Primitives*

6.1 *Three-Dimensional Objects, Modifiers, and Options*

By simply substituting **Graphics3D** for **Graphics** and adding a third dimension to each coordinate, we can produce 3D graphics. **Point**, **Line**, **Polygon**, and **Text** are as before. There is a new primitive geometric object, a **Cuboid**. If one is willing to calculate and type in the coordinates of the relevant points, then there is no limitation to the three-dimensional figures that can be constructed.

```
In[1]:= Show[Graphics3D[
        {PointSize[0.05], Point[{0, 2, 2}], Point[{4, 2, 2}],
        Thickness[0.02], Line[{{0, 2, 2}, {4, 2, 2}}],
        Polygon[{{0, 0, 0}, {1.5, 1.5, 1.5},
                {2.5, 1.5, 1.5}, {4, 0, 0}}],
        Polygon[{{4, 0, 0}, {2.5, 1.5, 1.5},
                {2.5, 2.5, 1.5}, {4, 4, 0}}],
        Polygon[{{4, 4, 0}, {2.5, 2.5, 1.5},
                {2.5, 2.5, 2.5}, {4, 4, 4}}],
        Polygon[{{1.5, 2.5, 2.5}, {2.5, 2.5, 2.5},
                {4, 4, 4}, {0, 4, 4}}],
        Polygon[{{0, 4, 4}, {1.5, 2.5, 2.5},
                {1.5, 2.5, 1.5}, {0, 4, 0}}],
        Polygon[{{0, 4, 0}, {1.5, 2.5, 1.5},
                {1.5, 1.5, 1.5}, {0, 0, 0}}],
        Cuboid[{1.5, 1.5, 1.5}, {2.5, 2.5, 2.5}] }]];
```

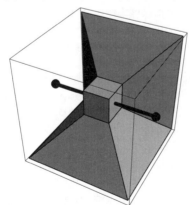

All of the modifiers for two-dimensional graphics are available along with three new ones:

```
EdgeForm[specification]
FaceForm[frontspec, backspec]
SurfaceColor[specification]
```

EdgeForm is simple to use. **EdgeForm[]** means no edges are to be drawn. Otherwise, **specification** can be any modification or list of modifications involving **Hue**, **RGBColor**, **CMYK-Color**, **GrayLevel**, or **Thickness**. **FaceForm** is only useful if one can see both the front and back faces of some list of similar polygons and one wants them colored or shaded differently. **frontspec** and **backspec** can be color or shading modifiers or a **SurfaceColor** object. It is more complicated and will be discussed below. There are many new options available for **Graphics3D**.

In[2]:= **Complement[Options[Graphics3D], Options[Graphics]]**

Out[2]= {AmbientLight → GrayLevel[0], AspectRatio → Automatic,
      AxesEdge → Automatic, Boxed → True, BoxRatios → Automatic,
      BoxStyle → Automatic, FaceGrids → None, Lighting → True,
      LightSources → {{{1., 0., 1.}, RGBColor[1, 0, 0]}, {{1., 1., 1.},
          RGBColor[0, 1, 0]}, {{0., 1., 1.}, RGBColor[0, 0, 1]}},
      Plot3Matrix → Automatic, PolygonIntersections → True,
      RenderAll → True, Shading → True,
      SphericalRegion → False, ViewCenter → Automatic,
      ViewPoint → {1.3, -2.4, 2.}, ViewVertical → {0., 0., 1.}}

The only one of these that is familiar is **AspectRatio**, which just has a different default value here. **AmbientLight** specifies the general overall illumination level of the graphics. Its value can be either a **GrayLevel**, **Hue**, or **RGBColor** specification. **AxesEdge** determines on which edges of the display the axes should be drawn. See The *Mathematica* Book for a description of its possible values. The three next options, **Boxed**, **BoxRatios**, and **BoxStyle**, refer to the enclosing box drawn around the graphics object. **Boxed** itself is either **True** or **False**. Changing **BoxRatios** can distort the graphics by making it fit in a strangely shaped box. **BoxStyle** can take a list of modifiers such as **GrayLevel**, **Hue**, **Thickness**, and **Dashing**. **FaceGrids** determines if the faces of the bounding box should have grids drawn on them. It is similar to the option **GridLines** for two-dimensional graphics. See The *Mathematica* Book for directions. The options **Lighting** and **LightSources** determine if the graphics should appear to be colored by reflecting light from the indicated point sources. **Plot3Matrix** has been replaced by **ViewCenter** and **ViewVertical**. **PolygonIntersections** and **RenderAll** affect how polygons are drawn and for which ones PostScript code is generated. **Spherical-Region** is mainly useful in creating graphics animations. The final three determine the relative position of the graphics object in the viewing area. **ViewPoint** can be set from a special graphics dialog box in notebook versions of *Mathematica*. Here is a slight modification of the preceding graphics using some of these.

```
In[3]:= Show[Graphics3D[
       {PointSize[0.05], Point[{0, 2, 2}], Point[{4, 2, 2}],
        Thickness[0.02], Line[{{0, 2, 2}, {4, 2, 2}}]],
        Polygon[{{0, 0, 0}, {1.5, 1.5, 1.5},
                {2.5, 1.5, 1.5}, {4, 0, 0}}],
        Polygon[{{4, 0, 0}, {2.5, 1.5, 1.5},
                {2.5, 2.5, 1.5}, {4, 4, 0}}],
        Polygon[{{4, 4, 0}, {2.5, 2.5, 1.5},
                {2.5, 2.5, 2.5}, {4, 4, 4}}],
        Polygon[{{1.5, 2.5, 2.5}, {2.5, 2.5, 2.5},
                {4, 4, 4}, {0, 4, 4}}],
        Polygon[{{0, 4, 4}, {1.5, 2.5, 2.5},
                {1.5, 2.5, 1.5}, {0, 4, 0}}],
        Polygon[{{0, 4, 0}, {1.5, 2.5, 1.5},
                {1.5, 1.5, 1.5}, {0, 0, 0}}],
        {EdgeForm[{Hue[1], Thickness[0.02]}],
          Cuboid[{1.5, 1.5, 1.5}, {2.5, 2.5, 2.5}]}
       }], Boxed -> False,
          ViewPoint->{1.091, -2.930, 1.294}];
```

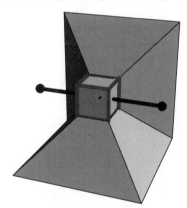

Of course, we prefer to use *Mathematica* itself to create the graphics objects rather than typing in coordinates. E.g., here is a five sided random folded polygon.

```
In[4]:= Show[Graphics3D[
       Polygon[Table[{Random[], Random[], Random[]}, {5}] ],
       ViewPoint->{1.711, -2.751, 0.975} ]];
```

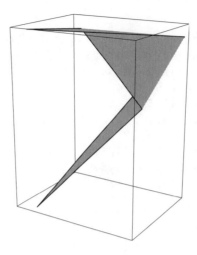

6.2 Three-Dimensional Objects in Packages

6.2.1 Shapes

The graphics packages supplied with *Mathematica* contain a number of extra three-dimensional geometrical objects in two different packages. The following are in **Graphics`-Shapes`**.

```
Cylinder[radius(1), height(1), number(20)]
Cone[radius(1), height(1), number(20)]
Torus[radius(1), radius(0.5), number(20), number(10)]
Sphere[radius(1), number(20), number(15)]
MoebiusStrip[radius(1), radius(0.5), number(20)]
Helix[radius(1), height(0.5), turns(2), number(20)]
DoubleHelix[radius(1), height(0.5), turns(2), number(20)]
```

The numbers in parentheses are the default values when the names are used without any specified arguments. These geometric objects are all displayed in a standard position centered on the vertical axis at the origin. In order to locate them differently, it is necessary to use the two operations **TranslateShape[shape, {x, y, z}]** and **RotateShape[shape, φ, θ, ψ]** that are found in the same package. **TranslateShape** is easy to understand; it just translates every coordinate by the given vector. **RotateShape** is more complicated since φ, θ and ψ refer to Euler angles and there are different conventions concerning them. First of all, *Mathematica* uses the European convention, which has the names φ and ψ interchanged with the American convention. However, it keeps them in the American order, which is confusing. Secondly, **RotateShape** makes use of **RotationMatrix3D**, which is in the package **Geometry`Rotations`** and which calculates the appropriate matrix for a rotation with given Euler angles. The matrix used refers to what are called "body coordinates" in physics.

However, *Mathematica* constructs everything with reference to a fixed set of "space coordinates", which means that the transpose of this matrix should have been used. To correct this inside **RotateShape**, it is necessary to use negative angles. Thus, we redefine these two important constructs as follows:

In[5]:=**Needs["Graphics`Master`"]**

In[6]:= **Needs["Geometry`Rotations`"]**

In[7]:=**rotationMatrix3D[φ_, Θ_, ψ_] :=**
        Transpose[RotationMatrix3D[φ, Θ, ψ]]

In[8]:=**rotateShape[shape_, φ_, Θ_, ψ_] :=**
        RotateShape[shape, -ψ, -Θ, -φ]

Then we can illustrate Euler angles by showing their effect on a standard coordinate system.

In[9]:=**axes =**
        {{Thickness[0.02], Line[{{0, 0, 0}, {1, 0, 0}}],
          Text["x", {1.1, 0, 0}]},
         {Thickness[0.0175], Line[{{0, 0, 0}, {0, 1, 0}}],
          Text["y", {0, 1.1, 0}]},
         {Thickness[0.015], Line[{{0, 0, 0}, {0, 0, 1}}],
          Text["z", {0, 0, 1.1}]} };

We also want to change the view point

In[10]:=**Show[Graphics3D[axes],**
          Boxed -> False, ViewPoint->{2.996, 0.318, 1.540}];

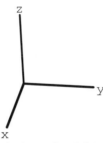

The effect of a rotation by Euler's angles (phi, theta, psi) = $(\pi/6, \pi/4, \pi/5)$ can be demonstrated by showing the successive positions of the axes under these rotations. We write three graphics commands with suppressed outputs and then show all three pictures by a **GraphicsArray**. In each case, the new position of the coordinate axes is shown in black, and the old position is in gray. The labels of the original **x**, **y**, and **z** axes remain in all three pictures.

```
In[11]:=pict1 = Show[Graphics3D[
        {{GrayLevel[0.5], axes},
         {rotateShape[axes, π/6, 0, 0],
          Text["x'",
               rotationMatrix3D[N[π/6], 0, 0].{1.2, 0, 0}],
          Text["y'",
               rotationMatrix3D[N[π/6], 0, 0].{0, 1.2, 0}],
          Text["z'",
               rotationMatrix3D[N[π/6], 0, 0].{0, 0, 1.3}]}}
      ], Boxed -> False,
         ViewPoint->{2.996, 0.318, 1.540},
         PlotLabel -> "Rotate about the \nz-axis by π/6",
         DisplayFunction -> Identity];

In[12]:=pict2 = Show[Graphics3D[
        {{GrayLevel[0.5],
          rotateShape[axes, π/6, 0, 0],
          Text["x'",
               rotationMatrix3D[N[π/6], 0, 0].{1.1, 0, 0}],
          Text["y'",
               rotationMatrix3D[N[π/6], 0, 0].{0, 1.1, 0}],
          Text["z'",
               rotationMatrix3D[N[π/6], 0, 0].{0, 0, 1.3}]},
         {rotateShape[axes, π/6, π/4, 0],
          Text["x''",
            rotationMatrix3D[N[π/6], N[π/4], 0].{1.4, 0, 0}],
          Text["y''",
            rotationMatrix3D[N[π/6], N[π/4], 0].{0, 1.2, 0}],
          Text["z''",
            rotationMatrix3D[N[π/6], N[π/4], 0].{0, 0, 1.1}]}}
      ], Boxed -> False,
          ViewPoint->{2.996, 0.318, 1.540},
          PlotLabel->"Rotate about the \nnew x'-axis by π/4",
          DisplayFunction -> Identity];

In[13]:=pict3 = Show[Graphics3D[
       {{GrayLevel[0.5],
         rotateShape[axes, π/6, π/4, 0],
         Text["x''",
           rotationMatrix3D[N[π/6], N[π/4], 0].{1.2, 0, 0}],
         Text["y''",
           rotationMatrix3D[N[π/6], N[π/4], 0].{0, 1.2, 0}],
         Text["z''",
           rotationMatrix3D[N[π/6], N[π/4], 0].{0, 0, 1.1}]},
        {rotateShape[axes, π/6, π/4, π/5],
         Text["x'''",
```

```
            rotationMatrix3D[N[π/6],N[π/4],N[π/5]].{1.4,0,0}],
          Text["y'''",
            rotationMatrix3D[N[π/6],N[π/4],N[π/5]].{0,1.2,0}],
          Text["z'''",
            rotationMatrix3D[N[π/6],N[π/4],N[π/5]].{0,0,1.3}]}}
      ], Boxed -> False,
        ViewPoint->{2.996, 0.318, 1.540},
        PlotLabel->"Rotate about the \nnew z''-axis by π/5",
        DisplayFunction -> Identity];
```

In[14]:= `Show[GraphicsArray[{pict1, pict2, pict3}],`
         `DisplayFunction -> $DisplayFunction];`

The first picture is the action of $\varphi = \pi/6$, which is a rotation about the **z**-axis by $\pi/6$. The new **z'** axis is the same as the **z**-axis. The second picture is the action of Θ, which is a rotation about the new **x'**-axis by $\pi/4$. The new **x'**-axis is the same as the **x'**-axis. Finally, the third picture is the action of ψ, which is a rotation about the new **z''**-axis by $\pi/5$. The new **z'''**-axis is the same as the **z''**-axis. Using translations and rotations, we can make a construction from the graphics objects in the **Shapes** package.

In[15]:= `Show[Graphics3D[`
```
        {Cuboid[{-1, -1, -1}, {1, 1, 1}],
         TranslateShape[Cylinder[], {0, 0, 2}],
         rotateShape[TranslateShape[Cone[], {0, 0, 2}], 0, Pi/2, 0],
         TranslateShape[Sphere[0.5], {0, -3, 0}],
         rotateShape[TranslateShape[Helix[], {0, 0, 2}],
                     Pi/2, Pi/2, 0]}
        ], Axes -> True];
```

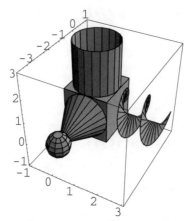

Here, the **Cylinder**, **Cone**, and **Helix** are all originally placed on top of the **Cuboid**. Then the **Cone** is rotated about the **x'**-axis = the **x**-axis (because $\varphi = 0$) by $\Theta = \pi/2$. (Note that the **x**-axis runs along the lower front of the box.) For the **Helix**, the new **x'**-axis is the **y**-axis (because $\varphi = \pi/2$) and it is rotated about this axis by $\Theta = \pi/2$.

6.2.2 Polyhedra

In the package **Graphics`Polyhedra`**, there are more shapes which are displayed in a somewhat different manner.

```
Tetrahedron
Cube
Octahedron
Dodecahedron
Icosahedron
Hexahedron
GreatDodecahedron
SmallStellatedDodecahedron
GreatStellatedDodecahedron
GreatIcosahedron.
```

These are actually the names of the lists of polygons making up the various shapes. they are converted into **Graphics3D** objects by affixing the head **Polyhedron**. In addition, **Polyhedron** can take two optional arguments specifying the center of the shape and a scaling number specifying its size. In order to try them out, load the **Graphics`Master`** package if it hasn't already been loaded. The default location of the center is {0, 0, 0}, with default size equal to 1.

In[16]:= **Show[Polyhedron[Tetrahedron], Axes -> True];**

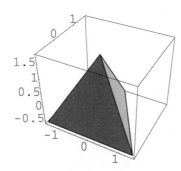

All of the regular solids are shown here, in different locations.

```
In[17]:= Show[Polyhedron[Cube, {0, 0, -1.5}],
         Polyhedron[Tetrahedron],
         Polyhedron[Octahedron, {0, -1.5, 1.5}, 0.8],
         Polyhedron[Dodecahedron, {1.5, 0.5, 1.5}, 0.8],
         Polyhedron[Icosahedron, {-1.5, 0.5, 1.5}, 0.8],
         Axes -> True];
```

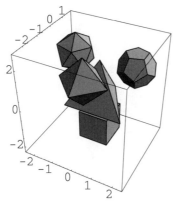

The polygons making up one of these shapes can be accessed by using **First[-Polyhedron[name]]**. Thus:

```
In[18]:= First[Polyhedron[Tetrahedron]]
```

Out[18]= {Polygon[{{0, 0, 1.73205},
        {0, 1.63299, -0.57735}, {-1.41421, -0.816497, -0.57735}}],
    Polygon[{{0, 0, 1.73205}, {-1.41421, -0.816497, -0.57735},
        {1.41421, -0.816497, -0.57735}}],
    Polygon[{{0, 0, 1.73205}, {1.41421, -0.816497, -0.57735},
        {0, 1.63299, -0.57735}}],
    Polygon[{{0, 1.63299, -0.57735}, {1.41421, -0.816497, -0.57735},
        {-1.41421, -0.816497, -0.57735}}]}}

Since a polyhedron consists of a list of polygons, this description can be used together with graphics modifiers to create polyhedra with other characteristics. For instance, a dodecahedron has 12 faces which can be colored by giving a list of 12 hues. In order to see these colors, we have to turn off the default lights.

```
In[19]:= Show[Graphics3D[
        Transpose[{
            Table[Hue[1 - i/12], {i, 12}],
            First[Polyhedron[Dodecahedron]]}]
        ], Lighting -> False];
```

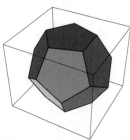

A cube just fits inside an octahedron with its vertices touching the faces of the octahedron. (Here, **WireFrame** is an operation from the package **Graphics`Shapes`** which removes the surfaces, just leaving the edges of the polygons.)

```
In[20]:= Show[Polyhedron[Cube],
            WireFrame[Polyhedron[Octahedron, {0, 0, 0}, 1.45]],
            Boxed -> False];
```

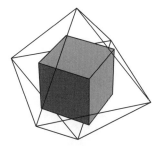

A rotated cube fits inside a dodecahedron with its vertices the same as some of the vertices of the dodecahedron and its edges lying in the faces of the dodecahedron.

```
In[21]:=Show[
        rotateShape[Polyhedron[Cube], 0.35, 0.54, 0],
        rotateShape[WireFrame[
          Polyhedron[Dodecahedron, {0, 0, 0}, 1.15]],
          0, 0, 0],
        Boxed -> False,
        ViewPoint->{2.222, -2.451, 0.713}];
```

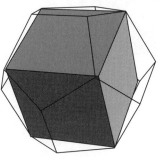

6.2.3 Color in three-dimensional graphics

```
In[22]:=Needs["Graphics`Master`"]

In[23]:=Show[Graphics3D[
        {SurfaceColor[GrayLevel[0.2], GrayLevel[0.8], 5],
         Sphere[]}],
        LightSources -> {{{1., 0., 1.}, GrayLevel[0.9]},
                         {{0., 1., 1.}, GrayLevel[0.9]}}];
```

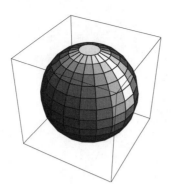

```
In[24]:= Show[Graphics3D[
        {SurfaceColor[RGBColor[0.9, 0.9, 0.9], White, 10],
         Sphere[]}],
        LightSources -> {{{1., 0., 0.3}, Red},
                         {{0., 1., 0.3}, Yellow},
                         {{-0.3, 0., 1.}, Blue}}];
```

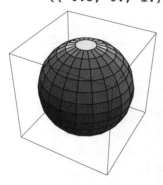

6.2.4 Combining three-dimensional graphics

The output of **Plot3D** is -**SurfaceGraphics**- so it cannot be used together with other **Graphics3D** constructions. The solution is that **Graphics3D[SurfaceGraphics[---]]** converts the **SurfaceGraphics** to a **Graphics3D** object. However, different **Surface-Graphics** can be combined with **Show**.

```
In[25]:= Show[Plot3D[Sin[x y], {x, 0, Pi}, {y, 0, Pi},
               DisplayFunction -> Identity],
          Plot3D[Cos[x y], {x, 0, Pi}, {y, 0, Pi},
               DisplayFunction -> Identity],
          DisplayFunction -> $DisplayFunction];
```

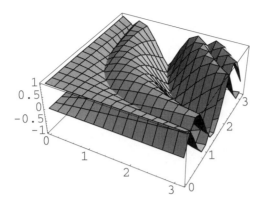

```
In[26]:= Show[Graphics3D[Plot3D[Sin[x y], {x, 0, Pi}, {y, 0, Pi},
              DisplayFunction -> Identity]],
         Polyhedron[Dodecahedron, {Pi/2, Pi/2, 0}, 1.5],
         DisplayFunction -> $DisplayFunction];
```

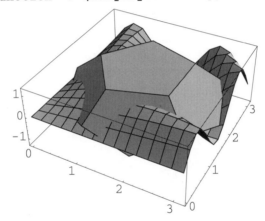

7 Exercises

1. Make pictures of the partial sums of the Maclaurin's series approximation to sin x.

2. Complete the discussion of Fourier series approximations. Use cos series and the full sin and cos series for given periodic functions.

3. Make pictures of the solutions of partial differential equations.

4. In the picture with the cube, cylinder, cone, and helix, where would the helix appear if the directions for it are replaced by

```
rotateShape[TranslateShape[Helix[], {0, 0, 2}], π/2, π/2, π/2]
```

Make a picture to see if your prediction is correct.

5. Make some of the other pictures of regular solids fitting nicely inside other regular solids.

Some Finer Points

1 Introduction

There are still many things to be learned about using *Mathematica* as a programming language. In this chapter six miscellaneous topics are collected together to help you fine-tune your programming abilities: packages, attributes, named optional arguments, evaluation, general recursive functions, and substitution and the lambda calculus. The first four are basic aspects of the *Mathematica* programming language, while the last two consider how more general programming issues are treated in *Mathematica*. There are several sources for further information; e.g., The *Mathematica* Journal and news features on computer networks. A good way to deepen your knowledge is to read other people's programs and try to decide why things are written the way they are. The packages supplied with *Mathematica* are a good place to start. When you find that you can do better by writing briefer, more transparent, more cogent, or faster programs, then you have begun to master *Mathematica*.

2 Packages

Packages are the final organizing ingredient in the *Mathematica* language. These are structures that enable one to completely isolate certain portions of code from the outside world. Not only are variables protected as in **Module**s, but function definitions are protected as well. The structural feature that permits this is the notion of contexts, which will require some explanation.

2.1 Contexts

Contexts make themselves evident in a setting familiar to most users of *Mathematica*. For example, let us try to evaluate a Laplace transform, forgetting that it is defined in a package.

In[1]:= **LaplaceTransform[2 t^3, t, s]**

Out[1]= LaplaceTransform[2 t^3, t, s]

As is to be expected, nothing happens because we haven't loaded the appropriate package. So, load the package.

In[2]:=**Needs["Calculus`LaplaceTransform`"]**

```
   LaplaceTransform::shdw :
    Symbol LaplaceTransform appears in multiple
       contexts {Calculus`LaplaceTransform`, Global`};
       definitions in context Calculus`LaplaceTransform` may
       shadow or be shadowed by other definitions.
```

What does this strange message about "multiple contexts" and "definitions . . . shadowed by other definitions" mean? Furthermore, **LaplaceTransform** still doesn't work.

In[3]:= **LaplaceTransform[2 t^3, t, s]**

Out[3]= LaplaceTransform[2 t^3, t, s]

Maybe the message means we have to clear **LaplaceTransform** before using it.

In[4]:=**Clear[LaplaceTransform]**

In[5]:= **LaplaceTransform[2 t^3, t, s]**

Out[5]= LaplaceTransform[2 t^3, t, s]

It still doesn't work, but there is a more powerful way to clear expressions, after which it finally works.

In[6]:=**Remove[LaplaceTransform]**

In[7]:= **LaplaceTransform[2 t^3, t, s]**

Out[7]= $\dfrac{12}{s^4}$

This is all very strange, but note well the remedy for functions refusing to work after they have been tried before the appropriate package is loaded: namely, **Remove** the offending function. What is going on? First we have to understand names in *Mathematica*.

2.2 Names

Names are an important aspect of *Mathematica*. Everything, in fact, depends on the way names are handled. The command **Names[string]** returns all names known to *Mathematica* at the time of its being run that match the given pattern.

```
In[8]:=Names["B*"]
```

```
Out[8]= {Background, Backward, BaseForm, Baseline, Before, Begin,
         BeginPackage, Below, BernoulliB, BesselI, BesselJ, BesselK,
         BesselY, Beta, BetaRegularized, BinaryGet, BinaryOp,
         Binomial, Blank, BlankForm, BlankNullSequence, BlankSequence,
         Block, Bold, BoldItalic, Bottom, BoxData, BoxDimensions,
         Boxed, BoxForm, BoxFrame, BoxMargins, BoxRatios, BoxRegion,
         BoxShift, BoxSizeAdjustments, BoxStyle, Break, Button,
         ButtonBox, ButtonCell, ButtonContents, ButtonData,
         ButtonEvaluator, ButtonExpandable, ButtonFrame, ButtonFunction,
         ButtonMargins, ButtonMinHeight, ButtonMnemonic, ButtonNote,
         ButtonNotebook, ButtonSource, ButtonStyle, Byte, ByteCount}
```

In this example, **"B*"** stands for all words beginning with B. The ***** is a wild card that matches anything. Certain names have values attached to them, either because they have built-in values or because values have been assigned in the current session. **Clear[name]** clears values assigned to **name**. As an experiment, give **aa** the value 5 by an assignment statement.

```
In[9]:=aa = 5
```

```
Out[9]= 5
```

Then, of course, **aa** has the value 5, as we can check.

```
In[10]:=aa
```

```
Out[10]= 5
```

Now clear **aa** and observe that it no longer has a value.

In[11]:=**Clear[aa]**

However, *Mathematica* still knows about **aa**, as the following demonstrates.

In[12]:=**Names["a*"]**

Out[12]= {aa}

Clear[name] or **Clear["nameform"]** removes values assigned to a particular name, or to all symbols whose names match a particular nameform. It does not remove the name, however; it just removes values assigned to a name. **Remove[aa]** actually removes the object itself from the context so that *Mathematica* no longer knows anything about it.

In[13]:=**Remove[aa]; Names["a*"]**

Out[13]= {}

But what is meant by removing an object from a context?

2.3 The Hierarchy of Contexts

Actually, everything in *Mathematica* has a much more complicated name which includes its context. Contexts form a hierarchy that lies behind everything that we have seen so far in our use of *Mathematica*. One can see these contexts by asking for the contexts of particular names.

In[14]:= **{Context[LaplaceTransform], Context[Sin], Context[aa]}**

Out[14]= {Calculus`LaplaceTransform`, System`, Global`}

The system variable **$ContextPath** describes the current state of this hierarchy and **$Context** tells where we are on this path at present.

In[15]:= **$ContextPath**

Out[15]= {Calculus`LaplaceTransform`, Calculus`Common`TransformCommon`,
          Calculus`DiracDelta`, Graphics`Animation`, Global`, System`}

In[16]:= **$Context**

Out[16]= Global`

Every name, either built-in or user defined, has a full name that includes the context in which it is defined. For instance, the context of all built-in system commands is **System`** and the context in which one normally works is **Global`**. Thus, the full name of **Sin** is **System`Sin** and the full name of **aa** is **Global`aa**. These names can always be used instead of their abbreviated forms. Note that context names always end with a tick, " ` ". The long context name **Calculus`LaplaceTransform`** indicates that **LaplaceTransform`** is a subcontext of the context **Calculus`**. We can get a list of all of the names that have been introduced in the **Global** context during the current session by the following command. Note that the output here depends on everything that has been done in the current session as well as certain system-dependent names.

In[17]:=**Names["Global`*"]**

Out[17]= {aa, s, t}

From one point of view, the value of **$ContextPath** should be viewed as a tree structure reflecting the directory structure of the **Packages** folder. Thus, ignoring the **Graphics`Animation`** context (whose presence is hard to account for), it looks like the following tree.

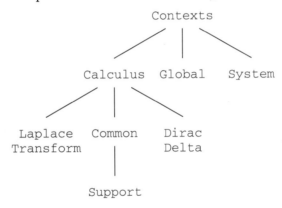

This tree will be searched from left to right by a depth first search to find names. Note that only the leaves of this tree are actual contexts. There is no context named just **Calculus`**. If more packages are added, then the simple directory structure disappears.

In[18]:=**Needs["Graphics`ImplicitPlot`"]**

In[19]:=**Needs["Calculus`Limit`"]**

In[20]:=**$ContextPath**

Out[20]= {Calculus`Limit`, Graphics`ImplicitPlot`,
        Utilities`FilterOptions`, Calculus`LaplaceTransform`,
        Calculus`Common`TransformCommon`, Calculus`DiracDelta`,
        Graphics`Animation`, Global`, System`}

Now there are two nodes named **Calculus`** separated by nodes named **Graphics`** and **Utilities`**. So, from this point of view it is better to view **$ContextPath** as a simple list which is searched from left to right. Notice that the most recently loaded packages will be searched first.

When a package is loaded using **Needs**, then *Mathematica* locates the appropriate file by looking in directories that start from multiple "roots" which are specified by **$Path**.

In[21]:= **$Path**

Out[21]= {Macintosh HD:System Folder:Preferences:*Mathematica*:3.0:Kernel,
        Macintosh HD:System Folder:Preferences:*Mathematica*:
            3.0:AddOns:Autoload, Macintosh HD:System Folder:
            Preferences:*Mathematica*:3.0:AddOns:Applications, :,
        Macintosh HD:*Mathematica* 3.0 Files:AddOns:StandardPackages,
        Macintosh HD:*Mathematica* 3.0
            Files:AddOns:StandardPackages:StartUp,
        Macintosh HD:*Mathematica* 3.0 Files:AddOns:Autoload,
        Macintosh HD:*Mathematica* 3.0 Files:AddOns:Applications,
        Macintosh HD:*Mathematica* 3.0 Files:AddOns:ExtraPackages,
        Macintosh HD:*Mathematica* 3.0
            Files:SystemFiles:Graphics:Packages}

This is a list of full path names of directories on a Macintosh hard disk named MacintoshHD. The ":" means the current working directory. On other systems the path names will be different.

When a new symbol is entered, *Mathematica* does the following:

i) It looks in the current context to see if the symbol belongs to the list of names in that context. If so, it returns the latest value it has for the symbol, or just the symbol itself if there is no value for it.

ii) If the symbol is not in the current context, it then searches the contexts on the current context path, and does the same with the first context in which it finds the symbol.

iii) If the symbol is nowhere on the current context path, it adds the symbol to the list of known names in the current context. Next time the symbol is used, it will be found in the current context.

The **LaplaceTransform** that was mistakenly typed at the beginning of this session was therefore placed in the **Global`** context. When the **LaplaceTransform** package was loaded, the context named **Calculus`LaplaceTransform`** was created. The next time **LaplaceTransform** was typed, it was found in the **Global`** context where it has no value, so it was returned unevaluated. The **LaplaceTransform** in the **Global`** context hid, or shadowed, the **LaplaceTransform** in the **Calculus`LaplaceTransform`** context so the real one couldn't be found. Only after **Remove[LaplaceTransform]** removed **Laplace-Transform** from the **Global`** context (only) could *Mathematica* find the real one in the **Calculus`LaplaceTransform`** context. *Mathematica* is supposed to warn one about the possibility of this happening, which is exactly what it did with the warning message.

2.4 To Make a New Context

Start a new session and put a name in the **Global`** context.

In[22]:= **a**

Out[22]= a

To create a new context, use the following command.

In[23]:= **Begin["newstuff`"];**

Check that we are now in a different context.

In[24]:= **{$Context, $ContextPath}**

Out[24]= {newstuff`, {Calculus`Limit`, Graphics`ImplicitPlot`,
         Utilities`FilterOptions`, Calculus`LaplaceTransform`,
         Calculus`Common`TransformCommon`, Calculus`DiracDelta`,
         Graphics`Animation`, Global`, System`}}

Notice that **newstuff`** has not been added to the context path. Now give a value to **a** using a new symbol.

In[25]:= **a = b + 5;**

The symbol **b** has been introduced in this new context, so its real name is **newstuff`b**.
In[26]:= **Names["newstuff`*"]**

Out[26]= {b}

One can also find it by the following command, which actually shows **b** with its complete name.

In[27]:=`??newstuff`*`

```
newstuff`b
```

In this context, **a** has its given value, even though **a** is in the `Global`` context.

In[28]:=`a`

Out[28]= $5 + b$

Context[a]

```
Global`
```

We can introduce a new symbol with complete name **newstuff`a** by using its complete name in the assignment statement.

In[29]:=`newstuff`a = c 7`

Out[29]= $7 c$

Now if we ask for **a** we get its value in the current context.

In[30]:=`a`

Out[30]= $7 c$

Nevertheless, we can still retrieve the previous **a** by using its complete name.

In[31]:=`Global`a`

Out[31]= $5 + b$

Finally, if we ask for the current names in **newstuff`** we get three entries.

In[32]:=`Names["newstuff`*"]`

Out[32]= $\{a, b, c\}$

To leave the context **newstuff`**, use the command **End[]**.

In[33]:=`End[];`

Finally, let's check the context path and the context. Now the current context is again **Global`**.

In[34]:= **{$Context, $ContextPath}**

Out[34]= {Global`, {Calculus`Limit`, Graphics`ImplicitPlot`,
        Utilities`FilterOptions`, Calculus`LaplaceTransform`,
        Calculus`Common`TransformCommon`, Calculus`DiracDelta`,
        Graphics`Animation`, Global`, System`}}

Now the current context is again **Global`** and the context **newstuff`** seemingly has disappeared. However, the following shows that it is still there, somewhere, in the background.

In[35]:= **newstuff`a**

Out[35]= 7 newstuff`c

2.5 Using Packages

Packages are a technique for

    i) setting up new contexts and adding them to the context path

    ii) exporting certain information to the current context

    iii) hiding the rest of the information

Here is a brief example of how this works. Instead of a **Begin** statement, use a **BeginPackage** statement.

In[36]:= **BeginPackage["newerstuff`"];**

Check the current context and context path.

In[37]:= **{$Context, $ContextPath}**

Out[37]= {newerstuff`, {newerstuff`, System`}}

Note that **Global`** and everything else is gone but **newerstuff`** has been added to the context path; the only other context on the path is **System`**. Now we give a usage message for the function that is to be exported to the outer context.

In[38]:= **gamma::usage =**
          "This function is to be exported to the Global context.";

Then, start another new context that is to be a subcontext of **newerstuff`**, and give it the standard name **`private`**. (Note the tick at the beginning and the end.)

In[39]:= `Begin["`private`"]`

Out[39]= newerstuff`private`

Check where we are.

In[40]:= `{$Context, $ContextPath}`

Out[40]= {newerstuff`private`, {newerstuff`, System`}}

There are several things to notice. The syntax `` `private` ``, with an additional tick at the beginning, means that this context is a subcontext of the current context, which is **newerstuff`**, so its actual name is the compound form **newerstuff`private`** given by **$Context**. Since we used **Begin**, this context was not added to the context path, which remains unchanged. Next, introduce an auxilary variable **beta**, give it a value, and use it to define the function **gamma**.

In[41]:= `beta = 57;`
        `gamma[x_] := beta^x`

Now end the private context.

In[43]:= `End[]`

Out[43]= newerstuff`private`

Check where we are again.

In[44]:= `{$Context, $ContextPath}`

Out[44]= {newerstuff`, {newerstuff`, System`}}

Finally, end the package.

In[45]:= `EndPackage[]`

Check where we are yet again.

In[46]:= `{$Context, $ContextPath}`

Out[46]= {Global`, {newerstuff`, Calculus`Limit`,
        Graphics`ImplicitPlot`, Utilities`FilterOptions`,
        Calculus`LaplaceTransform`, Calculus`Common`TransformCommon`,
        Calculus`DiracDelta`, Graphics`Animation`, Global`, System`}}

We are back in the **Global`** context, but the context **newerstuff`** has been added to the context path so symbols that exist in the **newerstuff`** context will be found without using their full names. We can now use the function **gamma**.

In[47]:= **gamma[3]**

Out[47]= 185193

But the constant **beta** is hidden because it is in the context **newerstuff`private`**, which is not on the context path. The idea is that **beta**, being in the private context, is not accessible to the user.

In[48]:= **beta**

Out[48]= beta

This is supposed to protect the definition of **gamma** from inadvertently changing it by giving some other value to **beta** in the **Global`** context. However, it's not really lost since we can still get it back by using the full context name.

In[49]:= **??newerstuff`private`beta**

```
newerstuff`private`beta
newerstuff`private`beta = 57
```

We can also reset **beta**, changing **gamma** along with it.

In[50]:= **newerstuff`private`beta = 100;**

In[51]:= **gamma[3]**

Out[51]= 1000000

We can find out what *Mathematica* knows about **gamma** using **??**.

In[52]:= **??gamma**

```
This function is to be exported to the Global context.
gamma[newerstuff`private`x_] :=
  newerstuff`private`beta^newerstuff`private`x
```

Notice that the variable **x** used in the definition of **gamma** also has a very long "real" name. See Maeder, *Programming in Mathematica* [Maeder 1], for a detailed treatment of contexts.

2.6 *Features of Packages*

2.6.1 The BeginPackage statement

The general structure of a package is as follows. First comes a **BeginPackage** statement containing the name of the new package in quotation marks.

```
BeginPackage["PackageName`"]
```

Notice that when we loaded the package **Calculus`LaplaceTransform`** above and then looked at the context path, there was also a context **Calculus`Common`TransformCommon`** there.

```
$ContextPath
```

```
{Calculus`LaplaceTransform`, Calculus`Common`TransformCommon`,
  Calculus`DiracDelta`, Graphics`Animation`, Global`, System`}
```

If you look at the Laplace transform package, then you will see that the **BeginPackage** statement contains a second argument.

```
BeginPackage["Calculus`LaplaceTransform`",
            "Calculus`Common`TransformCommon`"]
```

There can be as many additional arguments as desired which are the quoted names of packages containing operations that are required in the present package. They will all be automatically opened by the **BeginPackage** statement. Alternatively, one can also follow the **Begin-Package** statement with a **Needs["Package`"]** statement to read in further needed operations. (Note that the **Master** packages cannot be used either in the **BeginPackage** statement or in a **Needs** statement inside a package.) This means that there is an actual hierarchy of contexts determined by which contexts depend on other contexts by calling them when they are loaded. This hierarchy forms a directed graph since a given context may have more than one ancestor and of course more than one descendant.

2.6.2 The usage messages

The general format of the usage messages is

```
name::usage = "message";
```

Note the semicolon at the end. If it is omitted, then all of the usage messages will be printed if the package is read in as a notebook. Usage messages are not required. What is required is that the objects that are to be exported from the private part of the package must be mentioned before the **Begin["`private`"]** statement, so that when their names are mentioned in the private part of the package, they will be found outside in the main part of the package. There is no danger of these names conflicting with names in the **Global`** context, since that has been removed from the context path. It is sufficient to just list all the names before starting the **`private`** context, followed by semicolons. However, if a usage message is given, then once the package is loaded, typing **?name** will display the message just as it does for built-in operations. The desired format is to give a usage message for the name of the package itself first so the user can find out what it does. Then give usage messages for the exported objects in the form

```
"name[argument1, argument2, . . .] does something.";
```

where the arguments are given names that suggest their roles in the object. The idea is that if users read the usage message, then they will know how to use the operation. In particular, they will know how many arguments of what kinds the object expects.

2.6.3 The private part of the package

This is where all the work is done in constructing the required operations, defining rewrite rules, etc. Usually, in complicated situations, other auxiliary operations are needed to define the ones that will be exported. Because these constructions are given in the private context, they will not be available to the user. One justification for this is that the usage messages are specifications for the operations constructed in the package. All the user needs to know is what the usage messages promise the operations will do. How this is accomplished is up to the implementer, and the implementer may change his or her mind at some later point when the package is updated or improved. As long as the exported operations do what they are supposed to do, the details of the implementation shouldn't matter. Therefore, they should be kept hidden from the user. In particular, the implementer should be free to change the hidden auxiliary operations at any time without affecting the user's programs.

Of course, in *Mathematica*, these concerns are somewhat academic, since if you have access to a package at all, then you can look at the complete package and find out exactly how it is constructed. But there is still a point in using only exported operations with usage messages, precisely because packages do get updated. For instance, packages supplied with *Mathematica* itself are often updated when a new version of the program comes out.

2.7 An Alternative Form for Packages

Henry Cejtin and Theodore Gray have advocated an alternative form for packages, as discussed beginning on p. 259 of [Blachmann].

3 Attributes

There are 20 possible **Attributes** that a function can have, given by the following list.

`Constant`	`Flat`
`HoldAll`	`HoldAllComplete`
`HoldFirst`	`HoldForm`
`HoldRest`	`Listable`
`Locked`	`NHoldAll`
`NHoldFirst`	`NHoldRest`
`NumericFunction`	`OneIdentity`
`Orderless`	`Protected`
`ReadProtected`	`SequenceHold`
`Stub`	`Temporary`

Attributes have an important effect on the way in which functions are evaluated. (See Section 5 below.) There are several ways to manipulate **Attributes** of both built-in and user-defined functions. One can add attributes or change those that are already present by using the command **SetAttributes**.

In[1]:= **?SetAttributes**

```
SetAttributes[s, attr] adds attr
    to the list of attributes of the symbol s.
```

This is how to set attributes for user-defined functions, but of course, it works for built-in functions only if they are unprotected first. **Attributes** can be removed by using the command **ClearAttributes**.

In[2]:= **?ClearAttributes**

```
ClearAttributes[s, attr] removes attr
    from the list of attributes of the symbol s.
```

The command **Attributes[Symbol]** returns the current list of attributes for a symbol. It can be used to change this list just by assigning some new list of attributes to it. Again, unprotect built-in functions before doing this. The attributes **HoldAll**, **HoldFirst**, and **HoldRest** will be discussed in Section 5 below.

Let us look at some attributes that are involved in algebraic operations.

In[3]:=**Attributes[Plus]**

Out[3]= {Flat, Listable, NumericFunction,
        OneIdentity, Orderless, Protected}

Flat corresponds to associativity in the sense that (a + b) + c is the same as a + b + c. It is called **Flat** because in its general guise it means that **f[f[a, b], c] = f[a, b, c]** and this looks like flattening a list in case **f = List**. Similarly, **Orderless** corresponds to commutativity in the sense that a + b is the same as b + a. What it actually means is that the arguments to an orderless function are sorted according to the built-in **Sort** function before the function is applied. As we saw in the chapter on imperative programming, **Plus** is not actually commutative because the arguments are evaluated before they are sorted. (See also the section on Evaluation below.) You might think that **OneIdentity** has something to do with 0 being an identity for addition, but it doesn't. What it in fact means is that **Plus** of a single argument is the identity operation; i.e., **Plus[x] = x**. The identity for addition very nicely arises as the value of **Plus[]**, but this is controlled by a default value rather than by an attribute.

Listable is an important option that is possessed by many built-in functions. We can find all such functions by the following command.

In[4]:= **Select[Names["*"], MemberQ[Attributes[#], Listable]&]**

The output, which is omitted, is a long list which leads one to the conclusion that if it would make sense for a function to be **Listable**, then it probably is. It is clear what it means for a function of one variable to be **Listable**; it automatically maps itself down lists. But notice that **Plus**, **Power**, and **Times** are listable even though they are functions of two or more variables. Listability for functions of several variables includes the property of threadability as discussed in Chapter 5, Section 3.1.

Our technique for finding all **Listable** functions works for other attributes, too. For instance, we can find out which things are **Constant**.

In[5]:= **Select[Names["*"], MemberQ[Attributes[#], Constant]&]**

Out[5]= {Catalan, Degree, E, EulerGamma, GoldenRatio, Pi}

The output from **Names** consists of strings, so to see the values of these constants we have to first convert them to expressions. Note that both **N** and **ToExpression** are listable.

In[6]:= **N[ToExpression[%]]**

Out[6]= {0.915966, 0.0174533, 2.71828, 0.577216, 1.61803, 3.14159}

Something is **Locked** if you can't change it at all, even by unprotecting it.

In[7]:=`Select[Names["*"], MemberQ[Attributes[#], Locked]&]`

Out[7]= `{Fail, False, I, List, Symbol, TooBig, True, $Aborted,`
        `$BatchOutput, $CreationDate, $DumpDates, $DumpSupported,`
        `$InitialDirectory, $Input, $LinkSupported, $MachineType,`
        `$Off, $OperatingSystem, $PipeSupported, $PrintForms,`
        `$PrintLiteral, $ProcessorType, $ReleaseNumber,`
        `$Remote, $System, $TimeUnit, $Version, $VersionNumber}`

Something has the attribute **Stub** if, whenever its name is used, the appropriate package is loaded.

In[8]:=`Select[Names["*"], MemberQ[Attributes[#], Stub]&]`

Out[8]= `{Animate, MovieContourPlot, MovieDensityPlot, MovieParametricPlot,`
        `MoviePlot, MoviePlot3D, ShowAnimation, SpinShow}`

In earlier versions, nothing had this attribute, but now various animation commands have it. These commands are used in non-front-end environments. In general, **Master** packages assign it to operations in their directories. Presumably commands like **Integrate** have the attribute **Stub**, except that it is hidden from users. Finally, nothing has the attribute **Temporary**.

In[9]:=`Select[Names["*"], MemberQ[Attributes[#], Temporary]&]`

Out[9]= `{}`

We have to use a **Module** that exports its local variable to get a temporary name.

In[10]:=`Module[{t}, t];`
        `Select[Names["*"], MemberQ[Attributes[#], Temporary]&]`

Out[11]= `{t$1}`

According to The *Mathematica* Book, these names are removed "when they are no longer needed". What that means is that if they occur just within a **Module** and are never exported to the global context, then they disappear when the **Module** has finished evaluating. Otherwise, they are removed when nothing refers to them anymore.

4 Named Optional Arguments

Named optional arguments as found, for instance, in the plotting functions are very convenient to use. They are to be distinguished from positional arguments, which must always be present in order for a function to work and whose effect on the output is determined by their position in the function. Named optional arguments can be given in any order (usually only after the positional arguments, although this is only a convention) and may not be present at all. We'll give three illustrations here how to define your own named optional arguments using three different techniques.

4.1 The Gram–Schmidt Procedure Revisited

We shall rewrite the Gram–Schmidt procedure that was developed in Exercise 8.2 of Chapter 7 so that the inner product that is used becomes an optional argument. Also, whether the vectors should be normalized and what inner product to use for that will also be optional arguments. The format here is based on modifying the Gram–Schmidt package by John M. Novak that is distributed with *Mathematica*. Consider the problem of normalizing a vector. The default is to divide the vector by the square root of its **Dot** product with itself. If some other inner product is specified, then we want to replace **Dot** by that inner product. This is done by first giving a list of the options for a function **normalize** as substitutions. Here there is just one.

```
In[1]:=Options[normalize]   = {innerProduct -> Dot};
```

Thus, the default value of **innerProduct** is set to **Dot**. Other possible values are pure functions of two variables that can serve as inner products. Then we have to define **normalize** in such a way as to make use of this in the form of an optional argument that may or may not be present. The solution is based on the fact that **/.** associates to the left.

```
In[2]:=normalize[vec_, opts___] :=
    With[
      {innerp =
        innerProduct /. {opts} /. Options[normalize]},
        If[innerp[vec, vec]=!= 0,
          vec / Sqrt[innerp[vec, vec]],
          (*else*) 0 vec] ];
```

In the definition of **normalize**, the purpose of the local variable **innerp** is to pick up the desired inner product to use in normalizing vectors. Since **/.** associates to the left, the line

```
    innerProduct /. {opts} /. Options[normalize]
```

gives **innerProduct** the value specified in **opts** if there is one, in which case the expression **innerProduct** is no longer present, so the second **/. Options[normalize]** has no effect. Otherwise it gets its value from **Options[normalize]**. Try out **normalize** with a weighted dot product. Note: The round brackets are necessary here.

```
In[3]:=normalize[{2, -1, 4},
          innerProduct ->
              (Plus@@Thread[Times[#1, #2, {1, 2, 3}]]&)]
```

$$\text{Out[3]= } \left\{ \frac{\sqrt{\frac{2}{3}}}{3}, \ -\frac{1}{3\sqrt{6}}, \ \frac{2\sqrt{\frac{2}{3}}}{3} \right\}$$

The projection function works in exactly the same way.

```
In[4]:=Options[projection]  = {innerProduct -> Dot};
```

```
In[5]:=projection[v1_, v2_, opts___] :=
        With[
          {innerp =
             innerProduct /. {opts} /. Options[projection]},
          If[innerp[v2, v2] =!= 0,
             innerp[v1, v2] v2 / innerp[v2, v2],
             (*else*) 0]]
```

This is used to define a multiple projection operation as before.

```
In[6]:=multipleProjection[v1_, vecs_, opts___] :=
        Plus @@ Map[projection[v1, #, opts]&, vecs]
```

Finally, the Gram–Schmidt procedure itself has three possible optional arguments: which inner product to use; whether the vectors should be normalized, and if so how; and whether or not zero vectors are to be removed.

```
In[7]:=Options[gramSchmidt] = {innerProduct -> Dot,
                               normalized -> True,
                               deleteZeros -> False};
```

innerProduct will work as before. **normalized** is allowed to have three possible values: **True**, meaning that vectors are to be normalized by using the given inner product, **False**, meaning that they are not to be normalized at all, and some alternative inner product to use just for normalizing. **deleteZeros** also has three possible options: **False**, meaning they are not deleted; **True**, meaning vectors with zero components of the length of the input vectors are to be deleted; and finally, some other description of vectors to be deleted.

The `gramSchmidt` procedure has to be written to make use of all three optional arguments.

```
In[8]:= gramSchmidt [vecs_List, opts___] :=
    Module [
     {orthogs,
      norm    = normalized  /.{opts}/.Options [gramSchmidt],
      innerp = innerProduct/.{opts}/.Options [gramSchmidt],
      delete = deleteZeros /.{opts}/.Options [gramSchmidt]},
     orthogs =
     Fold [Join[#1,
          {#2 -
          multipleProjection [#2, #1,
                              innerProduct->innerp] }] &,
         {}, vecs];
     Which [
       norm === True,
         orthogs =
          Map [normalize [#, innerProduct->innerp] &, orthogs],
       norm === False, orthogs,
       True,
         orthogs =
          Map [normalize [#, innerProduct->norm] &, orthogs]];
     Which [
       delete === False, orthogs,
          delete === True,
             Select [orthogs,
                 (# =!= Table[0, {Length[ vecs[[1]] ]}])&],
          True, Select [orthogs, (# =!= delete)&] ] ]
```

The heart of this program is the **Fold** statement, which now includs a possible optional value for **innerProduct**. Then the output from that is processed two more times in the **Which** statements to take care of possible optional values for **normalized** and **deleteZeros**.

4.1.1 Examples

As the first example, consider six vectors in three-dimensional space.

```
In[9]:= vectors = {{1, 2, 3}, {2, -3, -4}, {3, -1, -1},
              {1, -5, -7},{-1, 5, 2}, {6, 2, -8}};
```

```
In[10]:= gramSchmidt [vectors]
```

Out[10]= $\{\{\frac{1}{\sqrt{14}}, \sqrt{\frac{2}{7}}, \frac{3}{\sqrt{14}}\}, \{\frac{22}{5\sqrt{21}}, -\frac{1}{\sqrt{21}}, -\frac{4}{5\sqrt{21}}\},$

$\{0, 0, 0\}, \{0, 0, 0\}, \{\frac{1}{5\sqrt{6}}, \sqrt{\frac{2}{3}}, -\frac{7}{5\sqrt{6}}\}, \{0, 0, 0\}\}$

In[11]:= **gramSchmidt [vectors, normalized -> False]**

Out[11]= $\{\{1, 2, 3\}, \{\frac{22}{7}, -\frac{5}{7}, -\frac{4}{7}\}, \{0, 0, 0\},$

$\{0, 0, 0\}, \{\frac{7}{30}, \frac{7}{3}, -\frac{49}{30}\}, \{0, 0, 0\}\}$

In[12]:= **gramSchmidt [vectors, deleteZeros -> True]**

Out[12]= $\{\{\frac{1}{\sqrt{14}}, \sqrt{\frac{2}{7}}, \frac{3}{\sqrt{14}}\},$

$\{\frac{22}{5\sqrt{21}}, -\frac{1}{\sqrt{21}}, -\frac{4}{5\sqrt{21}}\}, \{\frac{1}{5\sqrt{6}}, \sqrt{\frac{2}{3}}, -\frac{7}{5\sqrt{6}}\}\}$

One can also change the options by the usual built-in command.

In[13]:= **SetOptions [gramSchmidt, normalized -> False]**

Out[13]= $\{$ innerProduct $\to$ Dot, normalized $\to$ False, deleteZeros $\to$ False$\}$

In[14]:= **gramSchmidt [vectors, deleteZeros -> True]**

Out[14]= $\{\{1, 2, 3\}, \{\frac{22}{7}, -\frac{5}{7}, -\frac{4}{7}\}, \{\frac{7}{30}, \frac{7}{3}, -\frac{49}{30}\}\}$

We change them back for further use.

In[15]:= **SetOptions [gramSchmidt, normalized -> True];**
As the second example, define a new inner product using a 4 × 4 matrix.

In[16]:= **matrix =** $\begin{pmatrix} 8 & 3 & 0 & 0 \\ 3 & 2 & 1 & 2 \\ 0 & 1 & 2 & 2 \\ 0 & 2 & 2 & 14 \end{pmatrix}$ **;**

In[17]:= **gramSchmidt [{{1, 0, 0, 0}, {0, 1, 0, 0},**
                     {0, 0, 1, 0}, {0, 0, 0, 1}},
                     innerProduct -> (#1 . matrix . #2&)]

Out[17]= $\left\{\left\{\frac{1}{2\sqrt{2}}, 0, 0, 0\right\}, \left\{-\frac{3}{2\sqrt{14}}, 2\sqrt{\frac{2}{7}}, 0, 0\right\},\right.$

$\left.\left\{\sqrt{\frac{3}{14}}, -4\sqrt{\frac{2}{21}}, \sqrt{\frac{7}{6}}, 0\right\}, \left\{\frac{\sqrt{\frac{3}{7}}}{2}, -\frac{4}{\sqrt{21}}, \frac{1}{2\sqrt{21}}, \frac{\sqrt{\frac{3}{7}}}{2}\right\}\right\}$

As the last example, consider the Legendre polynomials again.

In[18]:= `gramSchmidt[{1, x, x², x³, x⁴},`

     `innerProduct → ((∫₋₁¹ #1 #2 dx) &),`

     `normalized → ((#1² /. x → 1) &)] // Together`

Out[18]= $\left\{1, x, \frac{1}{2}(-1 + 3 x^2), \frac{1}{2}(-3 x + 5 x^3), \frac{1}{8}(3 - 30 x^2 + 35 x^4)\right\}$

Here the zero vectors are just the expression "0", so this has to be explicitly specified in the option **deleteZeros** if dependent functions are included in the input list.

In[19]:= `gramSchmidt[{1, x, x², 2 x² - 3 x, x³, x⁴, x⁴ - x³},`

     `innerProduct → ((∫₋₁¹ #1 #2 dx) &),`

     `normalized → ((#1² /. x → 1) &),`

     `deleteZeros → 0] // Together`

Out[19]= $\left\{1, x, \frac{1}{2}(-1 + 3 x^2), \frac{1}{2}(-3 x + 5 x^3), \frac{1}{8}(3 - 30 x^2 + 35 x^4)\right\}$

4.2 Newton's Method Revisited

Here is another example of defining optional arguments. This time we want to have our own optional argument together with an optional argument that is passed on to a built-in command. First set the default options for Newton's method.

In[20]:= `Options[newtonsMethod] = {precision -> None,`
                                       `SameTest -> SameQ};`

The option **precision** is our own, user-defined optional argument, so we have to take care of its possible values ourselves. The intentions is that with the default value **None** for precision, the output will be the value of **N[number]**, whereas if **precision** is given a specific value, either a number or **$MachinePrecision**, then the output will be the value of **N[number, precision]**. The value for **SameTest** will just be passed to **FixedPoint-List**. It knows how to take care of the various possible optional values for **SameTest**, so we don't have to do anything about them.

```
In[21]:=newtonsMethod[expr_, {x_, x0_}, opts___] :=
     With[
       {prec = precision /. {opts} /. Options[newtonsMethod],
        test = SameTest  /. {opts} /. Options[newtonsMethod],
        fun  = Evaluate[
                 Simplify[x - expr/D[expr, x]]/. x -> #]},
       FixedPointList[
         Which[prec === None, N[fun],
               True,          N[fun, prec]]&,
          x0, SameTest -> test] ]
```

The **Which** clause inside of **FixedPointList** chooses which precision will be used for the final output.

4.2.1 Examples

In[22]:= **newtonsMethod[x^3 - 10, {x, 1}]**

Out[22]= {1, 4., 2.875, 2.31994, 2.16596, 2.1545, 2.15443, 2.15443, 2.15443}

In[23]:= **newtonsMethod[x^3 - 10, {x, 1}, precision ⟶30]**

Out[23]= {1, 4.00000000000000000000000000000,
         2.87500000000000000000000000000, 2.31994328922495274102079395l,
         2.16596155517792788479790169, 2.15449592515337473955275702,
         2.15443469177229294471607676, 2.15443469003188837231652421,
         2.15443469003188372l759294, 2.15443469003188372l759294}

In[24]:= **newtonsMethod[x^3 - 10, {x, 1},
         precision ⟶11, SameTest ⟶(Abs[#1 - #2] < 10^{-10} &)]**

Out[24]= {1, 4., 2.875, 2.3199432892, 2.1659615552,
         2.1544959252, 2.1544346918, 2.15443469, 2.15443469}

4.3 Solids of Revolution

In the third example, we illustrate a further problem that arises if we want to define a function with optional arguments, some of which are to be passed on to a built-in function, but we don't know which ones ahead of time: for instance, if the function being defined includes a built-in plotting command and we want to be able to specify optional arguments in our function that will be passed to the plotting command. This problem is solved by the package `Utilities`FilterOptions``. (Look at it to see how it works.)

In[25]:= **Needs ["Utilities`FilterOptions`"]**

We'll use this operation in a plotting routine that illustrates a surface of revolution together with cylindrical shells that show how the volume under the surface is approximated by such shells. As an option, we want to have a bounding cylinder for figures that require it, but we also want to pass on ordinary plotting options to **Show**. The operation will be called **shellPlot**. Its special optional argument is first given a default value.

In[26]:= **Options[shellPlot] = {boundingCylinder -> False};**

In[27]:= **Needs ["Graphics`Master`"]**

The operation **FilterOptions** occurs in the next to the last line of the following definition where it picks out from all given options those that apply to **Graphics3D**.

```
In[28]:= shellPlot[expr_, shells_List, {x_, x0_, x1_}, opts___]:=
    Module[
      {picture,
       shellvals = (expr/.x -> shells) / 2,
       val = (expr /. x -> x1) / 2,
       bound =
         boundingCylinder/.{opts}/.Options[shellPlot]},
      picture =
        {Map[Graphics3D[TranslateShape[
             Cylinder[#[[1]], #[[2]], 40],
                     {0, 0, #[[2]]}]]&,
           Transpose[{shells, shellvals}]],
         WireFrame[
           ParametricPlot3D[
             {x Cos[theta], x Sin[theta], expr},
             {x, x0, x1}, {theta, 0, 2 Pi},
             DisplayFunction -> Identity]]};
      If[bound, AppendTo[picture,
          WireFrame[Graphics3D[TranslateShape[
            Cylinder[x1, val, 40], {0, 0, val}]]]]];
      Show[Flatten[picture],
```

```
        FilterOptions[Graphics3D, opts],
        DisplayFunction -> $DisplayFunction]];
```

Here are two examples, using different options.

```
In[29]:=shellPlot[2 Sin[x], {Pi/6, Pi/3, Pi/2, 2 Pi/3, 5 Pi/6},
          {x, 0, Pi}, Boxed -> False];
```

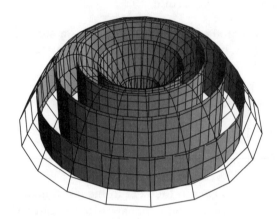

```
In[30]:=shellPlot[Sin[x], {Pi/6, Pi/3},
          {x, 0, Pi/2}, Boxed -> False,
          boundingCylinder -> True,
          ViewPoint -> {1.308, 1.738, 2.591}];
```

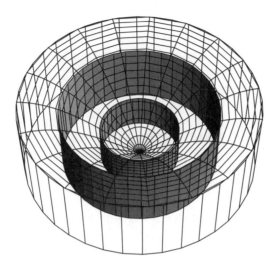

5 Evaluation

5.1 Kinds of values

When we regard *Mathematica* as a functional programming language, we think of each head as a function. When such a head is given appropriate arguments, it processes them and returns some value. Sometimes what is returned is just the head wrapped around the arguments, as with **List**; sometimes it is the word **Graphics**, as with **Plot**; sometimes it is a real number, as with **Sin** for real arguments, etc. But what is really going on is somewhat different. What really happens is that we type in some expression and then, using Enter or Shift-Return, send it to the evaluator. We have been calling the evaluator "*Mathematica*" when we speak of *Mathematica* doing something. The evaluator is a meta-function or meta-processor, sitting hidden behind everything, which takes simgle expressions as arguments and produces expressions as outputs. We can try to describe precisely what the evaluator does for certain classes of expressions, but as will be seen, the situation is rather complicated.

Let **Expr** denote the collection of all *Mathematica* expressions. The evaluator, **Eval**, is a function from **Expr** to itself; i.e., **Eval** : **Expr**   ←**Expr**. (Here we use an arrow to mean a function from its left-hand side to its right-hand side.) Can one say what **Eval** does to certain subsets of **Expr**? For instance, consider the subset of **Expr** consisting of expressions whose head is **Integer**. Clearly, **Eval** is the identity function on such expressions. In fact, there is a large class of expressions, including badly formed ones and those for which there are no rewrite rules, on which **Eval** acts as the identity operation, returning the input expression unchanged. Do expressions with head **List** belong to this class? Well, not exactly. What **Eval** does to an expression with head **List** is just to move inside it and evaluate the arguments. If **Eval** acts as the identity on the arguments, then it acts as the identity on the whole list. A more precise description would be that **Eval** commutes with the head **List** in the following sense. Regard **List** as a function from sequences of expressions to expressions, so if **Expr*** denotes the collection of all sequences of expressions, then **List: Expr*** ←**Expr**, where **List** applied to a sequence of expressions wraps itself around the expressions, separating them by commas. Also think of **Eval** as determining a function **Eval* : Expr*** ←**Expr*** by separately evaluating each expression in a sequence, one after the other. Then for a given sequence of expresions, **seq**,

$$\text{Eval}(\texttt{List}[\text{seq}]) = \texttt{List}[\text{Eval*}(\text{seq})]$$

This is what we mean by saying that **Eval** commutes with **List**. Actually, another concern is raised here because what you actually see as the result of such an evaluation is not something with head **List**, but something wrapped in curly brackets. These are produced by the formatter, **Formatter**, so there is an extra stage determining the actual appearance of the output. In fact, **Eval** only works on full forms of expressions so the real situation looks more like

FullForm followed by **Eval** followed by **Formatter**

In a certain sense, **Formatter** is the inverse to **FullForm**.

There is another large class of expressions which **Eval** takes to the constant **Null** and which **Formatter** then reduces to nothing at all. This class includes well-formed expressions with head **Do**, **For**, **While**, etc. Similarly, well-formed expressions with heads including **Plot** in some form are taken by the **Formatter** to the expression -Graphics-. This of course brings up another agent processing expressions, the *side-effector*, which acts on expressions with head **Print**, **Graphics**, **Set**, etc.

Now, how does **Eval** actually do its work? Viewed as a symbolic computation program, **Eval** does only two things: it calls C code to compute particular numerical functions, and it evaluates rewrite rules. There are many things to understand about evaluation, but perhaps the most important thing is the order in which parts of an expression are evaluated. Very detailed information about this can be found in [Withoff] and The *Mathematica* Book. We summarize some of this information here and then investigate the parts of it that are available for experimentation. As we know, *Mathematica* maintains tables of rules attached to symbols. There are in fact 10 kinds of such lists. The first four are:

DownValues[symbol]
    Rules for evaluating expressions of the form **symbol[-]**

SubValues[symbol]
> Rules for evaluating expressions such as **symbol[-][-]** with a *symbolic head* of **symbol**

OwnValues[symbol]
> A rule for evaluating **symbol** itself

UpValues[symbol]
> Rules for evaluating expressions such as **f[symbol]**, where **symbol** appears as an argument or the head of an argument

Here are some examples of these kinds of values. Each command returns a possibly empty list of the appropriate values.

In[1]:= **f[x_] := x^2**

In[2]:= **{DownValues[f], SubValues[f], OwnValues[f], UpValues[f]}**

Out[2]= $\{\{\text{HoldPattern}[f[x\_]] :\rightarrow x^2\}, \{\}, \{\}, \{\}\}$

In[3]:= **g[x_][y_] := x y**

In[4]:= **{DownValues[g], SubValues[g], OwnValues[g], UpValues[g]}**

Out[4]= $\{\{\}, \{\text{HoldPattern}[g[x\_][y\_]] :\rightarrow x\,y\}, \{\}, \{\}\}$

In[5]:= **a = 5;**

In[6]:= **{DownValues[a], SubValues[a], OwnValues[a], UpValues[a]}**

Out[6]= $\{\{\}, \{\}, \{\text{HoldPattern}[a] :\rightarrow 5\}, \{\}\}$

In[7]:= **h[x_] + h[y_] ^:= h[x y]**

In[8]:= **{DownValues[h], SubValues[h], OwnValues[h], UpValues[h]}**

Out[8]= $\{\{\}, \{\}, \{\}, \{\text{HoldPattern}[h[x\_] + h[y\_]] :\rightarrow h[x\,y]\}\}$

The other kinds of values have a slightly different character.

FormatValues[symbol]	Printing rules for **symbol**
NValues[symbol]	Rules used in evaluating **N[symbol]**
DefaultValues[symbol]	Default values for arguments in **symbol[-]**
Options[symbol]	Default options attached to **symbol**

>   Messages[symbol]   Messages attached to symbol
>   Attributes[symbol]   Attributes associated with symbol

For instance:

In[9]:=`Format[v[x_]] := Subscripted[v[x]]`

In[10]:=`FormatValues[v]`

Out[10]= {HoldPattern[MakeBoxes[$v_{x\_}$, FormatType_]] :>
          Format[v_x , FormatType], HoldPattern[$v_{x\_}$] :> v_x}

In[11]:=`N[e] = 2.7;`

In[12]:=`{OwnValues[e], NValues[e]}`

Out[12]= {{}, {HoldPattern[N[e]] :> 2.7}}

In[13]:=`DefaultValues[Plus]`

Out[13]= {HoldPattern[Default[Plus]] :> 0}

In[14]:=`mappingGraphics::codomainDimensions =`
     `" Codomain dimensions are too large for plotting.\n`
     `Dimensions should be 2 or 3.";`

In[15]:=`Messages[mappingGraphics]`

Out[15]= {HoldPattern[mappingGraphics::codomainDimensions] :>
           Codomain dimensions are too large for plotting.
        Dimensions should be 2 or 3.}

In[16]:=`Attributes[Plus]`

Out[16]= {Flat, Listable, NumericFunction,
          OneIdentity, Orderless, Protected}

Note: The **Messages** example above occurs in Chapter 14.

5.2 Normal order of evaluation

The normal order of evaluation of an expression is to evaluate first the head of the expression, then the arguments in order from left to right, and then the evaluated head is applied to the evaluated arguments. However, the situation is actually somewhat more complicated than this. In detail, according to [Withoff], the following steps are carried out recursively.

1. If the expression is a string, a number, a symbol with no **OwnValues**, or if no part of the expression has changed since the last evaluation, then return the expression.

2. Expressions which are symbols with **OwnValues** are evaluated.

3. The head of the expression is evaluated.

4. The arguments are evaluated from left to right, with several provisos: If **head** has attribute **HoldFirst**, **HoldRest**, or **HoldAll**, do not evaluate the corresponding arguments unless they have head **Evaluate**. (This means that all arguments have to be looked at, in any case, to see if they have the head **Evaluate**.) If an argument has head **Unevaluated**, replace it with the arguments of the argument and keep a record of the original expression. Flatten out nested expressions with head **Sequence**.

5. The attributes of the head are used next.

i) **Flat** means flatten out nested expressions.

ii) **Listable** means thread head over any arguments that are lists.

iii) **Orderless** means the evaluated arguments are to be sorted.

Note that these are applied only after the arguments have been evaluated.

6. UpValues attached to the symbolic heads of the arguments are applied, using user defined values before internally defined ones.

7. DownValues are applied if the head is a symbol, otherwise **SubValues** attached to the symbolic head are applied, using user defined values before internally defined ones.

8. The head **Unevaluated** is replaced if no applicable rules were found.

9. The head **Return** is discarded, if present, for expressions generated through application of user defined rules.

5.2.1 Normal evaluation

Let us see if we can persuade **Eval** to display the order of some evaluations. First introduce a shorthand for **Module[{t}, t]**, which is to be evaluated anew each time it is called. We have already seen that the result of this is to just output **t** with the current value of the evaluation counter appended to it.

In[17]:=**mod := Module[{t}, t]**

Then use this as head and arguments for a generic function.

In[18]:=**mod[mod, mod, mod]**

Out[18]= t$1[t$2, t$3, t$4]

This at least shows the order of evaluation of the head and the arguments. These rules are applied recursively to each argument in turn. Thus:

In[19]:=**mod[mod[mod, mod], mod[mod[mod], mod[mod]]]**

Out[19]= t$5[t$6[t$7, t$8], t$9[t$10[t$11], t$12[t$13]]]

In other words, viewing the expression as a tree, the nodes are evaluated by a depth-first traversal of the tree.

5.2.2 Hold

If the head has the attribute **HoldAll**, then the situation changes. Here is an example.

In[20]:=**SetAttributes[gg, HoldAll]**
        gg[x_] := {x, x}

In[22]:=**hh[x_] := {x, x}**

In[23]:=**{gg[mod], hh[mod]}**

Out[23]= {{t$14, t$15}, {t$16, t$16}}

In the case of **gg**, the argument **mod** is not evaluated until it is used in the right-hand side of the definition of **gg**. It then is used twice giving two successive values of t$. In the case of **hh**, the argument **mod** is evaluated before the operation **hh** is applied. Its single value is then used twice in the right-hand side.

An argument which is held is not evaluated. There are two ways to overcome this which were confused in earlier versions of *Mathematica*. Starting in Version 2, they have been separated. For instance, **gg** has the attribute **HoldAll**. If we want **gg** to evaluate its argument, one can replace **gg** by **ReleaseHold[gg[argument]]**, or we can use **gg[-Evaluate[argument]]**. Thus,

In[24]:= **{ReleaseHold[gg[mod]], gg[Evaluate[mod]]}**

Out[24]= {{t$17, t$18}, {t$19, t$19}}

Clearly, in the second version, the argument of **gg** is evaluated before **gg** is applied, whereas **ReleaseHold** has no effect on the evaluation. Thus, one should use **Evaluate[argument]** inside functions that have the attribute **HoldAll** or **HoldFirst**. On the other hand, if something is explicitly held, then **ReleaseHold** outside the function is the appropriate operation.

In[25]:= **Hold[2 + 2]**

Out[25]= Hold[2 + 2]

In[26]:= **ReleaseHold[%]**

Out[26]= 4

Note that **ReleaseHold** only removes one layer of holding.

In[27]:= **ReleaseHold[Hold[2 + Hold[2 + 2]]]**

Out[27]= 2 + Hold[2 + 2]

Furthermore, there is another similar operation, **HoldForm**, that does the same thing as **Hold** but prints the result without wrapping **Hold** around it. It is also removed using **ReleaseHold**.

In[28]:= **HoldForm[2 + 2]**

Out[28]= 2 + 2

5.2.3 HoldPattern

When we looked at various kinds of values, they were displayed with **HoldPattern** wrapped around the left-hand sides. To see what this is about, we'll first look at the built-in information about **Rule**, **RuleDelayed**, and **HoldPattern**. Note that in earlier versions of *Mathematica*, **HoldPattern** was called **Literal**.

In[29]:= **??Rule**

> lhs -> rhs represents a rule that transforms lhs to rhs.
> Attributes[Rule] = {Protected, SequenceHold}

In[30]:= **??RuleDelayed**

> lhs :> rhs represents a rule that transforms lhs
>   to rhs, evaluating rhs only when the rule is used.
> Attributes[RuleDelayed] = {HoldRest, Protected, SequenceHold}

In[31]:= **??HoldPattern**

> HoldPattern[expr] is equivalent to expr for pattern
>   matching, but maintains expr in an unevaluated form.
> Attributes[HoldPattern] = {HoldAll, Protected}

Thus, from the **Attributes** statements, we see that **Rule** evaluates both of its arguments while **RuleDelayed** evaluates only its first argument. **HoldPattern** is used to prevent **RuleDelayed** from evaluating its first argument, without changing the form of the pattern to be matched. Here is a nice example from The *Mathematica* Book.

In[32]:= **Hold[u[1 + 1]] /. HoldPattern[1 + 1] -> x**

Out[32]= Hold[u[x]]

HoldPattern cannot be replaced by **Hold** here, since **Hold** is a part of any pattern in which it appears, whereas, for purposes of pattern matching, **HoldPattern** is invisible.

5.2.4 Evaluation of conditions

To investigate the order of evaluation of conditions, consider the following function definition.

In[33]:= **f[x_Integer /; mod || EvenQ[x]] :=**
**   mod[mod, mod, mod] /; (mod; Positive[x])**

It is not clear how to apply the nine rules for evaluation to determine in what order an evaluation of **f[2]** will actually be carried out. However, **Trace** will show us explicitly what happens.

In[34]:= **Trace[f[2]]**

Out[34]= {f[2], {mod || EvenQ[2],
        {mod, Module[{t}, t], t$20}, {EvenQ[2], True}, True},
        {{mod; Positive[2], {mod, Module[{t}, t], t$21},

```
{Positive[2], True}, True}, RuleCondition[
 $ConditionHold[$ConditionHold[mod[mod, mod, mod]]], True],
 $ConditionHold[$ConditionHold[mod[mod, mod, mod]]]},
mod[mod, mod, mod], {mod, Module[{t}, t], t$22},
{mod, Module[{t}, t], t$23}, {mod, Module[{t}, t], t$24},
{mod, Module[{t}, t], t$25}, t$22[t$23, t$24, t$25]}
```

Thus, the first thing to be evaluated is the condition inside the definition of **f** for matching the pattern for the argument to **f**. Next the condition at the end of the definition for application of the rule is checked, and then the usual order of evaluation is followed.

The order of evaluation of substitutions is just what one would expect from the **FullForm** of a substitution.

In[35]:= **Trace[mod /. mod -> mod]**

Out[35]= {{mod, Module[{t}, t], t$27},
      {{mod, Module[{t}, t], t$28}, {mod, Module[{t}, t], t$29},
        t$28 → t$29, t$28 → t$29}, t$27 /. t$28 → t$29, t$27}

6 General Recursive Functions

6.1 . Primitive Recursive Functions

Recursive functions were introduced by Gödel, Kleene, and Church. We consider them in the form described by Kleene. They are defined in two stages. See [Rogers] or [Soare]. First of all, the functions under consideration are all functions of k natural number variables, for some k, and which take natural numbers as values. A *primitive recursive function* is one that can be constructed by using the following rules.

i) The identity function is primitive recursive; i.e., $\mathrm{id}(x) = x$ is primitive recursive.

ii) A constant function is primitive recursive; i.e., $f(x) = c$ is primitive recursive.

iii) The successor function is primitive recursive; i.e., $s(x) = x + 1$ is primitive recursive.

iv) A projection from a product onto one of the factors is primitive recursive; i.e., $\mathrm{pr}_i(x_1, \ldots$

$. x_n) = x_i$ is primitive recursive for all $i \le n$.

v) A composition of primitive recursive functions is primitive recursive.

vi) (Induction) If a function $f(x_1, \ldots, x_k)$ satisfies a pair of equations of the form

$$f(0, x_2, \ldots, x_k) = g(x_2, \ldots, x_k),$$
$$f(x_1 + 1, x_2, \ldots, x_k) = h(x_1, f(x_1, x_2, \ldots, x_k), x_2, \ldots, x_k),$$

where g and h are primitive recursive, then f is primitive recursive. In this way, standard arithmetic functions like addition, multiplication, and exponentiation can be shown to be primitive recursive. For instance, addition is described by the equations

$$\text{plus}(0, n) = n$$
$$\text{plus}(n + 1, m) = \text{plus}(n, m) + 1,$$

which is an instance of rule (vi) using rules (i) and (iii).

One can easily implement a version of these basic primitive recursive functions using either a functional style or an imperative style of programming. Actually, a very large class of functions are primitive recursive, and it is clear in principle how to implement them in computer programs. The first five cases are obvious, and the inductive rule in (vi) says that if one has implementations of g and h, then f is implemented by the indicated pair of rewrite rules.

6.2 Unbounded Search and Church's Thesis

Partial, or *general*, *recursive functions* are defined by adding one more scheme to the above procedures for constructing functions; namely, they are those partial functions which are definable using the preceding six schemes plus the scheme
   vii) (Unbounded search)

$$f(x_1, \ldots, x_n) = \mu y \, (g(x_1, \ldots, x_n, y)' = \text{True and } (\forall z \le y)(g(x_1, \ldots, x_n, z)')$$

Here g is a partial recursive function, $(g(x_1, \ldots, x_n, y)'$ means that g is defined for these values of the arguments, and $\mu y \, (g(x_1, \ldots, x_n, y))$ means the least y such that the predicate $g(x_1, \ldots, x_n, y)$ is satisfied. This means that $f(x_1, \ldots, x_n)$ is defined to be the least value of y such that $g(x_1, \ldots, x_n, z)$ is defined for $z = y$ and all smaller values of z and such that $g(x_1, \ldots, x_n, y) = \text{True}$. If $k = 1$, this can be paraphrased by the prescription

$$f(x) = \text{"the smallest value of } y \text{ such that } g(x, z) \text{ is defined for all } z \le y$$
$$\text{and } g(x, y) \text{ is true"}.$$

(Note that one can actually restrict the predicate g to be primitive recursive.) Church's thesis is the statement that effectively computable functions are the same as partial recursive functions. It is the word "effective" here that has no precise definition. The defender of Church's thesis is committed to showing that any proposed mechanical computation of a function produces a partial recursive function.

It is a theorem that partial recursive functions defined this way coincide with the functions that can be expressed in the (untyped) lambda calculus as well as those functions calculable by any Turing complete programming language. However, we need not resort to the lambda calculus to find implementations of such functions in *Mathematica*. Functions described by unbounded search have natural expressions both imperatively and functionally by structures that are intended just for this purpose: namely, `While` and `FixedPoint`. The abstract form of such an algorithm in *Mathematica* is

```
f[x_] := Module[{y = 1}, While[!g[y], y++]; y]

f[x_] := FixedPoint[If[g[#],#, # + 1], 1]
```

Note that these algorithms can fail to return a value either because `g[y]` does not return a value for some `y` that is reached or because it never happens that `g[y]` is True.

6.3 Examples

6.3.1 Fractionalize

Replace the built-in function `Rationalize` by a function that finds a best possible rational approximation to a real number whose numerator and denominator have at most a specified number of digits, with one of them having at least that many digits. (The problem of finding a functional program to do this was suggested by Charles Wells in an e-mail communication.) We want to use the built-in function in the form

```
Rationalize[N[r, y], 0.1^(y-1)]
```

and the problem is, given `r`, find the least value of `y` so that the result has the correct number of digits in its numerator and denominator. The solution is an unbounded search on values of `y`, checking the number of digits in the numerator and denominator as one goes.

```
In[1]:= fractionalize[number_, size_] :=
         With[{term = FixedPoint[
             With[{value = Rationalize[N[number, # + 1], 0.1^#]},
               If[Length[IntegerDigits[Numerator[value]]] ≤ size &&
                   Length[IntegerDigits[Denominator[value]]] ≤ size,
                 # + 1, #]] &, 1]},
             Rationalize[N[number, term], 0.1^(term-1)]]
```

Here are some results for π.

In[2]:= `Map[{#, fractionalize[Pi, #]} &, Range[5, 15]]`

Out[2]= $\{\{5, \frac{355}{113}\}, \{6, \frac{312689}{99532}\}, \{7, \frac{5419351}{1725033}\},$

$\{8, \frac{80143857}{25510582}\}, \{9, \frac{245850922}{78256779}\}, \{10, \frac{6167950454}{1963319607}\},$

$\{11, \frac{21053343141}{6701487259}\}, \{12, \frac{21053343141}{6701487259}\}, \{13, \frac{8958937768937}{2851718461558}\},$

$\{14, \frac{8958937768937}{2851718461558}\}, \{15, \frac{428224593349304}{136308121570117}\}\}$

6.3.2 Expressions for primes

Can a prime number p can be written in the form $2^n - 3^m$ or $3^m - 2^n$ for some choice of m and n? Note that m and n can be arbitrarily large. Given p we can conduct an unbounded search for m and n by using the usual reverse diagonal recursive enumeration of pairs of natural numbers $e(k) = \{p_1(k), p_2(k)\}$, given by the formula

In[3]:= `pair[k_] := pair[k] =`

$$\text{With}\left[\left\{r = \text{Floor}\left[N\left[\frac{\sqrt{1 + 8\ k} - 1}{2}\right]\right]\right\},\right.$$

$$\left.\left\{k - \frac{r\ (r + 1)}{2},\ \frac{r\ (r + 3)}{2} - k\right\}\right]$$

The following picture shows the values of **pair** for **k** between 0 and 20, starting from the origin.

In[4]:= `Show[Graphics[`
        `With[{points = Table[pair[k], {k, 0, 20}]},`
            `{Prepend[Map[Point, points], PointSize[0.03]],`
            `Line[points]}] ]];`

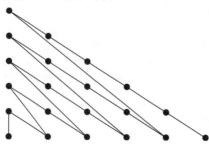

The functional version of unbounded search for pairs (i, j) such that $|2^i - 3^j| = $ `Prime[n]` is as follows.

```
In[5]:= findPair[n_] :=
    With[{y =
        FixedPoint[If[Abs[Thread[{2, 3}^pair[#]] . {1, -1}] =!= Prime[n],
            #+1, #] &, 1]},
    Print[SequenceForm["|", "2"^pair[y][[1]] _ "3"^pair[y][[2]],
        "| == ", Prime[n]]]; Null]
```

Calculate the values for the first 12 primes.

```
In[6]:= Map[findPair[#] &, Range[12]];
```

$$|1 - 3| == 2$$
$$|-1 + 2^2| == 3$$
$$|2^2 - 3^2| == 5$$
$$|2 - 3^2| == 7$$
$$|2^4 - 3^3| == 11$$
$$|2^4 - 3| == 13$$
$$|2^6 - 3^4| == 17$$
$$|2^3 - 3^3| == 19$$
$$|2^2 - 3^3| == 23$$
$$|2^5 - 3| == 29$$
$$|-1 + 2^5| == 31$$
$$|2^6 - 3^3| == 37$$

What about the 13th prime?

```
In[7]:= Timing[findPair[13]]
```

```
Out[7]= $Aborted
```

```
In[8]:= Prime[13]
```

```
Out[8]= 41
```

41 is conjectured to be the smallest prime which has no such representation. In order to find many primes which have such a representation, it is much faster to find all primes < 20,000 that have such a representation for $k <= 5050$; i.e., for $m + n \le 100$. Such a bounded search always terminates and is to be contrasted with the unbounded search above, which presumably would never terminate for $p = 41$.

```
In[9]:= goodPrimes =
    Select[Union[Select[
        Map[Abs[2^pair[#][[1]] - 3^pair[#][[2]]] &, Range[5050]],
        (# < 20000 &)]], PrimeQ]
```

Out[9]= {2, 3, 5, 7, 11, 13, 17, 19, 23, 29, 31, 37, 47, 61, 73, 79, 101, 127,
         139, 179, 211, 227, 229, 239, 241, 269, 431, 503, 509, 601, 727, 997,
         1021, 1163, 1319, 1931, 2039, 2179, 3299, 3853, 4093, 4513, 6529,
         6553, 7949, 8111, 8191, 11491, 14197, 16141, 16381, 19427, 19681}

Any prime < 20,000 which is not on this list is a candidate for a prime with no such representation.

6.4 WithRec

A seemingly more general form of unbounded search is given by the functional programming **letrec** construct. An expression of the form **letrec x = expr1 in expr2**, where x occurs in **expr1**, means substitute **expr1** for x in **expr2**. If the resulting expression contains x, then again substitute **expr1** for it, continuing this way until x no longer occurs in the expression. Thus, an unbounded search is being conducted for an iterated substitution that doesn't contain x. This behavior can be implemented very simply in *Mathematica* by a **FixedPoint** operation.

```
In[10]:= Attributes[withRec] = {HoldFirst};
```

```
In[11]:= withRec[{x_ = expr1_}, expr2_] :=
            FixedPoint[With[{x = expr1}, #]&, expr2]
```

The **HoldFirst** attribute is required in order for the first argument to be in the form {x = expr1}. For instance, here is what we hope will be the last version of a factorial computation and a Fibonacci computation.

```
In[12]:= withRec[{fac = If[# == 0, 1, # fac[# - 1]]&}, fac[10]]
```

Out[12]= 3628800

```
In[13]:= withRec[{fib =
                Which[# == 1, 1,
                      # == 2, 1,
                      True, fib[# - 1] + fib[# - 2]]&},
            fib[20]]
```

Out[13]= 6765

7 Substitution and the Lambda Calculus

There are two ways to substitute values for arguments in *Mathematica*, neither of which is completely satisfactory.

7.1 With versus /.

Recall the meaning of **/.**

In[1]:= **? /.**

> expr /. rules applies a rule or list of rules in an
>   attempt to transform each subpart of an expression expr.

As we have seen, **/.** is a very general mechanism for applying local rules. However, if the rules are of the form **expr /. x -> expr1**, then the effect is to substitute **expr1** for all occurrences of **x** in **expr**. This substitution is purely and relentlessly syntactical. If *Mathematica* sees an **x** as a separate symbol, it sticks in a copy of **expr1**. We used this kind of substitution, for instance, in checking solutions of equations where it works very well. However, sometimes **/.** does the wrong thing. Consider a pure function.

In[2]:= **f = Function[{x}, x + y];**

Applying this as a function to values works as it is supposed to:

In[3]:= **{f[2], f[x], f[y]}**

Out[3]= {2 + y, x + y, 2 y}

But now try substituting something for **x** and **y**. The result depends on what is substituted.

In[4]:= **{f /. y -> 3, f /. x -> 3}**

> Function::flpar :
>   Parameter specification {3} in Function[{3}, 3 + y]
>   should be a symbol or a list of symbols.

Out[4]= {Function[{x}, x + 3], Function[{3}, 3 + y]}

The first one is OK, but the second one makes no sense, as the warning message points out. The **x** in **Function[{x}, x + y]** is a bound variable and it should not be possible to substitute anything for it. However, the other built-in operation, **With**, does carry out substitutions (for variables only) in a way that is mostly correct.

In[5]:= **?With**

> With[{x = x0, y = y0, ... }, expr]
>   specifies that in expr occurrences of the symbols
>   x, y, ... should be replaced by x0, y0, ...

Note that in **With**, the left-hand side of the = expression has to be a symbol and not some more complicated pattern. For instance,

In[6]:= **With[{x = 2}, x²]**

Out[6]= 4

Also, **With** uses the call-by-value mode of evaluation.

In[7]:= **With[{x = Module[{t}, t]}, {x, x}]**

Out[7]= {t$1, t$1}

Try this on the function definition.

In[8]:=**With[{y = 3}, Function[{x}, x + y]]**

Out[8]= Function[{x$}, x$ + 3]

In[9]:=**With[{x = 3}, Function[{x}, x + y]]**

Out[9]= Function[{x}, x + y]

Thus, the substitution for **y** is carried out as it should be and the name of **x** is actually changed to a new **x$**. The substitution for **x** has no effect, which is also correct. Now consider a more complicated function whose value is again a function.

In[10]:=**g = Function[{x}, Function[{y}, x + y]];**

Try evaluating this at **a**.

In[11]:=**g[a]**

Out[11]= Function[{y$}, a + y$]

Now evaluate this at **y**.

In[12]:=**g[y]**

Out[12]= Function[{y$}, y + y$]

Note that there is no conflict because the name of the bound variable **y** has been changed to **y$**, so the **y** outside the function definition is completely separate from the one inside. However, this arrangement can be fooled if **g** somehow gets an argument of the form **y$**.

In[13]:=**g[y$]**

Out[13]= Function[{y$}, y$ + y$]

In the next section we give an example where this occurs. This would be much safer if the evaluation counter were used here. Also note that **Function[{x}, x + y]** properly does not depend on **x**.

In[14]:=**Function[{x}, Function[{x}, x + y]][a]**

Out[14]= Function[{x}, x + y]

7.2 The lambda calculus

Sorting out the relationships between pure functions (with named bound variables) like **Function[{x}, expr]**, function applications like **f[a]**, and substitutions like **With[{x = a}, expr]** is a non-trivial task. Fortunately, these relationships were all carefully worked out in the 1930s with the development of the lambda calculus. At present, the lambda calculus is more often regarded as an abstract prototype of a functional programming language. Functional programs deal with expressions which encode both the algorithm being processed and the input data. These expressions are evaluated by rewrite rules. Suppose **expr** is some expression and **part** is a part of **expr** which we write as **expr[part]**. A rewrite rule is a rule of the form **part -> part'**. Then **expr** is reduced by rewriting **expr[part] -> expr[part']**. This continues recursively until no more rewrite rules can be applied. The resulting expression is called the *normal form* of the original expression and is the output calculated by the original expression. At first it is hard to see how this procedure can be used to calculate anything, but our discussion of the lambda calculus should help explain how this works.

7.2.1 The syntax of the lambda calculus

The lambda calculus is deceptively simple to describe. There is an alphabet consisting of names (or variables or identifiers) that represent ambiguously values and functions; i.e., a given name can function both as a value and a function within the same expression. There is one binary operation, that of applying a function to a value, written **(f g)** (using LISP-like notation), called *application*, and one construction written λx . **expr**, called *abstraction*, whose intended interpretation is "the function of x given by the expression **expr**"; i.e., it is the same thing as **Function[{x}, expr]** in *Mathematica*. Here, **expr**, which is called the *body* of the abstraction, is some expression built up using the operations of application and abstraction. The name **x** in the abstraction is called the *bound variable*. Any other name **y** in **expr** that is not preceded by (actually, is not in the scope of) a λy is called a *free variable*. A lambda term of the form λx. **expr** can be thought of as a *canonical name* for the function of x described by **expr**. Finally, there is a single rewrite rule called *beta reduction* which says that an application of the form **((λx . expr) y)**, whose first argument is an abstraction, rewrites to the result of substituting **y** for **x** in **expr**, which is written here in *Mathematica* notation as **expr /. {x -> y}**; i.e., the value **y** is substituted for the bound variable **x** in **expr**. We will write this reduction step as

$$((\lambda x . \textbf{expr}) \; y) \quad \Leftarrow \textbf{expr} \; /. \; \{x \to y\}.$$

For instance, in ordinary algebra, if **f** = λx . x^2, then the result of applying **f** to the number 2 is given by

$$(\textbf{f} \; 2) = ((\lambda x . x^2) \; 2) \quad \Leftarrow x^2 /. \{x \to 2\} \quad \Leftarrow 2^2 = 4.$$

In *Mathematica*, one uses the syntax of ordinary function application. Thus,

In[1]:= **Function[{x}, x^2][2]**

Out[1]= 4

and this, of course, is the same as

In[2]:= x^2 **/. {x** ⟶ **2}**

Out[2]= 4

This very simple syntax is the starting point of a vast theory called the lambda calculus. It is intended to capture the two most fundamental properties of functions in a certain sense, namely, that of applying a function to an argument by substituting the argument for a bound variable in the body of the function and that of giving a name to the function determined by a rule. Here are some examples of terms in the lambda calculus.

i) (The identity function) λx . x. If this is applied to any expression, the result is the expression itself since

    ((λx . x) expr) ⇐ /. {x -> expr} ⇐ **expr**.

ii) (Self-application) λx . (x x). Applying this to any expression **expr** results in **(expr expr)** which may or may not make sense. Note that applying self-application to itself never terminates, since

  ((λx . (x x)) (λx . (x x))) ⇐ (x x) /. {x -> λx . (x x)} ⇐ λx . (x x) λx . (x x)) ⇐ .

This combination is called the *paradoxical combinator*. It is the model of a non-terminating computation.

iii) (The Church numerals) The *n*th Church numeral is the lambda expression λf . λx . (f (f . . . x) . . .), where there are *n* nested applications of *f* to *x*. E.g.,

    $1 = \lambda f . \lambda x . (f\ x),$
    $2 = \lambda f . \lambda x . (f\ (f\ x)\ ),$
    $3 = \lambda f . \lambda x . (f\ (f\ (f\ x)\ )\ ),$ etc.

The Church numerals serve as a model for the natural numbers inside the lambda calculus.

The lambda calculus is a surprisingly deep and subtle subject. One of its main results, called the Church–Rosser theorem, says that if beta reduction can be applied to more than one subexpression of a given expression leading to different results, then there are further beta reductions of each of these two terms that eventually lead to the same expression. Thus, a given expression can be reduced to at most one normal form. For an elementary introduction, see [Michaelson], and for a thorough treatment see [Curry and Feys] and [Barendregt].

7.2.2 Rewrite rules for the lambda calculus

A complete description of the operational semantics of the lambda calculus is given by three interrelated lists of rewrite rules; rules for the free variables function, various conversion rules, and rules for substitution. If **f** is any term built up recursively from symbols, applications, and abstractions, then **FV(f)** denotes the set of free variables, or symbols, of **f**. It is defined recursively by the following rules. (Here, **Var** denotes the set of variables, or symbols.)

We use " ⇐ here to mean the left hand side rewrites to the right-hand side in a single step.

i) If **x ∈ Var**, then **FV(x)** ⇐ {x}.

ii) **FV((f g))** ⇐ **FV(f)** ∪ **FV(g)**.

iii) **FV((λx . g))** ⇐ **FV(g)** - {x}.

A term **f** is called *closed* or a *program* if **FV(f)** is the empty list.

The main reduction rule in the operational semantics is beta reduction:

$$(\beta\text{–reduction}) \ (\lambda x . f) \ g \quad \twoheadleftarrow f /. \{x -> g\}. \ \text{(See below.)}$$

Besides this rule, there are implicit rewrite schemes:

a) $\dfrac{h \quad \twoheadleftarrow k}{(h \ g) \quad \twoheadleftarrow (k \ g)}$ b) $\dfrac{h \quad \twoheadleftarrow k}{(f \ h) \quad \twoheadleftarrow (f \ k)}$ c) $\dfrac{f \quad \twoheadleftarrow g}{(\lambda x . f) \quad \twoheadleftarrow (\lambda x . g)}$

The right-hand side of β-reduction depends on the notion of substituting **g** for **x** in **f**. This is defined by the following rules. Let $x \in$ **Var** and let **f** and **g** be terms. Then the operation **f** /. {x -> g} of substituting **g** for **x** in **f** is defined recursively by the rules:

i) $x /. \{x -> g\} \quad \twoheadleftarrow g$ if x is a variable

ii) $y /. \{x -> g\} \quad \twoheadleftarrow y$ if $y \neq x$ and y is a variable

iii) $(h \ k) /. \{x -> g\} \quad \twoheadleftarrow (h /. \{x -> g\}) \ (k /. \{x -> g\}))$

iv) $(\lambda x . f) /. \{x -> g\} \quad \twoheadleftarrow (\lambda x . f)$ (i.e., don't substitute for bound variables)

v) $(\lambda y . f) /. \{x -> g\} \quad \twoheadleftarrow (\lambda y . (f /. \{x -> g\}))$ if $x \neq y$ and $y \notin$ **FV(g)**

vi) $(\lambda y . f) /. \{x -> g\} \quad \twoheadleftarrow (\lambda z . (f /. \{y -> z\} /. \{x -> g\}))$ if $x \neq y$ and $y \in$ **FV(g)**, where $z \neq x, z \neq y$, and $z \notin$ **FV[f]** $\cup$ **FV[g]** (this prevents the capture of free variables)

The first three rules here are obviously required by any notion of substitution. It is the last three that contain the real content. Rule (iv) says in the strongest possible form what it means syntactically for a variable to be bound; namely, if you try to substitute something for it, nothing happens. Rules (v) and (vi) say that substituting something in a lambda term for a different variable means substituting it in the body of the term. However, this has to be done carefully. Consider the following example:

$$((\lambda x . ((\lambda y . (y \ x))) \ t) \ u) \quad \twoheadleftarrow ((\lambda y . (y \ x)) \ /. \{x -> t\}) \ u)$$
$$\twoheadleftarrow ((\lambda y . ((y \ x) \ /. \{x -> t\}))) \ u)$$
$$\twoheadleftarrow ((\lambda y . (y \ t)) \ u) \quad \twoheadleftarrow (u \ t)$$

If, instead of **t**, we use **y**, then using rule (v) we get

$$((\lambda x . ((\lambda y . (y \ x))) \ y) \ u) \quad \twoheadleftarrow (\lambda y . (y \ y)) \ u) \quad \twoheadleftarrow (u \ u)$$

which is incorrect. The problem is in the step

$$((\lambda y . (y \ x)) \ /. \{x -> y\}) \ u) \quad \twoheadleftarrow ((\lambda y . ((y \ x) \ /. \{x -> y\})) \ u)$$
$$\twoheadleftarrow ((\lambda y . (y \ y)) \ u)$$

One says that the free variable **y** outside the lambda term has been captured by the bound variable **y** in the lambda term when it is substituted for **x** there. That's clearly wrong if the bound variable **y** is really not there; i.e., if it could be replaced by some other variable. What we should do is first replace **y** in the lambda term by a new variable **z** that doesn't occur anywhere else in the expression, and then continue.

$$((\lambda y . (y\ x))\ /.\ \{x \rightarrow y\})\ u) \quad \Leftarrow ((\lambda z . (z\ x))\ /.\ \{x \rightarrow y\})\ u)$$
$$\Leftarrow (\lambda z . ((z\ x)\ /.\ \{x \rightarrow y\}))\ u)$$
$$\Leftarrow (\lambda z . (z\ y))\ u) \quad \Leftarrow (u\ y)$$

This is exactly what rule (vi) says to do. The problem of capture of free variables is a serious one in functional progrmming languages, where it is usually discussed under the heading of the *scope* of a bound variable. The scope of **y** in λy . **expr** is exactly **expr**. This topic was discussed in Chapter 8, where it arose in the treatment of local variables (which are another case of bound variables). Another way to solve the problem of capture of free variables is to find a different notation that never mentions variables at all. There are two ways to do this in the lambda calculus. One is to replace the lambda calculus by combinators and the other is to use de Bruijn variable free notation. See [Barendregt] for discussions of both of these. In *Mathematica*, it is done by the pure function notation using **#** and **&**.

In using these rewrite rules, substitution is usually considered as one step; i.e., after a β-reduction, the resulting substitution is propagated throughout the entire expression before any other rule is used. With this proviso, if a sequence of applying rules to a given expression terminates after finitely many steps in an expression to which no further rules apply, then the resulting expression is independent of the order in which the rules were applied. Such an expression is called the *normal form* of the original expression.

7.2.3 An Implementation of the lambda calculus

The lambda calculus is the place where the buck stops as far as substitution is concerned. It is there that the rules for substitution are precisely specified in the situation where the only operations are function application and function abstraction. Trying to write these rules in *Mathematica* is a non-trivial exercise. Fortunately, it is not too difficult; we just have to carefully follow the rules given above. As we have seen, we cannot use the *Mathematica* operations of **Function** and **ReplaceAll** since they do not work correctly together.

It is tempting to try to use **With** instead of **ReplaceAll** since as we saw above, that fixes some of the problems. Unfortunately, it does not fix all of them, so we have to implement all of the operations ourselves. Thus, we construct a basic operation that does not evaluate its arguments at all; λ**[x].body** for function abstraction, and we use *Mathematica* function application **f[a]** for application. Substitution, written above in *Mathematica* notation as **f/.{x -> g}**, is implemented by a **let** operation since *Mathematica* substitution doesn't work correctly. We write it in the form **let[{x = expr1}, expr2]**, as in the notation used with **With**. The other thing that is required is an operation to calculate the free variables in an expression, given here by **freeVars[expr]**. The basic relation between these notions is given by the beta reduction rule, which says that applying a function written in the form of a lambda expression to an argument should rewrite to the value of replacing the variable of the lambda expression in the argument by the body of the lambda expression.

In[3]:=**Attributes[let] = {HoldFirst};**

In order to use the **Dot** notation that is standard in the lambda calculus, we have to unprotect **Dot** first.

In[4]:= **Unprotect[Dot];**

The rule for beta reduction then reads

In[5]:= **(λ[x_] . expr2_)[expr1_] := let[{x = expr1}, expr2];**
  Protect[Dot];

Warning: This rule interferes with the usual use of **Dot**. It also interferes with the use of **Dot** in Chapters 9 and 13 involving object-oriented programming.

The key to all of this, of course, is given by the rules governing **let**. We implement the rules (i), . . . , (vi) above in as direct a way as possible. Notice how variable capture is avoided in the last rule by using **Unique**.

```
In[6]:= let[{x_ = expr_}, x_] := expr;
    let[{x_ = expr_}, y_Symbol] := y /; x =!= y;
    let[{x_ = expr1_}, expr2_[expr3_]] :=
        let[{x = expr1}, expr2][let[{x = expr1}, expr3]];
    let[{x_ = expr1_}, (λ[x_] . expr2_)] := (λ[x] . expr2);
    let[{x_ = expr1_}, (λ[y_] . expr2_)] :=
        (λ[y] . let[{x = expr1}, expr2]) /;
        (x =!= y) && Not[MemberQ[freeVars[expr1], y]];
    let[{x_ = expr1_}, (λ[y_] . expr2_)] :=
        let[{x = expr1}, ((λ[y] . expr2) /. y ⭠Unique["q"])] /;
        (x =!= y) && MemberQ[freeVars[expr1], y];
```

Finally, the free variable operation is specified by the following rules.

```
In[7]:= freeVars[x_] := {x};
    freeVars[expr1_[expr2_]] :=
        Union[freeVars[expr1], freeVars[expr2]];
    freeVars[(λ[x_] . expr_)] :=
      Select[freeVars[expr], (# =!= x &)]
```

These 10 rules constitute a complete implementation of the lambda calculus. Here is a simple example.

```
In[8]:= (λ[z] . z[a])[(λ[x] . x)]
```

Out[8]= a

The result of the first use of beta reduction replaces **z** by **λ[x].x** (which we saw above is just the identity operation), so one has the expression **(λ[x].x)[a]**. Beta reduction applies again, reducing to the output **a**. Here is a slightly more complicated example with three applications.

```
In[9]:= ((λ[x] . a)[x])[((λ[y] . b)[y])[c]]
```

Out[9]= a[b[c]]

In this example, the final result is an iterated application in which no lambda terms appear, so no further reduction is possible. Now consider an example which is a possible source of trouble.

```
In[10]:= ((λ[x] . (λ[y] . y[x]))[t])[u]
```

Out[10]= u[t]

If we use **y** instead of **t**, then variable capture is possible, as discussed above, but is avoided because **Unique** is used in the appropriate rule.

```
In[11]:= ((λ[x] . (λ[y] . y[x]))[y])[u]
```

Out[11]= u[y]

If the variable **y** had been captured, the second application of beta reduction would have been to the expression

```
In[12]:= (λ[y] . y[y])[u]
```

Out[12]= u[u]

which would have been incorrect. As remarked above, it would be tempting to replace all of
these rules by the simple definition

$$(\lambda[x].expr1\_)[expr2\_] := With[\{x = expr2\}, expr1]$$

If you clear all of the previous definitions and try it, then it fails exactly on this expression,
actually for reasons having to do with the way that **With** evaluates its arguments. The direct
translation of the lambda expression into a **With** expression works correctly.

In[13]:= **With[{x = y}, With[{y = u}, y[x]]]**

Out[13]= u[y]

However, it fails when used as the right-hand side of a rule for **let**. The point is that when **x**
is replaced by **y** in **y[x]**, then the outer **y** should be turned into some new symbol because it
is in the body of a **With** statement specifying a substitution for **y**.

Here is another quite intricate example that reduces to the symbol **a**.

In[14]:= **(λ[f] . f[f[a]]) [(λ[x] . x[x]) [(λ[y] . y) [(λ[y] . y)]]]**

Out[14]= a

The combination $\lambda[x].x[x]$ is called the *paradoxical combinator*. Applied to itself, it is the archetype of a non-terminating computation. It behaves as it should.

In[15]:= **(λ[x] . x[x]) [(λ[x] . x[x])]**

    $IterationLimit::itlim : Iteration limit of 4096 exceeded.

Out[15]= Hold[let[{x = λ[x] . x[x]}, x][let[{x = λ[x] . x[x]}, x]]]

Thus, beta reduction was carried out 4096 times, resulting in the same expression being held.
This is exactly what should happen. Now consider the following evaluation. It goes into an
infinite loop, as would be expected by call-by-name evalution, but when it hits the iteration
limit it succeeds in finishing the evaluation with the correct answer.

In[16]:= **(λ[y] . a) [(λ[x] . x[x]) [(λ[x] . x[x])]]**

    $IterationLimit::itlim : Iteration limit of 4096 exceeded.

Out[16]= a

7.2.4 Arithmetic in the lambda calculus

The lambda calculus as implemented here is actually a complete programming language in itself. Any calculation that can be done in any of the standard programming languages can also be done in the lambda calculus, although it might be very unwieldy to actually carry it out. We'll show here how to introduce arithmetic via the Church numerals, which represent numbers in the lambda calculus. First some preliminary definitions of standard terms.

```
In[17]:= true = (λ[x] . (λ[y] . x));
         false = (λ[x] . (λ[y] . y));
         if = (λ[p] . (λ[x] . (λ[y] . p[x][y])));
```

Check that **if**, **true**, and **false** fit together in the expected way.

```
In[18]:= {if[true][t][f], if[false][t][f]}
```

```
Out[18]= {t, f}
```

Now create some Church numerals.

```
In[19]:= zero = λ[f] . (λ[x] . x);
         one = λ[f] . (λ[x] . f[x]);
         two = λ[f] . (λ[x] . f[f[x]]);
         three = λ[f] . (λ[x] . f[f[f[x]]]);
         four = λ[f] . (λ[x] . f[f[f[f[x]]]]);
```

The general Church numeral **n** can be constructed using **Nest**.

```
In[20]:= churchN[n_] := λ[f] . (λ[x] . Nest[f[#] &, x, n])
```

There are standard formulas to define the usual arithmetic functions in terms of this representation of the natural numbers.

```
In[21]:= succ = λ[n] . (λ[f] . (λ[x] . (n[f])[f[x]]));
         iszero = λ[n] . (n[(λ[x] . false)])[true];
         add = λ[m] . (λ[n] . (λ[f] . (λ[x] . (m[f])[n[f][x]]))));
         mult = λ[m] . (λ[n] . (λ[f] . m[n[f]]));
         exp = λ[m] . (λ[n] . (λ[f] . (λ[x] . ((n[m])[f])[x]))));
```

For instance:

In[22]:= {iszero[zero], iszero[four]}

Out[22]= {λ[x] . λ[y] . x, λ[x] . λ[y] . y}

We recognize the output as being {true, false}. Next try out the successor function.

In[23]:= {succ[zero], succ[one], succ[two]}

Out[23]= {λ[f] . λ[x] . f[x], λ[f] . λ[x] . f[f[x]], λ[f] . λ[x] . f[f[f[x]]]}

We recognize the results as being **one**, **two**, and **three**. Finally try addition, multiplication, and exponentiation.

In[24]:= **add[two][two]**

Out[24]= λ[f] . λ[x] . f[f[f[f[x]]]]

In[25]:= **add[churchN[2]][churchN[2]]**

Out[25]= λ[f] . λ[x] . f[f[f[f[x]]]]

In[26]:= **mult[churchN[2]][churchN[2]]**

Out[26]= λ[f] . λ[x] . f[f[f[f[x]]]]

In[27]:= **exp[churchN[2]][churchN[2]]**

Out[27]= λ[f] . λ[x] . f[f[f[f[x]]]]

In each case we recognize that the answer is **four**. However, it is very inconvenient to have to count the number of f[-]'s to recognize what number the output represents. Instead, we add a formatting command that formats the output as a Church numeral.

In[28]:= **Format[(λ[f_] . (λ[x_] . expr_)) /; numberlike[expr, x, f]] :=**
        SequenceForm["churchN[", Depth[expr] - 1, "]"]

In[29]:= **numberlike[expr_, x_, f_] :=**
        (expr === x) || (expr === f[x]) ||
         ((Depth[expr] ≤ 2) &&
           (Head[expr] === f) && (Length[expr] === 1) &&
           numberlike[expr[[1]], x, f])

Try out some small examples.

In[30]:= **add[churchN[5]][churchN[12]]**

Out[30]= churchN[17]

In[31]:= **mult[churchN[5]][churchN[5]]**

Out[31]= churchN[25]

In[32]:= **exp[churchN[3]][churchN[3]]**

Out[32]= churchN[27]

Now we are ready to try some larger calculations. First, the recursion limit and iteration limit have to be increased since these operations are completely recursive.

In[33]:= **\$RecursionLimit = 16000;**
     \$IterationLimit = 16000;

In[34]:= **add[churchN[275]][churchN[275]]**

Out[34]= churchN[550]

In[35]:= **mult[churchN[22]][churchN[22]]**

Out[35]= churchN[484]

In[36]:= **exp[churchN[4]][churchN[4]]**

Out[36]= churchN[256]

If the values here are increased, then the program crashes on my machine for unknown reasons. Compare this implementation with the implementation of the lambda calculus in ML given in [Paulson].

Technical note: *Mathematica* will not allow a definition in the form **let[{x = expr1}, expr2]** unless the first argument is held. This forces **let** to use call-by-name evaluation. A call-by-value version can be implemented just by giving **let** three separate values; i.e., **let[x, expr1, expr2]** with none of them held. This form is marginally faster.

8 Exercises

1. Newton's method for finding a zero of several functions of the same number of variables is a generalization of the method for one function of one variable. It views the several functions as a single vector-valued function of one vector variable and tries to write the same formula. Given $g(x) = \{g_1 (x_1, \ldots x_n), \ldots g_n (x_1, \ldots x_n)\}$, then the formula for the next step in the approximation is

$$\texttt{x}_{\texttt{n+1}}\texttt{=(x - Inverse[jacobian[g, x]].g[x])/.x->x}_{\texttt{n}}$$

Here x and g represent n-dimensional vectors. Use this formula to define **oneNewtonZero-Step** and then use **Nest, NestList,** and **FixedPoint** to define various versions of a **NewtonZero** function.

2. Newton's method can be adapted to finding critical points of a function by taking g in Exercise 1 to be the gradient of a single function f of n variables. Since the Jacobian of the gradient of a function is the same as the Hessian of the function, this leads to a formula

$$\texttt{x}_{\texttt{n+1}}\texttt{=(x-Inverse[hessian[f, x]].gradient[f, x])/.x->x}_{\texttt{n}}$$

Use this formula to define **oneNewtonStep** and then as above to define various versions of a **NewtonCriticalPoint** function.

3. Carry out a similar discussion for method of steepest descent, for Broyden's zero method and for Broyden's method.

4. Construct a package called **minimization** to find a local minimum of a function of several variables, given a starting point. It should take one optional argument, **method**, whose possible values are **newton, steepestDescent, broydenZero,** and **broyden**. It exports a single function called **findMinimum**. Note: There is a built-in function called **FindMinimum,** so the spelling checker will object, but just ignore that. You may want to look at the options for it and try to include similar options in your function.

III

Mastering Knowledge Representation in Mathematica

Polya Pattern Analysis

1 Introduction

Polya's Pattern Inventory is concerned with the following combinatorial problem. Suppose there is a pattern consisting of n regions which are to be colored using m colors. The regions could be stripes on a flag, beads on a necklace, or sides (or edges) of a geometric figure, etc. (Polya's original problem concerned isomers of molecules in which given numbers of different atoms could be arranged in different ways in the molecule). For instance, suppose we want to make a necklace consisting of 5 beads and there are both red and blue beads available. Clearly there are $2^5 = 32$ possible necklaces. Now suppose that we decide to use exactly 2 reds and 3 blues. Then it is almost as immediate that there are Binomial[5, 2] = 10 such necklaces. Now suppose that we decide that we will consider two necklaces the same if one is a rotation of the other. Then the answer takes further thought, particularly if we want to find the principle that answers all such questions. Polya's Pattern Inventory answers the general question: Suppose there is a group of symmetries acting on the n regions, and two colorings by m colors are to be considered equivalent if one coloring is taken to the other by one of the symmetries. In our example of a necklace, the rotation group acts on the colorings of the necklace and two colorings are considered the same if they differ just by the action of some rotation of the necklace. Polya's Pattern Inventory will determine how many different necklaces there are all together of each kind, allowing for equivalence under rotations. Thus, given two colors and five beads, there are six possible choices for numbers of colors: 5 red, 4 red and 1 blue, 3 red and 2 blue, 2 red and 3 blue, 1 red and 4 blue, and 5 blue. For each choice, we will determine how many necklaces there are, considering two necklaces to be the same if they differ just by a rotation. For instance, there is only one necklace consisting of 5 red beads, but there is also only one necklace consisting of 4 red and 1 blue beads, since any two colorings with these colors differ by a rotation.

This question can be investigated from a geometrical or an algebraic point of view.

•   The geometric approach starts by constructing all colorings of the regions with the given colors and then determines the orbit of each coloring under the action of the specified symmetry group. Different colorings can determine the same orbit; namely, the colorings in a particular orbit all determine exactly that orbit. The geometric solution consists in extracting one representative coloring from each distinct orbit. For small values of n and m these can be illustrated by pictures.

•   The algebraic approach, discovered by Polya, consists of the construction of a polynomial from which one can read off how many colorings there are, modulo a group action, for specific numbers of regions of specified colors. It does not provide an actual description of the different colorings, but is the only feasible approach for large values of the parameters.

Both approaches require some sample groups to use as examples, so these will be constructed first.

2 *The Geometric Approach*

If the regions to be colored are numbered 1 through n, then the symmetry groups acting on the regions can always be regarded as permutations of $\{1, \ldots, n\}$, so it suffices to construct some examples of permutation groups.

In what follows, we discuss the constructions needed for this chapter, but they have been placed in non-evaluable cells and we only show the examples as evaluated. To try the examples and others, load the package **PolyaPatternAnalysis**.

In[1]:= **Needs["PolyaPatternAnalysis`"]**

2.1 *Permutations*

A permutation is an expression of the form

$$\begin{pmatrix} 1, 2, 3, 4, 5, 6, 7 \\ 3, 1, 2, 6, 5, 7, 4 \end{pmatrix}$$

This describes the permutation where 1 goes to 3, 2 goes to 1, 3 goes to 2, etc. Usually, and especially in *Mathematica*, the top row is omitted and this is written just as (3, 1, 2, 6, 5, 7, 4). The *Mathematica* operation **Part** allows one to apply such a permutation to any list of the same length. For example,

In[2]:= **{a, b, c, d, e, f, g}[[{3, 1, 2, 6, 5, 7, 4}]]**

Out[2]= {c, a, b, f, e, g, d}

In particular, a permutation can be applied to another permutation and the result is again a permutation.

```
In[3]:= {5, 3, 4, 2, 7, 6, 1}[[{3, 1, 2, 6, 5, 7, 4}]]
```

Out[3]= {4, 5, 3, 6, 7, 1, 2}

2.2 Permutation Groups

Permutations form a group under this operation, called composition of permutations; i.e., there is an identity element (= the identity permutation), composition of permutations is associative, and each permutation has an inverse. To keep things straight, elements of the group of all permutations of $1, - - -, n$ will be written in the form $ge[i_1, . . . , i_n]$ rather than as $\{i_1, . . . , i_n\}$. The head ge stands for "group element". The formula above for the composition of permutations, modified for group elements, is

```
    g1_ · g2_ := g1[[List@@g2]] /; Length[g1] == Length[g2]
```

Here we have used the symbol **CenterDot** for the multiplication of permutations. For instance:

```
In[4]:= g1 = ge[3, 1, 2, 6, 5, 7, 4];
        g2 = ge[5, 3, 4, 2, 7, 6, 1];
```

```
In[6]:= g2 · g1
```

Out[6]= ge[4, 5, 3, 6, 7, 1, 2]

The identity permutation of length **n** is given by

```
    id_n_ := ge @@ Range[n]
```

Composition with the identity element on either side has no effect.

```
In[7]:= {g1 · id_7, id_7 · g2}
```

Out[7]= {ge[3, 1, 2, 6, 5, 7, 4], ge[5, 3, 4, 2, 7, 6, 1]}

The inverse of a permutation is given by a simple formula.

```
    inv[p_ge] :=
        ge@@Flatten[Map[Position[p, #]&,
                        Range[Length[p]]]]
```

For instance:

In[8]:= **{inv[g1], g1 · inv[g1]}**

Out[8]= {ge[2, 3, 1, 7, 5, 4, 6], ge[1, 2, 3, 4, 5, 6, 7]}

A collection of permutations determines a subgroup of the group of all permutations consisting of all possible compositions of members of the list with each other. **Outer** of **composition** with a list of permutations gives all pairwise compositions of the permutations in the list. Applying **Union** to the flattening of this eliminates duplicates and puts the result in canonical order. If this operation is nested until there are no further changes; i.e., if **Fixed-Point** is used, then all possible compositions are obtained, which gives the group generated by the permutations. Instead of just getting a list of group elements, we change the head to **group** to remind ourselves that this is a group. (Note: sometimes it is necessary to include the identity group element with the generators and sometimes it can be omitted, so we include it always.)

```
generatedGroup[permutations_List] :=
  group@@ FixedPoint[Union[Flatten[Outer[#1 · #2 &, #, #]]] &,
    Prepend[permutations, id_Length[permutations[[1]]]]]]
```

The output from **generatedGroup** clearly contains the identity permutation of the appropriate length as well as the composition of any two entries it contains. A little thought shows that it also contains the inverse of any entry, but we provide a check operation for this anyway which verifies that the collection of inverses of entries coincides with the collection of entries themselves.

```
checkGroup[g_group] :=
  With[{gr = g}, Union[Map[inv, gr]] == gr]
```

2.2.1 The rotation group

The rotation group of size n is the group of all cyclic permutations of $1, \ldots, n$. It is generated by a single rotation, in the sense that every rotation is a composition of copies of this smallest rotation. (To see the generators of the following groups, it is necessary to evaluate the appropriate cells here, even though they are evaluated in the package.)

In[9]:= **rotationGenerator_n_ := ge@@ RotateLeft[Range[n], 1]**

For instance:

In[10]:= **rotationGenerator_5**

Out[10]= ge[2, 3, 4, 5, 1]

The composition of this with itself rotates left by two steps, etc.

In[11]:= **rotationGenerator₅ · rotationGenerator₅**

Out[11]= ge[3, 4, 5, 1, 2]

The rotation group of a given size consists of all compositions of this with itself and the identity permutation.

rotationGroup$_{n\_}$:= generatedGroup[{rotationGenerator$_n$}]

For instance:

In[12]:= **rotationGroup₅**

Out[12]= group[ge[1, 2, 3, 4, 5], ge[2, 3, 4, 5, 1],
        ge[3, 4, 5, 1, 2], ge[4, 5, 1, 2, 3], ge[5, 1, 2, 3, 4]]

In[13]:= **checkGroup[rotationGroup₅]**

Out[13]= True

Moderately large examples can be constructed and checked.

In[14]:= **checkGroup[rotationGroup₁₀₀]**

Out[14]= True

2.2.2 The tetrahedron edge group

The tetrahedron edge group is the group of symmetries of the six edges of a tetrahedron determined by all proper physical motions of the tetrahedron. It is generated by (i) rotating by 120 degrees around a vertex and the center of the opposite face and (ii) rotating by 180 degrees about the line joining the centers of two opposite edges. Number the edges 1, 2, 3 around a given vertex and then 4, 5, 6 around the opposite face, as illustrated.

In[15]:=**Needs["Graphics`Master`"]**

In[16]:= **Show[WireFrame[Polyhedron[Tetrahedron]],
        Graphics3D[
            {Text["1", {-0.6, 0, 1}],
            Text["2", {0.1, 0, 1}],
            Text["3", {0.1, 0.7, 1}],**

```
      Text["5", {0.8, 0.6, -0.6}],
      Text["6", {-0.5, 0.2, -0.6}],
      Text["4", {0, -0.6, -0.5}]}],
   Boxed -> False];
```

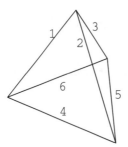

If the 120 degree rotation is about the vertex joining edges 1, 2, and 3, and the 180 degree rotation is about the line joining the centers of edges 1 and 5, then the generators are

In[17]:= **tetrahedronGenerator$_1$ = ge[2, 3, 1, 5, 6, 4];**
      tetrahedronGenerator$_2$ = ge[1, 6, 4, 3, 5, 2];

The whole group is generated by all compositions of these generators. Note that there is only one tetrahedron group rather than a family as with the rotation groups.

In[18]:= **tetrahedronGroup =**
      generatedGroup[{tetrahedronGenerator$_1$, tetrahedronGenerator$_2$}]

Out[18]= group[ge[1, 2, 3, 4, 5, 6], ge[1, 6, 4, 3, 5, 2],
         ge[2, 3, 1, 5, 6, 4], ge[2, 4, 5, 1, 6, 3],
         ge[3, 1, 2, 6, 4, 5], ge[3, 5, 6, 2, 4, 1],
         ge[4, 1, 6, 2, 3, 5], ge[4, 5, 2, 6, 3, 1], ge[5, 2, 4, 3, 1, 6],
         ge[5, 6, 3, 4, 1, 2], ge[6, 3, 5, 1, 2, 4], ge[6, 4, 1, 5, 2, 3]]

In[19]:= **checkGroup[tetrahedronGroup]**

Out[19]= True

2.2.3 The octahedron edge group

This is similar to the tetrahedron edge group. It consists of all symmetries of the 12 edges of a regular octahedron generated by rotating by 90 degrees about two adjacent vertices. If the edges are numbered 1, 2, 3, 4 around a given top vertex, 5, 6, 7, 8 around the middle square, and 9, 10, 11, 12 around the bottom vertex, then the generators are

```
In[20]:= octahedronGenerator₁ =
         ge[2, 3, 4, 1, 6, 7, 8, 5, 10, 11, 12, 9];
         octahedronGenerator₂ = ge[8, 4, 7, 12, 1, 3, 11, 9, 5, 2, 6, 10];
```

```
In[21]:= octahedronGroup =
         generatedGroup[{octahedronGenerator₁, octahedronGenerator₂}];
```

The output is suppressed since it is rather long.

```
In[22]:= Length[octahedronGroup]
```

```
Out[22]= 24
```

```
In[23]:= checkGroup[octahedronGroup]
```

```
Out[23]= True
```

2.2.4 The dodecahedron face group

A dodecahedron has pentagonal faces. The physical symmetries of a dodecahedron are generated by a rotation by 120 degrees about a vertex and by a rotation by 72 degrees of one face. For a suitable numbering of the faces, these are given by

```
In[24]:= dodecahedronGenerator₁ =
         ge[2, 3, 1, 5, 6, 4, 8, 9, 7, 11, 12, 10];
         dodecahedronGenerator₂ =
         ge[1, 3, 4, 7, 2, 9, 5, 6, 10, 11, 8, 12];
```

```
In[25]:= dodecahedronGroup = generatedGroup[
         {dodecahedronGenerator₁, dodecahedronGenerator₂}];
```

```
In[26]:= {Length[dodecahedronGroup], checkGroup[dodecahedronGroup]}
```

```
Out[26]= {60, True}
```

2.3. Group Actions and Orbit Spaces

Each time the kernel is restarted, the package has to be reloaded.

In[1]:= **Needs ["PolyaPatternAnalysis` "]**

If G is a group and X is a set, then a *group action* of G on X is a function act : $X \times G \rightarrow X$ satisfying two equations:

$$act(x, e) = x, \text{ where } e \text{ is the identity element of G, and}$$
$$act(act(x, g_1), g_2) = act(x, g_1 * g_2),$$

where $g_1 * g_2$ is the product of elements g_1 and g_2 from the group G. Starting with an element x_0 of X, the *orbit* of x_0 under the group action is the set $\{act(x_0, g) \mid g \in G\}$; i.e., it is the set of all elements of X that can be obtained from x_0 by acting on x_0 with some group element. If **act** is some group action on a set written in *Mathematica*, then the orbit of the element under the group action is expressed in *Mathematica* by the operation

```
orbit[setelement_, g_group, act_] :=
    List@@Union[Map[act[setelement, #]&, g]]
```

The purpose of **Union** here is to remove duplicate elements. Now, the orbits of two elements either coincide or are disjoint, so they determine a partition of X into disjoint subsets. The set of these subsets is called the *orbit space* X / G of X by the group action of G. Thus, an element of X / G is an orbit. We want to construct X / G in certain examples. The idea of the algorithm to be developed is to choose an element in X, find its orbit, add it to the set of orbits, delete its elements from X, choose another of the remaining elements, etc. This iterative procedure acting on pairs {A, X'}, where A is the partially constructed set of orbits and X' is the set of remaining elements in X, can be implemented by a one-step procedure which takes the first element of X', calculates its orbit, appends that orbit to A, and deletes its elements from X'.

```
oneStep[g_group, act_] :=
    If[#[[2]] =!= {},
        With[{orb = orbit[First[#[[2]]], g, act]},
            {Append[#[[1]], orb], Complement[#[[2]], orb]}],
        #]&
```

The first component of the fixed point of this operation starting with **{{}, X}** is the desired orbit space.

```
orbitSpace[set_, g_group, act_] :=
    First[FixedPoint[oneStep[g, act], {{}, set}]]
```

This general construction can be applied in many specific circumstances. We work out a simple example before turning to the case we are interested in.

2.3.1 Example

As a simple example, consider the following set of points.

In[2]:=**points = {a, b, c, d, e}**

Out[2]= {a, b, c, d, e}

Form the set of all pairs of points as the example of a set X.

In[3]:=**pairs = Flatten[Outer[List, points, points], 1]**

Out[3]= {{a, a}, {a, b}, {a, c}, {a, d}, {a, e}, {b, a},
        {b, b}, {b, c}, {b, d}, {b, e}, {c, a}, {c, b},
        {c, c}, {c, d}, {c, e}, {d, a}, {d, b}, {d, c},
        {d, d}, {d, e}, {e, a}, {e, b}, {e, c}, {e, d}, {e, e}}

Now, define an action of the group **rotationGroup$_5$** on this set by operating on both components of a pair. The components are permuted according to their position in the set **points**.

In[4]:=**actpair[{x_, y_}, groupelement_ge] :=**
        Flatten[
          Map[points[[List@@groupelement]][[
                   Flatten[Position[points, #]]]]]&,
              {x, y}]]

For instance:

In[5]:=**actpair[{a, b}, ge[5, 1, 2, 3, 4]]**

Out[5]= {e, a}

Using this action, the orbit space of **pairs** under the action can be constructed using the operation defined above.

In[6]:= **orbs = orbitSpace[pairs, rotationGroup$_5$, actpair]**

Out[6]= {{{a, a}, {b, b}, {c, c}, {d, d}, {e, e}},
         {{a, b}, {b, c}, {c, d}, {d, e}, {e, a}},
         {{a, c}, {b, d}, {c, e}, {d, a}, {e, b}},
         {{a, d}, {b, e}, {c, a}, {d, b}, {e, c}},
         {{a, e}, {b, a}, {c, b}, {d, c}, {e, d}}}

Each row here is an orbit. In order to make a picture of these orbits, turn the pairs of points into pairs of numbers, and join the pairs in each orbit by a line.

```
In[7]:= Show[Graphics[
          With[{numbs = orbs /. Thread[points - <Range[5]]},
            {Map[Line, numbs],
             PointSize[0.03], Map[Point, numbs, {2}]}]]];
```

That is, an orbit consists of the collection of points joined by a line in this picture. If all we are interested in is a representation of how many orbits there are, it is sufficient to choose a representative pair from each orbit, for instance, the first pair in each orbit. This gives a set which is isomorphic to the set of orbits.

```
In[8]:= orb1 = Map[First, orbs]
```

Out[8]= {{a, a}, {a, b}, {a, c}, {a, d}, {a, e}}

Finally, there is a "projection" function mapping the set **pairs** onto the orbit space **orb1** given by assigning to each pair the representation of its orbit in **orb1**.

```
In[9]:= proj[{x_, y_}] := First[orbit[{x, y}, rotationGroup_5, actpair]]
```

For instance:

```
In[10]:= proj[{e, b}]
```

Out[10]= {a, c}

The inverse image of an element in **orb1** under this mapping is the orbit containing that element.

In[11]:=**projInverse[{x_, y_}] :=**
        Select[pairs, (proj[#] == {x, y}&)]

For instance:

In[12]:=**projInverse[{a, c}]**

Out[12]= {{a, c}, {b, d}, {c, e}, {d, a}, {e, b}}

2.4 *The Pattern Array for a Group Action*

Suppose there are six regions to be colored with three colors, say, red, green, and blue, and we want to find the distinct patterns with respect to the action of some groups of permutations. There are several things that have to be constructed before we arrive at the final answer.

i) A possible choice of colors could be 2 reds, 1 green, and 3 blues. This choice is abbreviated as **pt[2, 1, 3]**, where **pt** stands for *partition*. Our first task is to generate the list of all such choices; i.e., all partitions of 6 into three summands (including 0 as a possible summand).

ii) One possible coloring of six regions using the partition **pt[2, 1, 3]** is represented by **pattern[red, red, green, blue, blue, blue]**. Our next task is to construct one such pattern for every possible choice of colors. These will be called *basic* patterns.

iii) The collection of all patterns is constructed by forming all permutations of each basic pattern. This collection is called **setOfPatterns**.

iv) Now all that has to be done is to specify how a permutation acts on a pattern. Then the **orbitSpace** construction from above will construct the set of orbits.

v) Finally, pick out a representative pattern from each distinct orbit. This is the pattern array we are seeking.

2.4.1 Partitions

The operations here are not exported from the package so we evaluate them in place here. With a little bit of experimentation, a procedure can be written that generates all partitions of n into m non-negative summands (and nothing else).

```
In[13]:=partitions[0, m_] := {pt@@(0 Range[m])};
     partitions[n_, 1] := pt[n];
     partitions[n_Integer?Positive, m_Integer?Positive] :=
        Table[Flatten[partitions[n-i, m-1]] //.
            pt[x___] /; Length[{x}] == m-1 :> Prepend[pt[x], i],
            {i, 0, n}]
```

For instance:

```
In[16]:=partitions[5, 3]
```

```
Out[16]= {{pt[0, 0, 5], pt[0, 1, 4],
       pt[0, 2, 3], pt[0, 3, 2], pt[0, 4, 1], pt[0, 5, 0]},
      {pt[1, 0, 4], pt[1, 1, 3], pt[1, 2, 2], pt[1, 3, 1], pt[1, 4, 0]},
      {pt[2, 0, 3], pt[2, 1, 2], pt[2, 2, 1], pt[2, 3, 0]},
      {pt[3, 0, 2], pt[3, 1, 1], pt[3, 2, 0]},
      {pt[4, 0, 1], pt[4, 1, 0]}, {pt[5, 0, 0]}}
```

2.4.2 All patterns

Rather than using the names red, green, etc., we denote colors by c[1], c[2], etc., and instead of putting the colors in a list, we use an expression with head **pattern**. Thus, one possible coloring of 6 regions using the choice of colors {2, 1, 3} is represented by **pattern[c[1], c[1], c[2], c[3], c[3], c[3]]**. Our next task is to construct one such pattern for every possible choice of colors. In the following construction we treat c as a variable since we will later want to replace it by an operation that actually does something. In these three rewrite rules, n represents the number of regions and m the number of colors.

```
In[17]:=oneEach[n_Integer?Positive, 0, c_] := {{}};
     oneEach[n_Integer?Positive, 1, c_] := pattern@@Table[c[1], {n}];
     oneEach[n_Integer?Positive, m_Integer?Positive, c_] :=
        Map[pattern@@Flatten[Table[Table[c[i], {#[[i]]}], {i, m}]]&,
            Flatten[partitions[n, m]]]
```

For instance, to see how this works, we treat the case of 5 regions and 2 colors, since for 3 colors the output to be calculated later becomes rather large. First of all,

```
In[20]:=Flatten[partitions[5, 2]]
```

```
Out[20]= {pt[0, 5], pt[1, 4], pt[2, 3], pt[3, 2], pt[4, 1], pt[5, 0]}
```

Each of the six partitions will determine a basic pattern.

```
In[21]:=Map[Flatten[Table[Table[c[i], {#[[i]]}], {i, 2}]]&,
           Flatten[partitions[5, 2]]]
```

```
Out[21]= {{c[2], c[2], c[2], c[2], c[2]}, {c[1], c[2], c[2], c[2], c[2]},
         {c[1], c[1], c[2], c[2], c[2]}, {c[1], c[1], c[1], c[2], c[2]},
         {c[1], c[1], c[1], c[1], c[2]}, {c[1], c[1], c[1], c[1], c[1]}}
```

All that has to be done is to change the head of each inner list to **pattern**, to get the resulting list of six basic patterns.

```
In[22]:=basicPatterns = oneEach[5, 2, c]
```

```
Out[22]= {pattern[c[2], c[2], c[2], c[2], c[2]],
         pattern[c[1], c[2], c[2], c[2], c[2]],
         pattern[c[1], c[1], c[2], c[2], c[2]],
         pattern[c[1], c[1], c[1], c[2], c[2]],
         pattern[c[1], c[1], c[1], c[1], c[2]],
         pattern[c[1], c[1], c[1], c[1], c[1]]}
```

To display this in a more condensed form, replace each pattern by a "*".

```
In[23]:=basicPatterns/._pattern -> "*"
```

```
Out[23]= {*, *, *, *, *, *}
```

The output of **oneEach** gives one basic pattern for each choice of colors. To find all colorings of the regions for a particular choice of colors, it is necessary to construct all permutations of the given pattern. For the first pattern in the example, all permutations are the same. For the second pattern, there are 5! permutations, but only 5 of them represent different colorings, because permutating $c[2]$'s amongst themselves produces no change. Notice that **Permutations** gives the correct result when some of the items are the same. For example,

```
In[24]:=Permutations[{a, b, b}]
```

```
Out[24]= {{a, b, b}, {b, a, b}, {b, b, a}}
```

To find all patterns, we just have to apply **Permutations** to each of the patterns given by **oneEach**. In the example this is done as follows:

```
In[25]:=setOfPatterns[n_, m_, c_] :=
        Map[Permutations, oneEach[n, m, c]];
```

For instance:

In[26]:=`setOfPatterns[5, 2, c];`

gives us all possible $2^5 = 32$ colorings of 5 regions using 2 colors. Its output is suppressed because it is long. A typical entry is `pattern[c[2], c[1], c[2], c[1], c[2]]`. To visualize the output, we again replace each pattern by "*".

In[27]:=`TableForm[setOfPatterns[5, 2, c]/._pattern :> "*",`
          `TableSpacing ←{1, 1}]`

```
   Out[27]//TableForm=
   *

   *    *    *    *    *

   *    *    *    *    *    *    *    *    *    *

   *    *    *    *    *    *    *    *    *    *

   *    *    *    *    *

   *
```

Each star here represents a permutation of a basic pattern and shows that each of the 6 basic patterns has been expanded to a number of permuted patterns. The first and last basic patterns have no permutations, so produce only 1 pattern each, the second and fifth have 5 permutations each, and the third and fourth have 10 permutations each. (If 3 colors were used instead of 2, then there would be 3^5 colorings broken up into similar groupings, etc.) These grouping are determined by the partitions of 5 into two summands and give rise to a *stratification* of the set `setOfPatterns[5, 2, c]` into 6 strata corresponding to the 6 rows in the preceding output. This stratification will turn out to be important.

2.4.3 Orbits

So far, no use has been made of a symmetry group which we assume here to be `rotationGroup`$_5$. It acts on each of these possible patterns by permuting the colors in them by rotations; i.e., we define the action of a group element on a pattern as follows:

In[28]:=`act[pattern_, groupelement_ge] :=`
          `pattern[[List@@groupelement]] /;`
          `Length[pattern] == Length[groupelement]`

This describes an action of the group G = `rotationGroup`$_5$ on the set X = `setOfPatterns`. For instance:

In[29]:=`act[pattern[c[2], c[1], c[2], c[1], c[2]],`
          `ge[5, 1, 2, 3, 4]]`

Out[29]= `pattern[c[2], c[2], c[1], c[2], c[1]]`

Furthermore, the strata are clearly invariant under this action since the orbit of a particular pattern consists of all the patterns formed by all rotations of the colors in the given pattern. That is,

```
In[30]:= orbit[pattern[c[2], c[1], c[2], c[1], c[2]],
             rotationGroup₅, act]
```

```
Out[30]= {pattern[c[1], c[2], c[1], c[2], c[2]],
         pattern[c[1], c[2], c[2], c[1], c[2]],
         pattern[c[2], c[1], c[2], c[1], c[2]],
         pattern[c[2], c[1], c[2], c[2], c[1]],
         pattern[c[2], c[2], c[1], c[2], c[1]]}
```

Since the strata are invariant under the group action, the orbit space of `setOfPatterns[5, 2, c]` is the union of the orbit spaces of the strata. Hence, the orbit space inherits a stratification. Thus, to find it, we map the function `orbitSpace` down `setOfPatterns[5, 2, c]`. As before, the result is too large to understand we replace each pattern by a "*".

```
In[31]:= MatrixForm[Map[orbitSpace[#, rotationGroup₅, act] &,
             setOfPatterns[5, 2, c]] /. _pattern - <"*"]
```

Out[31]//MatrixForm=

$$
\begin{pmatrix}
\{\{*\}\} \\
\{\{*,\ *,\ *,\ *,\ *\}\} \\
\{\{*,\ *,\ *,\ *,\ *\},\ \{*,\ *,\ *,\ *,\ *\}\} \\
\{\{*,\ *,\ *,\ *,\ *\},\ \{*,\ *,\ *,\ *,\ *\}\} \\
\{\{*,\ *,\ *,\ *,\ *\}\} \\
\{\{*\}\}
\end{pmatrix}
$$

Here we see the 32 patterns divided up into 8 orbits falling into 6 kinds of orbits. The first and last are patterns in which all colors are the same, so there is only one pattern of each type. The second and sixth consist of patterns with one region being a different color than the others. There is 1 kind of orbit for each color consisting of the 5 rotations of the basic pattern. Finally, there are 2 kinds of orbits consisting of patterns with 2 regions of one color and 3 of the other. Each kind consists of 2 types of orbit with 5 patterns each. Another way to display this information is to choose the first patterns from each orbit and just look at that. This helps to understand the kinds of orbits and their grouping in this display.

In[32]:= **Map[First, Map[orbitSpace[#, rotationGroup$_5$, act] &,**
          setOfPatterns[5, 2, c]], {2}]

Out[32]= {{pattern[c[2], c[2], c[2], c[2], c[2]]},
          {pattern[c[1], c[2], c[2], c[2], c[2]]},
          {pattern[c[1], c[1], c[2], c[2], c[2]],
           pattern[c[1], c[2], c[1], c[2], c[2]]},
          {pattern[c[1], c[1], c[1], c[2], c[2]],
           pattern[c[1], c[1], c[2], c[1], c[2]]},
          {pattern[c[1], c[1], c[1], c[1], c[2]]},
          {pattern[c[1], c[1], c[1], c[1], c[1]]}}

Finally, replacing patterns by stars as before shows the inherited stratification of the orbit space more clearly.

In[33]:= **TableForm[%/._pattern :> "*", TableSpacing ∢1,1}]**

Out[33]//TableForm=

*

*

*   *

*   *

*

*

We collect together the results of this analysis into a single operation to construct the Polya Pattern Array.

```
patternArray[g_group, act_, m_Integer?Positive, c_] :=
  Map[First,
      Map[orbitSpace[#, g, act]&,
          setOfPatterns[Length[g[[1]]], m, c]], {2}]
```

As a check, repeat the calculation we just stepped through.

In[34]:= **patternArray[rotationGroup$_5$, act, 2, c]**

Out[34]= {{pattern[c[2], c[2], c[2], c[2], c[2]]},
          {pattern[c[1], c[2], c[2], c[2], c[2]]},
          {pattern[c[1], c[1], c[2], c[2], c[2]],
           pattern[c[1], c[2], c[1], c[2], c[2]]},
          {pattern[c[1], c[1], c[1], c[2], c[2]],

```
          pattern[c[1], c[1], c[2], c[1], c[2]]},
        {pattern[c[1], c[1], c[1], c[1], c[2]]},
        {pattern[c[1], c[1], c[1], c[1], c[1]]}}
```

If we want a diagram representing this output with *'s replacing the patterns, it is useful to be able to take the transpose of a table with rows of unequal lengths. To do so, we have to pad all of the rows until they have the same length with something that doesn't appear in the final table. The following does it.

```
pad[list_] :=
  With[{len = Max[Map[Length, list]]},
       Map[Join[#, Table[" ", {len - Length[#]}]]&, list]]
```

The strata are now represented by the columns.

```
In[35]:= TableForm[
           Transpose[pad[patternArray[rotationGroup₅, act, 2, c] /.
             _pattern ↵"*"]], TableSpacing ↤{1, 1}]
```

```
Out[35]//TableForm=
  *   *   *   *   *   *
          *   *
```

In the same way, we can display the output for colorings of the corners of a square using three colors.

```
In[36]:= TableForm[Transpose[pad[patternArray[
             rotationGroup₄, act, 3, c] /. _pattern ↵"*"]],
           TableSpacing ↤{1, 1}]
```

```
Out[36]//TableForm=
  *   *   *   *   *   *   *   *   *   *   *   *   *   *   *
      *               *   *       *   *   *
                      *   *           *
```

The following is the result for coloring the edges of a tetrahedron with two colors.

```
In[37]:= TableForm[Transpose[pad[
             patternArray[tetrahedronGroup, act, 2, c]/.
               _pattern -> "*"]],
             TableSpacing ↤{1, 1}]
```

```
Out[37]//TableForm=
  *   *   *   *   *   *   *
          *   *   *
              *
              *
```

The corresponding result for the octahedron group takes somewhat longer to evaluate and has a much larger output, which is therefore omitted.

```
In[38]:= TableForm[Transpose[pad[
             patternArray[octahedronGroup, act, 2, c]/.
             _pattern -> "*"]],
             TableSpacing ←{1, 1}];
```

Finally, the dodecahedron group shows a different pattern, which is again too large to display here.

```
In[39]:= TableForm[Transpose[pad[
             patternArray[dodecahedronGroup, act, 2, c]/.
             _pattern -> "*"]],
             TableSpacing ←{1, 1}];
```

The number of entries in each stratum in the orbit space of a group action is just the length of the corresponding column. For the rotation group of size 5, the tetrahedron group and the octahedron group, using two colors, it is given by the commands:

```
In[40]:= Map[Length, patternArray[rotationGroup₅, act, 2, c]]
```

Out[40]= {1, 1, 2, 2, 1, 1}

```
In[41]:= Map[Length, patternArray[tetrahedronGroup, act, 2, c]]
```

Out[41]= {1, 1, 2, 4, 2, 1, 1}

```
In[42]:= Map[Length, patternArray[octahedronGroup, act, 2, c]]
```

Out[42]= {1, 1, 5, 13, 27, 38, 48, 38, 27, 13, 5, 1, 1}

```
In[43]:= Map[Length, patternArray[dodecahedronGroup, act, 2, c]]
```

Out[43]= {1, 1, 3, 5, 12, 14, 24, 14, 12, 5, 3, 1, 1}

These lists of numbers are what will be calculated by the *Polya Pattern Inventory* in the algebraic treatment of this subject.

2.5 The Picture Array for a Group Action

In[1]:= **Needs["PolyaPatternAnalysis`"]**

2.5.1 Picture array

Our ultimate goal is to make pictures of all the patterns modulo symmetries for a given design. In order to use the pattern arrays derived above in plots, the head **pattern** has to be replaced by **List** everywhere.

```
pictureArray[g_group, act_, m_Integer?Positive, c_] :=
    patternArray[g, act, m, c] /. pattern ←List
```

For instance:

In[2]:= **pictureArray[rotationGroup$_5$, act, 2, c]**

Out[2]= {{{c[2], c[2], c[2], c[2], c[2]}}, {{c[1], c[2], c[2], c[2], c[2]}},
    {{c[1], c[1], c[2], c[2], c[2]}, {c[1], c[2], c[1], c[2], c[2]}},
    {{c[1], c[1], c[1], c[2], c[2]}, {c[1], c[1], c[2], c[1], c[2]}},
    {{c[1], c[1], c[1], c[1], c[2]}}, {{c[1], c[1], c[1], c[1], c[1]}}}}

2.5.2 Rotation groups

First we treat the case of a rotation group. A necklace will be pictured as a circle with small disks equally spaced aroung the circle for the beads. **pictureArray** can be used to calculate how to color each bead in a necklace by replacing each abstract color name of the form **c[i]** with an actual color specification using **Hue[N[$\frac{\#}{2}$]]&** in place of **c** when evaluating the function. For instance:

In[3]:= **pictureArray$\left[$rotationGroup$_5$, act, 2, Hue$\left[$N$\left[\frac{\#}{2}\right]\right]$ &$\right]$**

Out[3]= {{{Hue[1.], Hue[1.], Hue[1.], Hue[1.], Hue[1.]}},
    {{Hue[0.5], Hue[1.], Hue[1.], Hue[1.], Hue[1.]}},
    {{Hue[0.5], Hue[0.5], Hue[1.], Hue[1.], Hue[1.]},
     {Hue[0.5], Hue[1.], Hue[0.5], Hue[1.], Hue[1.]}},
    {{Hue[0.5], Hue[0.5], Hue[0.5], Hue[1.], Hue[1.]},
     {Hue[0.5], Hue[0.5], Hue[1.], Hue[0.5], Hue[1.]}},
    {{Hue[0.5], Hue[0.5], Hue[0.5], Hue[0.5], Hue[1.]}},
    {{Hue[0.5], Hue[0.5], Hue[0.5], Hue[0.5], Hue[0.5]}}}}

Now all we have to do is combine these color specifications with a `Disk` description of the beads.

```
In[4]:= necklaces[n_, m_] :=
    Map[{Circle[{0, 0}, 1],
       Transpose[
          {#,
           Map[Disk[#, Min[0.25, 1/n]] &,
            Table[{Cos[N[2 k π/n]], Sin[N[2 k π/n]]}, {k, n}]]}]} &,
       pictureArray[rotationGroup_n, act, m, Hue[N[#/m]] &], {2}]
```

Here is a small example constructing the three necklaces consisting of two beads using two colors; namely, the two necklaces that use only one color and the necklace using one bead of each color.

```
In[5]:= Chop[necklaces[2, 2]]

Out[5]= {{{Circle[{0, 0}, 1], {{Hue[1.], Disk[{-1., 0}, 0.25]},
           {Hue[1.], Disk[{1., 0}, 0.25]}}}},
         {{Circle[{0, 0}, 1], {{Hue[0.5], Disk[{-1., 0}, 0.25]},
           {Hue[1.], Disk[{1., 0}, 0.25]}}}},
         {{Circle[{0, 0}, 1], {{Hue[0.5], Disk[{-1., 0}, 0.25]},
           {Hue[0.5], Disk[{1., 0}, 0.25]}}}}}
```

The plotting routine is now very simple using this geometric construction of the necklaces. The actual graphics objects are constructed by a routine called `polyaPictures`. In order to use particular groups as the first argument to `polyaPictures`, we have to give it the attribute `HoldFirst`.

```
Attributes[polyaPictures] = {HoldFirst};
```

We would also like to display the pictures here in the same orientation as the diagrams above. This time we have to be able to take the transpose of a matrix of graphics objects of unequal lengths. The following operation is what we need.

```
padGraphics[list_] :=
   With[{len = Max[Map[Length, list]]},
      Map[Join[#, Table[Graphics[{}], {len - Length[#]}]]&,
         list]];


polyaPictures[rotationGroup_n_ , m_] :=
  GraphicsArray[Transpose[padGraphics[
     Map[Graphics[#, AspectRatio →Automatic] &,
     necklaces[n, m], {2}] ]]];
```

We can finally illustrate all of the different necklaces with five beads using beads of two colors, red and blue, where necklaces are considered the same if they differ by a rotation.

In[6]:= **Show[polyaPictures[rotationGroup$_5$, 2]];**

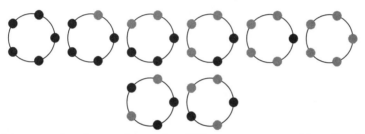

In this picture, each column shows the different patterns with a fixed distribution of colors. The numbers of necklaces of each kind are given by the command:

In[7]:= **Map[Length, pictureArray[rotationGroup$_5$, act, 2, c]]**

Out[7]= {1, 1, 2, 2, 1, 1}

For six beads and two colors, there are almost twice as many necklaces.

In[8]:= **Show[polyaPictures[rotationGroup$_6$, 2]];**

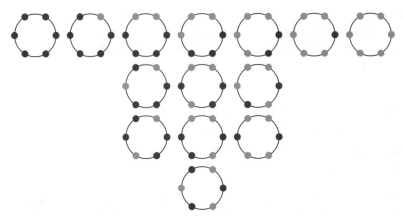

As before, the numbers of necklaces of each kind are given by the command

```
In[9]:= Map[Length, pictureArray[rotationGroup₆, act, 2, c]]
```

```
Out[9]= {1, 1, 3, 4, 3, 1, 1}
```

2.5.3 The tetrahedron group

Next we treat the case of the edges of a tetrahedron. In order to make a picture of a tetrahedron we need the list of the space coordinates of its four vertices; this is found in one of the graphics packages.

```
In[10]:= Needs["Graphics`Master`"]
```

```
In[11]:= Vertices[Tetrahedron]
```

```
Out[11]= {{0, 0, 1.73205}, {0, 1.63299, -0.57735},
          {-1.41421, -0.816497, -0.57735}, {1.41421, -0.816497, -0.57735}}
```

We also need a description of the edges of the tetrahedron which we have to calculate for ourselves.

```
In[12]:= edges[Tetrahedron] = Union[
            Map[Union[Take[#, 2]]&, Permutations[{1, 2, 3, 4}]]]
```

```
Out[12]= {{1, 2}, {1, 3}, {1, 4}, {2, 3}, {2, 4}, {3, 4}}
```

The actual graphical lines representing the edges of the tetrahedron are given by the operation

```
In[13]:= Map[Line[Vertices[Tetrahedron][[#]]]&,
            edges[Tetrahedron]]
```

```
Out[13]= {Line[{{0, 0, 1.73205}, {0, 1.63299, -0.57735}}],
     Line[{{0, 0, 1.73205}, {-1.41421, -0.816497, -0.57735}}],
     Line[{{0, 0, 1.73205}, {1.41421, -0.816497, -0.57735}}],
     Line[{{0, 1.63299, -0.57735}, {-1.41421, -0.816497, -0.57735}}],
     Line[{{0, 1.63299, -0.57735}, {1.41421, -0.816497, -0.57735}}],
     Line[{{-1.41421, -0.816497, -0.57735},
        {1.41421, -0.816497, -0.57735}}]}
```

Using this construction, the geometric tetrahedra are constructed as before.

```
In[14]:= tetrahedra[m_] :=
     With[{lines = Map[Line[Vertices[Tetrahedron][[#]]]&,
                    edges[Tetrahedron]]},
        Map[Prepend[#, Thickness[0.03]]&,
           Map[Transpose[{#, lines}]&,
              pictureArray[tetrahedronGroup, act, m,
                          GrayLevel[N[(# - 1)/m]]&],
              {2}],
           {3}]];
```

The plotting routine, using the geometric construction of the tetrahedra, is now almost exactly the same as for the rotation group.

```
     polyaPictures[tetrahedronGroup, m_] :=
        GraphicsArray[Transpose[padGraphics[
           Map[Graphics3D[#, Boxed -> False]&,
              tetrahedra[m],
              {2}]]] ];
```

```
In[15]:= Show[polyaPictures[tetrahedronGroup, 2]];
```

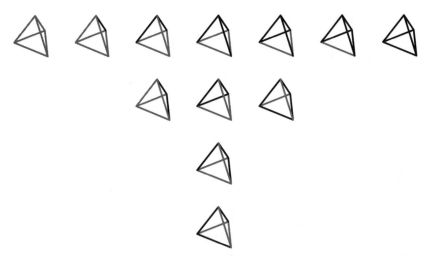

As before, the numbers of tetrahedra of each kind are given by the command

In[16]:=`Map[Length, pictureArray[tetrahedronGroup, act, 2, c]]`

Out[16]= `{1, 1, 2, 4, 2, 1, 1}`

The colorings for the octahedron group and the dodecahedron group are too large to illustrate.

3 The Algebraic Approach

The geometric approach in the preceding section succeeded through a brute force construction of the desired patterns. Counting how many patterns there are for each choice of colors (i.e., the number of entries in a column) is an incidental by-product of the construction. The algebraic approach, which is highly refined, concentrates solely on finding these numbers and never does spell out what the actual patterns are. These numbers will appear as coefficients in a polynomial whose variables represent the colors; e.g., in the polynomial for the rotation group of size 5 with 2 colors, the term $2\,c[1]^3\,c[2]^2$ will mean that there are 2 (equivalence classes of) necklaces using 3 beads of color c[1] and 2 of color c[2]. There are 3 steps in constructing this polynomial.

   i) First, we have to adopt a different representation of permutations in which they are given as products of cycles.

   ii) Next, given a permutation group G, a polynomial $P_G(x_1, \ldots, x_k)$, called the *cycle index* of G is constructed. Here k is the maximum length of a cycle in the cycle representations of the elements of G.

iii) Finally, the polynomial for m colors is given by substituting $\sum_{j=1}^{m} c[j]^i$ for x_i in P_G and dividing the result by the number of elements in G.

It is a non-trivial result of group theory that this construction gives the desired answer. For an introductory treatment of the theory, see [Tucker], and for the full story, see [Rotman] or [Biggs, L].

3.1 *The Cycle Representation of a Permutation*

In[1]:= **Needs["PolyaPatternAnalysis`"]**

There is another way to represent permutations; namely, as products of cycles. For instance, in the permutation {3, 1, 2, 6, 5, 7, 4}, 1 goes to 3, 3 goes to 2 and 2 goes to 1, so these three entries are cyclically permuted. This is represented by the cycle {3, 2, 1}, or equivalently {1, 3, 2}, or {2, 1, 3}. Notice that the *cycle* {3, 2, 1} is different from the *permutation* {3, 2, 1}, even though they are both written as the same list. It is a theorem that any permutation is equivalent to a product of cycles. There is a nice functional program which appeared years ago on the network from "The Gang of Four at Stanford" which calculates the cycle representation of a permutation. It is constructed as follows: First take a test permutation.

In[2]:= **perm = {3, 1, 2, 6, 5, 7, 4};**

As we have seen, following 1 to 3 to 2 to 1 leads to the cycle {1, 3, 2}. This can be calculated by the operation

In[3]:= **NestList[perm[[#]]&, 3, Length[perm]]**

Out[3]= {3, 2, 1, 3, 2, 1, 3, 2}

We have followed the sequence for eight terms because we don't know exactly how long the cycle is going to be. Of course, we only need the first three terms here, which are given by

In[4]:= **Take[%, Length[Union[%]]]**

Out[4]= {3, 2, 1}

Do this for every entry in the permutation.

In[5]:= **Map[NestList[perm[[#]]&, #, Length[perm]]&, perm]**

Out[5]= {{3, 2, 1, 3, 2, 1, 3, 2},
         {1, 3, 2, 1, 3, 2, 1, 3}, {2, 1, 3, 2, 1, 3, 2, 1},
         {6, 7, 4, 6, 7, 4, 6, 7}, {5, 5, 5, 5, 5, 5, 5, 5},
         {7, 4, 6, 7, 4, 6, 7, 4}, {4, 6, 7, 4, 6, 7, 4, 6}}

```
In[6]:=Map[Take[#, Length[Union[#]]]&, %]
```

```
Out[6]= {{3, 2, 1}, {1, 3, 2}, {2, 1, 3}, {6, 7, 4}, {5}, {7, 4, 6}, {4, 6, 7}}
```

Then pick out those cycles that start with the minimal entry, just to have a definite way to choose one representative of each cycle.

```
In[7]:=Select[%, First[#] == Min[#]&]
```

```
Out[7]= {{1, 3, 2}, {5}, {4, 6, 7}}
```

Putting these steps together gives the construction.

```
In[8]:=toCycles[perm_] :=
        Select[
            Map[Take[#, Length[Union[#]]]&,
                Map[NestList[perm[[#]]&, #, Length[perm]]&, perm]],
                First[#] == Min[#]&]
```

Check this on the example that was just worked out interactively.

```
In[9]:=toCycles[{3, 1, 2, 6, 5, 7, 4}]
```

```
Out[9]= {{1, 3, 2}, {5}, {4, 6, 7}}
```

3.2 The Cycle Index of a Group

Polya's Pattern Inventory is constructed from a polynomial called the cycle index. See the reference cited at the beginning of this section for further information. The cycle index for a group of symmetries is a polynomial $P_G(x_1, \ldots, x_k)$ in variables x_i, $i = 1, \ldots, k$, where k is the maximum length of a cycle in the cycle representations of the elements of the group G. We first work out an example of the construction interactively. Start with the tetrahedron group, rewrite it as a list of lists, and apply **toCycles** to each permutation in the group. This gives the following:

```
In[10]:=cycleList = Map[toCycles[List@@#]&,
                    List@@tetrahedronGroup]
```

Out[10]= {{{1}, {2}, {3}, {4}, {5}, {6}}, {{1}, {3, 4}, {5}, {2, 6}},
            {{1, 2, 3}, {4, 5, 6}}, {{1, 2, 4}, {3, 5, 6}},
            {{1, 3, 2}, {4, 6, 5}}, {{2, 5, 4}, {1, 3, 6}},
            {{1, 4, 2}, {3, 6, 5}}, {{2, 5, 3}, {1, 4, 6}},
            {{2}, {3, 4}, {1, 5}, {6}}, {{3}, {4}, {1, 5}, {2, 6}},
            {{1, 6, 4}, {2, 3, 5}}, {{1, 6, 3}, {2, 4, 5}}}

For instance, the first entry here at level two, $\{1\}$, $\{2\}$, $\{3\}$, $\{4\}$, $\{5\}$, $\{6\}$, is the representation of the identity permutation as a product of cycles. In general, each entry at level two here is a list of cycles. We now create new variables $x_1, \ldots, x_k$, where k is the maximum length of a cycle, and replace each cycle by the variable for its length.

In[11]:= **vars = Map[x$_{\text{Length}[\#]}$ &, cycleList, {2}]**

Out[11]= {{x_1, x_1, x_1, x_1, x_1, x_1}, {x_1, x_2, x_1, x_2}, {x_3, x_3},
            {x_3, x_3}, {x_3, x_3}, {x_3, x_3}, {x_3, x_3}, {x_3, x_3},
            {x_1, x_2, x_2, x_1}, {x_1, x_1, x_2, x_2}, {x_3, x_3}, {x_3, x_3}}

Thus, each cycle of the form $\{n\}$ is replaced by x_1, each one of the form $\{m, n\}$ by x_2, etc. Next, multiply together the variables in each sublist.

In[12]:= **terms = Apply[Times, vars, {1}]**

Out[12]= {x_1^6, $x_1^2 x_2^2$, x_3^2, x_3^2, x_3^2, x_3^2, x_3^2, x_3^2, $x_1^2 x_2^2$, $x_1^2 x_2^2$, x_3^2, x_3^2}

What has happened here is that each original permutation in the group has been replaced by a product of variables determined by the cycle structure of the permutation. The indexes of the variables give the lengths of the cycles and the exponents tell how many cycles there are of each length. Thus, the second group element ge[1, 6, 4, 3, 5, 2] has the cycle structure {{1}, {3, 4}, {5}, {2, 6}} with two cycles of length 1 and two of length two, so it yields the term $x_1^2 x_2^2$. The polynomial we want is the sum of all of these terms.

In[13]:= **Plus@@terms**

Out[13]= $x_1^6 + 3 x_1^2 x_2^2 + 8 x_3^2$

Combining these steps gives the general operation.

```
cycleIndex[g_group, x_] :=
  Plus @@ Apply[Times, Map[x_Length[#] &,
    Map[toCycles[List@@#] &, List@@g], {2}], {1}]
```

It is now easy to calculate the cycle index for the groups we have constructed.

In[14]:= **cycleIndex[rotationGroup$_5$, x]**

Out[14]= $x_1^5 + 4\ x_5$

In[15]:= **cycleIndex[tetrahedronGroup, x]**

Out[15]= $x_1^6 + 3\ x_1^2\ x_2^2 + 8\ x_3^2$

In[16]:= **cycleIndex[octahedronGroup, x]**

Out[16]= $x_1^{12} + 6\ x_1^2\ x_2^5 + 3\ x_2^6 + 8\ x_3^4 + 6\ x_4^3$

In[17]:= **cycleIndex[dodecahedronGroup, x]**

Out[17]= $x_1^{12} + 15\ x_2^6 + 20\ x_3^4 + 24\ x_1^2\ x_5^2$

We can't help pointing out that constructing this polynomial by hand seems like a daunting task. Furthermore, when Polya [Polya] discovered it, there were no symbolic computation programs to make its construction so remarkably simple.

3.3 Polya's Pattern Inventory for a Group Action

The Polya Pattern Inventory is constructed from the cycle index by evaluating the polynomial P_G for the arguments

$$P_G\left(\sum_{j=1}^m c[j],\ \sum_{j=1}^m c[j]^2, \ldots, \sum_{j=1}^m c[j]^k\right)$$

and dividing the result by the number of elements in the group. Here m is the number of colors and k is the maximum length of a cycle in a group element. In generating the list of substitutions in the following procedure, no harm is done if possibly too many powers are calculated, so we can ignore the problem of determining what k is and just use the maximum value it could possibly have. It is non-trivial to prove that this gives the answer to our question.

```
polyaPatternInventory[g_group, m_Integer? Positive, c_] :=
  Cancel[Expand[cycleIndex[g, c] /.
    Table[c_i -> Sum[c_j^i, {j, m}], {i, Length[g[[1]]]}]] /
    Length[g]]
```

For a necklace consisting of 5 beads, using 2 colors, this gives the polynomial:

In[18]:= **polyaPatternInventory[rotationGroup$_5$, 2, c]**

Out[18]= $c_1^5 + c_1^4 c_2 + 2 c_1^3 c_2^2 + 2 c_1^2 c_2^3 + c_1 c_2^4 + c_2^5$

Each term here corresponds to one way of coloring the necklace. The exponents correspond to the number of beads of each color and the coefficient gives the number of colorings, module rotations, using that choice of beads. For instance, the term $2 \ c_1{}^2 \ c_2{}^3$ means that there are 2 ways to construct a necklace using 2 beads of color c_1 and 3 beads of color c_2. We would like to compare the coefficients in this polynomial with the geometrically determined numbers of colorings of each kind. The following routine will extract them.

```
polyaCoefficients[g_group, m_Integer? Positive, c_] :=
    Select[Flatten[CoefficientList[polyaPatternInventory[g, m, c],
        Table[c_i, {i, m}]]], # ≠ 0 &]
```

For instance:

In[19]:= **polyaCoefficients[rotationGroup$_5$, 2, c]**

Out[19]= {1, 1, 2, 2, 1, 1}

We can let *Mathematica* do the comparison with the geometrical results.

In[20]:= **polyaCoefficients[rotationGroup$_5$, 2, c] ==**
        Map[Length, pictureArray[rotationGroup$_5$, act, 2, c]]

Out[20]= True

For 5 beads and 3 colors there are many more necklaces.

In[21]:= **polyaPatternInventory[rotationGroup$_5$, 3, c]**

Out[21]= $c_1^5 + c_1^4 c_2 + 2 c_1^3 c_2^2 + 2 c_1^2 c_2^3 + c_1 c_2^4 + c_2^5 + c_1^4 c_3 + 4 c_1^3 c_2 c_3 +$
    $6 c_1^2 c_2^2 c_3 + 4 c_1 c_2^3 c_3 + c_2^4 c_3 + 2 c_1^3 c_3^2 + 6 c_1^2 c_2 c_3^2 + 6 c_1 c_2^2 c_3^2 +$
    $2 c_2^3 c_3^2 + 2 c_1^2 c_3^3 + 4 c_1 c_2 c_3^3 + 2 c_2^2 c_3^3 + c_1 c_3^4 + c_2 c_3^4 + c_3^5$

In[22]:= **polyaCoefficients[rotationGroup$_5$, 3, c] ==**
        Map[Length, pictureArray[rotationGroup$_5$, act, 3, c]]

Out[22]= True

Here are several more examples.

In[23]:= **polyaPatternInventory[rotationGroup$_6$, 2, c]**

Out[23]= $c_1^6 + c_1^5 \, c_2 + 3 \, c_1^4 \, c_2^2 + 4 \, c_1^3 \, c_2^3 + 3 \, c_1^2 \, c_2^4 + c_1 \, c_2^5 + c_2^6$

In[24]:= **polyaCoefficients[rotationGroup$_6$, 2, c] ==**
      Map[Length, pictureArray[rotationGroup$_6$, act, 2, c]]

Out[24]= True

In[25]:=**polyaPatternInventory[tetrahedronGroup, 2, c]**

Out[25]= $c_1^6 + c_1^5 \, c_2 + 2 \, c_1^4 \, c_2^2 + 4 \, c_1^3 \, c_2^3 + 2 \, c_1^2 \, c_2^4 + c_1 \, c_2^5 + c_2^6$

In[26]:=**polyaCoefficients[tetrahedronGroup, 2, c] ==**
      Map[Length, pictureArray[tetrahedronGroup, act, 2, c]]

Out[26]= True

In[27]:=**polyaPatternInventory[tetrahedronGroup, 3, c]**

Out[27]= $c_1^6 + c_1^5 \, c_2 + 2 \, c_1^4 \, c_2^2 + 4 \, c_1^3 \, c_2^3 + 2 \, c_1^2 \, c_2^4 + c_1 \, c_2^5 + c_2^6 + c_1^5 \, c_3 +$
    $3 \, c_1^4 \, c_2 \, c_3 + 6 \, c_1^3 \, c_2^2 \, c_3 + 6 \, c_1^2 \, c_2^3 \, c_3 + 3 \, c_1 \, c_2^4 \, c_3 + c_2^5 \, c_3 + 2 \, c_1^4 \, c_3^2 +$
    $6 \, c_1^3 \, c_2 \, c_3^2 + 9 \, c_1^2 \, c_2^2 \, c_3^2 + 6 \, c_1 \, c_2^3 \, c_3^2 + 2 \, c_2^4 \, c_3^2 + 4 \, c_1^3 \, c_3^3 + 6 \, c_1^2 \, c_2 \, c_3^3 +$
    $6 \, c_1 \, c_2^2 \, c_3^3 + 4 \, c_2^3 \, c_3^3 + 2 \, c_1^2 \, c_3^4 + 3 \, c_1 \, c_2 \, c_3^4 + 2 \, c_2^2 \, c_3^4 + c_1 \, c_3^5 + c_2 \, c_3^5 + c_3^6$

Notice that there are 9 ways, up to symmetries, to color the edges of a tetrahedron using 2 edges each of 3 colors.

In[28]:=**polyaCoefficients[tetrahedronGroup, 3, c] ==**
      Map[Length, pictureArray[tetrahedronGroup, act, 3, c]]

Out[28]= True

In[29]:=**polyaPatternInventory[octahedronGroup, 2, c]**

Out[29]= $c_1^{12} + c_1^{11} \, c_2 + 5 \, c_1^{10} \, c_2^2 + 13 \, c_1^9 \, c_2^3 + 27 \, c_1^8 \, c_2^4 + 38 \, c_1^7 \, c_2^5 +$
    $48 \, c_1^6 \, c_2^6 + 38 \, c_1^5 \, c_2^7 + 27 \, c_1^4 \, c_2^8 + 13 \, c_1^3 \, c_2^9 + 5 \, c_1^2 \, c_2^{10} + c_1 \, c_2^{11} + c_2^{12}$

The numbers here are much bigger. Thus, for instance, there are 48 ways to color the edges of an octahedron using equal numbers of 2 colors. We can check that the geometrical description agrees with the theory.

```
In[30]:=polyaCoefficients[octahedronGroup, 2, c] ==
        Map[Length, pictureArray[octahedronGroup, act, 2, c]]
```

Out[30]= True

```
In[31]:= polyaPatternInventory[dodecahedronGroup, 2, c]
```

Out[31]= $c_1^{12} + c_1^{11} c_2 + 3 c_1^{10} c_2^2 + 5 c_1^9 c_2^3 + 12 c_1^8 c_2^4 + 14 c_1^7 c_2^5 +$
$24 c_1^6 c_2^6 + 14 c_1^5 c_2^7 + 12 c_1^4 c_2^8 + 5 c_1^3 c_2^9 + 3 c_1^2 c_2^{10} + c_1 c_2^{11} + c_2^{12}$

```
In[32]:= polyaCoefficients[dodecahedronGroup, 2, c] ==
        Map[Length, pictureArray[dodecahedronGroup, act, 2, c]]
```

Out[32]= True

3.4 The Burnside Number for a Group

The Burnside number for a permutation group G and a number of colors m is the total number of colorings of the design by m colors modulo the symmetries in the group. It is given by evaluating the polynomial P_G with all variables set equal to m and dividing by the number n of elements in the group; i.e., $P_G[m, \ldots, m] / n$. Of course, it is also the sum of the numbers given by **polyaCoefficients**.

```
    burnsideNumber[g_group, m_Integer? Positive] :=
     Module[{c},
      cycleIndex[g, c] / Length[g] /.
       Table[c_i ←m, {i, Length[g]}]]
```

Here are the numbers of various necklaces and the tetrahedron, octahedron, and dodecahedron colorings using just two colors.

```
In[33]:= {burnsideNumber[rotationGroup_5, 2],
         burnsideNumber[rotationGroup_10, 2],
         burnsideNumber[rotationGroup_40, 2],
         burnsideNumber[tetrahedronGroup, 2],
         burnsideNumber[octahedronGroup, 2],
         burnsideNumber[dodecahedronGroup, 2]}
```

Out[33]= {8, 108, 27487816992, 12, 218, 96}

Object-Oriented Graph Theory

1 Introduction

A graph consists of a finite set of vertices, some of which are joined by edges. Here are several examples that will be constructed later.

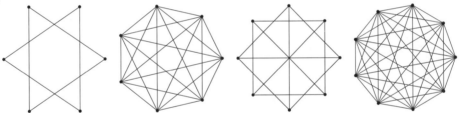

The vertices are indicated by heavy dots. When the edges are drawn in the plane they sometimes intersect, but these intersection points are not considered as part of the graph. The important thing is whether or not there is an edge joining two vertices. These kinds of graphs are sometimes called undirected, simple graphs to distinguish them from directed graphs that have arrows on their edges and from multigraphs that can have several edges joining two vertices. Sometimes edges are allowed from a vertex to itself, but we rule that out here.

Mathematically, a graph can be considered as a relation between vertices. Two vertices are related if and only if they are joined by an edge. This relation is clearly symmetric: if x is joined to y then y is joined to x. We assume explicitly that it is anti-reflexive; i.e., a vertex is not joined to itself. In other words, there are no loops in the graph. Clearly, any symmetric, anti-reflexive relation can be pictured by such a graph, so it doesn't matter if we talk about graphs or about such relations. Now there are various ways to describe relations, each of which corresponds to a way to describe graphs. A relation on a set V (of vertices) can be considered as a subset of the Cartesian product V × V; i.e., as a set of ordered pairs of elements of V, so one can simply make a list of those pairs that belong to the relation. This will be one of our basic representations of a graph, called the *ordered pair* representation. Alternatively, such a subset can be described by its characteristic function, that is, a function on V × V with values 0 and 1 which is 1 exactly for those pairs belonging to the subset. If we name the elements of V by the numbers 1 through n, where n is the number of elements of V, then this characteristic function can be described by an $n \times n$ matrix of 0's and 1's, in which the (i, j)th entry is 1 if and only if the pair (i, j) belongs to the subset; i.e., if and only if there is an edge from vertex i to vertex j. This matrix is called the *adjacency matrix* of the graph, since a 1 in position (i, j) is interpreted as meaning that vertex i is adjacent to vertex j. The relation being symmetric is equivalent to the adjacency matrix being symmetric, and the relation being anti-reflexive means that the diagonal entries are all 0. This will be another of our basic representations of graphs called the *adjacency matrix* representation. The third basic representation is simply to list, for each vertex, the other vertices to which it is connected. This is called the *edge list* representation.

Approximately two-thirds of S. Skiena's book, Implementing Discrete Mathematics [Skiena], is concerned with graph theory. We present here a somewhat different treatment of graph theory as an illustration of systematically developing a part of mathematics in the object-oriented style developed in Chapter 9.

There are many aspects to graph theory. There must be thousands or possibly tens of thousands of algorithms concerning properties of graphs. Many are to be found in Skiena's book. Each algorithm expects its input in a particular form and works most conveniently or most efficiently in that form, which is one reason why there are many different representation of graphs. As the preceding pictures show, one can make drawings of graphs and try to understand them through these drawings, so such illustrations are an intrinsic feature of graph theory. Skiena's representation of graphs includes instructions for making a drawing of each graph. We omit this feature for simplicity and just have a few general plotting routines that are suitable for all graphs, and a few special ones for particular kinds of graphs. In this study we concentrate on the construction of new graphs from old ones, and leave the study of graph algorithms mainly to Skiena's book.

2 Representations of Graphs

2.1 The Class Hierarchy

As a first simple example, consider the complete graph on three vertices, K[3]. It consists of three vertices, labeled 1, 2, 3, each of which is connected by an edge to the other two.

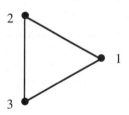

As described above, the adjacency matrix of a graph is the $n \times n$ matrix G in which

$$G[i, j] == 1 \text{ iff vertex } i \text{ is connected to vertex } j.$$

Our assumptions are that this matrix is symmetric with 0's on the main diagonal. In particular, the adjacency matrix for **K[3]** is

```
0   1   1
1   0   1
1   1   0
```

The edge list representation is a list of lists in which the ith list is the list of vertices connected to the ith vertex. E.g., for **K[3]**, we have the list of lists:

$$\{\{2, 3\}, \{1, 3\}, \{1, 2\}\}$$

Here, the first entry, {2, 3}, means that the first vertex is connected to vertices 2 and 3, etc. Clearly, to be the list of edge lists for a graph requires that the individual lists are increasing, that the largest entry is less than or equal to the number of lists, that i does not occur in the ith list, and that if j occurs in the ith list, then i occurs in the jth one.

Finally, the ordered pair representation is the list of pairs of vertices that are connected by edges. E.g., for **K[3]**, we have the list of pairs:

```
{{1, 2}, {1, 3}, {2, 1}, {2, 3},
 {3, 1}, {3, 2}}
```

Here the first entry {1, 2} means that there is an edge from 1 to 2, etc. The only restrictions on a list of pairs to be the list of ordered pairs of a graph are that there are no pairs of the form {i, i}, and that if {i, j} occurs in the list then so does {j, i}. Note that the number of vertices of a graph cannot be determined from its ordered pair representation since there may be isolated vertices; i.e., vertices that are not connected to any others.

The situation here is very similar to that of points in the plane treated in Chapter 11. There are three different ways to represent graphs, so we have to have operations translating between them and the whole situation should be embedded in a small hierarchy of classes consisting of an abstract top class **graph**, under the class **Object**, followed by three subclasses, one for each way of representing graphs. We call these subclasses **adjacents**, **edges**, and **ordereds**, just to have names that won't conflict with the operations to be constructed for them. Actually, there will be a number of subsubclasses as well, as indicated in the following picture.

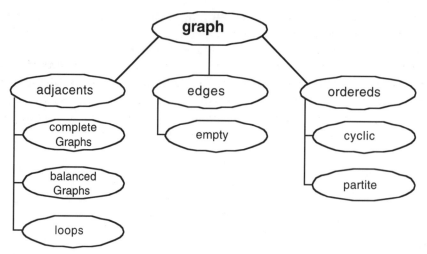

Altogether the hierarchy contains 10 classes, but there could be many more.

2.2 *Outline Versions of the Classes*

The structure to be set up is complicated both in the number and detail of operations to be implemented and in their object-oriented organization. To get started, we'll first look at the classes in outline form just to see what's to go in them. First of all is the top class **graph**, just under **Object**. It has no instance variables and no **new** method as well as three methods with default second components.That is, like the class **point** in Chapter 11, there are no objects belonging to this class; it is just there to organize the classes below it. However, just as before, it will turn out that this class contains almost all of the knowledge about graphs. Note: Many of the input cells that follow are not intended to be evaluated. They are just for illustrative purposes. They have had the **Evaluatable** property turned off in the **Cell** menu. This is indicated by the short horizontal line at the top of the cell bracket. However, all specific examples can be evaluated.

```
Class[
    graph,                         (* name of the class*)
    Object,                        (* super class*)
    {},                            (* an abstract class*)
    {{graphQ, ---},                (* methods*)
     {adjacencyMatrix,  NIM[self, adjacencyMatrix]&},
     {edgeLists,        NIM[self, edgeLists]&},
     {orderedPairs,     NIM[self, orderedPairs]&},
     {numberOfVertices, ---},
     {numberOfEdges,    ---},
     --- }]
```

Actually, there will be many more methods in this class. The methods with default second components have to be implemented in the subclasses of graph, and in fact that is about all that is implement in them.

Consider the first subclass, `adjacents`. It has one instance variable called `matrix` and the idea is that when a new object of the class is constructed, `matrix` will be set equal to the adjacency matrix of some graph. The method `adjacencyMatrix` will just return this matrix, while the methods `edgeLists` and `orderedPairs` will have to carry out some computation to find the edge lists and ordered pairs corresponding to the adjacency matrix.

```
Class[
    adjacents,                      (* name of the class *)
    graph,                          (* super class *)
    {matrix},                       (* the adjacency matrix *)
    {{new, super.new[]; (matrix = #)&},     (* methods *)
     {adjacencyMatrix,  matrix&},
     {edgeLists,        "calculate the edge lists"},
     {orderedPairs,     "calculate the ordered pairs"}} ]
```

The class **edges** is similar, except this time the single instance variable expects to be given the list of edge lists of some graph. The method **edgeLists** just returns this list of lists while the other two methods involve computations.

```
Class[
    edges,                          (* name of the class *)
    graph,                          (* super class *)
    {eds},                          (* the edge lists *)
    {{new, super.new[]; (eds = #)&},        (* methods *)
     {adjacencyMatrix, "calculate the adjacency matrix"},
     {edgeLists,        eds&},
     {orderedPairs,     "calculate the ordered pairs"}} ]
```

The pattern is now clear. For the class **ordereds**, the single instance variable is set equal to the list of ordered pairs of some matrix, which is returned by the method **orderedPairs**, and the other two methods require computations.

```
Class[
    ordereds,                       (* name of the class *)
    graph,                          (* super class *)
    {ords},                         (* the ordered pairs *)
    {{new, super.new[]; (ords = #)&},       (* methods *)
     {adjacencyMatrix, "calculate the adjacency matrix"},
     {edgeLists,        "calculate the edge lists"},
     {orderedPairs,     ords&}} ]
```

The subsubclasses will be treated later.

2.3 The Subclasses in Detail

Before looking at the class **graph**, which is rather large, we discuss the three subclasses in detail since that is where graphs are actually created.

2.3.1 The class `adjacents`

For the class **adjacents**, the list of edge lists and the list of ordered pairs has to be calculated from the adjacency matrix . Conceptually, it is simple to see how to find the list of edge lists. Consider the first row of the adjacency matrix. The places in that row where there are 1's correspond to the vertices to which the first vertex is connected. The operation **Position** can find those places. Suppose we were just implementing this in our usual functional style. Then we could define a function as follows:

```
edgeListsFromAdjacencyMatrix[matrix_] :=
   Map[(Flatten[Position[#, edge_ /; (edge != 0)]])&,
       matrix]
```

The reason for the **Flatten** is because **Position** returns its results wrapped in extra parentheses. For example,

In[1]:= `Position[{0, 1, 1}, edge_ /; (edge != 0)]`

Out[1]= `{{2}, {3}}`

In object-oriented programming, this function gets replaced by a method, named **edgeLists** here, with the body of the definition as second argument. Thus, we could just write a method for **adjacents** as follows:

```
{edgeLists,
    Map[(Flatten[Position[#, edge_ /; (edge != 0)]])&,
        matrix]
```

This works because the instance variable for **adjacents** is named **matrix**. However, this method, written this way, is not inherited correctly by subclasses which will be constructed later. Instead of **matrix**, we have to write **adjacencyMatrix[self]**, or rather **self.adjacencyMatrix[]** in message sending style, so that the correct matrix from an object of the subclass will be used. Thus, the actual method is

```
{edgeLists,
    Map[(Flatten[Position[#, edge_ /; (edge != 0)]])&,
        self.adjacencyMatrix[]]&}
```

We also need a way to calculate the list of ordered pairs corresponding to the adjacency matrix of a graph. This is even simpler than finding the edge lists since the desired ordered pairs are exactly the positions in the adjacency matrix where there is a non-zero entry. In a functional style, we would just write

```
orderedPairsFromAdjacencyMatrix[matrix_] :=
    Position[matrix, edge_ /; (edge != 0)]]
```

As before, in object-oriented style, this becomes the method:

```
{orderedPairs,
    Position[self.adjacencyMatrix[],
            edge_/; (edge != 0)]&}
```

Here, **matrix** is replaced by **self.adjacencyMatrix[]** for the same reason as before. Putting this all together gives the following class definition.

```
Class[adjacents, graph, {matrix},
   {{new, (super.new[]; matrix = #)&},
    {adjacencyMatrix, matrix&},
    {edgeLists,
       Map[(Flatten[Position[#, edge_ /; (edge != 0)]])&,
           self.adjacencyMatrix[] ]&},
    {orderedPairs,
       self.adjacencyMatrix[].Position[
               edge_ /; (edge != 0)]&}
   }];
```

If the package **GraphTh** is loaded, then Maeder's package **Classes** is also evaluated and we can try out examples to be sure that everything works correctly.

In[2]:=**Needs["GraphTh`"]**

When the package is loaded, the following commands are evaluated so the notation introduced in Chapter 9 can be used.

```
Unprotect[Dot];
SetAttributes[Dot, HoldAll];
Dot[object_, method_[parameters___]] :=
  method[object, parameters];
Protect[Dot];
```

Start with the complete graph on three vertices, entered "by hand" and given a different name so it doesn't clash with the complete graphs in the package.

In[3]:= `KK[3] = adjacents.new[{{0, 1, 1}, {1, 0, 1}, {1, 1, 0}}]`

Out[3]= `-adjacents-`

Check the calculation of edge lists and ordered pairs.

In[4]:= `{KK[3].edgeLists[], KK[3].orderedPairs[]}//MatrixForm`

Out[4]//MatrixForm=

$$\begin{pmatrix} \{\{2, 3\}, \{1, 3\}, \{1, 2\}\} \\ \{\{1, 2\}, \{1, 3\}, \{2, 1\}, \{2, 3\}, \{3, 1\}, \{3, 2\}\} \end{pmatrix}$$

2.3.2 The class `edges`

Given the edge lists of a graph, we have to find the corresponding adjacency matrix and list of ordered pairs. The first calculation is the inverse of the first calculation in the preceding section: the calculation of the adjacency matrix of a graph from the edge lists for the graph. Clearly, the edge lists tell us where, in each row of the adjacency matrix, there is to be a 1. In functional style, this gives the following operation.

```
adjacencyMatrixFromEdgeLists[edges_]:=
  Map[ReplacePart[0 Range[Length[edges]], 1, #]&,
      Map[Partition[#, 1]&, edges]]
```

In object-oriented style, this becomes the method

```
{adjacencyMatrix,
   With[{edges = self.edgeLists[]},
      Map[ReplacePart[0 Range[Length[edges]], 1, #]&,
         Map[Partition[#, 1]&, edges]]]&}
```

The **With** construction is used to avoid calculating **self.edgeLists[]** twice.

The second calculation needed here is a new conversion, this time from edge lists of a graph to the list of ordered pairs of the same graph. If the ith entry in the edge lists is $(i_1, \ldots, i_n)$, then there should be ordered pairs of the form $(i, i_1), \ldots, (i, i_n)$ in the list of ordered pairs of the graph. It is simpler to construct all pairs $\{i, j\}$ and then select the ones that we want. This requires an auxilary expression to extract the diagonal from a matrix using a clever method found by Allan Hayes. (This has to be evaluated for use later.)

In[5]:= `diagonal[matrix_List] := Transpose[matrix, {1, 1}]`

In functional form, the required conversion operation is

```
orderedPairsFromEdgeLists[edges_] :=
 Flatten[
   diagonal[Outer[{#1, #2}&,
                  Range[Length[edges]], edges] ],
   1]
```

This works because of the way that **Outer** organizes its output. Turned into an object-oriented message, the operation becomes the last method in the class **edges**.

```
Class[edges, graph, {eds},
 {{new, (super.new[]; eds = #)&},
  {adjacencyMatrix,
   With[{edges = self.edgeLists[]},
     Map[ReplacePart[0 edges.Length[].Range[], 1, #]&,
       Map[Partition[#, 1]&, edges]]]&},
  {edgeLists, eds&},
  {orderedPairs,
    With[{edges = self.edgeLists[]},
      Flatten[
        diagonal[Outer[{#1, #2}&,
                       edges.Length[].Range[], edges]],
     1]]&}
 }];
```

Try this out using the edge lists from KK[3].

In[6]:= **edg[3] = edges.new[KK[3].edgeLists[]]**

Out[6]= -edges-

Check that the two important messages work correctly.

In[7]:= **{edg[3].adjacencyMatrix[], edg[3].orderedPairs[]}//MatrixForm**

Out[7]//MatrixForm=
$$\begin{pmatrix} \{\{0, 1, 1\}, \{1, 0, 1\}, \{1, 1, 0\}\} \\ \{\{1, 2\}, \{1, 3\}, \{2, 1\}, \{2, 3\}, \{3, 1\}, \{3, 2\}\} \end{pmatrix}$$

2.3.3 The class `ordereds`

Starting with the list of ordered pairs of a graph, this class will find the corresponding adjacency matrix and edge lists. Both operations are inverse to operations considered in the preceding two sections. We calculate the adjacency matrix first. Conceptually, this is very simple since the list of ordered pairs describes the positions of the 1's in the adjacency matrix. So just start with a matrix of 0's of the correct size and use the ordered pairs to replace appropriate 0's by 1's. In functional form, this looks like

```
adjacencyMatrixFromOrderedPairs[pairs_] :=
    ReplacePart[0 IdentityMatrix[Max[Flatten[pairs]]],
                1, pairs]
```

As a method, it becomes

```
{adjacencyMatrix,
    With[{pairs = self.orderedPairs[]},
        ReplacePart[0 IdentityMatrix[Max[Flatten[pairs]]],
                    1, pairs]]&}
```

Again, we use a **With** construction to avoid calculating **self.orderedPairs[]** twice.

Lastly, we find the edge lists in terms of the ordered pairs. The edge list for the vertex *i* consists of all second entries of ordered pairs whose first entry is *i*. In functional form, this is given by the operation

```
edgeListsFromOrderedPairs[pairs_] :=
    Table[Cases[pairs, {i, x_} -> x],
          {i, Max[Flatten[pairs]]} ]
```

As a message, it becomes the second method in the class **ordereds**.

```
Class[ordereds, graph, {ords},
  {{new, (super.new[]; ords = #)&},
   {adjacencyMatrix,
      With[{pairs = self.orderedPairs[]},
          ReplacePart[0 IdentityMatrix[Max[Flatten[pairs]]],
                      1, pairs]]&},
   {edgeLists,
      With[{pairs = self.orderedPairs[]},
          Table[Cases[pairs, {i, x_} -> x],
                {i, Max[Flatten[pairs]]}]]&},
   {orderedPairs, ords&}
  }];
```

As a simple example, consider

In[8]:=`orp[3] = ordereds.new[KK[3].orderedPairs[]]`

Out[8]= `-ordereds-`

In[9]:= `{orp[3].adjacencyMatrix[], orp[3].edgeLists[]}//MatrixForm`

Out[9]//MatrixForm=

$$\begin{pmatrix} \{0, 1, 1\} & \{1, 0, 1\} & \{1, 1, 0\} \\ \{2, 3\} & \{1, 3\} & \{1, 2\} \end{pmatrix}$$

2.3.4 Discussion

These three classes make it possible to construct graphs by specifying the adjacency matrix, the edge pairs, or the ordered pairs of the graph. Once the graph is made, these three messages, **adjacencyMatrix**, **edgeLists**, and **orderedPairs**, can be sent without worrying about how the graph was originally created This kind of object-oriented polymorphism is a powerful idea. As long as we describe all further operations in terms of these three constructs, we never have to be concerned with what a graph actually is; graphs in this sense are abstract, which is the intuitive reason why the top class **graph** will be an abstract class.

2.4 The Top Class graph: *Basic Structure*

We're now ready to begin the discussion of the class **graph** itself. Ultimately, almost all of our knowledge about graphs will be contained in the messages for this graph, except for the three calculations in the subclasses. Recall that the outline version of this class looks like

```
Class[graph, Object, {},
      {{graphQ, ---},
       {adjacencyMatrix,    self.NIM[adjacencyMatrix]&},
       {edgeLists,          self.NIM[edgeLists]&},
       {orderedPairs,       self.NIM[edgeLists]&},
       {numberOfVertices, ---},
       {numberOfEdges,    ---},
       --- }]
```

The default methods are all the same: **self.NIM[nameOfMethod]**. The meaning of this is that these methods must be implemented in all of the subclasses of **graph**. If they are not, then a message to that effect is returned. Now consider the method with name **graphQ**. Its second component should be a predicate that returns `True` if and only if the object under consideration is a graph. It is easy to describe in functional form when a matrix is the adjacency matrix of a graph.

```
In[10]:= adjacencyMatrixOfGraph[matrix_] :=
        MatrixQ[matrix, (# === 0 || # === 1 &)] &&
          (matrix === Transpose[matrix])          &&
          (Plus @@ (diagonal[matrix]²) === 0)
```

The first clause says that matrix is a matrix of 0's and 1's, the second that it is symmetric (and in particular square), and the third that the diagonal elements are all 0. For instance:

```
In[11]:= adjacencyMatrixOfGraph[KK[3].adjacencyMatrix[]]
```

```
Out[11]= True
```

As a message, it becomes the first method in the class **graph**. The methods for the numbers of vertices and edges are simple to write, so, as a start, the class **graph** is given as follows:

```
Class[graph, Object, {},
  {{graphQ,
      With[{matrix = self.adjacencyMatrix[]},
          MatrixQ[matrix, (#===0 || #===1)&] &&
          (matrix === Transpose[matrix])     &&
          (Plus@@(diagonal[matrix]^2)===0)]&},
  {adjacencyMatrix,    self.NIM[ adjacencyMatrix]},
  {edgeLists,          self.NIM[edgeLists]},
  {orderedPairs,       self.NIM[orderedPairs]},
  {numberOfVertices,   self.edgeLists[].Length[]&},
  {numberOfEdges,
      self.edgeLists[].Flatten[].Length[] / 2 &}
  }]
```

For instance:

```
In[12]:= {K[3].graphQ[], K[3].numberOfVertices[], K[3].numberOfEdges[]}
```

```
Out[12]= {True, 3, 3}
```

The actual class **graph** as described at the end of the chapter contains all of these methods as well as many others. Some of the most important additions are methods to make pictures of graphs. There are three that are contained in the top class, called **randomImmersion**, **circularImmersion**, and **centerCircularImmersion**. The underlying principle of all of them is the same. If the graph has *n* vertices, then *n* points in the plane are given explicitly and the list of ordered pairs of the graph is used to determine lines between appropriate pairs of these points. In doing this we don't need all of the ordered pairs, just those whose first coordinate is less than the second coordinate. For instance, the method **randomImmersion** is given as follows:

```
{randomImmersion,
  With[{verts =
         Table[{Random[], Random[]},
               {self.numberOfVertices[]}]},
     Graphics[Join[
       {PointSize[0.035]}, Map[Point, verts],
         Map[Line[verts[[#]]]&,
             self.orderedPairs[].Select[(#[[1]] < #[[2]])&]]
         ]] ]&}
```

Here **verts** is set equal to a table of *n* random points in the plane. Then **Show[-Graphics[---]]** is called with an argument consisting of a chosen point size, a point for each entry in **verts**, and a line for each pair of entries in **verts** that corresponds to an ordered pair of vertices of the graph. For instance:

In[13]:=**K[3].randomImmersion[].Show[];**

The principle is the same for the other two drawing messages. The argument of **Graphics[---]** is always the same. What differs is the construction of **verts**. Thus:

```
{circularImmersion,
   Module[{n = self.numberOfVertices[], verts},
     verts = Table[{N[Cos[2 Pi i/n]/ 2],
                    N[Sin[2 Pi i/n]/2]},
                   {i, 0, n - 1}];
     Graphics[Join[
         {PointSize[0.035]}, Map[Point, verts],
         Map[Line[ verts[[#]] ]&,
             self.orderedPairs[].Select[(#[[1]] < #[[2]])&]]
         ]] ]&}
```

Here the vertices are located uniformly around the unit circle, with the first vertex at the point 1 on the x-axis. For instance:

In[14]:=**K[3].circularImmersion[].Show[AspectRatio -> Automatic];**

The third message, `centerCircularImmersion`, works in the same way, except the last vertex is located at the origin. This doesn't work well for K[3], since all of the vertices come out colinear.

In[15]:=`K[3].centerCircularImmersion[].Show[AspectRatio -> 1];`

What we need at this point is some more graphs to experiment with. That's the purpose of the subsubclasses.

2.5 Some Classes of Special Graphs

All of the special kinds of graphs to be treated here will be constructed as subclasses of the three main subclasses **adjacents**, **edges**, and **ordereds**. Mostly all that has to be done is to overwrite the method **new**.

2.5.1 Complete graphs

A complete graph is one in which every vertex is connected to every other vertex. Its adjacency matrix therefore consists entirely of 1's except for 0's on the main diagonal. Cameron Smith (personal communication) had the nice idea of using **Listability** of subtraction to describe such a matrix as `1 - IdentityMatrix[n]`. We use this to construct a subclass of **adjacents** for complete graphs.

```
Class[completeGraph, adjacents, {int},
      {{new, super.new[(1 - IdentityMatrix[int = #])]&}}];
```

The standard notation for the complete graph on n vertices is `K[n]`, which is introduced as an abbreviation.

```
K[n_] := completeGraph.new[n]
```

This gives us a large collection of graphs which will be useful both in themselves and for use in other constructions. For instance:

In[16]:=`K[6].edgeLists[]`

Out[16]= `{{2, 3, 4, 5, 6}, {1, 3, 4, 5, 6}, {1, 2, 4, 5, 6},`
`{1, 2, 3, 5, 6}, {1, 2, 3, 4, 6}, {1, 2, 3, 4, 5}}`

In[17]:=`K[4].orderedPairs[]`

Out[17]= `{{1, 2}, {1, 3}, {1, 4}, {2, 1}, {2, 3}, {2, 4},`
`{3, 1}, {3, 2}, {3, 4}, {4, 1}, {4, 2}, {4, 3}}`

In[18]:=`K[11].circularImmersion[].Show[AspectRatio -> Automatic];`

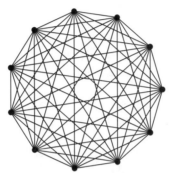

2.5.2 Balanced graphs

One can think of a complete graph as one for which all vertices look the same; namely, each vertex is connected to every other one. In a balanced graph, again all vertices look the same, but a given vertex is connected only to certain others given by taking every second or every third, or in general every kth vertex. The rows in the adjacency matrix of a complete graph on n vertices can be described as starting with the table

```
Prepend[Table[1, {n - 1}], 0]
```

and rotating it to the right successively to fill out the adjacency matrix. For a balanced graph, we can do essentially the same thing putting in 1's and 0's depending on whether the vertex number is divisible by k or not. Unfortunately, just rotating such a row to the right may not produce a symmetric matrix, so we have to symmetrize it, keeping the entries 0's and 1's. This is done by a general auxiliary function, which we regard as being outside the class system.

```
adjust[matrix_ /;MatrixQ[matrix, MatchQ[#, _Integer]&]]:=
  Module[{mnew = matrix},
    mnew = mnew -
      diagonal[mnew] IdentityMatrix[Length[mnew]];
    mnew = (mnew + Transpose[mnew]);
```

```
    mnew = Map[If[(# != 0), 1, 0]&, mnew, {2}]
  ] /; Length[matrix] == Length[Transpose[matrix]]
```

In three steps, this first sets all of the diagonal entries to 0, then makes the matrix symmetric, and finally turns all non-zero entries into 1's. There are two built-in checks: that the matrix is square and that its entries are integers.

```
Class[balancedGraph, adjacents, {n, k},
  {{new,
    (n = #1; k = #2;
    super.new[
      Module[{i, edges},
        edges = Table[
          If[Mod[i, k] == 0, 1, 0], {i, 0, n - 1}];
        adjust[Table[RotateRight[edges, i],
                {i, 0, n - 1}]]])&}}];
```

The table **edges** constructed here always has 1's on the main diagonal and sometimes is not symmetric, which is why the **adjust** operation is required.

```
In[19]:=balancedGraph.new[9, 3].circularImmersion[].Show[
        AspectRatio -> Automatic];
```

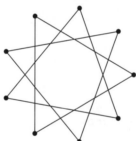

Here are pictures of the balanced graphs between (6, 2) and (9, 3). The top row consists of the graphs that are shown at the beginning of this chapter.

```
In[20]:=Map[balancedGraph.new[#[[1]], #[[2]] ].circularImmersion[].
       Show[AspectRatio -> Automatic,
           DisplayFunction ->Identity]&,
             Table[{i, j}, {j, 2, 3}, {i, 6, 9}], {2}].Graphics-
     Array[].Show[
       DisplayFunction -> $DisplayFunction];
```

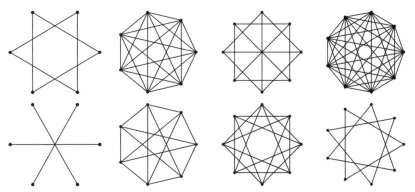

If we want to write this in a more object-oriented form, then something has to be done about **Map**, which is the archetype of a functional command, since it "sends" to a function the data on which to operate. This can be rewritten as our own **map** method which sends to a list the instructions as to what to do with its elements.

```
In[21]:= map[object_, function_, opts___] := Map[function, object, opts]
```

```
In[22]:= Table[{i, j}, {j, 2, 3}, {i, 6, 9}] .
          map[balancedGraph . new[#[[1]], #[[2]]] .
            circularImmersion[] .
             Show[AspectRatio - <Automatic,
                DisplayFunction - <Identity] &, {2}] .
          GraphicsArray[DisplayFunction - $DisplayFunction] .
            Show[];
```

The picture is omitted since it is the same as before.

2.5.3 Loops

Although we have stipulated that our graphs have no loops, it will be convenient when we define tensor products of graphs below to pretend that there are graphs with loops. In particular, we need the "graph" consisting of n vertices with a loop on each vertex but no other edges. Its adjacency matrix is an identity matrix of the appropriate size.

```
    Class[loops, adjacents, {int},
       {{new, super.new[IdentityMatrix[int = #]]&}}];
```

```
In[23]:= loops.new[5].orderedPairs[]
```

```
Out[23]= {{1, 1}, {2, 2}, {3, 3}, {4, 4}, {5, 5}}
```

As expected, the program does not think that a loop is a graph.

In[24]:= `loops.new[8].graphQ[]`

Out[24]= `False`

2.5.4 Empty graphs

We also need another seemingly strange collection of graphs; namely, those with no edges at all. It is easiest to consider this as a subclass of **edges**.

```
Class[empty, edges, {int},
      {{new, super.new[Table[{}, {#}]]&}}];
```

In[25]:= `emp = empty.new[5];`

In[26]:= `emp.edgeLists[]`

Out[26]= `{{}, {}, {}, {}, {}}`

In[27]:= `emp.adjacencyMatrix[]`

Out[27]= `{{0, 0, 0, 0, 0}, {0, 0, 0, 0, 0},`
`        {0, 0, 0, 0, 0}, {0, 0, 0, 0, 0}, {0, 0, 0, 0, 0}}`

In[28]:= `emp.orderedPairs[]`

Out[28]= `{}`

2.5.5 Cyclic graphs

A cyclic graph is one in which each vertex is connected only to the preceding and succeeding ones. These graphs can be constructed as a subclass of **ordereds**.

```
Class[cyclicGraph, ordereds, {int},
   {{new,
        (int = #;
        super.new[
            Union@@
              Map[Function[{place},
                    {{place, Mod[place, int] + 1},
                     {Mod[place, int] + 1, place}}],
                    Range[int]])&}}];
```

```
In[29]:= cyclicGraph.new[4].edgeLists[]
```

```
Out[29]= {{2, 4}, {1, 3}, {2, 4}, {1, 3}}
```

```
In[30]:= cyclicGraph.new[10].circularImmersion[].
         Show[AspectRatio -> Automatic];
```

2.5.6 Partite graphs

A Kpartite graph is specified by a list of numbers $n_1, \ldots, n_k$. It has $n = \Sigma\, n_i$ vertices grouped in blocks of sizes n_i. Each vertex in a block is connected to all of the vertices not in the block. One way to construct this graph is by starting with the complete graph on n vertices and removing the complete graphs on each of the blocks. This requires that we be able to construct the disjoint union (= coproduct) of the complete graphs on each of the blocks. This is implemented in the class **graph** and will be discussed below. We also want a special way to display Kpartite graphs. If there are k blocks, we locate k equal length segments symmetrically around the unit circle and place n_i points in the ith segment.

```
Class[partite, ordereds, {list},
  {{new, (list = #;
     super.new[
       Complement[completeGraph.new[Plus@@list].orderedPairs[],
         coproduct[
           Sequence@@
           Map[Function[{value},
                         completeGraph.new[value]],
               list]
         ].orderedPairs[]]])&},
   {partiteImmersion,
     Module[
       {n = Length[list], p, verts},
       p[i_] := {Cos[(2 i + 1) Pi/(2n)],
                 Sin[(2i+1) Pi/(2n)]};
       verts =
         Flatten[
           Table[
             (p[2i-1] +
```

```
                (j/(list[[i]]-1)) (p[2i] - p[2i-1])),
              {i, n}, {j, 0, list[[i]]-1}], 1];
          Graphics[
           Join[{PointSize[0.035]},
              Map[Point, verts],
              Map[
                Line[ verts[[#]] ]&,
                Select[self.orderedPairs[],
                       (#[[1]] < #[[2]])&]] ]] ]&} }]
```

Here is the standard abbreviation for these graphs.

```
    Kpartite[numbers__Integer] := partite.new[ {numbers}]
```

A Kpartite graph can of course be displayed by a circular immersion, since **partite** is a subsubclass of **graph**.

In[31]:=**Kpartite[4, 3].circularImmersion[].**
           Show[AspectRatio -> Automatic];

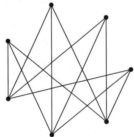

But the special drawing routine makes a nicer picture.

In[32]:=**Kpartite[4, 3].partiteImmersion[].**
           Show[AspectRatio -> Automatic];

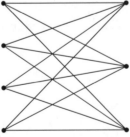

Here are two larger examples.

In[33]:=**Kpartite[3, 2, 4].partiteImmersion[].**
           Show[AspectRatio -> Automatic];

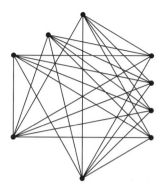

```
In[34]:= Kpartite[3, 2, 4, 6, 5].partiteImmersion[].
         Show[AspectRatio -> Automatic];
```

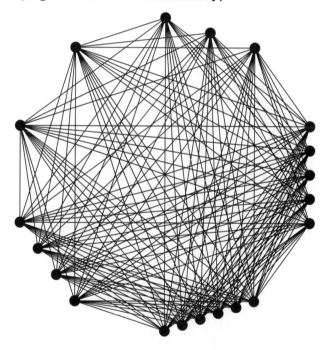

3 Products

Products are operations on graphs that depend on more than one graph. As discussed in Chapter 9, in a strict object-oriented programming language this can only be implemented by sending a message to the first graph telling it to construct the product using the other graphs given as parameters to the message. The coproduct of graphs is implemented in this way as a method for the class **graph**. Of course, in *Mathematica*, we can perfectly well write functional programs depending on several graphs provided we access them through other methods to which they can respond. That is, we can implement *multimethods* in the outside language, independent of the class structure. This technique is used for the other two products: Cartesian products and tensor products.

3.1 Coproducts

A coproduct of graphs means their disjoint union; i.e., place them side by side with no vertices or edges overlapping. One way to construct the coproduct of two graphs is join together their edge lists after adding the number of vertices of the first graph to every entry in every edge list of the second graph. This operation can be iterated for several graphs by using **Fold**. In functional form, this looks like

```
coproduct[graphs___?graphQ] :=
  new[edges,
      Fold[Join[#1, #2 + Length[#1]]&, {},
          Map[edgeLists,{graphs}] ] ]
```

However, we have chosen to add **coproduct** as a message to the class **graph**, as discussed above, where it becomes the method:

```
{coproduct,
      edges.new[
          Fold[Join[#1, #2 + Length[#1]]&,
              self.edgeLists[],
              Map[edgeLists, {##}]]]&}
```

The main difference is that the **Fold** operation in the method starts with the edge lists of the graph to which the message is sent rather than with {}. For instance:

In[1]:=**Needs["GraphTh`"]**

In[2]:=`K[3].coproduct[K[5], K[4]]. circularImmersion[].Show[AspectRatio`
`    -> Automatic];`

A common notation for the coproduct of two graphs is **g1⊕g2**. We can easily implement this notation by the definition

In[3]:= `g1_?graphQ⊕g2_?graphQ := coproduct[g1, g2]`

This gives us a convenient way to construct graphs.

In[4]:= `(K[2]⊕K[3]).adjacencyMatrix[] // MatrixForm`

Out[4]//MatrixForm=
$$\begin{pmatrix} 0 & 1 & 0 & 0 & 0 \\ 1 & 0 & 0 & 0 & 0 \\ 0 & 0 & 0 & 1 & 1 \\ 0 & 0 & 1 & 0 & 1 \\ 0 & 0 & 1 & 1 & 0 \end{pmatrix}$$

Normally, ⊕ is regarded as being associative and is written without bracketing. For this to work here, we have to tell *Mathematica* to rewrite **g1⊕g2⊕g3** as **((g1⊕g2)⊕g3**. This is not so easy to do, although there are facilities making it possible. Fortunately, many situations of this sort are covered by a new package **Notations** which provides convenient templates for introducing such rules. First, load the package.

In[5]:= `Needs["Utilities`Notation`"]`

This puts up a palette from which we choose the Standard Form implication template and fill it in as follows.

In[6]:= `Notation[ g1_⊕g2_⊕g3_ ⟹ ((g1_⊕g2_)⊕g3_) ]`

Try this out.

In[7]:= `(K[2]⊕K[3]⊕K[2]).adjacencyMatrix[] // MatrixForm`

Out[7]//MatrixForm=

$$\begin{pmatrix} 0 & 1 & 0 & 0 & 0 & 0 & 0 \\ 1 & 0 & 0 & 0 & 0 & 0 & 0 \\ 0 & 0 & 0 & 1 & 1 & 0 & 0 \\ 0 & 0 & 1 & 0 & 1 & 0 & 0 \\ 0 & 0 & 1 & 1 & 0 & 0 & 0 \\ 0 & 0 & 0 & 0 & 0 & 0 & 1 \\ 0 & 0 & 0 & 0 & 0 & 1 & 0 \end{pmatrix}$$

This in fact implements the general associativity law.

In[8]:= **(K[2] ⊕ K[3] ⊕ K[2] ⊕ K[2] ⊕ K[2]) . adjacencyMatrix[] // MatrixForm**

Out[8]//MatrixForm=

$$\begin{pmatrix} 0 & 1 & 0 & 0 & 0 & 0 & 0 & 0 & 0 & 0 & 0 \\ 1 & 0 & 0 & 0 & 0 & 0 & 0 & 0 & 0 & 0 & 0 \\ 0 & 0 & 0 & 1 & 1 & 0 & 0 & 0 & 0 & 0 & 0 \\ 0 & 0 & 1 & 0 & 1 & 0 & 0 & 0 & 0 & 0 & 0 \\ 0 & 0 & 1 & 1 & 0 & 0 & 0 & 0 & 0 & 0 & 0 \\ 0 & 0 & 0 & 0 & 0 & 0 & 1 & 0 & 0 & 0 & 0 \\ 0 & 0 & 0 & 0 & 0 & 1 & 0 & 0 & 0 & 0 & 0 \\ 0 & 0 & 0 & 0 & 0 & 0 & 0 & 0 & 1 & 0 & 0 \\ 0 & 0 & 0 & 0 & 0 & 0 & 0 & 1 & 0 & 0 & 0 \\ 0 & 0 & 0 & 0 & 0 & 0 & 0 & 0 & 0 & 0 & 1 \\ 0 & 0 & 0 & 0 & 0 & 0 & 0 & 0 & 0 & 1 & 0 \end{pmatrix}$$

3.2 Cartesian Products

The Cartesian product of graphs is described as follows: Given graphs G and H, form the Cartesian product of the sets of vertices. If G has n vertices and H has m vertices, this will be a set with $n\,m$ elements described as pairs (v, w) where v is a vertex of G and w is a vertex of H. There is an edge in the Cartesian product of G and H from (v, w) to (v', w') if and only if there is a edge in G from v to v' *and* an edge in H from w to w'. This is an inconvenient description to use with our representations of graphs because the vertices ultimately have to be ordered from 1 to n. However, there is a well-known operation on matrices called the Kronecker product which does exactly the right thing to the adjacency matrices of the graphs. We define a more general operation which is just a rearrangement and flattening of **Outer**. The actual operation we want will then be given by taking the first argument to be **Times**.

```
In[9]:=kronecker[f_, p_List, q_List] :=
       Flatten[Map[Flatten, Transpose[Outer[f, p, q],
            {1, 3, 2}], {2}], 1];
```

Here is an example.

```
In[10]:= (mat1 = Kpartite[2, 2].adjacencyMatrix[]) // TableForm
```

Out[10]//TableForm=

0	0	1	1
0	0	1	1
1	1	0	0
1	1	0	0

```
In[11]:= (mat2 = balancedGraph.new[4, 2].adjacencyMatrix[]) // TableForm
```

Out[11]//TableForm=

0	0	1	0
0	0	0	1
1	0	0	0
0	1	0	0

```
In[12]:=kronecker[Times, mat1, mat2]
```

```
Out[12]= {{0, 0, 0, 0, 0, 0, 0, 0, 0, 0, 1, 0, 0, 0, 1, 0},
          {0, 0, 0, 0, 0, 0, 0, 0, 0, 0, 0, 1, 0, 0, 0, 1},
          {0, 0, 0, 0, 0, 0, 0, 0, 1, 0, 0, 0, 1, 0, 0, 0},
          {0, 0, 0, 0, 0, 0, 0, 0, 0, 1, 0, 0, 0, 1, 0, 0},
          {0, 0, 0, 0, 0, 0, 0, 0, 0, 0, 1, 0, 0, 0, 1, 0},
          {0, 0, 0, 0, 0, 0, 0, 0, 0, 0, 0, 1, 0, 0, 0, 1},
          {0, 0, 0, 0, 0, 0, 0, 0, 1, 0, 0, 0, 1, 0, 0, 0},
          {0, 0, 0, 0, 0, 0, 0, 0, 0, 1, 0, 0, 0, 1, 0, 0},
          {0, 0, 1, 0, 0, 0, 1, 0, 0, 0, 0, 0, 0, 0, 0, 0},
          {0, 0, 0, 1, 0, 0, 0, 1, 0, 0, 0, 0, 0, 0, 0, 0},
          {1, 0, 0, 0, 1, 0, 0, 0, 0, 0, 0, 0, 0, 0, 0, 0},
          {0, 1, 0, 0, 0, 1, 0, 0, 0, 0, 0, 0, 0, 0, 0, 0},
          {0, 0, 1, 0, 0, 0, 1, 0, 0, 0, 0, 0, 0, 0, 0, 0},
          {0, 0, 0, 1, 0, 0, 0, 1, 0, 0, 0, 0, 0, 0, 0, 0},
          {1, 0, 0, 0, 1, 0, 0, 0, 0, 0, 0, 0, 0, 0, 0, 0},
          {0, 1, 0, 0, 0, 1, 0, 0, 0, 0, 0, 0, 0, 0, 0, 0}}
```

A careful look at this 16×16 table shows that if it is divided into 4×4 blocks, then in each position where there is a 1 in **mat1** there is a copy of **mat2** in the Kronecker product. Using **kronecker**, the Cartesian product of many graphs is constructed in the same form as the coproduct involving **Fold**. The differences are that the function which is folded is **kronecker[Times, #1, #2]&** and the starting value is **{{1}}** rather than **{}**, and the operation is applied to adjacency matrices rather than edge lists. Note that we omit the predicate **graphsQ** in the following because we want to use the construction for graphs with loops, which would be ruled out by the predicate.

```
In[13]:= cartesianProduct[graphs___] :=
        adjacents . new[Fold[kronecker[Times, #1, #2] &,
                {{1}},
                Map[adjacencyMatrix, {graphs}]]]
```

This could, but won't, be turned into a method for the class **graph**. Here is an example.

```
In[14]:= cartesianProduct[K[3], K[4]] .
        circularImmersion[] . Show[AspectRatio ←Automatic];
```

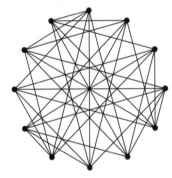

Just as with coproducts, there is a standard infix notation for Cartesian products; namely, **g1 ✘ g2**. In order to use it we first have to unprotect **Cross**.

```
In[15]:= Unprotect[Cross];
```

```
In[16]:= g1_ ? graphQ ✘ g2_ ? graphQ := cartesianProduct[g1, g2]
```
As before, we want to write a Cartesian product of several graphs without using brackets.

```
In[17]:= Notation[ g1_ × g2_ × g3_  ⟹  ((g1_ × g2_) × g3_) ]
```
For instance:

```
In[18]:= (K[2] ✘ K[3] ✘ K[4]) . circularImmersion[] .
        Show[AspectRatio ←Automatic];
```

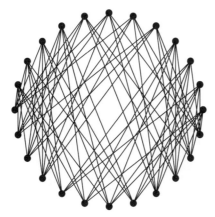

3.3 Tensor Products

The tensor product of two graphs has the same vertices as their Cartesian product, but edges are introduced in a different way. If G has *n* vertices and H has *m* vertices, then the set of vertices has *n m* elements described as pairs (*v*, *w*) where *v* is a vertex of G and *w* is a vertex of H. There is an edge in the tensor product of G and H from (*v*, *w*) to (*v'*, *w'*) if and only if there is a edge in G from *v* to *v'* and *w = w'*, or there is an edge in H from *w* to *w'* and *v = v'*. This construction is why we want to have the illegitimate class of loops, because the edges of the first kind look like the edges in the Cartesian product of G with loops on the number of vertices of H, and conversely for the edges of the second kind. This leads to the following simple implementation.

```
In[19]:= tensorProduct[g_?graphQ, h_?graphQ] :=
    ordereds.new[Union[
        cartesianProduct[g, loops.new[h.numberOfVertices[]]].
         orderedPairs[],
        cartesianProduct[loops.new[g.numberOfVertices[]], h].
         orderedPairs[]]]
```

We have implemented the case of a tensor product of two graphs. The general case can be handled in a generic way.

```
In[20]:= tensorProduct[graphs___?graphQ] :=
        Fold[tensorProduct, empty.new[1], {graphs}];
```

As with the Cartesian product, this could be turned into a method for the class **graph**, but we forego doing so. Here is an example.

In[21]:= **tensorProduct[K[3], K[4]] .**
        circularImmersion[] . Show[AspectRatio →Automatic];

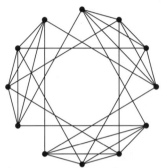

As with the other two products, there is a standard infix notation for the tensor product of two graphs; namely, **g1⊗g2**. We implement it and the required associativity rule as before.

In[22]:= **g1_ ? graphQ ⊗ g2_ ? graphQ := tensorProduct[g1, g2]**

In[23]:= **Notation[g1_⊗g2_⊗g3_ ⟹ ((g1_⊗g2_)⊗g3_)]**

It is instructive to compare the Cartesian product and the tensor product of graphs which consist just of a single edge.

In[24]:= **GraphicsArray[{**
        (K[2] × K[2]) . circularImmersion[] .
          Show[AspectRatio →1, DisplayFunction →Identity],
        (K[2] ⊗ K[2]) . circularImmersion[] .
          Show[AspectRatio →1, DisplayFunction →Identity]}] .
        Show[DisplayFunction →$DisplayFunction];

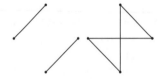

Finally, here is an example of a tensor product of three graphs.

In[25]:= **(K[2] ⊗ K[3] ⊗ K[4]) . circularImmersion[] .**
        Show[AspectRatio →Automatic];

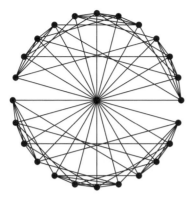

4 Other Graph Constructions in the Class `graph`

The top class **graph** knows about a number of other constructions for graphs. Those presented here are all constructions that start with a single graph and produce another related graph.

4.1 Complement of a Graph

The complement of a graph is the graph on the same vertices as the original graph which has an edge wherever the original graph does not have an edge. In terms of ordered pairs, the ordered pairs of the original graph are subtracted from the ordered pairs of a complete graph of the same size. This is implemented directly as a method in the class **graph**.

```
{complement,
 ordereds.new[
   Complement[
     completeGraph.new[self.numberOfVertices[]].orderedPairs[],
    self.orderedPairs[]]]&}
```

Recall that the complement construction is used in the definition of partite graphs.

To evaluate the examples here, reload the package if necessary and recall the notations introduced in the preceding section.

In[1]:=**Needs["GraphTh`"]**

In[2]:= **g1_?graphQ⊕g2_?graphQ := coproduct[g1, g2];**
    Unprotect[Cross];

```
g1_?graphQ ✘ g2_?graphQ := cartesianProduct[g1, g2];
Protect[Cross];
g1_?graphQ ⊗ g2_?graphQ := tensorProduct[g1, g2];
```

Here is an example that has interesting threefold symmetry

```
In[3]:= (K[3] ⊗ K[4]) .complement[] .
    circularImmersion[] .Show[AspectRatio →Automatic];
```

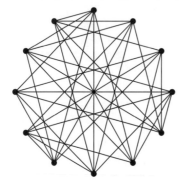

4.2 Cones, Stars, and Wheels

The cone on a graph is the graph given by joining a new vertex to all the vertices of the original graph. In terms of ordered pairs, it consists of the ordered pairs of the original graph together with all pairs consisting of an original vertex together with a fixed new vertex. This is also implemented directly as a method for the class **graph**.

```
{cone,
    Module[{n = self.numberOfVertices[], i},
        ordereds.new[
            Union[self.orderedPairs[],
                Table[{i, n + 1}, {i, n}],
                Table[{n + 1, i}, {i, n}]]]]&}
```

Here are some examples.

```
In[4]:= K[5] .cone[] .centerCircularImmersion[] .Show[AspectRatio →1];
```

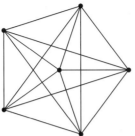

A brief command can produce a graph with striking symmetries.

```
In[5]:= (K[3] × K[3]) . cone[] .
        centerCircularImmersion[] . Show[AspectRatio → 1];
```

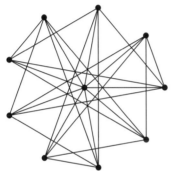

We use the cone construction to describe stars and wheels. Namely, stars are cones on empty graphs and wheels are cones on cyclic graphs. Both of these could be defined as new classes, but it is easy enough to just describe them directly.

```
star[n_] := empty . new[n] . cone[]
```

```
In[6]:= star[5] . centerCircularImmersion[] . Show[AspectRatio → 1];
```

```
wheel[n_] := cyclicGraph . new[n] . cone[]
```

```
In[7]:= wheel[12] . centerCircularImmersion[] . Show[AspectRatio → 1];
```

4.3 Induced Subgraphs

Given a graph and a subset of the vertices, there is an induced subgraph on the subset in which there is an edge between two vertices in the subset if and only if there is an edge between them in the original graph. This is implemented as a method in the class **graph** just by selecting the appropriate rows and columns of the adjacency matrix.

```
{inducedSubgraph,
    Function[{subset},
        adjacents.new[
            Transpose[Transpose[
                (self.adjacencyMatrix[])[[subset]] ][[subset]]]]]}
```

For instance, the graph induced on any subset of a complete graph is again a complete graph.

In[8]:= **K[5].inducedSubgraph[{1, 3, 5}].adjacencyMatrix[]**

Out[8]= {{0, 1, 1}, {1, 0, 1}, {1, 1, 0}}

Since the subset is actually given as a list, this can be used to generate a permutation of a graph.

In[9]:= **star[5].inducedSubgraph[{6, 5, 4, 3, 2, 1}].adjacencyMatrix[] //**
        TableForm

Out[9]//TableForm=

0	1	1	1	1	1
1	0	0	0	0	0
1	0	0	0	0	0
1	0	0	0	0	0
1	0	0	0	0	0
1	0	0	0	0	0

Now the first, rather than the last, vertex is connected to all the other vertices.

4.4 Incidence Matrix of a Graph

The incidence matrix of a graph is a (normally) non-square matrix whose rows correspond to the edges of the graph and whose columns correspond to the vertices. A 1 in position (i, j) means that the jth vertex is one of the ends of the ith edge. Thus, the number of rows is the number of edges, the number of columns is the number of vertices, each row has exactly two 1's in positions corresponding to its two ends, and a column has as many 1's as there are edges meeting at that vertex. This is implemented as a method of the class **graph**.

```
{incidenceMatrix,
    Map[
        ReplacePart[
            Table[0, {self.numberOfVertices[]}],
            1, Partition[#, 1]]&,
        self.orderedPairs[].Select[
                (#[[1]] < #[[2]])&] ]&}
```

To see how this works, consider the complete graph on four vertices. It has six edges, as the following shows.

In[10]:= **K[4].orderedPairs[].Select[(#[[1]] < #[[2]])&]**

Out[10]= {{1, 2}, {1, 3}, {1, 4}, {2, 3}, {2, 4}, {3, 4}}

Since there are 4 vertices and 6 edges, the incidence matrix is a 6×4 matrix. The code produces rows of 0's of length 4 and replaces certain entries by 1's. The first ordered pair {1, 2} means that the first row of this matrix should have 1's in positions 1 and 2, etc.

In[11]:= **K[4].incidenceMatrix[] // TableForm**

Out[11]//TableForm=

1	1	0	0
1	0	1	0
1	0	0	1
0	1	1	0
0	1	0	1
0	0	1	1

It would be possible to treat the incidence matrix as another way to represent graphs and add it as a fourth subclass under the class **graph**. The other three subclasses would then have to have methods to calculate the incidence matrix from their data and the incidence matrix class would have to know how to calculate the other three representations from its data. We leave this as an exercise.

4.5 Line Graphs

The line graph L(G) of a graph G has a vertex for each edge of G and there is an edge in L(G) from a vertex v to a vertex w if and only if the two edges of G corresponding to v and w share a common vertex in G. Thus, the number of vertices of L(G) is the number of rows of the incidence matrix of G, and row i and row j are joined by an edge in L(G) if and only if there is some column of the incidence matrix that has a 1 in both of these rows. The way to detect when that occurs is to multiply the incidence matrix of G by its transpose and look to see if the (i, j) entry there is non-zero. This leads to a square integer matrix which, after adjustment, is the adjacency matrix of the desired graph. The algorithm is implemented as a method of the class **graph**.

```
{lineGraph,
    With[{im = self.incidenceMatrix[]},
        adjacents.new[adjust[im . Transpose[im]]]]&}
```

Note: **Dot** cannot actually be used here because of the new rule given for it. Instead, **Inner[-Times, im, Transpose[im], Plus]** is used in the implementation.
The complete graph on n vertices has $n(n - 1)/2$ edges, so the line graphs for these graphs grow quickly in size.

In[12]:= **K[5] . lineGraph[] . circularImmersion[] .**
  Show[AspectRatio ←Automatic];

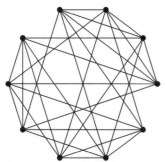

The line graph of a graph of the form **Kpartite[m, n]** has **m n** edges. Pictures of them have interesting symmetries.

In[13]:= **Kpartite[3, 4] . lineGraph[] .**
  circularImmersion[] . Show[AspectRatio ←Automatic];

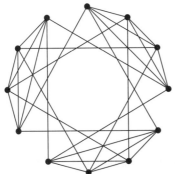

The line graph of a cycle is a cycle of the same length, except that two of the vertices are permuted.

```
In[14]:= cyclicGraph . new[5] . lineGraph[] .
           circularImmersion[] . Show[AspectRatio ↔Automatic];
```

5 Some Graph Algorithms

There are thousands of algorithms that use graphs in one way or another. Some of them are just concerned with properties of graphs, and that is all that we care about here. In principle, such algorithms belong in the class **graph**, but then the class becomes unwieldy, so we have put a few of them there and implemented a few others outside the class in functional style.

5.1 Graph Isomorphism

In principle, every class should have a method to determine when two objects of the class are isomorphic. In our case, although objects may belong to different classes, everything is ultimately a graph, so it is sufficient to be able to decide if two graphs are isomorphic. We cannot compare them directly as objects since they will have different identifying numbers and may belong to different subclasses, but we can, for instance, compare their adjacency matrices.

5.1.1 Isomorphism predicate

Isomorphism testing will be implemented in functional style first, to clarify what is going on. Graphs are isomorphic if there is some bijection between their vertices that preserves the property of being connected by an edge. This is equivalent to saying that graphs G and H are isomorphic if and only if there is some permutation of the vertices of H such that the adjacency matrix of G is the same as the adjacency matrix of the graph induced from H by the permutation. Here is a function that checks if a given permutation yields such an isomorphism.

```
In[15]:=isomorphismQ[g1_?graphQ, g2_?graphQ, p_List] :=
        SameQ[g1.adjacencyMatrix[],
             inducedSubgraph[g2, p].adjacencyMatrix[]] /;
             Length[p] == g2.numberOfVertices[]
```

For instance:

```
In[16]:=isomorphismQ[K[5], K[5], {1, 5, 3, 2, 4}]
```

Out[16]= True

```
In[17]:=isomorphismQ[Kpartite[2, 2], Kpartite[2, 2],
               {1, 3, 2, 4}]
```

Out[17]= False

5.1.2 Finding isomorphisms

To determine if two graphs are isomorphic, it is necessary to search through all possible permutations of the vertices to see if some permutation yields an isomorphism. Of course, the number of permutations grows exponentially with the number of vertices, so such a search should be avoided if possible. Clearly, if the numbers of vertices of the two graphs are different, then they cannot be isomorphic, so the number of vertices is an isomorphism invariant of graphs. There are many other such invariants. We consider just one of them here: the degree sequence. The degree of a vertex is the number of edges that meet at that vertex. In our representation, the degree of vertex i is just the number of 1's in the ith row of the adjacency matrix. The degree sequence of a graph is the decreasing sequence of degrees of vertices of the graph.

```
    degreeSequence[g_graph] :=
        Map[(Apply[Plus, #])&,
            g.adjacencyMatrix[]].Sort[].Reverse[]
```

Actually, this is implemented as a method in the class **graph**.

```
In[18]:=Kpartite[2, 3, 4].degreeSequence[]
```

Out[18]= {7, 7, 6, 6, 6, 5, 5, 5, 5}

The output means that there are two vertices of degree 7, three of degree 6, and four of degree 5. This sequence is clearly an isomorphism invariant.

Since searching for an isomorphism is a lengthy procedure, some safeguards are built in to check beforehand if the graphs are clearly non-isomorphic. Any number of invariants could be used, but we only consider the two mentioned above: namely, the number of vertices and the degree sequence. The procedure checks if these two invariants are the same before it embarks on searching through all permutations to try to discover an isomorphism. If the graphs are not isomorphic, a message giving the reason is printed and the empty list is returned. If they are, then a specific permutation is returned which realizes the isomorphism. Functionally, we can implement this using the usual message reporting facilities.

```
Graph::vertices = "Different numbers of vertices";

Graph::degreeSequence = "Different degree sequences.";

Graph::isomorphism = "The graphs are not isomorphic.";
```

The function **findIso** will first test if the number of vertices and the degree sequences are the same, using a **Which** clause to report failure of these tests. If they both succeed, then it proceeds to look at all permutations of the vertices of the first graph and **Scan** them using the (written out) predicate from above, returning the first permutation it finds that works. If none of them work, it reports that the graphs are not isomorphic.

```
findIso[g1_?graphQ, g2_?graphQ]:=
  Module[{iso},
    Which[
       g1.numberOfVertices[] =!= g2.numberOfVertices[],
         Message[Graph::vertices];{},
       g1.degreeSequence[] =!= g2.degreeSequence[],
         Message[Graph::degreeSequence];{},
       True,
         iso =
          Scan[If[
                  g1.adjacencyMatrix[] ===
                    g2.inducedSubgraph[#].adjacencyMatrix[
                    ],
                  Return[#]]&,
              g1.numberOfVertices[].Range[].Permutations[
              ]];
          If[iso =!= Null, iso,
            Message[Graph::isomorphism];{} ]] ]
```

The **findIsomorphism** method of the class **graph** is just the object-oriented version of this operation. It doesn't use the **Message** mechanism, but just prints the appropriate string.

In[19]:=**K[4].findIsomorphism[K[3]]**

     Different numbers of vertices

Out[19]= {}

In[20]:=**Kpartite[6, 4].findIsomorphism[Kpartite[5, 5]]**

     Different degree sequences.

Out[20]= {}

In[21]:=**Kpartite[3, 2].findIsomorphism[Kpartite[2, 3]]**

Out[21]= {3, 4, 5, 1, 2}

This result means that if the vertices of the second graph are rearranged in the order {3, 4, 5, 1, 2} then it becomes isomorphic to the first graph (which, of course, is obvious).

In[22]:=**Kpartite[3, 3].findIsomorphism[K[2]⊗K[3]]**

     The graphs are not isomorphic.

Out[22]= {}

In this example, both graphs have 6 vertices of degree 3. The search space consists of all 720 permutations of {1, 2, 3, 4, 5, 6}. There may be other graphs that have 6 vertices of degree 3. Here is an attempt to construct one.

In[23]:=**newgraph =**
          With[{edge = {0, 0, 1, 1, 1, 0}},
              adjacents.new[
                  Table[RotateRight[edge, i], {i, 0, 5}]]];

We can make a picture of all three graphs to see if two of them are obviously isomorphic.

In[24]:= **GraphicsArray[**
          Map[#.circularImmersion[] .
              Show[AspectRatio →Automatic, DisplayFunction →Identity] &,
              {newgraph, Kpartite[3, 3], K[2]⊗K[3]}]] .
          Show[DisplayFunction →$DisplayFunction];

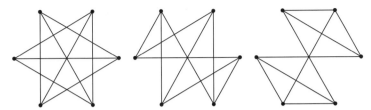

Clearly, these graphs all have six vertices of degree 3. (In all cases, the center is not a vertex.) But are any of them isomorphic? The first and the third both contain two triangles, so we check if they are isomorphic.

In[25]:= (K[2] ⊗ K[3]) . findIsomorphism[newgraph]

Out[25]= {1, 3, 5, 4, 6, 2}

5.2 Maximum Cliques

A *clique* of size k in a graph G is a subset of k vertices of G which induce a complete graph. Finding the largest clique in a graph is very similar to finding isomorphisms between graphs. One can define a predicate that checks if a given subset is a clique.

```
In[26]:= cliqueQ[g_?graphQ, clique_List] :=
        SameQ[K[Length[clique]].adjacencyMatrix[],
            g.inducedSubgraph[clique].adjacencyMatrix[]]
```

For instance:

In[27]:= cliqueQ[K[5], {1, 2, 3, 4, 5}]

Out[27]= True

In[28]:= cliqueQ[Kpartite[3, 3], {1, 2, 3, 4}]

Out[28]= False

To find a maximum clique in a graph one has to scan all subsets of the vertices, ordered by decreasing size, to find the first one which is a clique. Recall the construction of all subsets of a set.

```
subsets[list_List] :=
  Sort[Map[Flatten,
      Distribute[ Map[({{}, {#}})&, list], List]]]
```

Then the following will find a maximum clique. Note that this algorithm always succeeds since a single vertex is a clique.

```
maximumClique[g_graph] :=
  Scan[(If[cliqueQ[g, #], Return[#] ])&,
       Reverse[subsets[vertices[g] ] ] ]
```

MaximumClique is actually implemented as a method for the class **graph**. In doing so, we have spelled out the predicate explicitly, although it could have been included as a separate method.

```
{maximumClique,
  Scan[
    If[SameQ[
         completeGraph.new[Length[#]].adjacencyMatrix[],
         self.inducedSubgraph[#]. adjacencyMatrix[]],
       Return[#]]&,
    self.numberOfVertices[].Range[].subsets[].Reverse[]]&}
```

Another way to see that the graph **newgraph** constructed in the preceding section is not isomorphic to the graph **Kpartite[2, 3]** is to find maximum cliques in each. Since they have different sizes, the graphs must be non-isomorphic.

```
In[29]:= {Kpartite[2, 3].maximumClique[],
          newgraph.maximumClique[]}
```

Out[29]= {{2, 5}, {2, 4, 6}}

However, **K[2]⊗K[3]** also has a maximum clique with three vertices (as it must).

```
In[30]:= (K[2] ⊗ K[3]) . maximumClique[]
```

Out[30]= {4, 5, 6}

5.3 Minimum Vertex Covers

A vertex cover of a graph G is a subset of the vertices of G such that every edge has one of its vertices in the subset. A moment's reflection should convince you that two vertices not in the cover cannot be connected by an edge in G, since in that case, the subset would not be a cover. Thus, in the complementary graph of G, the vertices not in a clique define a cover. A minimum vertex cover of G therefore is the complement of a maximum clique in the complementary graph of G.

```
minimumVertexCover[g_?graphQ]  :=
    Complement[g.numberOfVertices[].Range[],
                g.complement[].maximumClique[]]
```

For instance:

In[31]:= (K[2] ⊗ K[3]) . minimumVertexCover[]

Out[31]= {1, 2, 4, 6}

This, and the following algorithms have not been added as methods to the class **graph**. But they are functions in the package.

5.4 *Maximum Independent Sets of Vertices*

An *independent set* of vertices in a graph G is a subset of the vertices such that no two vertices in the subset are joined by an edge. If V is a vertex cover of a graph, then the complement of V is an independent set. Hence, a maximum independent set is the compliment of a minimum vertex cover.

```
maximumIndependentSet[g_?graphQ]  :=
    Complement[g.numberOfVertices[].Range[],
                g.minimumVertexCover[]]
```

For instance:

In[32]:= (K[2] ⊗ K[3]) . maximumIndependentSet[]

Out[32]= {3, 5}

5.5 *Hamiltonian Cycles*

A *Hamiltonian cycle* of a graph G is a cycle in G which visits every vertex in G exactly once. As with finding isomorphisms and maximum cliques, we first need a predicate to determine if a particular permutation of the vertices determines a Hamiltonian cycle. All that is necessary is that there be edges in G between the successive vertices of the permutation together with an edge from the last entry of the permutation to the first.

```
hamiltonianCycleQ[g_?graphQ, vert_List] :=
  Complement[
    Partition[Append[vert, First[vert]], 2, 1],
    g.orderedPairs[]] == {} /;
  Sort[vert] === g.numberOfVertices[].Range[]
```

Consider the graph `Kpartite[3, 3]`.

In[33]:= `Kpartite[3, 3] . partiteImmersion[] . Show[AspectRatio → 1];`

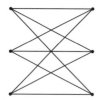

The order of vertices here is up the right side and down the left.

In[34]:=`Kpartite[3, 3].hamiltonianCycleQ[{6, 5, 4, 3, 2, 1}]`

Out[34]= `False`

In[35]:=`Kpartite[3, 3].hamiltonianCycleQ[{1, 4, 2, 5, 3, 6}]`

Out[35]= `True`

So there is a Hamiltonian cycle in `Kpartite[3, 3]`. Next we want to construct a procedure that yields such a cycle if there is one, and otherwise returns the empty list.

```
findHamiltonianCycle[g_?graphQ] :=
  With[
    {path = Scan[(If[g.hamiltonianCycleQ[#], Return[#]])&,
              g.numberOfVertices[].Range[].Permutations[]]},
    If[path === Null, {}, path] ]
```

Now instead of guessing the cycle above, we can use the program to find it.

In[36]:=`Kpartite[3, 3].findHamiltonianCycle[]`

Out[36]= `{1, 4, 2, 5, 3, 6}`

Many graphs don't have Hamiltonian cycles.

In[37]:=`Kpartite[2, 3].findHamiltonianCycle[]`

Out[37]= { }

But many do.

In[38]:= (K[2] ⊗ K[3]) . findHamiltonianCycle[]

Out[38]= {1, 2, 3, 6, 5, 4}

Six vertices have 720 permutations, which is a feasible number to search, but seven vertices have 5040, all of which would have to be searched for a graph which doesn't have a Hamiltonian cycle, like **Kpartite[3, 4]**. For those that do, like **K[3]⊗ K[3]**, it might be possible for the program to find one if it didn't have to first build the list of all 362,880 permutations and then search that. What is needed is an operation "**nextPermutation**" to generate permutations and test them one at a time.

6 *Exercises*

1. Define the class **empty** as a subclass of **adjacents**.

2. Add a fourth subclass, **incidents**, to the class graph and fill in all of the required details, as described in Section 2.3.

3. Describe the tensor product of several graphs by a single operation analogous to the descriptions of the coproduct and Cartesian product.

4. Look up Skiena's method to represent graphs, which includes instructions for drawing each graph. Consider the three constructions, coproduct, Cartesian product, and tensor product, in this light. Given the drawing instructions for graphs G and H, what should the drawing instructions be for **coproduct[G, H]**, **cartesianProduct[G, H]**, and **tensorProduct[G, H]** ?

5. What happens to the three kinds of products for the case of reflexive graphs, i.e., graphs in which it is assumed that there is always at least a loop on every vertex? The whole theory can be redone for this case. In particular, empty graphs for this case would be what we have called loops here. Work out a way to make drawings of such graphs.

7 Implementation

A complete package implementing all of the commands developed here will be found on the CD-ROM distributed with this book. It is called **GraphTh.m**. Together with its parent file **GraphTh.nb**, it should be located in the **Applications** subdirectory of the **AddOns** subdirectory of the *Mathematica* **3.0 Files** directory. Make sure that the package **Classes.m** is also located in this directory. Essentially everything there has been discussed here already except for the complete form of the class graph, so we include that here.

```
Class[graph, Object, {},
  {{graphQ,
      Module[{matrix = self.adjacencyMatrix[]},
        MatrixQ[matrix, (#===0 || #===1)&]    &&
        (matrix === Transpose[matrix])        &&
        And@@Map[(# === 0)&,
                   Transpose[matrix, {1, 1}]] ]&},
   {numberOfVertices, self.edgeLists[].Length[]&},
   {numberOfEdges, self.edgeLists[].Flatten[].Length[]/2&},
   {incidenceMatrix,
      Map[
        ReplacePart[
          Table[0, {self.numberOfVertices[]}],
          1, Partition[#, 1]]&,
        self.orderedPairs[].Select[
              (#[[1]] < #[[2]])&] ]&},
   {coproduct,
      edges.new[
        Fold[Join[#1, #2 + Length[#1]]&,
            edgeLists[self], Map[edgeLists[#]&, {##}]]]&},
   {complement,
    ordereds.new[
            Complement[completeGraph.new[
            self.numberOfVertices[]].orderedPairs[],
              self.orderedPairs[] ]]&},
   {cone,
     Module[{n = self.numberOfVertices[], i},
       ordereds.new[
         Union[self.orderedPairs[],
             Table[{i, n + 1}, {i, n}],
             Table[{n + 1, i}, {i, n}]]]]&},
   {lineGraph,
     With[{im = self.incidenceMatrix[]},
       adjacents.new[Inner[Times,im, Transpose[im],
         Plus].adjust[]]]&},
```

```
{randomImmersion,
   With[{verts =
       Table[{Random[], Random[]},
              {self.numberOfVertices[]}]},
     Graphics[
       Join[
        {PointSize[0.035]}, Map[Point, verts],
         Map[Line[{verts[[#[[1]]]], verts[[#[[2]]]]}]&,
           self.orderedPairs[].Select[(#[[1]] < #[[2]])&]]
     ]] ]&},
{circularImmersion,
   Module[{n = self.numberOfVertices[], verts},
     verts = Table[{N[Cos[2 Pi i/n]/ 2],
                     N[Sin[2 Pi i/n]/2]},
                    {i, 0, n - 1}];
     Graphics[
       Join[
        {PointSize[0.035]}, Map[Point, verts],
         Map[Line[{verts[[#[[1]]]], verts[[#[[2]]]]}]&,
           self.orderedPairs[].Select[(#[[1]] < #[[2]])&]]
        ]] ]&},
  {centerCircularImmersion,
    Module[{n = self.numberOfVertices[], verts},
     verts =
       Append[Table[{N[Cos[2 Pi i/(n-1)]/2],
                     N[Sin[2 Pi i/(n-1)]/2]},
                    {i, (n-1)}],
              {0, 0}];
     Graphics[
       Join[
        {PointSize[0.035]},
        Map[Point, verts],
        Map[Line[{verts[[#[[1]]]], verts[[#[[2]]]]}]&,
           self.orderedPairs[].Select[(#[[1]] < #[[2]])&]]
     ]] ]&},
  {degreeSequence,
       Map[(Apply[Plus, #])&,
           self.adjacencyMatrix[]].Sort[].Reverse[]&},
  {inducedSubgraph,
    Function[{subset},
      adjacents.new[
        Transpose[Transpose[
          (self.adjacencyMatrix[])[[subset]] ][[subset]]
        ]]]},
  {maximumClique,
```

```
      Scan[
        If[SameQ[
          completeGraph.new[Length[#]].adjacencyMatrix[],
          self.inducedSubgraph[#]. adjacencyMatrix[]],
        Return[#]]&,
      self.numberOfVertices[].Range[].subsets[].Reverse[]]&},
    {findIsomorphism,
      Function[{gh},
        Module[{iso},
          Which[
            self.numberOfVertices[] =!= gh.numberOfVertices[],
              Print["Different numbers of vertices"];{},
            self.degreeSequence[] =!= gh.degreeSequence[],
              Print["Different degree sequences."];{},
            True,
              iso =
                Scan[
                  (If[self.adjacencyMatrix[] ===
                      gh.inducedSubgraph[#].adjacencyMatrix[],
                      Return[#]])&,
                  self.numberOfVertices[].Range[].Permutations[
                  ]];
            If[iso =!= Null, iso,
                Print[
                "The graphs are not isomorphic."];{} ]] ]]}
  }];
```

$$CHAPTER \quad \boldsymbol{14}$$

Differentiable Mappings

1 Introduction

The preceding two chapters covering Polya's pattern inventory and graph theory involve finite, discrete mathematical structures whose representations in terms of numbers in a computer are probably as concrete a presentation of these structures as can be given. Such finite structures don't require *Mathematica*'s symbolic powers; they can be and have been programmed in lower level languages. In this chapter and the next one we will treat infinite, continuous structures associated with differentiable mappings. There is no way to directly realize such constructs in a computer other than in terms of symbolic representations of the basic entities. This is exactly the way *Mathematica* handles topics like symbolic differentiation, integration, and differential equations, and this is what will be used here.

In this chapter we will be concerned with differentiable mappings between finite-dimensional, real vector spaces. We symbolize such a mapping as $f : R^m \to R^n$, where R stands for the real numbers. Such a mapping is described analytically by n real-valued functions of m real variables. We will consider several special cases.

1. $f : R^2 \to R^2$. This will be treated as the generic case, where m and n are both 2. These values are small enough so that we can make pictures sometimes.

2. $f : R^1 \to R^1$. This seems too degenerate to involve any geometry, but in connection with the tangent space construction, it has some interest.

3. $f : R^m \to R^1$; i.e., one function of m variables. Here we will be concerned with the classification of the critical points of such a function.

4. $f : R^1 \to R^n$. Such a function represents a curve in R^n. If $n = 2$ or 3, then this concerns the classical differential geometry of curve in the plane or in space. If $n = 3$, then one can investi-

gate the curvature and torsion of the curve, and the tangent, normal, and binormal vector fields associated with it, find its arclength, etc. We leave this development for others to investigate.

5. $f : R^2 \to R^n$. Such a mapping represents a surface in R^n. If $n = 3$, then this concerns the classical differential geometry of surfaces in space. Here, we construct the first and second fundamental forms of such a parametric surface and use them to calculate the Gaussian and mean curvatures of a number of examples.

In all of these examples, our main interest is to represent the theory in *Mathematica* in a clean and consistent fashion. *Mathematica* is a sufficiently high-level language to allow one to do this in a way that looks almost exactly the way the theory looks in a textbook. This is important because differential geometry is perhaps unique in having two widely different aspects: on the one hand, the theory can be described in an abstract conceptual way which is very beautiful, while on the other hand, actual computations with examples are filled with symbols having multiple sub- and superscripts making them virtually opaque. If the computations can be described on the theoretical level, then one has made a clear conceptual gain.

The main theme of this chapter is the Jacobian of a differentiable mapping and its use in the tangent mapping associated with a differentiable mapping. In Chapters 3 and 5 there were problems concerning Jacobians of differentiable transformations. These will be investigated here in a much more systematic fashion.

1.1 Types in Mathematica

Types were discussed earlier in Chapter 5 where the basic types **Symbol**, **Integer**, and **Real** were identified, as well as the built-in type **List**. As discussed there, any head of an expression can be regarded as a type. This is certainly the case for those heads that just hold their arguments together without processing them, such as **List**, **Graphics**, and **Graphics3D**, as well as the user defined types that were introduced in Chapter 12 such as **group** for groups and **ge** for group elements. *Mathematica* uses the term *object* to refer to expressions with given heads; e.g., graphics objects are expressions with head **Graphics**. Thus, it makes sense to regard a group, for instance, as an object of type "group". In type theory functions have types determined by the types of their arguments and the type of their output. Thus, a function whose argument is of type A and whose output is of type B is said to have type A $\to$ B. Note that *Mathematica* does provide the facility to restrict the application of a function to arguments with a given head by the construction **f[x_head] := expr**. If **expr** has some other head **head1**, then we can say that **f** has type **head** $\to$ **head1**. In many parts of mathematics, there are two kinds of entities under consideration: some kind of geometric, algebraic, or analytic structure, and some appropriate kind of mapping or transformation from one such structure to another one of the same kind. For any particular kind of mathematics, our type theory will provide one type for the objects (or structures) and another type for the mappings.

2 *Differentiable Mappings*

The differentiable mappings we are concerned with are mappings between domains or regions in finite dimensional, real vector spaces. There are various ways in which one might try to describe domains in an n-dimensional space (for instance, by inequalities), but any such treatment leads inevitably to great complications. Thus, we assume here that our objects are just the whole n-dimensional spaces themselves. Such a space will be described by a list of n coordinates, $\{x, y, z, \ldots\}$. The type of the objects under discussion therefore is **List**. To describe a differentiable mapping between two such spaces we have to specify what the spaces are and then give rules telling how points in the first space are mapped to points in the second. Differentiable mappings will therefore be a expressions of the form

```
mapping[oldvariables, rule, newvariables]
```

We regard **mapping** as a type and expressions of this form as objects of type **mapping**. For instance,

```
mapping[{x, y}, {x^2 - y^2, 2 x y}, {u, v}]
```

represents the mapping from the x-y-plane to the u-v-plane given by the coordinate functions

$$u = x^2 - y^2 \text{ and } v = 2 \, x \, y.$$

Thus, the **rule** is a list of expressions describing how the new variables are functions of the old variables. Since a mapping consists of three lists, one could say that its type is the product type

$$\text{List} \times \text{List} \times \text{List},$$

except that there are implicit restrictions that the lengths of the second and third components should be the same and the first and third components should consist just of variables, so **mapping** is actually a subtype of this product type.

2.1 *Differentiable Mappings*

If the package **DiffMaps** is loaded using the following command, then all of the operations introduced in this chapter are automatically made available. Alternatively, they can be evaluated one at a time here.

In[1]:=**Needs["DiffMaps`"]**

The three components of a mapping can be extracted by functions called **dom** (for the list of domain variables), **rule**, and **cod** (for the list of codomain variables). We expect rules to be a list of expressions in the domain variables, whose length equals the length of the list of codomain variables, thought of as determining each codomain variable as a function of the domain variables, as in the example above. These extractors are defined in the obvious way.

```
dom[m_mapping]      := m[[1]];
rule[m_mapping]     := m[[2]];
cod[m_mapping]      := m[[3]];
```

In the preceding chapter on graph theory we were able to construct a predicate **graphQ** which served as a formal definition of a graph in *Mathematica*, but in the case of differentiable mappings it does not seem possible to write down anything more than the most trivial clauses in such a check. Certainly, a formal definition seems unattainable.

```
mappingQ[map_mapping]:=
    Length[map]  == 3                               &&
    Depth[dom[map]] == Depth[cod[map]] == 2  &&
    Length[rule[map]]  == Length[cod[map]]
```

An important feature of mappings is that they can be composed if the codomain of the first mapping is the same as the domain of the second. Furthermore, for any list of variables, there is an identity mapping from the domain represented by those variables to itself which serves as an identity for composition. **composition** is an operation taking a pair of mappings and returning a mapping, while **identityMapping** is an operation taking a list as argument and returning a mapping. The type of **identityMapping** clearly is **List** → **mapping**, while the type of **composition** is approximately **mapping** × **mapping** → **mapping**. Actually, its domain type is the subtype of **mapping** × **mapping** consisting of those pairs such that the codomain of the first equals the domain of the second.

```
composition[map1_mapping, map2_mapping]  :=
  mapping[
    dom[map1],
    rule[map2]  /. Thread[cod[map1]->rule[map1]],
    cod[map2]]  /; cod[map1]  === dom[map2]


identityMapping[var_List]  := mapping[var, var, var]
```

To understand the composition rule, consider the example

```
In[2]:=map1 = mapping[{x, y}, {x^2 + y^2, -2 x y}, {u, v}];
    map2 = mapping[{u, v}, {u + v, u - v}, {r, s}];
```

Then

In[4]:= `rule[map2]`

Out[4]= {u + v, u - v}

In[5]:= `Thread[cod[map1] -> rule[map1]]`

Out[5]= {u → x² + y², v → -2 x y}

In[6]:= `rule[map2] /. Thread[cod[map1] -> rule[map1]]`

Out[6]= {x² - 2 x y + y², x² + 2 x y + y²}

Thus, the rule for a composed mapping is given by substituting the formulas for **map1** into the formulas for **map2**.

In[7]:= `composition[map1, map2]`

Out[7]= mapping[{x, y}, {(x - y)², (x + y)²}, {r, s}]

In[8]:= `mappingQ[composition[map1, map2]]`

Out[8]= True

In[9]:= `identityMapping[{x, y, z}]`

Out[9]= mapping[{x, y, z}, {x, y, z}, {x, y, z}]

Our main interest in this section is in constructions on mappings. One such construction is the operation **inversemap**, which returns an object of the same type which is the inverse of the given mapping (if it exists). Thus, **inversemap** has type **mapping** → **mapping**. Actually, from the implementation using **Solve**, a given mapping may have several inverses so it would be more accurate to describe the type as **mapping** → **List of mappings**. In fact, what is actually constructed is a list of left inverses with respect to the composition operation defined below.

```
inversemap[m_mapping] :=
  With[
    {answers = Solve[Thread[rule[m] == cod[m]],dom[m]]},
    Map[mapping[cod[m], #, dom[m]]&,
        dom[m] /. answers] ]
```

As a very simple illustrative example (not requiring any extra simplification), we find the inverse mappings for the squaring mapping between one-dimensional spaces.

In[10]:= `mp = mapping[{x}, {x²}, {u}];`

In[11]:=**mpInv = inversemap[mp]**

Out[11]= $\{\mathrm{mapping}[\{u\},\ \{-\sqrt{u}\,\},\ \{x\}]$, $\mathrm{mapping}[\{u\},\ \{\sqrt{u}\,\},\ \{x\}]\}$

Thus, if $u = x^2$, then there are two inverses given by $x = \pm\sqrt{u}$.

Now, whenever possible, we want an equality test that determines if two objects of a given type are the same. The test for differentiable mappings is called **intentionalEqualQ**, the term *intentional* suggesting that we do not compare values of two mappings, but rather compare the expressions determining the mappings.

```
intentionalEqualQ[m1_mapping, m2_mapping] :=
    (dom[m1] == dom[m2]) &&
    (cod[m1] == cod[m2]) &&
    (Simplify[rule[m1] - rule[m2]] ===
        Table[0, {Length[cod[m1]]} ] ) )
```

intentionalEqualQ is an operation returning an object of type Boole (although there is no type by this name in *Mathematica*), i.e., **True** or **False**, so **intentionalEqualQ** has type **mapping×mapping** → **Boole**.

We can use these operations to check that the two mappings in **mapInv** are left inverses to **mp** with respect to composition.

```
In[12]:=Map[intentionalEqualQ[
            composition[#, mp],
            identityMapping[cod[mp]]]&,
        mpInv]
```

Out[12]= $\{\mathrm{True},\ \mathrm{True}\}$

However, only one of them is a right inverse to **mp**; namely, the second one.

```
In[13]:=Map[intentionalEqualQ[
            composition[mp, #]//PowerExpand,
            identityMapping[dom[mp]]]&,
        mpInv]
```

Out[13]= $\{\mathrm{False},\ \mathrm{True}\}$

2.2 *The Jacobian and the Tangent Map*

The Jacobian of a mapping at a point is a linear map (represented by a matrix) at each point of the domain of the mapping. It is an operation that applies to objects of class **mapping**.

```
jacobian[map_mapping] := Outer[D, rule[map], dom[map]]
```

For instance:

In[14]:= `jacobian[map1]`

Out[14]= $\{\{2\,x,\ 2\,y\},\ \{-2\,y,\ -2\,x\}\}$

Thus, at some fixed point $\{x_0,\ y_0\}$ this is the linear mapping represented by the matrix

$$\begin{pmatrix} 2\,x_0 & 2\,y_0 \\ -2\,y_0 & -2\,x_0 \end{pmatrix}.$$

One of the problems to be faced here is that there is no obvious type for the value of **jacobian**. One could say that given a mapping and a point in the domain of the mapping, it produces a linear transformation, but then this linear transformation has to have its own domain and codomain. This is not only a programming problem but a mathematical problem as well. In order to fit the Jacobian into our type system, we have to construct a domain and codomain for its values. In fact, starting from a given mapping, a new mapping will be constructed, called the *tangent mapping* whose rule makes use of the Jacobian of the original mapping. The tangent mapping of a given mapping will be a new mapping with a larger domain and codomain, called the tangent spaces of the original domain and codomain. In order to describe this in terms of lists of variables, we need a new head, **v**, that just puts the head **v** on each of the old variables, displayed in subscripted form; e.g., $\mathbf{v_x}$. Think of $\mathbf{v_x}$ as a vector coordinate in the direction of the **x** variable. The most convenient way to do this is to make **v** listable, with the stipulation that it print in subscripted form, and apply it to the old domain and codomain.

```
Attributes[v] = {Listable};

Format[v[x_]] := Subscripted[v[x]]

tangentSpace[list_List] := Join[list, v[list]]
```

For instance,

In[15]:= `tangentSpace[{x, y, z}]`

Out[15]= $\{x,\ y,\ z,\ v_x,\ v_y,\ v_z\}$

Thus, the tangent space of a real vector space is a real vector space of twice the dimension with new coordinates given by **v** applied to the old coordinates. The tangent mapping between the tangent spaces of the domain and codomain of a mapping uses the Jacobian. It solves the mathematical and programming problem of providing a type for the value of the Jacobian.

```
tangentMapping[m_mapping] :=
  mapping[tangentSpace[dom[m]],
          Join[rule[m], jacobian[m] . v[dom[m]]],
          tangentSpace[cod[m]]]
```

Try this on our main example.

In[16]:= `tangentMapping[map1]`

Out[16]= mapping$[\{x, y, v_x, v_y\},$
$\qquad \{x^2 + y^2, -2\,x\,y, 2\,x\,v_x + 2\,y\,v_y, -2\,y\,v_x - 2\,x\,v_y\}, \{u, v, v_u, v_v\}]$

The rules for the **tangentMapping** of **map1** are the same as those of **map1** as far as the variables **u** and **v** are concerned. For fixed values of **x** and **y**, the new variables v_u and v_v (in the tangent space of the codomain) are given in terms of the new variables v_x and v_y (in the tangent space of the domain) by multiplying them by the value of the Jacobian matrix at the point $\{x, y\}$. In terms of the equations

$$u = x^2 - y^2 \text{ and } v = 2\,x\,y$$

v_u and v_v are given by the matrix equation

$$\begin{pmatrix} v_u \\ v_v \end{pmatrix} = \begin{pmatrix} 2\,x & 2\,y \\ -2\,y & -2\,x \end{pmatrix} \cdot \begin{pmatrix} v_x \\ v_y \end{pmatrix}$$

In particular, the type of **tangentMapping** is **mapping** $\rightarrow$ **mapping**.

2.3 Tangent Vector Fields

The tangent mapping can be used to construct the vector fields tangent to the coordinate lines of a mapping by composing it with the unit tangent vector fields in the tangent space of the domain of the mapping. These are given by the operation

```
(*unitVectors[dom_List] :=
  Map[mapping[dom, Join[dom, #], tangentSpace[dom]]&,
      IdentityMatrix[Length[dom]]]*)
```

```
(*tangentVectorFields[map_mapping] :=
    Map[composition[#, tangentMapping[map]]&,
        unitVectors[dom[map]]]*)
```

We have commented out these operations because, although they are very attractive geometrically, in practice they turn out to be very slow. A much more efficient way to find the tangent vectors fields is to use the following version.

```
tangentVectorFields[map_mapping] :=
    Map[
      mapping[dom[map], Join[rule[map], #],
            tangentSpace[cod[map]]]&,
      Transpose[jacobian[map]]];
```

In[17]:= **tangentVectorFields[map1]**

Out[17]= $\{mapping[\{x, y\}, \{x^2 + y^2, -2xy, 2x, -2y\}, \{u, v, v_u, v_v\}],$
$mapping[\{x, y\}, \{x^2 + y^2, -2xy, 2y, -2x\}, \{u, v, v_u, v_v\}]\}$

This works because the transpose of the Jacobian has as its *i*th row the partial derivatives of the rule for the mapping with respect to the *i*th domain variable.

2.4 The Chain Rule

The chain rule for functions of several variables says that the Jacobian of a composed map is the matrix product of the Jacobians of the factors (expressed in the correct variables). This is the content of the exercises in Chapters 3 and 5. The problem of course is to get the expressions in terms of the correct variables. Once we have the concept of the tangent mapping of a mapping as well as the concept of the composition of two mappings, then everything takes care of itself very nicely. The proper theorem does not talk directly about the Jacobian at all, but just says that the tangent mapping of a composition of mappings is the composition of the tangent mappings of the given mappings. The only place one has to talk about substitution of expressions for variables is in the definition of composition. Once composition is given, then everything else follows. We express this as a theorem about a pair of composable mappings.

```
TheoremT[map1_mapping, map2_mapping] :=
    intentionalEqualQ[
      tangentMapping[ composition[map1, map2] ],
      composition[ tangentMapping[map1],
                  tangentMapping[map2] ]
    ] /; cod[map1] === dom[map2]
```

For instance:

In[18]:=**TheoremT[map1, map2]**

Out[18]= True

An auxiliary result says that the tangent mapping of an identity mapping is an identity mapping.

```
TheoremI[var_List] :=
    intentionalEqualQ[
        tangentMapping[ identityMapping[var] ],
        identityMapping[ tangentSpace[var] ] ]
```

These are called theorems because, in fact, we know that they are true. However, verifying **theoremT** in any particular case can be a very extensive calculation. In the examples below, we let *Mathematica* do such verifications.

2.5 Making Plots of Differentiable Mappings

The graphics routines implemented here are part of the package **DiffMaps**.

The graphical operations in this package illustrate mappings whose domain has dimension 1 or 2 and whose codomain has dimension 2 or 3. The only operation exported by the package is called **mapGraphics**. Its output is a graphics object, to be displayed by **Show**, which makes pictures of the domain and the codomain of a mapping, showing how the domain is transformed into the codomain. The domain is shown by a rectangular grid and the codomain by the image of that grid. To indicate the direction of the transformation, a named arrow is included between the two. Each of these ingredients is in a rectangle for assembly in the final graphics operation, so that both the grid and its image are displayed in the same picture. The operation is used in the form **mapGraphics[mapping, "name", range(s)]**, where the range is either an interval {a, b} or a pair of intervals with step sizes $\{a_1, a_2, step_a\}$, $\{b_1, b_2, step_b\}$. The use of this operation is illustrated by the following four mappings. The first two are mappings from a 1-dimensional space into a 2- and a 3-dimensional space, respectively.

In[19]:=**mapline = mapping[{t}, {t^2, t^3}, {x, y}];**

In[20]:=**mapcurve = mapping[{t},**
                          {Sin[t], Cos[t], Sin[t]^2}, {x, y, z}];

The second two mappings are mappings from a 2-dimensional space into a 2- and 3-dimensional space, respectively.

In[21]:=**map2d = mapping[{x, y}, {x^2 + y^2, -2 x y}, {u, v}];**

```
In[22]:= map3d = mapping[{u, v},
                {Cos[v] Cos[u], Cos[v] Sin[u], Sin[v]},
                {x, y, z}];
```

These mappings produce the following pictures.

```
In[23]:= Show[mapGraphics[mapline, "mapline", {-1, 1}]];
```

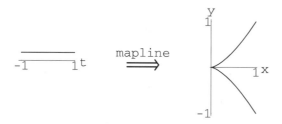

```
In[24]:= Show[mapGraphics[mapcurve, "mapcurve", {0, 2 Pi}]];
```

```
In[25]:= Show[mapGraphics[map2d, "map2d",
                {-1, 0, 0.1}, {0, 1, 0.1}]];
```

In[26]:= **Show[mapGraphics[map3d, "map3d",**
                  {0, Pi, Pi/10}, {0, 2Pi, Pi/5}]];

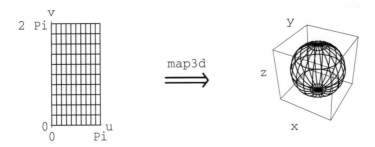

If the dimension of the codomain is too large, then an error message is displayed.

In[27]:= **toobig = identityMapping[{x, y, z, w}];**

In[28]:= **Show[mapGraphics[toobig, "toobig", {0, 1}]];**

```
mappingGraphics::codomainDimensions :
  Codomain dimensions are too large for plotting.
  The codomain should have dimension 2 or 3.
```

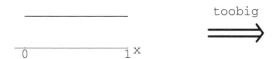

2.6 Examples

2.6.1 Example 1

Let us now look at the theoretical computations associated with **map2d** illustrated above. This is the mapping that was treated in the exercises in Chapters 3 and 4.

In[29]:=**map2d = mapping[{x, y}, {x^2 + y^2, -2 x y}, {u, v}];**

In[30]:=**jacobian[map2d]**

Out[30]= $\{\{2\,x,\,2\,y\},\,\{-2\,y,\,-2\,x\}\}$

In[31]:=**tangentMapping[map2d]**

Out[31]= mapping$[\{x,\,y,\,v_x,\,v_y\},$
$\qquad \{x^2 + y^2,\, -2\,x\,y,\, 2\,x\,v_x + 2\,y\,v_y,\, -2\,y\,v_x - 2\,x\,v_y\},\, \{u,\,v,\,v_u,\,v_v\}]$

In[32]:=**inverses = inversemap[map2d];**

The output, which has been suppressed because it is quite large, consists of four mappings. If **map2d** is composed with these different inverse mappings, the result is the identity mapping for the space **{x, y}** only in the first case.

In[33]:=**Map[composition[map2d, #]&, inverses] // FullSimplify // PowerExpand**

Out[33]= $\{$mapping$[\{x,\,y\},\,\{x,\,y\},\,\{x,\,y\}],$ mapping$[\{x,\,y\},\,\{-x,\,-y\},\,\{x,\,y\}],$
$\qquad$ mapping$[\{x,\,y\},\,\{-y,\,-x\},\,\{x,\,y\}],$ mapping$[\{x,\,y\},\,\{y,\,x\},\,\{x,\,y\}]\}$

However, each inverse map followed by **map2d** does give the identity map.

In[34]:= `Map[composition[#, map2d] &, inverses] // FullSimplify`

Out[34]= `{mapping[{u, v}, {u, v}, {u, v}], mapping[{u, v}, {u, v}, {u, v}],`
        `mapping[{u, v}, {u, v}, {u, v}], mapping[{u, v}, {u, v}, {u, v}]}`

Thus, each of the mappings found by **inversemap** is a left inverse to **map2d** but only the first one is a right inverse. **TheoremT** holds in both directions for **map2d** and all of its inverses. The first computation generalizes the result found in Exercise 13 of Chapter 3.

In[35]:= `Map[TheoremT[map2d, #] &, inverses]`

Out[35]= `{True, True, True, True}`

In[36]:= `Map[TheoremT[#, map2d]&, inverses]`

Out[36]= `{True, True, True, True}`

These computations are remarkably uneventful compared with the same computations in Version 2.2.

2.6.2 Example 2

The theorems concerning the tangent mapping work for generic functions of given numbers of variables. For instance, let **mapA** and **mapB** be generic mappings between two-dimensional spaces.

In[37]:= `mapA = mapping[{x, y}, {f[x, y], g[x, y]}, {u, v}];`
    `mapB = mapping[{u, v}, {r[u, v], s[u, v]}, {w, z}];`

Composition and identity mappings work correctly.

In[39]:= `composition[mapA, mapB]`

Out[39]= `mapping[{x, y}, {r[f[x, y], g[x, y]], s[f[x, y], g[x, y]]}, {w, z}]`

In[40]:= `composition[identityMapping[dom[mapA]], mapA] === mapA`

Out[40]= `True`

Furthermore, *Mathematica* is able to evaluate **TheoremsT** and **TheoremI** in this generality.

In[41]:= `TheoremT[mapA, mapB] // Simplify`

Out[41]= True

In[42]:=**TheoremI[dom[mapA]]**

Out[42]= True

The result for **TheoremT[mapA, mapB]** can be regarded as a proof of the theorem for the composition of two mappings between two-dimensional spaces. Clearly the same thing could be done for mappings between spaces of any fixed dimensions that fit together properly. However, we cannot think of any way to reformulate the theorem so that *Mathematica* can prove it for all possible dimensions in one step.

2.7 Example: A Phase Portrait

Here we look at a simple example, where the domain and codomain have dimension one. In this case the tangent mapping is a mapping between two-dimensional spaces. From it we will extract a mapping from one-dimensional space to two-dimensional space; i.e., a plane curve.- This example shows the phase portrait for the curve $x = \sin t$; i.e., for a particle undergoing simple harmonic motion as a function of time. The phase plane is the plane with coordinates position and velocity. It is the same as the tangent space to the one-dimensional coordinate space. The curve there given by $\{x(t), x'(t)\}$ is called the *phase portrait* of the motion. In *Mathematica*, this looks as follows:

In[43]:=**sincurve = mapping[{t}, {Sin[t]}, {x}];**

In[44]:=**phasecurve = First[tangentVectorFields[sincurve]]**

Out[44]= mapping[{t}, {Sin[t], Cos[t]}, {x, v_x}]

In[45]:=**Show[mapGraphics[phasecurve, "phase", {0, 2 Pi}]];**

2.8 Example: Damped Harmonic Motion

Here is another, more interesting example.

```
In[46]:= damped = mapping[{t}, {E^(-0.1 t) Sin[t]}, {x}];
```

```
In[47]:= phasecurve = First[tangentVectorFields[damped]]
```

Out[47]= mapping[{t},
         {E^{-0.1 t} Sin[t], E^{-0.1 t} Cos[t] - 0.1 E^{-0.1 t} Sin[t]}, {x, v_x}]

```
In[48]:= Show[mapGraphics[phasecurve, "phase", {0, 4 Pi}]];
```

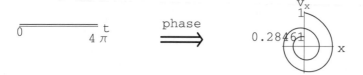

CHAPTER 15

Critical Points and Minimal Surfaces

This chapter continues the discussion of the preceding one. As before, all of the operations in this chapter are automatically made available by loading the following package.

```
In[1]:=Needs["DiffMaps`"]
```

In the two sections here, two other special cases of differentiable maps are considered:

i) Dimension[codomain] = 1; i.e., the mapping is determined by a single function of possibly many variables.

ii) Dimension[domain] = 2 and dimension[codomain] = 3; i.e., the mapping is a parametric surface in 3-dimensional space.

The Jacobian plays a central role in both cases. In the first case, the Jacobian of a single function is just the gradient of the function and the zeros of the gradient determine the critical points of the function. In the second case, the Jacobian determines the first fundamental form of a parametric surface. In addition to the Jacobian there is a new theoretical ingredient, the Hessian. In the first case, the Hessian is what classifies the critical points of a function. In the second case, the Hessian, in vector form, determines the second fundamental form of a surface, and the two fundamental forms together determine the Gaussian and mean curvatures of a surface. A minimal surface is one whose mean curvature is zero.

1. Critical points

1.1 The Mathematical Problem

If f is a differentiable function of several variables, then the critical points of f are the points where the gradient of f is 0. Such points are local minima, local maxima, or saddle points. The behavior of f at each critical point is determined by the Hessian matrix of f, evaluated at that critical point. This is the square matrix of all second partial derivatives of f with respect to the variables:

$$\mathbf{hessian[f]} = \left[\frac{\partial^2 \mathbf{f}}{\partial x_i\, \partial x_j}\right].$$

The analysis of the critical points involves looking at the values of the principal minors of this matrix at the critical points; namely, the determinants of the square submatrices running down the main diagonal in the upper left-hand side of the Hessian. For a 4×4 matrix, there are 4 such determinants.

i) The matrix is called *positive definite* if all of these determinants are positive. If the Hessian at a critical point is positive definite, then the critical point is a local minimum.

ii) The matrix is called *negative definite* if these determinants strictly alternate in sign, starting with the upper left-hand entry being negative. If the Hessian at a critical point is negative definite, then the critical point is a local maximum.

iii) If the matrix is neither positive nor negative definite, then the critical point will be called a *saddle point* here. (We are ignoring the case where some principal minor is 0, although that occurs in some of the examples). In the case of a saddle point, it is necessary to calculate the eigenvalues and eigenvectors of the matrix in order to understand the behavior of the function near such a point. If an eigenvalue is positive (respectively, negative), then the function increases (respectively, decreases) in the direction of the corresponding eigenvector. Positive (respectively, negative) definite corresponds to all eigenvalues being positive (respectively, negative).

1.2. The Mathematica Formulation

The commands implemented here are all included in the package **DiffMaps**.

In[1]:= **Needs["DiffMaps`"]**

1.2.1 Find the critical points

The gradient of a function is the vector of first partial derivatives of the function. (This is the same as the Jacobian of a single function with respect to several variables.)

```
grad[expr_, var_List] :=
    grad[expr, var] = D[expr, #]& /@ var
```

We want to use this for mappings whose codomain has dimension 1. We characterize such mappings as functions here. The default name for the single coordinate in the codomain space will be `lt` (for line type).

```
function[old_, rule_, new_] :=
   mapping[old, rule, new] /; Length[new] == 1

grad[fun_mapping] := grad[rule[fun], dom[fun]]
```

This can also be viewed more intrinsically as a mapping into the vector half of the tangent space of the domain of the function.

```
gradientMapping[fun_mapping] :=
   mapping[dom[fun], grad1[fun], v[dom[fun]]]
```

For instance, consider a generic function.

In[2]:=`genericFun = function[{x, y}, {f[x, y]}, {lt}];`

In[3]:=`grad[genericFun]`

Out[3]= $\{f^{(1,0)}[x, y], f^{(0,1)}[x, y]\}$

In[4]:=`gradientMapping[genericFun]`

Out[4]= $\text{mapping}[\{x, y\}, \{f^{(1,0)}[x, y], f^{(0,1)}[x, y]\}, \{v_x, v_y\}]$

The critical points of a function are the points where the gradient is zero. We are not interested in multiple solutions or in complex solutions, so we apply **Union** to the list of solutions and then select those that don't have complex entries. The operation **criticalPoints** is programmed dynamically in the package since it is a lengthy computation that is involved in all further calculations.

```
criticalPoints[fun_mapping] :=
   Select[Union[Solve[grad[fun] == 0, dom[fun],
                      VerifySolutions -> True]],
          FreeQ[#, Complex]& ]
```

1.2.2 Analyze the critical points

The Hessian matrix of a function is the matrix of all second partial derivatives of the function. As with the gradient, it is programmed in two forms, one in terms of functions and variables and one for functions as mappings whose codomain is 1-dimensional.

```
hessian[funs_, vars_] :=
        Outer[D[funs, #1, #2]&, vars, vars];

hessian[fun_mapping] :=
        hessian[First[rule[fun]], dom[fun]]/;
            Length[rule[fun]] == 1;
```

For our generic function this gives the following result.

In[5]:=**hessian[genericFun] // TableForm**

Out[5]//TableForm=

$f^{(2,0)}[x, y]$ $f^{(1,1)}[x, y]$

$f^{(1,1)}[x, y]$ $f^{(0,2)}[x, y]$

To find the principal minors, first define one step in the process of decreasing the size of a matrix by dropping the last row and column.

```
oneMinor[matrix_] := Map[Drop[#, -1]&, Drop[matrix, -1]]
```

The principal minors are given by nesting this operation.

```
principalMinors[matrix_] :=
    Det/@NestList[oneMinor, matrix, (Length[matrix] - 1)]
```

A matrix is positive definite if all principal minors are positive.

```
positiveDefiniteQ[matrix_] :=
        And@@Positive[principalMinors[matrix]]
```

A matrix is negative definite if the principal minors alternate in sign; equivalently, if −1 times the matrix is positive definite.

```
negativeDefiniteQ[matrix_] := positiveDefiniteQ[-matrix]
```

If the Hessian is positive definite at a critical point, then the critical point is a local minimum. If it is negative definite, then the critical point is a local maximum.

```
localMinima[fun_mapping] :=
    Select[criticalPoints[fun],
            positiveDefiniteQ[hessian[fun]/.#]&]

localMaxima[fun_mapping] :=
    Select[criticalPoints[fun],
            negativeDefiniteQ[hessian[fun]/.#]&]
```

If a critical point is neither positive nor negative definite, then we consider it here to be a saddle point. The output of the operation **saddlePoints** is an expression with head **criticalDirections** whose arguments are pairs consisting of a critical point and the eigensystem of the Hessian evaluated at that point.

```
otherCriticalPoints[fun_mapping]  :=
    Complement[criticalPoints[fun],
              localMinima[fun], localMaxima[fun] ]

saddlePoints[fun_mapping] :=
 With[{others = otherCriticalPoints[fun]},
  If[others == {}, {},
    Thread[criticalDirections[others,
      Transpose[Eigensystem[#]]&/@
            (hessian[fun]/.others)]]]]
```

1.2.3 Numerical versions of the commands

In the implementation package there are numerical versions of all of the preceding commands. They have the same names preceded by an N; i.e., **NcriticalPoints**, **NlocalMinima**, **NlocalMaxima**, **NotherCriticalPoints**, **NsaddlePoints**.

1.3 Examples

1.3.1 Example 1

```
In[6]:= function1 =
      function[{x, y, z},
          {3 x^2 + 2 y^2 + 2 z^2 + 2 x y + 2 x z + 2 y z},
          {1t}];
```

```
In[7]:= localMinima[function1]
```

```
Out[7]= {{x → 0, y → 0, z → 0}}
```

```
In[8]:= localMaxima[function1]
```

```
Out[8]= {}
```

In[9]:= **saddlePoints[function1]**

Out[9]= { }

1.3.2 Example 2

In[10]:= **function2 =**
        function[{x, y}, {x^4 + y^4 - x^2 - y^2 + 1}, {1t}];

In[11]:= **localMinima[function2]**

Out[11]= $\left\{ \left\{ x \to -\dfrac{1}{\sqrt{2}}, \ y \to -\dfrac{1}{\sqrt{2}} \right\}, \ \left\{ x \to -\dfrac{1}{\sqrt{2}}, \ y \to \dfrac{1}{\sqrt{2}} \right\}, \right.$

$\left. \left\{ x \to \dfrac{1}{\sqrt{2}}, \ y \to -\dfrac{1}{\sqrt{2}} \right\}, \ \left\{ x \to \dfrac{1}{\sqrt{2}}, \ y \to \dfrac{1}{\sqrt{2}} \right\} \right\}$

In[12]:= **localMaxima[function2]**

Out[12]= { {x → 0, y → 0} }

In[13]:= **saddlePoints[function2]**

Out[13]= $\left\{ \text{criticalDirections} \left[\right. \right.$

$\left\{ x \to 0, \ y \to -\dfrac{1}{\sqrt{2}} \right\}, \ \{\{-2, \{1, 0\}\}, \ \{4, \{0, 1\}\}\} \right],$

$\text{criticalDirections} \left[\left\{ x \to 0, \ y \to \dfrac{1}{\sqrt{2}} \right\}, \ \{\{-2, \{1, 0\}\}, \ \{4, \{0, 1\}\}\} \right],$

$\text{criticalDirections} \left[\left\{ y \to 0, \ x \to -\dfrac{1}{\sqrt{2}} \right\}, \right.$

$\{\{-2, \{0, 1\}\}, \ \{4, \{1, 0\}\}\} \right],$

$\text{criticalDirections} \left[\left\{ y \to 0, \ x \to \dfrac{1}{\sqrt{2}} \right\}, \ \{\{-2, \{0, 1\}\}, \ \{4, \{1, 0\}\}\} \right] \right\}$

In the case of a function of two variables, we can plot the function as a surface.

```
In[14]:= Plot3D[Evaluate[First[rule[function2]]],
         {x, -1, 1}, {y, -1, 1},
         PlotPoints -> 40, Mesh -> False];
```

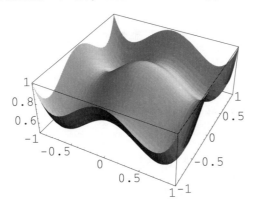

1.3.3 Example 3

```
In[15]:= function3 = function[{x, y, z},
                     {x^2 + y^2 + z^2 - 4 x z}, {lt}];
```

```
In[16]:= localMinima[function3]
```

Out[16]= { }

```
In[17]:= localMaxima[function3]
```

Out[17]= { }

```
In[18]:= saddlePoints[function3]
```

Out[18]= {criticalDirections[{x → 0, y → 0, z → 0},
         {{-2, {1, 0, 1}}, {2, {0, 1, 0}}, {6, {-1, 0, 1}}}]}

Thus, at the origin the function decreases in one direction and increases in the other two directions.

1.3.4 Example 4

```
In[19]:= function4 =
           function[{x, y, z},
           {x^4 + y^4 + z^4 - x^2 - y^2 - z^2 + 4 x y + 4 x z + 1},
           {1t}];
```

The exact symbolic program is unable to handle this function, so we have to use the numerical versions of the commands.

This time we display the critical points together with the value of **expr4** at each critical point.

```
In[20]:= With[{mins = NlocalMinima[function4]},
              Transpose[{mins,
                         val -> First[rule[function4]]/.mins}]]
```

```
Out[20]= {{{y → -1.28785, z → -1.28785, x → 1.49207}, val → -9.45797},
          {{y → 1.28785, z → 1.28785, x → -1.49207}, val → -9.45797}}
```

```
In[21]:= NlocalMaxima[function4]
```

```
Out[21]= {}
```

```
In[22]:= With[{sads = NsaddlePoints[function4]},
              Transpose[{sads,
                val ->
                First[rule[function4]]/.Map[(#[[1]])&, sads]}] ]
```

```
Out[22]= {{criticalDirections[{y → -0.707107, z → 0.707107, x → 0},
              {{7.40312, {0.515499, 0.605913, 0.605913}},
               {-5.40312, {-0.85689, 0.364513, 0.364513}},
               {4., {0, -0.707107, 0.707107}}}], val → 0.5},
          {criticalDirections[{y → 0, z → 0, x → 0},
              {{-2., {0, -1., 1.}}, {-7.65685, {-1.41421, 1., 1.}},
               {3.65685, {1.41421, 1., 1.}}}], val → 1},
          {criticalDirections[{y → 0.707107, z → -0.707107, x → 0},
              {{7.40312, {0.515499, 0.605913, 0.605913}},
               {-5.40312, {-0.85689, 0.364513, 0.364513}},
               {4., {0, -0.707107, 0.707107}}}], val → 0.5}}
```

There are two local minima and three saddle points. Note that the saddle points happen for **x** = 0, in which case, **function4** is the same as **function2** with one less domain variable.

1.3.5 Example 5

In this example, there is one local minimum and one saddle point. *Mathematica* is unable to find the exact eigenvectors, so we use the numerical version to find the saddle points.

```
In[23]:= function5 =
          function[{x, y, z},
                   {x^3 + y^3 + z^3 - 4 x z - 4 y z + 2}, {lt}];
```

```
In[24]:= localMinima[function5]
```

$$\text{Out[24]= } \left\{\left\{x \to \frac{4 \, 2^{1/3}}{3}, \; y \to \frac{4 \, 2^{1/3}}{3}, \; z \to \frac{4 \, 2^{2/3}}{3}\right\}\right\}$$

```
In[25]:= localMaxima[function5]
```

Out[25]= {}

Mathematica is unable to find the exact eigenvectors, so we use the numerical version to find the saddle points.

```
In[26]:= NsaddlePoints[function5]
```

Out[26]= {criticalDirections[{x → 0, y → 0, z → 0},
            {{0, {-1., 1., 0}}, {-5.65685, {0.707107, 0.707107, 1.}},
            {5.65685, {-0.707107, -0.707107, 1.}}}]}

Here, we see that there are positive, negative, and zero eigenvalues. The direction of the eigenvector corresponding to the negative eigenvalue should take us to the local minimum. If we restrict to a 2-dimensional subspace perpendicular to the direction of the null space of the Hessian, given by setting **y** equal to **x**, then we can make a picture of this situation. In the picture, we see the saddle point at $(0, 0)$ and the minimum at $(1.68, 2.12)$.

```
In[27]:= ContourPlot[Evaluate[First[rule[function5/.y -> x]]],
           {x, -1, 2.5}, {z, -1, 3},
           Contours -> 30, ContourShading -> False];
```

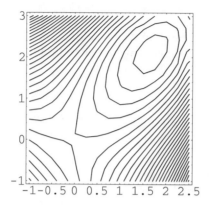

1.3.6 Example 6

For this function of four variables, there are no local minima or maxima and just one (real) saddle point which turns out to have two zero eigenvalues.

```
In[28]:= function6 =
        function[{x, y, z, w},
                {(x + 10 y)^2 + 5 (z - w)^2 +
                 (y - 2 z)^4 + 10 (x - w)^4},
                {1t}];
```

```
In[29]:= localMinima[function6]
```

```
Out[29]= {}
```

```
In[30]:= localMaxima[function6]
```

```
Out[30]= {}
```

```
In[31]:= saddlePoints[function6]
```

```
Out[31]= {criticalDirections[{y → 0, w → 0, x → 0, z → 0}, {{0, {0, 0, 1, 1}},
        {0, {-10, 1, 0, 0}}, {20, {0, 0, -1, 1}}, {202, {1, 10, 0, 0}}}]}
```

Notice that two eigenvalues are 0 and the other two are positive; in other words, the Hessian is positive semi-definite, so this is not really a saddle point. Analyze this as in Example 5 by looking at the function restricted to the orthogonal complement of the null space of the Hessian at the origin.

```
In[32]:= orthocomp = Solve[{-10 w + z == 0, x + y == 0}, {z, y}]
```

Out[32]= $\{\{z \rightarrow 10\,w,\; y \rightarrow -x\}\}$

In[33]:= **function66 =**
  function[{x, w}, rule[function6] /. orthocomp[[1]],
    {lt}]

Out[33]= $\mathrm{mapping}[\{x,\,w\},\;\{405\,w^2 + (-20\,w - x)^4 + 81\,x^2 + 10\,(-w + x)^4\},\;\{lt\}]$

This is obviously a convex function with a minimum at the origin. The symbolic critical points functions fail, but the numerical ones succeed.

In[34]:= **NlocalMinima[function66]**

Out[34]= $\{\{x \rightarrow 0,\; w \rightarrow 0\}\}$

In[35]:= **NlocalMaxima[function66]**

Out[35]= $\{\}$

In[36]:= **NsaddlePoints[function66]**

Out[36]= $\{\}$

In[37]:= **Show[**
  GraphicsArray[
   {Graphics3D[
    Plot3D[Evaluate[First[rule[function66]]],
     {x, -0.0001, 0.0001}, {w, -0.0001, 0.0001},
     DisplayFunction -> Identity]],
   Graphics3D[
    Plot3D[Evaluate[First[rule[function66]]],
     {x, -1, 1}, {w, -1, 1},
     DisplayFunction -> Identity]]}],
  DisplayFunction -> $DisplayFunction];

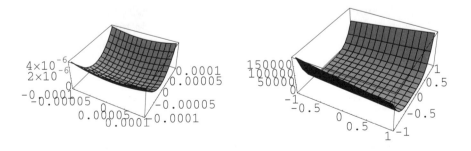

The pictures show that even in this subspace, the minimum is very flat.

1.3.7 Example 7

Neither procedure is able to deal with the last one.

```
In[38]:= function7 = function[{x, y, z},
                    {100 (z - (10/(2 N[Pi])) ArcTan[y/x])^2 +
                    (Sqrt[x^2 + y^2] - 1)^2 + z^2}, {1t}];
```

```
In[39]:= NcriticalPoints[function7]
```

Out[39]= $\{\{z \to 0,\ x \to -1.,\ y \to 0\}\}$

One critical point is found. However, if we proceed by hand using **FindRoot**, then we can find another one.

```
In[40]:= gradient = grad[function7];
```

```
In[41]:= solution =
              FindRoot[gradient == 0, {x, 1}, {y, 0}, {z, 0}]
```

Out[41]= $\{x \to 1.,\ y \to 0.,\ z \to 0.\}$

```
In[42]:= positiveDefiniteQ[hessian[function7] /. solution]
```

Out[42]= True

Thus, we conclude that the point $(1, 0, 0)$ is a local minimum of **function7**.

2 Minimal Surfaces

If the dimension of the domain of a mapping is 2, then the rule for the mapping consists of possibly many functions of 2 variables. Geometrically, the mapping is a parametric surface in some, possibly higher dimension, space. We consider just the case where the dimension of the codomain is 3; i.e., parametric surfaces in 3-space. The concepts analogous to the curvature of a curve are the Gaussian and mean curvatures of a surface. These in turn are determined by a pair of "forms" called the first and second fundamental forms of the (parametric) surface. A surface is called a minimal surface if its mean curvature is 0. (For a detailed treatment, see, e.g., Struik, Classical Differential Geometry, Addison-Wesley 1950 [Struik].)

2.1 The Differential Geometry of Minimal Surfaces: Mathematica Formulation

2.1.1 Differentiable surfaces

To use the commands implemented here, reload the package **DiffMaps`** if necessary.

In[1]:=**Needs["DiffMaps`"]**

Minimal surfaces are differentiable surfaces in 3-dimensional space whose mean curvature is zero. Interesting new minimal surfaces have been discovered recently. Pictures of such surfaces are often very attractive. For our purposes, a (parametric) surface is a vector valued function of two variables. It can be represented as a list of three ordinary differentiable functions of two variables:

$$X(u, v) = \{f(u, v), g(u, v), h(u, v)\}.$$

The goal here is to construct two functions of the form

gaussianCurvature[surface]
meanCurvature[surface]

which calculate the Gaussian and mean curvature functions of such a surface. These curvatures are defined in the following way. (For a detailed treatment, see, e.g., O'Neill, *Elementary Differential Geometry*, Academic Press 1966.)

A (parametric) surface is a differentiable mapping from a 2-dimensional space to a 3-dimensional space, so it can be characterized as a subtype of the general type of mapping.

```
surface[dom_, rule_, cod_] := mapping[dom, rule, cod] /;
        Length[dom] == 2 && Length[cod] == 3
```

For instance, here is a generic surface.

In[2]:=**generic =**
        surface[{u, v}, {f[u, v], g[u, v], h[u, v]}, {x, y, z}]

Out[2]= mapping[{u, v}, {f[u, v], g[u, v], h[u, v]}, {x, y, z}]

The rule for such a surface is given by a list of three functions of two variables, X(u, v) = {f(u, v), g(u, v), h(u, v)}. The vector fields on this surface along the coordinate lines are given by the partial derivatives with respect to u and v:

$$\frac{\partial X}{\partial u} = \left\{ \frac{\partial f}{\partial u}, \frac{\partial g}{\partial u}, \frac{\partial h}{\partial u} \right\}, \quad \frac{\partial X}{\partial v} = \left\{ \frac{\partial f}{\partial v}, \frac{\partial g}{\partial v}, \frac{\partial h}{\partial v} \right\}.$$

In *Mathematica*, these are the rows of the transpose of the Jacobian of the mapping.

In[3]:= `Transpose[jacobian[generic]]`

Out[3]= $\{\{f^{(1,0)}[u, v], g^{(1,0)}[u, v], h^{(1,0)}[u, v]\},$
    $\{f^{(0,1)}[u, v], g^{(0,1)}[u, v], h^{(0,1)}[u, v]\}\}$

We also have a more intrinsic representation of these vector fields as mappings to the tangent space of the codomain of the surface.

In[4]:= `tangentVectorFields[generic]`

Out[4]= $\{$mapping$[\{u, v\}, \{f[u, v], g[u, v], h[u, v], f^{(1,0)}[u, v],$
        $g^{(1,0)}[u, v], h^{(1,0)}[u, v]\}, \{x, y, z, v_x, v_y, v_z\}],$
    mapping$[\{u, v\}, \{f[u, v], g[u, v], h[u, v], f^{(0,1)}[u, v],$
        $g^{(0,1)}[u, v], h^{(0,1)}[u, v]\}, \{x, y, z, v_x, v_y, v_z\}]\}$

2.1.2 The first fundamental form of a surface

The "coefficients" of the "first fundamental form", $I(X)$, for a given surface by definition are the dot products of the tangent vectors:

$$E(X(u, v)) = \frac{\partial X}{\partial u} \cdot \frac{\partial X}{\partial u}, \quad F(X(u, v)) = \frac{\partial X}{\partial u} \cdot \frac{\partial X}{\partial v}, \quad G(X(u, v)) = \frac{\partial X}{\partial v} \cdot \frac{\partial X}{\partial v}.$$

The first fundamental form itself can be thought of as the matrix

$$\begin{pmatrix} E(X(u, v)) & F(X(u, v)) \\ F(X(u, v)) & G(X(u, v)) \end{pmatrix} = \begin{pmatrix} f_u & g_u & h_u \\ f_v & g_v & h_v \end{pmatrix} \begin{pmatrix} f_u & f_v \\ g_u & g_v \\ h_u & h_v \end{pmatrix}$$

Thus, in *Mathematica*, it is given by the operation

```
firstFundamentalForm[surf_mapping]  :=
    With[{partials = jacobian[surf]},
         Transpose[partials] . partials] // Simplify;
```

For the generic case, the result looks rather complicated.

In[5]:= `firstFundamentalForm[generic]`

Out[5]= $\{\{f^{(1,0)}[u, v]^2 + g^{(1,0)}[u, v]^2 + h^{(1,0)}[u, v]^2, f^{(0,1)}[u, v] f^{(1,0)}[u, v] +$
$g^{(0,1)}[u, v] g^{(1,0)}[u, v] + h^{(0,1)}[u, v] h^{(1,0)}[u, v]\},$
$\{f^{(0,1)}[u, v] f^{(1,0)}[u, v] + g^{(0,1)}[u, v] g^{(1,0)}[u, v] +$
$h^{(0,1)}[u, v] h^{(1,0)}[u, v],$
$f^{(0,1)}[u, v]^2 + g^{(0,1)}[u, v]^2 + h^{(0,1)}[u, v]^2\}\}$

2.1.3 Normal vectors and the second fundamental form

A normal vector field on the surface is given by the cross product of the tangent vectors

$$\text{Normal}(X(u, v)) = \frac{\partial X}{\partial u} \times \frac{\partial X}{\partial v},$$

and the unit normal vector field `UnitNormal(X(u, v))` is given by dividing this vector field by its length. In *Mathematica*, we need a formula for the length of a vector.

```
length[vect_] := Sqrt[vect . vect]//Simplify;
```

Then the unit normal vector field is given by the formula

```
unitNormal[surf_mapping] :=
    With[{vect = Cross@@Transpose[jacobian[surf]]},
        vect / length[vect] // Simplify // PowerExpand];
```

More intrinsically, we can define an associated mapping into the tangent space of the codomain of the surface.

```
normalVectorField[surf_mapping] :=
    mapping[dom[surf],
        Join[rule[surf], Cross@@Transpose[jacobian[surf]]],
        tangentSpace[cod[surf]]];
```

Both of these lead to large expressions if evaluated for **generic**. The "coefficients" of the "second fundamental form", $II(X)$, for a given surface, by definition, are the dot products of the second partial derivatives of X with the unit normal vector.

$$L(X(u, v)) = \frac{\partial^2 X}{\partial u^2} \cdot \text{unitNormal}(X(u, v))$$

$$M(X(u, v)) = \frac{\partial^2 X}{\partial u \, \partial v} \cdot \text{unitNormal}(X(u, v))$$

$$N(X(u, v)) = \frac{\partial^2 X}{\partial v^2} \cdot \text{unitNormal}(X(u, v)).$$

These coefficients can also be thought of as entries in a 2 × 2 symmetric matrix, but the formula to calculate it is more complicated:

$$\begin{pmatrix} L(X(u, v)) & M(X(u, v)) \\ M(X(u, v)) & N(X(u, v)) \end{pmatrix} = \begin{pmatrix} X_{u,u} & X_{u,v} \\ X_{v,u} & X_{v,v} \end{pmatrix} \cdot \frac{X_u \times X_v}{|X_u \times X_v|}.$$

The entries in the matrix on the right are vectors and the dot product means take the dot product of each of these vectors with the unit normal vector. The matrix of second partial derivatives is just the Hessian matrix in vector form as in the preceding section. For our generic surface, this is a 2 × 2 matrix whose entries are vectors.

In[6]:=**hessian[generic]**

Out[6]= $\{\{\{f^{(2,0)}[u, v], g^{(2,0)}[u, v], h^{(2,0)}[u, v]\},$
$\{f^{(1,1)}[u, v], g^{(1,1)}[u, v], h^{(1,1)}[u, v]\}\},$
$\{\{f^{(1,1)}[u, v], g^{(1,1)}[u, v], h^{(1,1)}[u, v]\},$
$\{f^{(0,2)}[u, v], g^{(0,2)}[u, v], h^{(0,2)}[u, v]\}\}\}$

The second fundamental form then has a very simple description.

```
secondFundamentalForm[surf_mapping]  :=
    Dot[hessian[surf], unitNormal[surf]] // Simplify;
```

Again, if this is evaluated for **generic**, then the result is very large.

2.1.4 The Gaussian and mean curvatures of a surface

It is shown in books like [O'Neill] and [Struik] cited above that

$$\text{gaussianCurvature}[X(u, v)] = (L N - M^2) / (E G - F^2)$$

$$\text{meanCurvature}[X(u, v)] = (E N - 2 F M + G L) / (E G - F^2).$$

These values can be derived directly from the first and second fundamental forms by forming the polynomial

$$\det(I(X) - x \; II(X))$$

and dividing by the leading coefficient. The constant term of the resulting polynomial is the Gaussian curvature and the coefficient of x is the negative of twice the mean curvature. In *Mathematica*, this is given by the operations

```
curvatureDet[surf_mapping, x_] :=
    curvatureDet[surf, x] =
        With[{det = Det[secondFundamentalForm[surf] -
                        x firstFundamentalForm[surf]]},
             Expand[det/Coefficient[det, x^2]]];

gaussianCurvature[surf_mapping] :=
    Module[{x},
            Coefficient[curvatureDet[surf, x], x, 0] ];

meanCurvature[surf_mapping] :=
    Module[{x},
            Coefficient[curvatureDet[surf, x], x] / 2 ];
```

The results of these operations applied to **generic** are huge expressions, so we only evaluate them for selected examples. Note that some of the following calculations take a long time.

2.2 Examples

2.2.1 Plane

Any parametric surface given by linear rules is a plane. We chose an arbitrary one and find, as expected, that both curvatures are 0.

```
In[7]:=plane =
    surface[{u, v},
            {a u + b v, c u + d v, p u + q v}, {x, y, z}];

In[8]:=gaussianCurvature[plane]
```

Out[8]= 0

In[9]:=`meanCurvature[plane]`

Out[9]= 0

2.2.2 Torus

Consider the pinched torus given by rotating a circle about an axis tangent to the circle.

```
In[10]:= torus = surface[{phi, theta},
                    {2 Sin[phi] Sin[phi] Cos[theta],
                     2 Sin[phi] Sin[phi] Sin[theta],
                     2 Sin[phi] Cos[phi]},
                    {x, y, z}];
```

About all that we can expect here is that the curvatures are independent of theta, but the exact forms are surprisingly brief.

In[11]:=`gaussianCurvature[torus]`

Out[11]= $-\dfrac{1}{2} \, \text{Cos[2 phi] Csc[phi]}^2$

In[12]:=`meanCurvature[torus]//PowerExpand//Together`

Out[12]= $\dfrac{1}{4} \left(2 - \text{Cos[2 phi] Csc[phi]}^2 \right)$

```
In[13]:= ParametricPlot3D[Evaluate[rule[torus]],
                       {phi, 0, Pi}, {theta, 0, 2Pi}];
```

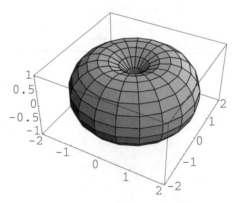

2.2.3 Sphere

Consider a parametric sphere of radius r.

```
In[14]:= sphere = surface[{u, v},
                  {r Cos[v] Cos[u], r Cos[v] Sin[u], r Sin[v]},
                  {x, y, z}];
```

```
In[15]:= gaussianCurvature[sphere]
```

$$Out[15]= \frac{1}{r^2}$$

```
In[16]:= meanCurvature[sphere]//PowerExpand
```

$$Out[16]= \frac{1}{r}$$

2.2.4 Catenoid

The only minimal surfaces of revolution are the catenoids.

```
In[17]:= catenoid =
      surface[{u, v},
              {a Cosh[u/a] Cos[v], a Cosh[u/a] Sin[v], u},
              {x, y, z}];
```

```
In[18]:= gaussianCurvature[catenoid]
```

$$Out[18]= -\frac{Sech[\frac{u}{a}]^4}{a^2}$$

```
In[19]:= meanCurvature[catenoid]
```

$$Out[19]= 0$$

```
In[20]:= ParametricPlot3D[Evaluate[a = 1; rule[catenoid]],
                  {u, -1, 1}, {v, 0, 2 Pi},
                  PlotPoints -> {15, 30}];
```

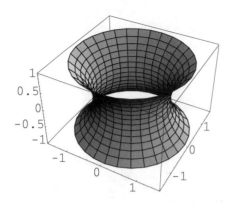

2.2.5 Helicoid

A right conoid is a surface generated by moving a straight line parallel to a plane and intersecting a line perpendicular to this plane. The only minimal right conoids are the helicoids.

```
In[21]:=helicoid = surface[{u, v},
                           {u Cos[v], u Sin[v], b v},
                           {x, y, z}];
```

```
In[22]:=gaussianCurvature[helicoid]
```

$$\text{Out[22]}= -\frac{b^2}{(b^2 + u^2)^2}$$

```
In[23]:=meanCurvature[helicoid]
```

Out[23]= 0

```
In[24]:=b = .3;
        ParametricPlot3D[Evaluate[b = .3; rule[helicoid]],
                         {u, 0, 2}, {v, -Pi, 4Pi},
                         PlotPoints -> {15, 40},
                         ViewPoint->{1.463, -2.702, 1.418}];
```

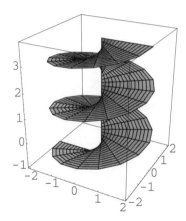

2.2.6 Sherk's minimal surface 1

Sherk's minimal surface "was the first minimal surface discovered after Meusnier's discovery of the catenoid and the helicoid" [Struik].

```
In[26]:=sherk1 = surface[{x, y},
                         {x, y, Log[Cos[y]] - Log[Cos[x]]},
                         {x, y, z}];
```

```
In[27]:=gaussianCurvature[sherk1]
```

$$Out[27]= -\frac{Sec[x]^2 Sec[y]^2}{(Sec[y]^2 + Tan[x]^2) (Sec[x]^2 Sec[y]^2 - Tan[x]^2 Tan[y]^2)}$$

```
In[28]:=meanCurvature[sherk1]
```

```
Out[28]= 0
```

```
In[29]:=Plot3D[Evaluate[rule[sherk1][[3]]],
               {x, -Pi/2, Pi/2}, {y, -Pi/2, Pi/2},
               PlotPoints -> 25, Mesh -> False];
```

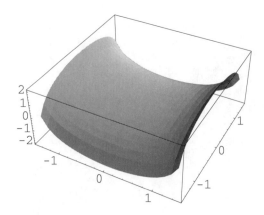

In[30]:= **Needs["Graphics`ContourPlot3D`"]**

In[31]:=**ContourPlot3D[Cos[x] E^z - Cos[y], {x, -Pi/2, Pi/2},**
            {y, -Pi/2, Pi/2}, {z, -3, 3}];

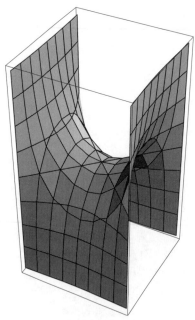

2.2.7 Sherk's second minimal surface

In[32]:=**sherk2 = surface[{x, y},**
                    {x, y, ArcSin[Sinh[x] Sinh[y]]},
                    {x, y, z}];

In[33]:=**gaussianCurvature[sherk2]//Simplify**

Out[33]= $-\text{Sech}[x]^2\,\text{Sech}[y]^2$

In[34]:=**meanCurvature[sherk2]//Simplify**

Out[34]= 0

In[35]:=**Plot3D[ArcSin[Sinh[x] Sinh[y]], {x, -1, 1}, {y, -1, 1},**
          BoxRatios -> {1, 1, 1}];

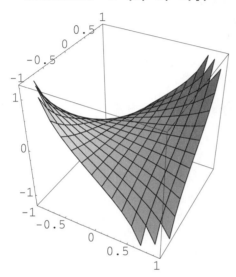

In[36]:=**ContourPlot3D[Sin[z] - Sinh[x] Sinh[y],**
          {x, -2, 2}, {y, -2, 2}, {z, -6, 2},
          PlotPoints->{5, 7}, PlotRange->All,
          Boxed->False];

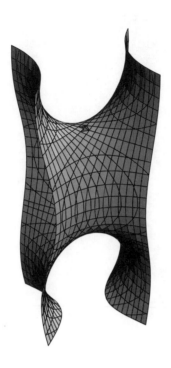

2.2.8 No name surface

I don't know the name of this minimal surface.

```
In[37]:= noname = surface[{x, y},
                          {x, y, ArcTan[y/x]},
                          {x, y, z}];
```

```
In[38]:= gaussianCurvature[noname]//Together
```

$$\text{Out[38]} = -\frac{1}{\left(1 + x^2 + y^2\right)^2}$$

```
In[39]:= meanCurvature[noname]//Together
```

Out[39]= 0

```
In[40]:= Plot3D[Evaluate[rule[noname][[3]]],
              {x, -1, 1}, {y, -1, 1},
              PlotRange -> All, PlotPoints -> 40,
              ViewPoint->{-2.5, -1.5, 1.7}];
```

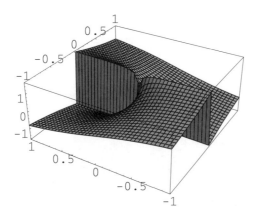

2.2.9 Monge surfaces

A Monge parametric surface is one of the form X[x, y] = {x, y, h[x, y]}. The last two examples are such surfaces. For these surfaces, there is a simpler formula for the mean curvature given in terms of the function h[x, y].

```
meanCurv[h_, {x_, y_}] :=
  With[{den = Sqrt[1 + D[h, x]^2 + D[h, y]^2]},
       (1/2)(D[D[h, x]/den, x] + D[D[h, y]/den, y])//
       Simplify]
```

This can be calculated for a generic function of two variables.

In[43]:=**mean1 = meanCurv[h[x, y], {x, y}]**

$$
\text{Out[43]=} \left(h^{(0,2)}[x, y] \left(1 + h^{(1,0)}[x, y]^2 \right) - 2\, h^{(0,1)}[x, y]\, h^{(1,0)}[x, y] \right.
$$

$$
\left. h^{(1,1)}[x, y] + h^{(2,0)}[x, y] + h^{(0,1)}[x, y]^2\, h^{(2,0)}[x, y] \right) \Big/
$$

$$
\left(2 \left(1 + h^{(0,1)}[x, y]^2 + h^{(1,0)}[x, y]^2 \right)^{3/2} \right)
$$

Minimal surfaces over a region in the x-y-plane are described by functions h[x, y] satisfying the partial differential equation given by setting this expression equal to 0. We can check that this formula is equivalent to our general formula applied to the special case of Monge surfaces by calculating the mean curvature for such a surface by our usual method.

In[44]:=**monge = surface[{x, y}, {x, y, h[x, y]}, {x, y, z}];**

In[45]:=**mean2 = meanCurvature[monge]//Together**

Out[45]= $\left(-h^{(0,2)}[x, y] - h^{(0,2)}[x, y] h^{(1,0)}[x, y]^2 + \right.$
$\qquad 2 h^{(0,1)}[x, y] h^{(1,0)}[x, y] h^{(1,1)}[x, y] -$
$\qquad \left. h^{(2,0)}[x, y] - h^{(0,1)}[x, y]^2 h^{(2,0)}[x, y]\right) \Big/$
$\qquad \left(2 \left(1 + h^{(0,1)}[x, y]^2 + h^{(1,0)}[x, y]^2\right)^{3/2}\right)$

Finally, check that these two expressions for the mean curvature of a Monge surface differ just by a minus sign.

In[46]:=**mean1 + mean2//Simplify**

Out[46]= 0

Of course, once one has this result, it is sufficient to set the numerator of **mean1** equal to 0 to describe a minimal surface in Monge form. This expression,

In[47]:= **Numerator[meanCurv[h[x, y], {x, y}]]**

Out[47]= $h^{(0,2)}[x, y] \left(1 + h^{(1,0)}[x, y]^2\right) - 2 h^{(0,1)}[x, y] h^{(1,0)}[x, y] h^{(1,1)}[x, y] +$
$\qquad h^{(2,0)}[x, y] + h^{(0,1)}[x, y]^2 h^{(2,0)}[x, y]$

is just the Euler–Lagrange equation for the area functional of a surface, which shows the connection between asking for the mean curvature to be equal to 0 and minimizing the area bounded by a curve.

IV

Answers

Chapter 1: Answers

Problem 1

```
Table[Factor[1 - x^n], {n, 1, 5}]//TableForm
```

Problem 2

```
Simplify[(1 - Cos[2 t])/(1 + Cos[2 t]) - Tan[t]^2]
```

Problem 3

```
expression1 = (x^2 + 5)/(x^5 + x^4 - x - 1);

integral1 = Integrate[expression1, x]//Simplify

derivative1 = D[integral1, x]//Simplify

derivative1 == expression1
```

Problem 4

```
expression1 = (a+b) ((c + d x) x + e x^2)

Expand[expression1]

Collect[expression1, {x, c}]

Factor[expression1]
```

Problem 5

```
Needs["Graphics`ImplicitPlot`"]
```

```
ImplicitPlot[9 x^2 + 4 x y + 6 y^2 == 1, {x, -0.5, 0.5},
            AspectRatio -> Automatic];
```

Problem 6

```
goodIntegrals =
    Table[{n, Integrate[(1 - 1/u)^(4/3)/u^n, u]}, {n, 0, 5}]

D[goodIntegrals, u] -
       Table[{0, (1 - 1/u)^(4/3)/u^n}, {n, 0, 5}] // Factor
```

Problem 8

```
Integrate[Sin[x^3]/Cos[x^3], {x, 0, 1}]//Timing

N[Integrate[Sin[x^3]/Cos[x^3], {x, 0, 1}]]//Timing

NIntegrate[Sin[x^3]/Cos[x^3], {x, 0, 1}]//Timing
```

Problem 10

```
expression6 = (x^3 + 6 x^5) / (2 (1 - x^3));

derivative6 = D[expression6, x] // Simplify

integral6 = Integrate[derivative6, x] // Simplify

(integral6 - expression6) // Simplify
```

Problem 11

```
num = (2 + 5 I)^12

N[num^(1/12)]

twelthroots = N[Solve[z^12 == num]]

Show[Graphics[{PointSize[0.03],
    Map[Point[{Re[#], Im[#]}]&, z/.twelthroots],
    Text["num^(1/12)", {8.5, 0.9}],
    Text["2 + 5 I", {4.5, 5}]}],
```

```
        Axes -> True,
        AspectRatio -> Automatic,
        PlotRange -> {{-7, 11.5}, Automatic}];

    rootsOne = N[Solve[z^12 == 1], 20]

    2. + 5. I == N[num^(1/12), 20] z/.rootsOne[[8]]
```

Problem 13

```
    matrix = {{1, 2, 3}, {4, 5, 6}, {7, 8, 9}};
```

Solution 1

```
    Transpose[Eigensystem[matrix]]//MatrixForm

    N[%]//MatrixForm
```

Solution 2

```
    Transpose[Eigensystem[N[matrix]]]//MatrixForm

    N[Eigensystem[matrix]][[2]] /
        Eigensystem[N[matrix]][[2]][[{3, 2, 1}]]
```

The check

```
    P = Transpose[N[Eigensystem[matrix]][[2]]]

    Inverse[P] . matrix . P

    Chop[%]
```

We can also carry out the check using the exact symbolic result.

```
    P = Transpose[Eigensystem[matrix][[2]]]

    Inverse[P] . matrix . P // Simplify
```

Problem 15a

```
hilbert[n_] := Table[1/(i + j - 1), {i, n}, {j, n}]

Timing[hilbert[20] . Inverse[hilbert[20]] ==
        IdentityMatrix[20]]
```

Problem 15b

$$\text{Factor}\left[\sum_{i=1}^{n} i^{30}\right]$$

Problem 15d

```
vanDerMonde[n_, x_] := Table[x_i^{j-1}, {j, n}, {i, n}]

Short[Det[vanDerMonde[6, x]], 2]

Det[vanDerMonde[6, x]]//Factor
```

Chapter 3: Answers

In some of the later problems here, we use commands that haven't been introduced yet. This is just to make the presentation of the answers as concise as possible. Everything that is used will be explained eventually.

```
equation1 = -3/7 + (37 x)/14 - (31 x^2)/7 + (17 x^3)/14 + x^4 == 0;

solution1 = Solve[equation1, x]

equation1 /. solution1
```

Problem 2

```
equation2 = x^5 - x^2/2740 - 3/9704700 == 0;

solution2 = Solve[equation2, x]
```

```
nsolution210 = N[solution2, 10]

equation2/.nsolution210

equation2[[1]]/.nsolution210//Chop

nsolution216 = N[solution2, 16]

equation2/.nsolution216

nsolution217 = N[solution2, 17]

equation2/.nsolution217
```

Seventeen digits is one more than machine accuracy.

Problem 3

```
equations3 = {x^2 y + y == 2, y - 4x == 8};

solution3 = Solve[equations3, {x, y}]//Simplify;

equations3/.solution3//Simplify
```

This last evaluation takes a long time. We are still searching for a similar example involving only polynomials in Version 3 where *Mathematica* can find the answer but can't verify it.

Problem 4

```
equations4 = {a x + b y -      z  ==  3 b,
              x - 4 y - 5 c z  ==  0,
              x + a y -    b z  ==  c};

solution4xyz = Solve[equations4, {x, y, z}]//Simplify

equations4/.solution4xyz//Simplify

solution4abc = Solve[equations4, {a, b, c}]//Simplify

equations4/.solution4abc//Simplify
```

Problem 5

```
equation5 = Sqrt[1 - x] + Sqrt[1 + x] == 3;

solution5 = Solve[equation5, x]

equation5/.solution5//Simplify
```

Note, however, the following anomalous result.

```
Reduce[%]
```

Problem 6

We illustrate the procedure using part (ii). There are two forms of **DSolve**; one solves for **y[x]** and the other for **y** as a pure function.

```
diffEquation = y'[x] - y[x] Tan[x] == Sec[x];

solution1 = DSolve[diffEquation, y[x], x]

diffEquation/.solution1/. D[solution1, x]//Simplify

solution2 = DSolve[diffEquation, y, x]

diffEquation /. solution2 // Simplify
```

Problem 9

These kinds of equations are called Lotka–Volterra systems. They describe things like predator–prey situations. For appropriate choices of the coefficients, the solution tends towards a fixed point and one is interested in stability properties of this fixed point. Here is a short routine to create the equations from a list of the right-hand sides together with the list of variables, the list of initial values, and an iterator to give the range for the desired solution.

```
diffEqSystem =
    {x'[t] == 2 x[t] - x[t] y[t] - 2 x[t]^2,
     y'[t] == y[t] - (1/2) x[t] y[t] - y[t]^2,
     x[0] == 2, y[0] == 2};
```

DSolve doesn't work.

```
DSolve[diffEqSystem, {x, y}, t]
```

Instead, we have to find a numerical solution and then use `ParametricPlot` to see what the answer looks like.

```
systemSol = NDSolve[diffEqSystem,
                    {x[t], y[t]}, {t, 0, 100}]

ParametricPlot[Evaluate[{x[t], y[t]}/.systemSol],
              {t, 0, 100}, PlotRange -> All];
```

Add an `Epilog` to give points on the curve at equal time intervals to show convergence to a fixed point. Also fix the plot range and the axes. Equibrium is actually reached for **t** ≈ 6.

```
ParametricPlot[
  Evaluate[{x[t], y[t]}/.systemSol], {t, 0, 6},
  PlotRange -> All, AxesOrigin -> {0, 0},
  Epilog ->
     {PointSize[0.02],
      Map[Point,
          Table[{x[t], y[t]}/.systemSol[[1]],
               {t, 0, 6}]]}];
```

Problem 11

```
pascaltrianglerow[n_] := Table[Binomial[n, i], {i, 1, n}]
```

Alternatively:

```
pascalTriangleRow[n_] := Binomial[n, Range[0, n]]
```

```
pascalTriangleRow[6]
```

`pascalTriangleRow` is itself `Listable` since it is built from `Listable` ingredients.

```
pascalTriangle[n_] :=
    TableForm[pascalTriangleRow[Range[0, n]],
            TableSpacing -> {1, 1}]
```

```
pascalTriangle[10]
```

Problem 12

There are several ways to construct the "complete the square" operation.

Solution 1

```
aa[expr_, x_] := Coefficient[expr, x, 2];
bb[expr_, x_] := Coefficient[expr, x, 1];
cc[expr_, x_] := Coefficient[expr, x, 0];

completeTheSquare[expr_, x_] :=
    aa[expr, x] (x + bb[expr, x]/(2 aa[expr, x]))^2 +
    cc[expr, x] - bb[expr, x]^2 / (4 aa[expr, x])

expr1 = 2 x^2 + 3 x + 4;

completeTheSquare[expr1, x]

expr2 = a x^2 + b x + c;

completeTheSquare[expr2, x]
```

There is no reasonable way (i.e., not using string operations) to get *Mathematica* to reverse the order in which it displays this result.

Solution 2

It is better to use local variables for this problem. We haven't discussed them yet, but they occur inside **Module** expressions.

```
Clear[completeTheSquare]

completeTheSquare[expr_, x_] :=
    Module[{a, b, c},
        {c, b, a} = CoefficientList[expr, x];
        a (x + b / (2a))^2 + c - b^2 / (4a)]

completeTheSquare[expr2, x]

Expand[%]
```

Problem 13

The point of the exercise is to check the generalized chain rule for functions of several variables. Here is the straightforward way to define the Jacobian matrix of a list of functions of many variables. See Problem 1 in the Answers to Chapter 5 for a better way.

```
jacobian[fun_List, var_List] :=
        Simplify[Table[D[fun[[i]],var[[j]]],
          {i, 1, Length[fun]}, {j, 1, Length[var]}]]

fun = {x^2 + y^2, -2 x y}; var = {x, y};

jak = jacobian[fun, var]
```

We view this pair of functions in **x** and **y** as a mapping from the x–y-plane to the u–v-plane, and we want to calculate the inverse mapping. The given mapping is not one-to-one and there are four "inverse" functions. The Jacobian matrix of such a mapping is the best linear approximation to the mapping at any given point; i.e., at some point (a, b) in the domain of the transformation, the evaluation of the Jacobian matrix at (a, b) is the matrix of the best linear approximation to the mapping at the point (a, b).

```
invexp = Solve[fun == {u, v}, {x, y}] // Simplify
```

We can check that the third pair of functions here is an actual inverse transformation by showing that

```
x(u(x, y), v(x, y)) = x,
y(u(x, y), v(x, y)) = y,
u(x(u, v), y(u, v)) = u,
v(x(u, v), y(u, v)) = v.
```

The first pair of equations are verified as follows:

```
{x, y} /. invexp[[1]] /. Thread[{u, v} -> fun] //
        Simplify // PowerExpand // Expand // PowerExpand
```

Here, `{x, y} /. invexp[[1]]` gives x(u, v} and y{u, v} for the third pair of functions above. Following this by the substitution `Thread[{u, v} -> fun]` gives x(u(x, y), v(x, y)) and y(u(x, y), v(x, y)). The output of this computation shows that the result eventually simplifies to `{x, y}`. The other check is done analogously. In this case, a simple `Simplify` suffices.

```
{u, v} /. Thread[{u, v} -> fun] /. invexp[[1]] // Simplify
```

Now we concentrate on the inverse mapping given by the third solution and call it **invFun**. The variables for this are **u** and **v**.

```
invFun = ( {x, y} /. invexp[[1]] );
invVar = {u, v};
```

Now construct the Jacobian of **invFun** in terms of **invVar**.

```
invJac = jacobian[invFun, invVar]
```

Our original Jacobian matrix **jak** is in terms of the variables **x** and **y**, while **invJac** is in terms of **u** and **v**. In order to check the generalized chain rule, we have to express both of them in terms of **x** and **y**. So define **jak'** to be **invJac** with **u** and **v** replaced by their values in terms of **x** and **y** from the original mapping **fun**. A lot of simplification is required to reduce **jak'** to its simplest form. This is done interactively until a reasonable form is arrived at.

```
jak' = invJac /. Thread[{u, v} -> fun] //
        Simplify // PowerExpand //
        ExpandAll // PowerExpand // Simplify

Simplify[jak . jak'] // TableForm
```

This result shows that the Jacobian matrix of a composition of transformations (in this case equal to the identity transformation so its Jacobian matrix is an identity matrix) is equal to the matrix product of the Jacobian matrices of the factors.

Pictures of the transformation $u = x^2 + y^2$, $v = -2xy$.

We construct a graphics function which constructs the image under a transformation of a rectangular grid in the x–y-plane. This is adapted from Roman Maeder's Complex Map construction.

```
cartesianMap[expr_, {x_, x0_, x1_, dx_},
                {y_, y0_, y1_, dy_}, options___] :=
   Module[{coords, lines},
      coords = Table[N[expr],
                    {x, x0, x1, dx}, {y, y0, y1, dy}];
      lines = Map[ Line, Join[coords, Transpose[coords]]];
      Show[ Graphics[lines],
           AspectRatio->Automatic, Axes->Automatic,
           options] ]
```

Here is a picture of the images of the upper half-plane and the lower half-plane.

```
Show[GraphicsArray[
     {cartesianMap[{x^2 + y^2, -2 x y},
                    {x, -4, 4, 0.2}, {y, 0.1, 3.6, 0.2},
                    DisplayFunction -> Identity],
      cartesianMap[{x^2 + y^2, -2 x y},
                    {x, -4, 4, 0.2}, {y, -3.7, -0.1, 0.2},
                    DisplayFunction -> Identity]}],
     DisplayFunction -> $DisplayFunction];
```

Discussion: The mapping f[x, y] is singular along the lines x = y and x = −y as one sees because the Jacobian is zero there. The mapping folds the first quadrant along the line x = y and covers the indicated region in the fourth quadrant twice. The second quadrant is mapped onto the first quadrant in the same way. Finally, the lower half-plane is mapped just like the upper half-plane, so every point in the image is covered four times. That's why there are four "inverse" functions.

Problem 14 i)

```
TableForm[Table[
            {n, AccountingForm[N[E^(Pi Sqrt[163]), n]]},
            {n, 30, 32}],
        TableHeadings -> {None, {"Precision", "Value"}},
        TableSpacing -> {1, 3}]
```

Thus, 31 digits of precision are required to show that this number is not an integer.

Problem 14 ii)

```
Infinity - Infinity

Infinity/Infinity

0^Infinity

1^Infinity
```

Problem 14 iii)

```
eqn = Sqrt[x] == 1 - x;

sol = Solve[eqn, x]

eqn/.sol//Simplify

Map[#^2&, %, {2}]//Simplify
```

Problem 14 iv)

```
Integrate[1/x^2, {x, -3, 2}]
```

Mathematica 3.0 handles this correctly.

```
Limit[Integrate[1/x^2, {x, -3, t}],
    t -> 0, Direction -> 1] +
Limit[Integrate[1/x^2, {x,  t, 2}],
    t -> 0, Direction -> -1]
```

Chapter 5: Answers

Problem 1

```
jacobian[fun_List, var_List] :=
    Simplify[Outer[D, fun, var]]

fun = {x^2 - y^2, - 2 x y};
var = {x, y};
newvar = {u, v};
```

Answer 1

```
jak = jacobian[fun, var];
inveqs[fun_, var_, newvar_] :=
    Solve[fun == newvar, var]//Simplify;
newfuns[n_] := inveqs[fun, var, newvar][[n]];
invfun[n_] := ({x, y}/.newfuns[n]);
invjak[n_] := jacobian[invfun[n], newvar]//Simplify

Timing[
   Table[
      (jak/.newfuns[n]).(invjak[n])//Simplify//Together,
            {n, Length[inveqs[fun, var,newvar]]}] //
         TableForm]
```

Answer 2

The following is much faster. The only difference is that it calculates the inverse functions immediately, whereas the first version calculates them four times.

```
jak = jacobian[fun, var];
inveqs[fun_, var_, newvar_] :=
    Solve[fun == newvar, var]//Simplify;
newfuns = inveqs[fun, var, newvar];
invfun[n_] := ({x, y}/.newfuns[[n]]);
invjak[n_] := jacobian[invfun[n], newvar]//Simplify
```

```
Timing[
  Table[
    (jak/.newfuns[[n]]).(invjak[n])//Simplify//Together,
         {n, Length[newfuns]}]// TableForm]
```

Problem 2

```
pascal[n_] := Binomial[n, Range[1, n - 1]]

Attributes[gcd] = {Listable};

gcd[n_] := Apply[GCD, pascal[n]]

Thread[{Range[100], gcd[Range[100]]}]

Prime[Range[50]] == gcd[Prime[Range[50]]]
```

Examining the table, we conjecture that the gcd of the p^nth row is also p and the gcd of the nth row where n has at least two prime factors is 1. These are somewhat harder to prove. We check the result for prime powers for a few primes.

```
Timing[And@@Flatten[
     Outer[(gcd[Prime[#1]^#2] == Prime[#1])&,
              Range[4], Range[4]]]]
```

We can check all three conjectures (which really are only two) in one step for values up to 300 in a reasonably short time.

```
Timing[And@@Table[If[Length[FactorInteger[n]] > 1,
            (gcd[n] == 1),
            gcd[n] == FactorInteger[n][[1, 1]]],
        {n, 2, 300}]]
```

Problem 4

We also want b[n, r] = f[n - r + 1, r] (turn this arround to define f[m, r] = b[m + r - 1, r]. This rotates the table by 45 degrees.

```
f[m_, r_] := Binomial[m + r - 1, r]
```

Pascal's original triangle looked as follows:

```
TableForm[Table[f[i, j], {i, 1, 12}, {j, 0, 11 - i}],
TableSpacing->{1, 1}]
```

Pascal's corollary 4 says that each entry is equal to all the sum of all the entries to the north-west of it plus 1.

```
cor4[m_, r_] :=
    f[m, r] == Sum[Sum[f[i, j], {j, 0 , r - 1}],
                   {i, 1, m - 1}]+1

Timing[
  And@@Map[And@@#&,
           Table[cor4[p, q], {p, 1, 25}, {q, 1, 25}]]]
```

Problem 5

To orthogonalize a list of vectors we first of all need to be able to project a vector onto another vector. Thus we construct a function to calculate the projection **projection[a, v]** of a vector **a** on a vector **v**.

```
projection[a_, v_] := ((a . v) / (v . v)) v ;
```

The general procedure of the Gram–Schmidt method applied to a list $v = \{v_1, v_2, v_3\}$ of vectors is

```
orthogonalize1[vectors_] :=
  {vectors[[1]],
   vectors[[2]] -
       projection[vectors[[2]], vectors[[1]] ],
   vectors[[3]] -
       projection[vectors[[3]], vectors[[1]] ] -
         projection[vectors[[3]],
           vectors[[2]] -
             projection[vectors[[2]], vectors[[1]]]]}


vects1 = {{1, 2, 3}, {2, -3, -4}, {-1, 5, 2}};

newvects1 = orthogonalize1[vects1]
```

A solution for n-dimensional space

This only works for three vectors in 3-dimensional space. Here is a recursive procedure to calculate this for any number of vectors, using the **Sum** function.

```
newvectors[i_, vectors_] :=
    vectors[[i]] -
        Sum[projection[vectors[[i]], newvectors[j, vectors]],
            {j, i - 1}]

orthogonalize2[vectors_] :=
    Table[newvectors[i, vectors], {i, Length[vectors]}]

orthogonalize2[vects1]

vects2 = {{1, 2, 3, 4}, {2, -3, -4, 5},
          {-1, 5, 2, -4}, {-2, -3, 4, 2}};

newvects2 = orthogonalize2[vects2]
```

Normalization

```
length[vector_] := Sqrt[vector . vector]

normalize[vectors_] :=
    Table[vectors[[i]] / length[vectors[[i]]],
          {i, Length[vectors]}]

orthonormvects1 = normalize[newvects1]
```

If this really is an orthonormal basis, then this matrix must be an orthogonal matrix. But that's easy to check, since then its transpose must be its inverse.

```
Transpose[orthonormvects1] == Inverse[orthonormvects1]

orthonormvects2 = normalize[newvects2]

Transpose[orthonormvects2] == Inverse[orthonormvects2]
```

Chapter 6: Answers

Problem 1

```
jacobian[fun_List, var_List] :=
    Simplify[Outer[D, fun, var]];
```

```
fun = {x^2 - y^2, - 2 x y};
var = {x, y}; newvar = {u, v};

jak = jacobian[fun, var]

newfuns = Solve[fun == newvar, var]//Simplify;

invfun = {x, y}/.newfuns;

invjaks = matrices@@Map[jacobian[#, newvar]&, invfun]//Simplify;

newjaks = matrices@@(jak/.newfuns);

List@@Map[Together,
            Thread[Dot[newjaks, invjaks], matrices] //
                Simplify // Together, {3}] // TableForm
```

Problem 2

The basic construction

```
newton[expr_, {x_, x0_, n_}] :=
   Nest[
     Evaluate[Simplify[x - expr/D[expr, x]]/. x -> #]&,
     N[x0], n]

newtonList[expr_, {x_, x0_, n_}] :=
   NestList[
     Evaluate[Simplify[x - expr/D[expr, x]]/. x -> #]&,
     N[x0], n]

newton[x^2 - 3, {x, 1.0, 10}]

newtonList[x^2 - 3, {x, 1.0, 10}]
```

Note that **Nest** and **NestList** require pure functions as their first arguments. We have achieved this by substituting **#** for **x** in the formula and appending an **&**.

The picture

```
newtonPicture[expr_, {x_, xmin_, xmax_}, {x0_, n_}] :=
    Show[Plot[expr, {x, xmin, xmax},
            DisplayFunction ->Identity],
        ListPlot[
          Flatten[
            Map[{{#, 0}, {#, expr/.x -> #}}&,
                newtonList[expr, {x, x0, n}]],
            1],
          PlotJoined -> True,
          PlotRange -> All,
          PlotStyle -> {Thickness[0.008]},
          DisplayFunction -> Identity],
        DisplayFunction -> $DisplayFunction]

newtonPicture[Cos[x^3], {x, 0.8, 1.5}, {.8788, 6}];
```

The fixed point construction

```
newton[expr_, {x_, x0_}] :=
    FixedPoint[
      N[Simplify[x - expr/D[expr, x]]]/. x -> #]&, x0]

newton[x^2 - 3, {x, 1.0}]

%^2 - 3
```

Finally, define a version of **newton** with three arguments to give an option to **FixedPoint-List**. (Note that this same optional argument could also be given to **FixedPoint**.)

```
newtonList[expr_, {x_, x0_}, opt___] :=
    FixedPointList[
      Evaluate[(x - expr/D[expr, x])/. x -> #]&, x0, {opt}]

newtonList[x^2 - 3, {x, 1.0},
    SameTest -> (Abs[#1 - #2] < 10^-3 &)]

newtonList[x - Cos[x], {x, 0.5},
        SameTest -> (Abs[#1 - #2] < 10^-5 &)]

Last[%] - Cos[Last[%]]
```

Another solution

Here is a different way to organize this construction that makes clear that a certain process is being repeated and gives us a function that can easily be converted to a pure function in `newton1`.

```
oneStepNewton[f_, {x_, x0_}] :=
    Simplify[ x - f/D[f, x] ] /. x -> x0

newton1[f_, {x_, x0_, n_}] :=
    NestList[oneStepNewton[f, {x, #}]&, x0, n]

newton1[x^2 - 3, {x, 1.0, 10}]
```

Different form of the answer

```
newtonSub[f_, {x_, x0_}] :=
  {x -> FixedPoint[oneStepNewton[f, {x, #}]&, N[x0]]}

expr1 = x^2 + x^3  - 13;

newtonSub[expr1, {x, 2}]
```

Problem 3

```
Fold[(#1/(1 + #2))&, a, {b, c, d}]

Fold[(#2/(1 + #1))&, a, {b, c, d}]
```

The only problem is that the arguments are in the wrong order and **a** is treated differently.

```
continuedFraction[list_List] :=
    Fold[(#2/(1 + #1))&,
        First[Reverse[list]],
        Rest[Reverse[list]] ]

continuedFraction[{a, b, c, d, e, f}]
```

If numbers rather than symbols are used, then *Mathematica* insists on evaluating the expression.

```
continuedFraction[{1, 1, 1, 1, 1, 1, 1}]
```

```
numfr =
    continuedFraction[Map[ToString, {1, 1, 1, 1, 1, 1, 1}]]
```

Another solution

```
continuedFraction1[list_List] :=
    Fold[(#2/(1 + #1))&, 0, Reverse[list] ]

continuedFraction1[{a, b, c, d, e, f}]
```

Problem 5

The projection function from the answers to Exercise 5 in Chapter 6 is easily modified to work with an arbitrary inner product.

```
projection[a_, v_, innerProduct_] :=
    (innerProduct[a, v] / innerProduct[v, v]) v;

newvectors[i_, vectors_, innerProduct_] :=
    vectors[[i]] -
      Sum[projection[vectors[[i]], newvectors[j, vectors],
                    innerProduct],
         {j, i - 1}];

orthogonalize2[vectors_, innerProduct_] :=
    Table[newvectors[i, vectors, innerProduct],
    {i, Length[vectors]}];
```

However, both of these violate the fundamental dictum, so they have to be replaced. A certain amount of reorganization is required to get satisfactory functional programs. First, we separate out the projection of a vector on a sum of orthogonal vectors as a new operation.

```
multiProjection[a_, basis_, innerProduct_] :=
    Apply[Plus, Map[projection[a, #, innerProduct]&, basis]]
```

The idea of the Gram–Schmidt process is to start with the empty list of vectors and successively feed in new vectors from the given list, modifying them and building up a list of orthogonal vectors. This sounds just like **Fold**. The problem is to figure out how to use it. Here is an elegant solution from John Novak of WRI.

```
orthogonalize[vectors_, innerProduct_] :=
  Fold[Join[#1,
      {Chop[#2 - multiProjection[#2, #1, innerProduct]]}]&,
      {}, vectors];
```

The action of **Fold** is to build up the basis by folding one vector at a time from the given list of vectors into the basis until there are no more vectors left. The picture is

$$basis <- vectors.$$

At the beginning, **basis** is empty. At each step one vector is removed from **vectors** and, in a suitably modified form, added to **basis**. At the end, **vectors** is empty and **basis** is the desired orthogonal basis. **Chop** is added to take care of vectors with real entries, since on such vectors **gramSchmidt** can fail due to tiny spurious components. We still have to worry about normalization, but that is easy to rewrite.

```
normalize[list_, innerProduct_] :=
  Map[Expand[# / Sqrt[innerProduct[#, #]]] &, list]
```

Finally, we need an operation to check that a set of vectors is actually orthonormal. The check we used before was **Transpose[vectors] == Inverse[vectors]**. This clearly won't work for a different inner product. We would like to replace it by something like

```
Outer[innerProduct, vectors, vectors] ==
    IdentityMatrix[length[vectors]
```

This doesn't work because **Outer** of matrices produces something of depth 4. (Try it and see.) However, **Distribute** does work for ordinary vectors and an arbitrary inner product. It won't work for functions.

```
orthoNormalQ[vectors_, innerProduct_] :=
  Simplify[
    Distribute[innerProduct[vectors, Transpose[vectors]],
            List, innerProduct, List, innerProduct]] ===
    IdentityMatrix[Length[vectors]]
```

Example: 3-dimensional space with the usual inner product

```
vects1 = {{1, 2, 3}, {2, -3, -4}, {-1, 5, 2}};

orthovects1 = orthogonalize[vects1, Dot]

orthonormvects1 = normalize[orthovects1, Dot]
```

```
orthoNormalQ[orthonormvects1, Dot]
```

Example: 4-dimensional space with a different inner product

The following matrix is positive definite and symmetric.

$$\text{matrix} \;=\; \begin{pmatrix} 8 & 3 & 0 & 0 \\ 3 & 2 & 1 & 2 \\ 0 & 1 & 2 & 2 \\ 0 & 2 & 2 & 14 \end{pmatrix};$$

Use it to define an inner product.

```
innerProduct4[v_List, w_List] := v . matrix . w;
```

Orthonormalize the four standard unit vectors with respect to this new inner product.

```
newvects = orthogonalize[IdentityMatrix[4], innerProduct4]

newbasis = normalize[newvects, innerProduct4]

orthoNormalQ[newbasis, innerProduct4]
```

Example: Legendre polynomials

In this example, our "vectors" are functions defined on the interval $-1 \le x \le 1$, and the inner product is given by integrating the product of the functions over this interval. Our test example consists of the powers of x.

$$\text{legendre}[f\_, \; g\_] := \int_{-1}^{1} f \; g \; dx$$

```
powers = {1, x, x^2, x^3, x^4};

legendrePowers = orthogonalize[powers, legendre]//Expand

notlegendrePolys = normalize[legendrePowers, legendre]
```

A check that these are orthonormal can use **Outer** as suggested above.

```
Outer[legendre, notlegendrePolys, notlegendrePolys] ==
    IdentityMatrix[5]
```

This does not give the Legendre polynomials because they are not usually normalized by making their length equal to one, but rather by making their value at the point 1 equal to 1.

We can achieve this by the small trick of defining a new "inner product" that isn't really an inner product.

```
atone[z_, w_] := z^2 /. x -> 1

legendrePolys = normalize[legendrePowers, atone]//Together

Plot[Evaluate[legendrePolys], {x, -1, 1}];
```

Note: these polynomials can be given by the formula

```
P[n_, x_] := (1/(2^n n!)) D[(x^2 -1)^n, {x, n}]//
                Simplify // Expand

Table[P[n, x], {n, 0, 4}]
```

They are also given by the recursion relations

```
P1[0, x_]   = 1;
P1[1, x_]   = x;
P1[n_, x_] := (1/n) ((2n - 1) x P1[n-1, x] -
                (n - 1) P1[n-2, x])//Expand

Table[P1[n, x], {n, 0, 4}]//Simplify
```

They are also built-in.

```
Table[LegendreP[n, x], {n, 0, 4}]
```

Problem 6

```
{ToCharacterCode[ToString[p]][[1]],
 ToCharacterCode[ToString[z]][[1]]}

applyVarsOnly[fun_, expr_] :=
  MapAt[fun, expr,
    Flatten[
    Map[
      Position[expr, #]&,
      Select[
        Level[expr, {-1}],
          (Not[NumberQ[#]] &&
            112 <=
            ToCharacterCode[ToString[#]][[1]]<=122)&]],
    1]]
```

```
applyVarsOnly[Sin, (3 + a) q + (1 - b x z)^3]
```

Problem 7

i) One can define **foldLeft** in terms of the built-in function **Fold**.

```
foldLeft[f_, list_List, seed_] :=
    Fold[f[#2, #1]&, seed, Reverse[list]]

foldLeft[f, {a, b, c}, d]
```

ii) Alternatively, it can be defined from scratch recursively.

```
foldLeftR[f_, {}, seed_] := seed;

foldLeftR[f_, list_List, seed_] :=
    f[First[list], foldLeftR[f, Rest[list], seed]]

foldLeftR[f, {a, b, c}, d]

{Timing[foldLeft[Plus, Range[250], 0]],
 Timing[foldLeftR[Plus, Range[250], 0]]}
```

Problem 8

```
Position[
  Partition[IntegerDigits[Floor[N[Pi, 1000] 10^1000]],
            6, 1], Table[9, {6}]]

Table[{n, Position[
  Partition[IntegerDigits[Floor[N[Pi, 1000] 10^1000]],
            3, 1],  Table[n, {3}]]}, {n, 9}]//MatrixForm
```

Chapter 7: Answers

Problem 1

```
Map[First,
    Select[Table[{n, Integrate[(1 - 1/u)^(4/3)/u^n, u]},
              {n, -10, 10, 1/3}],
          FreeQ[ #[[2]] , Integrate]& ]]
```

Problem 2

The following modifications are required for the Gram–Schmidt procedure. Two rules are required for the projection function to take care of projecting onto a 0 vector.

```
projection[a_, v_, innerProduct_] :=
    (innerProduct[a, v] / innerProduct[v, v]) v /;
        innerProduct[v, v] =!= 0;
projection[a_, v_, innerProduct_] := 0 v /;
        innerProduct[v, v] == 0;


multiProjection[a_, basis_List, innerProduct_] :=
    Plus@@((projection[a, #, innerProduct])& /@ basis);

orthogonalize[vectors_List, innerProduct_] :=
  Fold[
    Join[#1, {Chop[#2 -
              multiProjection[#2, #1, innerProduct]]}]&,
    {}, vectors];

normalize[vectors_List, innerProduct_]:=
    (Expand[# / Sqrt[innerProduct[#, #]] ] )& /@ vectors;

nozeros[vectors_List, zero_] := DeleteCases[vectors, zero];
```

Example: Too many vectors in 3-dimensional space

```
morevectors = {{1, 2, 3}, {2, -3, -4}, {3, -1, -1},
              {1, -5, -7},{-1, 5, 2}, {6, 2, -8}};

moreorthogonals = orthogonalize[morevectors, Dot]

result = nozeros[moreorthogonals, {0, 0, 0}]
```

```
randomvects =
    Table[{Random[], Random[], Random[]}, {100}];

nozeros[orthogonalize[randomvects, Dot], {0, 0, 0}]
```

Example: Polynomials

$$\text{legendre}[f\_,\ g\_] := \int_{-1}^{1} f\,g\,dx$$

```
morepowers =
 {1, x, x^2, 2 x^2 - 3, x^3, 5 x^3 - 3 x^2 + x,
  x^4, x^4 - x^3};

orthogonalize[morepowers, legendre]//Expand

nozeros[%, 0]
```

Problem 3

The algebraic expression predicate

This solution is based on using rewrite rules recursively.

```
algexpQ[u_ + v_]:= algexpQ[u] && algexpQ[v]
algexpQ[u_ v_]  := algexpQ[u] && algexpQ[v]
algexpQ[u_^v_]  := algexpQ[u] && algexpQ[v]
algexpQ[w_]      := MemberQ[{Symbol, Integer, Rational,
                            Real, Complex}, Head[w]]
```

```
{algexpQ[x^2 + (y + 2)^3],
 algexpQ[x^2 + (Sin[y] + 2)^3],
 algexpQ[(5 x y)^(z + w)],
 algexpQ[Sqrt[5 x y]^(z + w)],
 algexpQ[x^(x^(x^(x^x)))],
 algexpQ[(y + w)^(x + 2)],
 algexpQ[(x + 2 I) (3 + y I)^(5 + 4I)],
 algexpQ[(2x + y) + I (z w + u)],
 algexpQ[Tan[x^2 + y^2]]}
```

The type function

Next, we define the function type by giving conditional rules.

```
type[expr_Symbol]    := -1;
type[expr_Integer]   := 0;
type[expr_Rational]  := 1/2;
type[expr_Real]      := 1;
type[expr_Complex]   := 2;
type[expr_]          := 10 /;
        algexpQ[expr] && FreeQ[expr, Complex];
type[expr_]              := 20 /; algexpQ[expr]
type[expr_]              := Infinity

{type[anything], type[24], type[3/7], type[3.64],
 type[(5 + 3 I)], type[-(x + y z)^(z - 3 w)],
 type[(x + 2 I) (3 + y I)^(5 + 4I)],
 type[Sin[anything] + 4]}
```

Problem 5

Reference: [R. Maeder 2].

The recursive version of Fibonacci

This method uses the usual recursive definition of the Fibonacci numbers.

```
fibr[1] = 1; fibr[2] = 1;
fibr[n_] := fibr[n-1] + fibr[n-2]


fibrValues = Table[{2 m,
              Timing[fibr[2 m]][[1]]/Second}, {m, 1, 12}]

fibrFit = Fit[fibrValues, {1, x, x^2}, x]

fibrTimeToAMillion =
    (fibrFit/. x -> 1000000)/(60 60 24 356) years

fibrPlot = Plot[fibrFit, {x, 0, 24},
            PlotRange -> {{0, 24}, {-1, 10}},
            PlotLabel -> "Recursion",
            Epilog -> {PointSize[0.025], Map[Point, fibrValues]}];
```

Actually, it is known that the theoretical complexity of this algorithm is exponential (see later discussion). One way to find a suitable base is to take the limit of Fibonacci[n + 1] / Fibonacci[n] as n -> ∞, which is easily shown to be the golden ratio.

```
gr = N[GoldenRatio]

grfit = Fit[fibrValues, {1, gr^x}, x]

Show[GraphicsArray[
    {grplot = Plot[grfit, {x, 1, 24}, PlotRange -> All,
                    DisplayFunction -> Identity,
                    PlotLabel -> "grplot"],
     Show[{fibrPlot, grplot}, DisplayFunction -> Identity]}],
    DisplayFunction -> $DisplayFunction];
```

The dynamic programming version of Fibonacci

This version uses the usual recursive definition, but programmed dynamically.

```
fibd[1] = 1; fibd[2] = 1;
fibd[n_] := fibd[n] = fibd[n-1] + fibd[n-2]

fibdValues =
    Table[{100 m, Timing[fibd[100 m]][[1]]/Second},
          {m, 1, 20}]
```

This is essentially constant time since at each stage, 100 more values are calculated. If one tries individual values, then the recursion limit is exceeded exactly at **fibd[256]**. To do this, it is necessary to clear **fibd** before each calculation.

```
Clear[fibd];
fibd[1] = 1; fibd[2] = 1;
fibd[n_] := fibd[n] = fibd[n-1] + fibd[n-2];
fibd[256]

Clear[fibd];
fibd[1] = 1; fibd[2] = 1;
fibd[n_] := fibd[n] = fibd[n-1] + fibd[n-2];
fibd[257]
```

Thus, **fibd** uses one recursion step for each value and so it runs out of space after 256 steps. If one resets the recursion depth, then it will go farther. The correct thing to do is to clear **fibd** and redefine it at each step.

```
$RecursionLimit = 2000;

fibdValues =
  Table[{2^m,
          Clear[fibd];
```

```
            fibd[1] = 1; fibd[2] = 1;
            fibd[n_] := fibd[n] = fibd[n-1] + fibd[n-2];
            Timing[fibd[2^m]][[1]]/Second}, {m, 1, 9}]

    $RecursionLimit = 256;

    fibdFit = Fit[fibdValues, {1, x}, x]

    fibdTimeToAMillion = (fibdFit/.x -> 1000000)/(60 60) hours

    fibdPlot = Plot[fibdFit, {x, 0, 550},
                PlotLabel -> "Dynamic",
                Epilog -> {PointSize[0.025], Map[Point, fibdValues]}];
```

Analysis of recursion vs. dynamic programming

Why does the recursive program give up at a bit over 20, while the dynamic program goes up to 100 with no trouble? Use **Trace** to see how the tree is actually searched.

```
    Trace[fibr[6], fibr]

    Clear[fibd];
    fibd[1] = 1; fibd[2] = 1;
    fibd[n_] := fibd[n] = fibd[n-1] + fibd[n-2];
    Trace[fibd[6], fibd]
```

The definition of the Fibonacci numbers builds a tree of values to be calculated. For **n** = 6, it looks as follows:

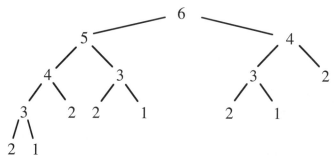

In the recursive version, every node of this tree is visited in a depth-first traversal, going first down the left-hand side, then recursively coming up until it is possible to descend again. The order is {6, 5, 4, 3, 2, 1, 2, 3, 2, 1, 4, 3, 2, 1, 2}, as one sees from the **Trace** of **fibr**. In the dynamic version, again the left-hand side is traversed, but that results in all of the required values being calculated, so only the tops of the rest of the subtrees are visited. The order is {6, 5, 4, 3, 2, 1, 2, 3, 4}. Note that the size of the tree satisfies the recursive equation tree[n] = tree[n - 1] + tree[n - 2] + 1, with tree[1] = tree[2] = 1, so tree[6] = 15. These sizes grow somewhat faster than the Fibonacci numbers and I don't know their limiting ratio, but this at least gives some justification for using the limiting ratio of the Fibonacci numbers as the exponential base in fitting a curve to their timing in the recursive case.

The iteration version of Fibonacci

Here we use a simple iteration repeated n times to calculate the nth Fibonacci number.

```
fibi[n_] :=
    Module[{an1 = 1, an2 = 1},
        Do[{an1, an2} = {an1 + an2, an1}, {i, 3, n}]; an1]

fibiValues = Table[{2^m, Timing[fibi[2^m]][[1]]/Second},
                {m, 1, 16}]

fibiFit = Fit[fibiValues, {1, x, x^2}, x]

fibiTimeToAMillion =(fibiFit/.x -> 1000000)/(60 60) hours

fibiPlot = Plot[fibiFit, {x, 0, 65600},
            PlotLabel -> "Iteration",
            Ticks ->{{{0, "0"}, {20000, "20000"},
                {40000, "40000"},{60000, "60000"}},
                Automatic},
            Epilog -> {PointSize[0.025],
                    Map[Point, fibiValues]}];
```

The symbolic formula version of Fibonacci

See the Maeder article referred to at the beginning for a derivation of these constants and this formula for the Fibonacci numbers.

```
e1 = (1 + Sqrt[5])/2;
e2 = (1 - Sqrt[5])/2;
b1 = (5 + Sqrt[5])/10;
b2 = (5 - Sqrt[5])/10;

fibf[n_] := Simplify[b1 e1^(n - 1) + b2 e2^(n - 1)]

fibfValues = Table[{2^m, Timing[fibf[2^m]][[1]]/Second}, {m, 1,
12}]

fibfFit = Fit[fibfValues, {1, x, x^2}, x]
```

```
fibfTimeToAMillion =
    (fibfFit/.x -> 1000000)/(60 60 24) days

fibfPlot = Plot[fibfFit, {x, 0, 4100},
                PlotLabel -> "Symbolic Function",
                Epilog -> {PointSize[0.025],
                            Map[Point, fibfValues]}];
```

The numeric formula version of Fibonacci

```
N[(1/Sqrt[5])(1 - Sqrt[5])/2]
```

```
{Log[N[1/Sqrt[5]]], Log[N[(1 + Sqrt[5])/2]]}
```

Since `Log[a]` + 1 is the number of digits of `a`, the nth Fibonacci number has at most $n/2$ digits, so it is sufficient to calculate the numerical value to $n/2$ digits of accuracy. (Actually, we will see below that $n/4$ would be sufficient.)

```
fibfn[n_] :=
    Round[N[(1/Sqrt[5]) ((1 + Sqrt[5])/2)^n, Round[n/2]]]
```

The following calculation checks our derivation.

```
Table[fibfn[2^n] - fibi[2^n], {n, 1, 14}]
```

```
fibfnValues =
    Table[{2^m, Timing[fibfn[2^m]][[1]]/Second},
          {m, 1, 16}]
```

```
fibfnFit = Fit[fibfnValues, {1, x, x^2}, x]
```

```
fibfnTimeToAMillion =
    (fibfnFit/.x -> 1000000)/(60 60 24) days
```

```
fibfnPlot = Plot[fibfnFit, {x, 0, 65600},
                PlotLabel -> "Numeric Function",
                Ticks ->{{{0, "0"}, {20000, "20000"},
                {40000, "40000"}, {60000, "60000"}},
                    Automatic},
                Epilog -> {PointSize[0.025],
                            Map[Point, fibfnValues]}];
```

Maeder gives a different analysis and a different algorithm. In the algorithm, a numerical approximation of the nth Fibonacci number is calculated along with the number of its digits. This is increased by 10 and used as the number of digit in the approximation of `Sqrt[5]`.

```
fibfnum[n_] :=
    Module[{digits, approx},
          approx = N[b1 e1^n];
          digits = Ceiling[Log[10, approx]] + 10;
          approx = N[b1, digits] N[e1, digits]^n;
          Round[approx]]
```

For some reason there is a shift in the values of the argument. Thus, `fibfnum[n] = fibfn[n + 1]`.

```
{fibfnum[9], fibfn[10]}

fibfnumValues =
    Table[{2^m, Timing[fibfnum[2^m-1]][[1]]/Second},
          {m, 1, 16}]

fibfnumFit = Fit[fibfnumValues, {1, x, x^2}, x]

fibfnumTimeToAMillion =
    (fibfnumFit/.x -> 1000000)/(60 60) hours

fibfnumPlot = Plot[fibfnumFit, {x, 0, 65600},
              PlotLabel -> "Maeder Function",
              Ticks ->{{{0, "0"}, {20000, "20000"},
              {40000, "40000"}, {60000, "60000"}},
                      Automatic},
              Epilog -> {PointSize[0.025],
                      Map[Point, fibfnumValues]}];
```

The matrix version of Fibonacci

This method uses **MatrixPower** to calculate the Fibonacci numbers. It is much faster than the other methods.

```
mat = {{1, 1}, {1, 0}};

fibm[n_] := MatrixPower[mat, n-1][[1, 1]]

fibmValues =
    Table[{2^p, Timing[fibm[2^p]][[1]]/Second}, {p, 1, 16}]

fibmFit = Fit[fibmValues, {1, x, x^2}, x]
```

```
fibmTimeToAMillion = (fibmFit/.x -> 1000000)/(60 60) hours

fibmPlot = Plot[fibmFit, {x, 0, 66000},
              PlotLabel -> "Matrix",
              Ticks ->{{{0, "0"}, {20000, "20000"},
              {40000, "40000"}, {60000, "60000"}},
                    Automatic},
              Epilog -> {PointSize[0.025],
                          Map[Point, fibmValues]}];
```

This method is fast enough to make it easily possible to calculate the millionth Fibonacci number on a PowerPC 8500.

```
{2^20, Timing[N[fibm[2^20]]]}
```

The time it takes is very close to the calculated time.

```
51.6833 / (60 60) hours
```

These results also show that the length of the nth Fibonacci number is smaller than $n/4$.

Comparison

The theoretical complexity of the different methods varies from exponential to apparently n^2 and there is a vast difference in the value of Fibonacci[n] that can be calculated in approximately one minute.

```
Show[GraphicsArray[{{fibrPlot, fibfPlot},
                    {fibdPlot, fibiPlot},
                    {fibfnPlot, fibfnumPlot},
                    {fibmPlot}}]];
```

Here is a picture of the last six methods. To get a better display, change all of the individual plots to include their name as a suitably located graphics element rather than a **PlotLabel**.

```
Show[{fibfPlot, fibdPlot, fibiPlot,
      fibfnPlot, fibfnumPlot, fibmPlot}];
```

Problem 6

```
maximafun[list_List] :=
    Union[Rest[FoldList[Max, -Infinity, list]]]
```

```
list = {-1.4, 3.2, 2.5, -5, 2.6, 7.3, 5, 3, 8, 6, 4};

maximafun[list]
```

Here is the rule based operation.

```
maxima[list_List] :=
    list //. {a___, x_, y_, b___} /; y <= x -> {a, x, b}
```

Construct a test to compare timings.

```
test[n_] := Table[Random[Integer, 100], {n}];

experiment =
 Module[{tt},
    Table[(tt = test[2^m]);
          {{2^m, Timing[maxima[tt];]},
           {2^m, Timing[maximafun[tt];]}},
          {m, 5, 10}]]
```

Extract the two sets of data.

```
listing[i_] := Map[{#[[i, 1]], #[[i, 2, 1]]/Second}&,
                     experiment]
```

Fit curves to the data.

```
fit[i_] := Fit[listing[i], {1, x, x^2}, x]
```

Plot everything together.

```
Show[Join[
     Map[Plot[Evaluate[fit[#]], {x, 32, 1024},
             PlotLabel -> "maxima test",
             Ticks ->{{{32, "32"}, {64, "64"}, {128, "128"},
                       {256, "256"}, {512, "512"},
                       {1024, "1024"}},
                        Automatic},
                 DisplayFunction -> Identity]&, {1, 2}],
     Map[ListPlot[listing[#],
                   PlotStyle -> {PointSize[0.025]},
                   DisplayFunction -> Identity]&, {1, 2}]],
     DisplayFunction -> $DisplayFunction,
     PlotRange -> All];
```

Chapter 8: Answers

Problem 1

The definition of deal

N elements are to be selected at random without replacement from a given population. A procedural version of the function can be written with either a **While** loop or a **Do** loop. We prefer the latter.

The procedural definition

```
dealProc[population_List, n_Integer]:=
  Module[{list = population, selection = {}, rand },
    Do[ rand = Random[Integer, {1, Length[list]}];
        AppendTo[selection, list[[rand]]];
        list = Delete[list, rand],
        {n}
    ]; selection] /; 0 <= n <= Length[population]
```

Rewrite rule definition 1

A student found the following elegant recursive rewrite rule version. This version, in effect, repeats dealing one element from **population** n times; i.e., it calls itself **n** times.

```
dealRew1[population_List, 0] := {};
dealRew1[population_List, n_Integer?Positive] :=
  Module[
      {rand = Random[Integer, {1, Length[population]}]},
      PrependTo[dealRew1[Delete[population, rand], n - 1],
                population[[rand]] ]
      ]/; n <= Length[population]
```

Rewrite rule definition 2

Here is another rewrite rule version that works in a different way.

```
oneStepRule = {hand_List, deck_List} :>
  Module[{choice = Random[Integer, {1, Length[deck]}]},
      {Append[hand, deck[[choice]]], Delete[deck, choice]}];
```

Next, we need a function to apply this rule.

```
oneStep[{hand_List, deck_List}]:=
  {hand, deck} /. oneStepRule
```

We can now use this to apply the rule a fixed number of times.

```
dealRew2[population_List, n_Integer] :=
    dealRew2[{}, population, n][[1]];
dealRew2[hand_List, deck_List, n_Integer] :=
    Nest[oneStep[#]&, {hand, deck}, n]
```

An attempted functional definition

Here is another version that works in a completely different way. It is more in the spirit of the fundamental dictum of functional programming. However, there seems to be no way to avoid generating a sequence of **n** random elements from scratch somewhere in the program.

```
dealRandom[population_List, n_Integer] :=
  Module[{i = n - 1, hand = {}},
    While[Length[hand] < n,
        i++;
        hand = population[[
          Union[
            Table[Random[Integer,
                         {1, Length[population]}],
                  {i}]] ]]];
        hand = Take[hand, n] ]
```

A one-liner

The following one-liner was posted to the mathgroup mailbox by Richard Gaylord, in response to our challenge to find such a version. Note that it also sorts the output, which makes it unusable in the final game deal function below.

```
dealFun[population_List, n_Integer] :=
    Complement[population,
        Nest[Delete[#, Random[Integer, {1, Length[#]}]]&,
            population, n]] /;
            0 <= n <= Length[population]
```

Examples

```
dealProc[{a, b, c, d, e, f, g, h, i, j, k, l, m, n}, 6]
```

The elements of **population** don't have to be all different. The following had to be run several times to get a nice output.

```
dealProc[{a, b, a, b, a, b}, 2]
```

Next we try some larger values and time them.

```
{Timing[dealRew1[Range[200], 60];],
 Timing[dealRew2[Range[200], 60];],
 Timing[dealProc[Range[200], 60];],
 Timing[dealFun[ Range[200], 60];]}
```

The performance of **dealRandom** is harder to measure, since it depends on how far one has to go to get enough different terms. We try averaging 10 runs.

```
Apply[Plus,
    Table[Timing[dealRandom[Range[200], 60];], {10}]]/10
```

This compares very favorably with **dealProc**, and sometimes it will have been significantly faster. However, if **n** approaches the size of **population**, then this method becomes very slow.

```
Timing[dealRandom[Range[50], 49];]
```

If all the entries in a population are dealt, then in effect a random permutation of the entries has been generated, providing the output is not sorted. Thus all methods except **dealRandom** and **dealFun** will generate such a random permutation.

The bridge deal

Several versions are given, starting with a very primitive one and progressing to a fairly nice one. The problem is to figure out how to combine the deal function in a efficient way to distribute the desired hands of cards.

The first version

Since **dealFun** is clearly the fastest procedure, we use it for the next few definitions and rebaptize it **deal**.

```
deal[population_List, n_Integer] :=
    Complement[population,
        Nest[Delete[#, Random[Integer, {1, Length[#]}]]&,
            population, n]] /;
            0 <= n <= Length[population]
```

First create a standard deck of cards.

```
deck = Flatten[Outer[List, {c, d, h, s},
                    Join[Range[2, 10], {J, Q, K, A}]], 1]
```

Try dealing a sample hand of 13 cards.

```
deal[deck, 13]
```

Note that this is automatically sorted. Here is our first try at defining **bridgeDeal**.

```
Clear[bridgeDeal]
```

```
bridgeDeal[deck_] :=
  Module[{hand1, hand2, hand3, hand4},
    hand1 = deal[deck, 13];
    hand2 = deal[Complement[deck, hand1], 13];
    hand3 = deal[
        Complement[deck, Join[hand1, hand2]], 13];
    hand4 = Complement[deck, Join[hand1, hand2, hand3]];
    TableForm[{hand1, hand2, hand3, hand4},
    TableHeadings -> {{"hand1","hand2","hand3", "hand4"},
                        None},
    TableSpacing -> {1, 1}] ]
```

```
bridgeDeal[deck]
```

A better version

Here is a better version where *Mathematica* does more of the work.

```
Clear[bridgeDeal]
```

```
bridgeDeal[deck_] :=
  Module[{hand},
    hand[i_] := hand[i] =
        deal[Complement[deck,
            Join[Sequence@@Table[hand[j], {j, i-1 }]]], 13];
    TableForm[Table[hand[i], {i, 1, 4}],
```

```
        TableHeadings ->
            {Table["hand["<>ToString[i]<>"]", {i, 4}],
             None},
        TableSpacing -> {1, 1}] ]

  bridgeDeal[deck]
```

A more general solution

The following generalization deals a given number of cards to a given number of players from a given deck using essentially the same strategy as the preceding version.

```
dealCards[deck_,
          numberOfPlayers_Integer?Positive,
          cardsPerPlayer_Integer?Positive] :=
  Module[{hand},
     hand[i_] := hand[i] =
         deal[Complement[deck,
                Join[Sequence@@
                     Table[hand[j], {j, i-1 }]]],
                 cardsPerPlayer];
     TableForm[
       Table[hand[i],{i, numberOfPlayers}],
          TableHeadings ->
              {Table["hand["<>ToString[i]<>"]",
                    {i, numberOfPlayers}],
               None},
          TableSpacing -> {1, 1}]]
```

Here is a sample poker deal to six players.

```
  dealCards[deck, 6, 5]
```

A still better solution

In the previous versions, the required number of cards are dealt in a block to each player. The new version here is based on nesting the operation of dealing one round of cards to each player. A **Transpose** operation is then required to see the cards dealt to each player; i.e., the view of the player is the transpose of the view of the dealer. Because the fastest version of **deal** sorts the cards that are dealt, it cannot be used here. We replace it by the procedural version. Also, the cards are sorted according to the usual ranking of cards.

```
  deal[population_List, n_Integer]:=
     Module[{list = population, selection = {}, rand },
```

```
           Do[rand = Random[Integer, {1, Length[list]}];
              AppendTo[selection, list[[rand]] ];
              list = Delete[list, rand ], {n}];
           selection] /; 0 <= n <= Length[population]


   oneRound[{alreadyDealt_List, remainingCards_List},
              noOfPlayers_Integer]:=
      Module[{round = deal[remainingCards, noOfPlayers]},
         {Append[alreadyDealt, round],
          Complement[remainingCards, round]}];


   gameDeal[deck_List, noOfPlayers_Integer, noOfCards_Integer] :=
      Map[Sort[#, cardOrderQ]&,
          Transpose[
              Nest[oneRound[#, noOfPlayers]&,
                   {{}, deck}, noOfCards][[1]] ] ] /;
          0 <= noOfPlayers noOfCards <= Length[deck]
```

The sorting routine is lexicographical in terms of suits and values.

```
   suits = {s, h, d, c};
   values = Join[Range[2, 10], {J, Q, K, A}];


   suitOrderQ[card1_, card2_] :=
      Position[suits, card1[[1]] ][[1, 1]] <
          Position[suits, card2[[1]] ][[1, 1]]


   valueOrderQ[card1_, card2_] :=
      Position[values, card1[[2]] ][[1, 1]] <
          Position[values, card2[[2]] ][[1, 1]]


   cardOrderQ[card1_, card2_] :=
      suitOrderQ[card1, card2] ||
      (card1[[1]] === card2[[1]] && valueOrderQ[card1, card2])
```

We also want to display the deal in a nice form.

```
   displayDeal[deck_List, noOfPlayers_Integer, noOfCards_Integer]:=
      TableForm[gameDeal[deck, noOfPlayers, noOfCards],
         TableHeadings -> {Table["hand["<>ToString[i]<>"]",
                                 {i, noOfPlayers}],
                           None},
         TableSpacing -> {1, 1}] /;
       0 <= noOfPlayers noOfCards <= Length[deck]
```

First try dealing the cards for a bridge game.

```
displayDeal[deck, 4, 13]
```

Now try dealing the cards for a poker game.

```
displayDeal[deck, 6, 5]
```

If we try to deal too many cards, it doesn't work.

```
displayDeal[deck, 4, 14]
```

A one-liner for gameDeal

Modify the ordering of cards to reflect the observation that the ordering of suits is the reverse of canonical ordering. Then define different card orderings for different games.

```
values = Join[Range[2, 10], {J, Q, K, A}];

valueOrderQ[card1_, card2_] :=
   Position[values, card1[[2]] ][[1, 1]] <
        Position[values, card2[[2]] ][[1, 1]]

bridgeOrderQ[card1_, card2_] :=
   (!SameQ[card2[[1]], card1[[1]]] &&
        OrderedQ[{card2[[1]], card1[[1]]} ]) ||
   (SameQ[card1[[1]], card2[[1]] ] &&
        valueOrderQ[card1, card2])

pokerOrderQ[card1_, card2_] :=
   (SameQ[card2[[2]], card1[[2]]] &&
        OrderedQ[{card2[[1]], card1[[1]]} ]) ||
   (!SameQ[card1[[2]], card2[[2]] ] &&
        valueOrderQ[card1, card2])
```

The following operation does everything.

```
gameDeal[deck_, noOfPlayers_, noOfCards_, gameOrderQ_]:=
  TableForm[
    Map[Sort[#, gameOrderQ]&,
        Partition[deal[deck, noOfPlayers noOfCards], noOfCards]],
    TableHeadings -> {Map[hand, Range[noOfPlayers]], None},
    TableSpacing -> {1, 1}] /;
    0 <= noOfPlayers noOfCards <= Length[deck]
```

```
gameDeal[deck, 4, 13, bridgeOrderQ]

gameDeal[deck, 6, 5, pokerOrderQ]
```

Problem 2

The algebraic expression predicate

There are many ways to define this predicate. We give seven of them and compare their speeds.

Solution 1

```
algheads[1] = {Plus, Times, Power,
                Integer, Rational, Real, Complex, Symbol};

algexp[1][exp_]:=
  MemberQ[algheads[1], Head[exp]] &&
    If[Length[exp] > 0, And@@Map[algexp[1], List@@exp],
        True]
```

Solution 2

Separate the allowed heads into the heads for leaves and the heads for internal nodes in the tree structure. Then use a **Which** clause to separate out the different cases.

```
algheads[2]  = {Plus, Times, Power};

algleaves[2] =
      {Symbol, Integer, Rational, Real, Complex};

algexp[2][exp_] :=
      Which[
        Length[exp] === 0,
            MemberQ[algleaves[2], Head[exp]],
        MemberQ[algheads[2], Head[exp]],
            And@@Map[algexp[2], List@@exp],
        Head[exp] === Rational, True,
            Head[exp] === Complex, True,
        True, False]
```

Solution 3

This solution is a more organized way to do the same thing.

```
algexp[3][exp_] :=
    Which[
        Map[(Head[exp] === #)&, Plus || Times || Power],
            And@@Map[algexp[3], List@@exp],
        (Head[exp] === Symbol) || NumberQ[exp], True,
        True, False]
```

Solution 4

This solution is based on a **Switch** clause that just looks at the head of the expression. It uses | to combine patterns.

```
algexp[4][exp_] :=
    Switch[exp,
        (_?NumberQ | _Symbol),
            True,
        (_Plus | _Times | _Power),
            And@@Map[algexp[4], List@@exp],
        _,   False]
```

Solution 5

This is another version using **Switch**.

```
algexp[5][exp_] :=
    Switch[
        Head[exp],
        (Symbol | Integer | Real | Rational | Complex),
            True,
        (Plus | Times | Power),
            And@@Map[algexp[5], List@@exp],
        _,   False]
```

Note that there does not seem to be any way to form Boolean combinations of patterns.

Solution 6

This solution is based on using rewrite rules recursively.

```
algexp[6][u_ + v_] := algexp[6][u] && algexp[6][v]
algexp[6][u_ v_]   := algexp[6][u] && algexp[6][v]
algexp[6][u_^v_]   := algexp[6][u] && algexp[6][v]
algexp[6][w_]       := MemberQ[{Symbol, Integer, Rational,
                                Real, Complex}, Head[w]]
```

Solution 7

This is the "power" solution. It just looks at all the head of all the subtrees of the expression, including the expression itself, and insists that they belong to the appropriate list.

```
algexp[7][exp_] :=
    Complement[Head /@ Append[Level[exp, Infinity], exp],
        {Plus, Power, Times, Symbol,
        Integer, Real, Rational, Complex}] == {}
```

Test of the solutions

```
testAlgExp[n_] :=
   {algexp[n][x^2 + (y + 2)^3],
    algexp[n][x^2 + (Sin[y] + 2)^3],
    algexp[n][(5 x y)^(z + w)],
    algexp[n][Sqrt[5 x y]^(z + w)],
    algexp[n][x^(x^(x^(x^x)))],
    algexp[n][(y + w)^(x + 2)],
    algexp[n][(x + 2 I) (3 + y I)^(5 + 4I)],
    algexp[n][(2x + y) + I (z w + u)],
    algexp[n][Tan[x^2 + y^2]]}


testAlgExp[1]


Table[{method[n],
        Timing[Do[testAlgExp[n], {40}]][[1]]}, {n, 1, 7}]
```

Thus, the first five methods are approximately the same, except for method 3 using **Witch**, which is definitely the worst one. Method 7 is clearly the winner, being half an order of magnitude faster than the slower methods. The pure rewrite rule method 6 is supprisingly fast.

One can get the same timing result by using a very large, deeply nested algebraic expression.

```
exp = Nest[((x^#) #&), z^2, 8];


Table[Timing[algexp[n][exp]], {n, 1, 7}]
```

Now method 7 appears to be about 10 times faster than the slower methods. Actually, of course, the first five methods probably are exponential while method 7 may be linear. The surprise is method 6, which should be at least quadratic.

Problem 3

```
countTheCharacters[text_String] :=
  With[{chars = Select[Characters[ToLowerCase[text]],
                       LetterQ[ToString[#]]&]},
    Sort[Map[
        {#, N[Count[chars, #] 100/Length[chars] "%", 3]}&,
        Union[chars] ], (#1[[2]]/"%" > #2[[2]]/"%")&] ]

text = "Pascal is for building pyramids - imposing,
breathtaking, static structures built by armies pushing heavy
blocks into place.  Lisp is for building organisms - imposing,
breathtaking, dynamic structures built by squads fitting fluctuat-
ing myriads of simpler organisms into place.";

countTheCharacters[text]
```

Problem 4

Part 1

Here are all three forms of the translation of the Pascal program into *Mathematica*, using a **Do** loop, a **For** loop, and a **While** loop. In order to avoid repeating the same fragment of code three times we put it into a separate **Module**.

```
ifStatement[trial_] :=
  Module[{divided},
      If[Mod[trial, 3] == 1,
          divided = 2 Floor[trial / 3];
          If[Mod[divided, 3] == 1,
              divided = 2 Floor[divided / 3];
              If[Mod[divided, 3] == 1,
                  divided = 2 Floor[divided / 3];
                  If[Mod[divided, 3] == 1,
                      Print[
                          PaddedForm[trial, 3],
                          " is a solution."]]]]]]
```

We compare timings for the three versions, editing out the **Print** statements from the second and third versions.

```
Timing[Module[{TrialNumber},
    Do[ifStatement[TrialNumber], {TrialNumber, 1, 500} ]]]

Timing[Module[
   {TrialNumber},
   For[TrialNumber = 1,
       TrialNumber <= 500,
       TrialNumber++,
       ifStatement[TrialNumber] ] ]]

Timing[Module[
   {TrialNumber = 0},
   While[TrialNumber++; TrialNumber <= 500,
       ifStatement[TrialNumber] ] ]]
```

As one might suspect, there is slightly less overhead in a **Do** loop than in the other version.

Part 2

To make the one-liner print its solution, just wrap **Print[-, "is a solution"]** around the given one-liner and put a semicolon at the end.

```
Timing[Map[Print[PaddedForm[#, 3], " is a solution"]&,
    Select[Range[500],
           And@@Map[(# == 1)&,
           Mod[NestList[2 Floor[#/3]&, #, 3], 3]]&]];]
```

Problem 5

First define a(n), which is used in both of the strange sums.

```
a[n_] :=
   Module[{intdig = IntegerDigits[n]},
       Length[
         Select[
           Delete[Reverse[intdig],
               Map[List, 2 Range[Floor[Length[intdig]/2]]]],
           OddQ]]];
```

For instance,

```
a[324234501]
```

The first series

```
series[k_] := N[Sum[a[n]/10^n, {n, k}], k]
```

Check its value for various values of **k**. Since the series consists of positive terms, if the value ever exceeds the value of 10/99, then the claimed result is false. This first occurs at $k = 100$.

```
series[100]
```

```
N[10/99, 100]
```

```
test[m_] := series[m] <= N[10/99, m]
```

```
test[100]
```

Clearly, the answer to the first series is wrong.

The second series

Here is the second series.

```
newSeries[k_] := N[Sum[a[2^n]/2^n, {n, k}], k/3]
```

No matter how far we go, the value is always slightly less than the value of 1/99. (This takes a long time.)

```
news = newSeries[1000]
```

```
N[1/99, 100]
```

```
news <= N[1/99, 250]
```

A third attempt

We make an attempt at another of the series discussed in Borwein and Borwein.

```
e[n_] :=
    Module[{intdig = IntegerDigits[n]},
        Reverse[intdig] . Table[(1/10)^j,
                                {j, Length[intdig]}]];
```

```
e[1234567891011121314151617181921]
```

```
N[%, 100]
```

```
newnewSeries[k_] :=
    N[Sum[e[n]/(n (n + 1)), {n, k}], k/100]
```

```
newnewSeries[500]
```

```
newnewSeries[1000]
```

```
newnewSeries[1500]
```

```
N[10/99 Log[10]]
```

For 500 terms, the first two digits were correct. For 1000 terms, three digits are correct. The answer for 5000 terms never returned. It seems that this is computationally infeasible. Apparently it is necessary to keep the sum in exact form until the last step in order to avoid roundoff errors. According to Borwein and Borwein, the answer is correct.

Problem 6

OutShuffle and inShuffle
First define `outShuffle` and `inShuffle`.

```
outShuffle[deck_List /; EvenQ[Length[deck]]] :=
    Flatten[Thread[{Take[deck,  Length[deck]/2],
                    Take[deck, -Length[deck]/2]}]]
```

```
inShuffle[deck_List /; EvenQ[Length[deck]]] :=
    Flatten[Thread[{Take[deck, -Length[deck]/2],
                    Take[deck,  Length[deck]/2]}]]
```

For instance,

```
outShuffle[Range[16]]
```

```
inShuffle[Range[16]]
```

The experimental orders of outShuffle and inShuffle

If outShuffle or inShuffle is iterated often enough, the deck must ultimately be brought back to its original order, since the group of all permutations is a finite group. Find the order of outShuffle in a deck with 2n cards. Since we don't know what the order is, we use a While loop that continues until we find the identity permutation. It is known that the group generated by outShuffle and inShuffle acts transitively on the deck and is the same as the group of symmetries of the n-octahedron.

```
outOrder[n_Integer/; Positive[n]] :=
    Module[{num = 1, out = outShuffle[Range[2 n]]},
           While[out =!= Range[2 n],
                   out = outShuffle[out]; num++]; num];

inOrder[n_Integer/; Positive[n]] :=
    Module[{num = 1, in = inShuffle[Range[2 n]]},
           While[in =!= Range[2 n],
                   in = inShuffle[in]; num++]; num];
```

Calculate the answers for the case of an ordinary deck of cards where **n** = 26; i.e., **2n** = 52.

```
{outOrder[26], inOrder[26]}
```

Thus, if a perfect out shuffle is performed eight times, the deck is returned to its original order. Now calculate out orders and in orders for even numbers of cards up to 100.

```
outOrdersUpTo[m_] :=
    Map[Flatten, Table[{2 n, outOrder[n]}, {n, m}]];

ListPlot[outOrdersUpTo[50], PlotJoined -> True];

inOrdersUpTo[m_] :=
    Map[Flatten, Table[{2 n, inOrder[n]}, {n, m}]];

ListPlot[inOrdersUpTo[50], PlotJoined -> True];
```

For both, the order apparently is less than or equal to 2n. From the values it seems clear that the orders of inShuffle are equal to those of outShuffle shifted by 2. It is known, in fact, that the order of outShuffle is less than or equal to 2n – 2 and the order of inShuffle is less than or equal to 2n.

Some experiments

The isomorphism with the symmetries of the **n** octahedron is based on the fact that every such permutation is centrally symmetric, as illustrated below.

```
Clear[a]

symdeck = Join[Table[a[i], {i, 10}],
                Reverse[Table[b[i], {i, 10}]]]]

outShuffle[symdeck]

outShuffle[%]
```

The theoretical orders of outShuffle and inShuffle

These calculations are done easily using a `While` loop.

```
outOrderCalc[n_] :=
    Module[{k = 1},
            If[n == 1, 1,
                While[Mod[2^k, 2n - 1] =!= 1, k++]; k]];

inOrderCalc[n_] :=
    Module[{k = 1},
        While[Mod[2^k, 2n + 1] =!= 1, k++]; k];

{outOrderCalc[26], inOrderCalc[26]}
```

They can also be written in functional form using `FixedPoint`.

```
outOrderCalcFun[n_] :=
    FixedPoint[If[Mod[2^#, 2n - 1] =!= 1, #+1, #]&, 1]

inOrderCalcFun[n_] :=
    FixedPoint[If[Mod[2^#, 2n + 1] =!= 1, #+1, #]&, 1]

{outOrderCalcFun[26], inOrderCalcFun[26]}
```

To check if this agrees with the experimental results, calculate many values.

```
outOrdersCalcUpTo[m_] :=
    Map[{2 #, outOrderCalc[#]}&, Range[m]];
```

```
inOrdersCalcUpTo[m_] :=
    Map[{2 #, inOrderCalc[#]}&, Range[m]];

outTest[m_] := outOrdersUpTo[m] === outOrdersCalcUpTo[m]

inTest[m_]  := inOrdersUpTo[m]  === inOrdersCalcUpTo[m]
{outTest[50], inTest[50]}
```

Problem 7

Find one representation

The procedure

The following procedure is optimized to find one representation of **n** as a sum of four squares. Originally it was written with a **For** loop, but a **Do** loop seems to be simpler. The purpose of the **Return** statement is to break out of the loop as soon as a solution is found. The program is developed in three steps. First find a representation of an integer **n** as a sum of two squares, if it exists. For this, it is sufficient to search for an integer **i** between **Floor[N[Sqrt[n/2]]]** and **Floor[N[Sqrt[n]]]** such that **n - i²** is the square of an integer. It is most efficient to start at the bigger value and step down. Then find a representation of **n** as a sum of three squares, if it exists, by searching for an integer **i** between **Floor[N[Sqrt[n/3]]]** and **Floor[N[Sqrt[n]]]** such that **n - i²** is the sum of two squares. Finally, find a representation of **n** as a sum of four squares by searching for an integer **i** between **Floor[N[-Sqrt[n/4]]]** and **Floor[N[Sqrt[n]]]** such that **n - i²** is the sum of three squares. This is guaranteed to exist.

```
sumOfTwoSquares[n_Integer] :=
  Module[{i, trial},
    Do[trial = Sqrt[n - i^2];
       If[IntegerQ[trial],
          Return[{i, trial}]],
          {i, Floor[N[Sqrt[n]]], Floor[N[Sqrt[n/2]]], -1}]];

sumOfThreeSquares[n_Integer] :=
  Module[{i, trial},
    Do[trial = sumOfTwoSquares[n - i^2];
       If[trial =!= Null,
          Return[Flatten[{i, trial}]]],
          {i, Floor[N[Sqrt[n]]], Floor[N[Sqrt[n/3]]], -1}]];
```

```
sumOfFourSquares[n_Integer] :=
  Module[{i, trial},
    Do[trial = sumOfThreeSquares[n - i^2];
      If[trial =!= Null,
        Return[Flatten[{i, trial}]] ],
      {i, Floor[N[Sqrt[n]]], Floor[N[Sqrt[n/4]]], -1}]];
```

Examples

```
sumOfTwoSquares[5]
```

```
sumOfThreeSquares[14]
```

```
sumOfFourSquares[1000]
```

```
Table[{i, sumOfFourSquares[i]}, {i, 150, 183, 3}]
```

```
Timing[sumOfFourSquares[16720845]]
```

This procedure is very fast, but the results are boring for small numbers since the first entry is almost always the largest integer whose square is less or equal to **n**. The following finds all integers between 1 and 1000 that are not sums of three squares.

```
Map[#[[1]]&, Select[Table[{i, sumOfThreeSquares[i]},
                 {i, 1, 1000}], #[[2]] === Null&]]
```

Find all representations

The procedure

These functions work in a somewhat different way. It takes much longer to find all representations. In this case we have written functional programs, but the strategy is the same as before. The procedure **twoSquares** is implemented using **Fold**, but the other two seem to be possible only by mapping an operation down the list of relevant values. The output from **fourSquares** is a list consisting of all representation of **n** as a sum of four squares. A procedure for checking the output is provided.

```
twoSquares[n_Integer] :=
  Fold[If[IntegerQ[Sqrt[n - #2^2]],
        Append[#1, {#2, Sqrt[n - #2^2]}], #1]&,
      {}, Range[Floor[N[Sqrt[n/2]]], Floor[N[Sqrt[n]]]]]];
```

```
    threeSquares[n_Integer] :=
      Union[Flatten[
        Cases[Map[{#, twoSquares[n - #^2]}&,
              Range[Floor[N[Sqrt[n/3]]], Floor[N[Sqrt[n]]]]],
                  {a_Integer, b_List} /; b =!= {} ] /.
          {a_Integer, b_List} :>
              Map[Sort[Flatten[{a, #}]]&, b], 1]];

    fourSquares[n_Integer] :=
      Union[Flatten[
        Cases[Map[{#, threeSquares[n - #^2]}&,
              Range[Floor[N[Sqrt[n/4]]], Floor[N[Sqrt[n]]]]],
                  {a_Integer, b_List} /; b =!= {} ] /.
          {a_Integer, b_List} :>
                Map[Reverse[Sort[Flatten[{a, #}]]]&, b], 1]];

    checkRep[list_List] := Map[Plus@@(#^2)&, list];
```

Examples

```
    twoSquares[25]

    checkRep[%]

    fourSquares[102]

    checkRep[%]

    Table[fourSquares[n], {n, 71, 75}]//MatrixForm

    Timing[fourSquares[3456]]
```

Sums of distinct squares

Some, but not all, numbers are the sum of four distinct, non-zero squares. Our representations are always in decreasing order, so the following predicate picks out the distinct representations.

```
    distinctQ[list_List] :=
        list[[1]] > list[[2]] > list[[3]] > list[[4]] > 0;
```

```
distinctSquares[n_Integer] :=
    Select[fourSquares[n], distinctQ]

distinctSquares[102]
```

The following takes a long time to calculate. The output is suppressed since it is over 50 pages long.

```
distinctRepresentations =
    Table[{n, distinctSquares[n]}, {n, 1, 1000}];
```

Once all distinct representations have been calculated, then information can be extracted from the table without actually displaying it all. The following finds all numbers between 1 and 1000 that have no representation as a sum of four distinct non-zero squares.

```
noRepresentations =
    Map[#[[1]]&,
        Select[distinctRepresentations, (#[[2]]==={})&]]
```

```
{1, 2, 3, 4, 5, 6, 7, 8, 9, 10, 11, 12, 13, 14, 15, 16, 17, 18, 19,
 20, 21, 22, 23, 24, 25, 26, 27, 28, 29, 31, 32, 33, 34, 35, 36,
 37, 38, 40, 41, 42, 43, 44, 45, 47, 48, 49, 52, 53, 55, 56, 58,
 59, 60, 61, 64, 67, 68, 69, 72, 73, 76, 77, 80, 82, 83, 88, 89,
 92, 96, 97, 100, 101, 103, 104, 108, 112, 115, 124, 128, 132,
 136, 144, 148, 152, 157, 160, 168, 172, 176, 188, 192, 208, 220,
 224, 232, 240, 256, 268, 272, 288, 292, 304, 320, 328, 352, 368,
 384, 388, 400, 412, 416, 432, 448, 496, 512, 528, 544, 576, 592,
 608, 640, 672, 688, 704, 752, 768, 832, 880, 896, 928, 960}
```

Problem 8

Here are the results of a student's investigations of magic squares. No attempt has been made to improve the procedures. Can these be replaced by functional programs?

Odd order magic squares

```
oddMagicSquare[n_]:=
  Module[{BasicList = Table[0, {n}, {n}], k},
    Do[BasicList[[
        Mod[1 - k + 2 Floor[(k-1)/n], n] + 1,
        Mod[1/2 (n-3) + k - Floor[(k-1)/n], n] + 1]] = k,
      {k, n^2}]; BasicList//TableForm];

magicSum[n_] := n (n^2 + 1) / 2;

oddMagicSquare[5]

Sum[oddMagicSquare[5][[1, 1, i]], {i, 5} ] == magicSum[5]
```

Double even order magic squares

```
doubleEvenMagicSquare[n_]:=
 Module[
  {BasicSquare = Table[0, {n}, {n}], i, j, k, p, q,
     AuxiliarySquare, Square},
   Do[BasicSquare[[1, i]] =
           {i, n + 1 - i}[[Random[Integer, {1, 2}] ]],
     {i, n/2} ];
   Do[BasicSquare[[1,j]] = n+1-BasicSquare[[1, n+1-j]],
     {j, n/2 + 1, n}];
   Do[BasicSquare[[k]] = BasicSquare[[1]],
     {k, n/2 - 1}];
   Do[BasicSquare[[p]] = n+1-BasicSquare[[p-1]],
     {p, 2, n/2, 2}];
   Do[BasicSquare[[q]] = BasicSquare[[n+1-q]],
     {q, n/2 + 1, n}];
   AuxiliarySquare =
     Flatten[Map[{(# - 1) n}&,
             Transpose[BasicSquare]], 1];
   Square[i_, j_]:=
        BasicSquare[[i, j]] + AuxiliarySquare[[i, j]];
   Table[Square[i, j], {i, n}, {j, n}] // TableForm];
```

```
doubleEvenMagicSquare[4]

Sum[%[[1, i]], {i, 4}] == magicSum[4]

doubleEvenMagicSquare[8]

Sum[%[[1, i]], {i, 8}] == magicSum[8]
```

References

[Abell 1] Abell, M. L. and Braselton, J. P., *Mathematica by Example*, Academic Press, New York, 1992.

[Abell 2] Abell, M. L. and Braselton, J. P., *Differential Equations with Mathematica*, Academic Press, New York, 1993.

[Abelson] Abelson, H. and Sussman, G. J. with Sussman, J., *Structure and Interpretation of Computer Programs*, The MIT Press, Cambridge, 1985.

[Barendregt 1] Barendregt, H. P., *The Lambda Calculus: Its Syntax and Semantics*, Studies in Logic and the Foundations of Mathematics 103, Second Edition, North-Holland, Amsterdam, 1984.

[Barendregt 2] Barendregt, H. P., "Functional Programming and Lambda Calculus", in *Handbook of Theoretical Computer Science*, Vol. B, Formal Models and Semantics, Ed. J. van Leeuwen, Elsevier Science Publishers, 1990, 321–363.

[Biggs] Biggs, N. L., *Discrete Mathematics*, Oxford University Press, Oxford, 1989.

[Blachman 1] Blachman, N., *Mathematica, a Practical Approach*, Prentice Hall, New Jersey, 1992.

[Blachman 2] Blachman, N., *Quick Reference Cards*, Prentice-Hall, New Jersey, 1992. Technical Report, Wolfram Research, Inc., 1992.

[Borwein] Borwein, J. M., and Borwein, P. B., Strange Series and High Precision Fraud, Amer. Math. Mon. 99 (1992), 622–640.

[Budd] Budd, T., *An Introduction to Object-Oriented Programming*, Addison-Wesley, New York, 1991.

[Church] Church, A., An Unsolvable Problem of Elementary Number Theory, Amer. J. Math 58, 1936, 354–363.

[Cooper] Cooper, D. and Clancey, M., *Oh! Pascal!*, W. W. Norton and Co., New York, 1982.

[Crandall] Crandall, R. E. *Mathematica for the Sciences*, Addison-Wesley, New York, 1991.

[Curry]   Curry, H. and Feys, R., *Combinatory Logic*, Vol I., North-Holland, Amsterdam, 1958.

[Dershowitz]   Dershowitz, N. and Jouannaud, J.-P., "Rewrite Systems", in *Handbook of Theoretical Computer Science*, Vol. B, Formal Models and Semantics, Ed. J. van Leeuwen, Elsevier Science Publishers, New York, 1990, 243–320.

[Ellis]   Ellis, W., and Lodi, E., *A Tutorial Introduction to Mathematica*, Brooks/Cole Pub. Co., Pacific Grove, California, 1991.

[Fitch]   J. Fitch, "A Survey of Symbolic Computation in Physics", in *Symbolic and Algebraic Computation, Proc. EUROSAM '79*, Lecture Notes in Computer Science 72, Springer-Verlag, New York, 1979, 30–41.

[Gray A]   Gray, A., *Modern Differential Geometry of Curves and Surfaces*, CRC Press, Boca Raton, 1993.

[Gray T1]   Gray, T. and Glynn, J., *Exploring Mathematics with Mathematica: Dialogs Concerning Computers and Mathematics*, Addison-Wesley, New York, 1991.

[Gray T2]   Gray, T. and Glynn, J., *The Beginners Guide to Mathematica* 2.0., Addison Wesley, New York, 1992.

[Hearn]   Hearn, A., "Reduce, A Case Study in Algebra System Development", in *Computer Algebra*, Lecture Notes in Computer Science 144, Springer-Verlag, New York, 1982, 263–272.

[Hindley]   Hindley, J. R. and Seldin, J. P., *Introduction to Combinators and λ-calculus*, Cambridge University Press, Cambridge, 1986.

[Horowitz]   Horowitz, E., *Programming Languages: A Grand Tour*, Third Edition, Computer Science Press, Rockville, MD, 1987

[Hoffman]   Hoffman, D., The Computer-Aided Discovery of New Embedded Minimal Surfaces, The Mathematical Intelligencer, Vol. 9, No. 3 (1987), 8 - 21.

[vanHulzen]   van Hulzen, J. A. and Calmet, J, *Computer Algebra: Symbolic and Algebraic Computation, Computer Algebra Applications*, Second Edition, Springer-Verlag, New York, 1983.

[Maeder 1]   Maeder, R., *Programming in Mathematica*, Second Edition, Addison-Wesley, New York, 1991.

[Maeder 2]   Maeder, R., *The Mathematica Programmer II*, Academic Press Professional, Cambridge, 1996.

[McCarthy]   McCarthy, J., Abrahams, P. W., Edwards, D. J., Hart, T. P., and Levin, M., LISP 1.5 Programmer's Manual, reprinted in *Programming Languages: A Grand Tour*, Third Edition, Ed., E. Horowitz, Computer Science Press, Rockville, MD, 1987, 215–239.

[Meyer]   Meyer, B., Object-Oriented Software Construction, Prentice-Hall, New York, 1986.

[Michaelson]   Michaelson, G., *An Introduction to Functional Programming through Lambda Calculus*, Addison-Wesley, New York, 1989.

[Miller]   Miller, L. H. and Quilici, A. E., *Programming in C*, Wiley & Sons Inc., New York, 1986.

[Mitchell]   Mitchell, J., "Type Systems for Programming Languages", in *Handbook of Theoretical Computer Science*, Vol. B, Formal Models and Semantics, Ed. J. van Leeuwen, Elsevier Science Publishers, 1990, 365–458.

[Ng]   Ng, E., Symbolic-Numeric Interface: A Review, Symbolic and Algebraic Computation, Proc. EUROSAM '79, Lecture Notes in Computer Science 72, Springer Verlag, New York, 1979, 330–345.

[O'Neill]   O'Neill, B., *Elementary Differential Geometry*, Academic Press, New York, 1966.

[Paulson]   Paulson, L, C., *ML for the Working Programmer*, Cambridge University Press, Cambridge, 1991.

[Rogers]   Rogers, H., *Theory of Recursive Functions and Effective Computability*, The MIT Press, Cambridge, 1987.

[Rotman]   Rotman, J., *An Introduction to the Theory of Groups*, Fourth Edition, Springer-Verlag, New York, 1994.

[Schmidt]   Schmidt, D. A., *Denotational Semantics, A Methodology for Language Development*, Allyn and Bacon, Boston, 1986.

[Simon 1]   Simon, B., Four Computer Mathematical Environments, Computers and Mathematics, Amer. Math. Soc. Notices, 37 (7), 1990, 861–868.

[Simon 2]   Simon, B., Comparitive CAS Review, Computers and Mathematics, Amer. Math. Soc. Notices, 39 (7), 1992, 700–710.

[Skeel]   Skeel, R. and Keiper, J., *Elementary Numerical Computing with Mathematica*, McGraw-Hill, New York, 1993.

[Skiena]   Skiena, S., *Implementing Discrete Mathematics: Combinatorics and Graph Theory with Mathematica*, Addison-Wesley, New York, 1990.

[Soare]   Soare, R. I., *Recursively Enumerable Sets and Degrees*, Springer-Verlag, New York, 1987.

[Stoutemyer]   Stoutemyer, D., Crimes and Misdemeanors in the Computer Algebra Trade, Computers and Mathematics, Amer. Math. Soc. Notices, 38 (7), 1992, 778–785.

[Struik]   Struik, D., *Classical Differential Geometry*, Addison-Wesley 1950.

[Tucker]   Tucker, A., *Applied Combinatorics*, Second Edition, John Wiley & Sons, New York, 1984.

[Vardi]   Vardi, I., *Computational Recreations in Mathematica*, Addison-Wesley, New York, 1991.

[Varian]   Varian, H., *Economic and Financial Modeling with Mathematica*, TELOS/Springer-Verlag, New York, 1993.

[Wagon]   Wagon, S., *Mathematica in Action*, W. H. Freeman, San Franscisco, 1991.

[Wei]   Wei, Sha Xin, *Mathematica* 2.0, Amer. Math. Soc. Notices, 39 (5), 1992, 428–435.

[Withoff]   *Mathematica* Internals, Technical Report, Wolfram Research, Inc. 1992.

[Wolfram 1]   Wolfram, S., *The Mathematica Book*, 3rd Edition, Wolfram Media/Cambridge University Press, 1996.

[Wolfram 2]   Wolfram Research, *Mathematica 3.0 Standard Add-On Packages*, Wolfram Media/Cambridge University Press, 1996.

Index

MATHEMATICA®

WOLFRAM RESEARCH

THE WORLD'S ONLY FULLY INTEGRATED TECHNICAL COMPUTING

Try Mathematica 3.0 for 30 days!

One of the best ways to explore the power of *Mathematica* is to use it with the book you have just bought and other *Mathematica* books published by Academic Press.

If you don't have *Mathematica* on your computer yet, now is your chance to try the revolutionary new *Mathematica* 3.0. You'll join over a million scientists, engineers, researchers, educators, students, and many other professionals who use *Mathematica* in their work everyday.

To get your copy of *Mathematica* 3.0 today*, contact Wolfram Research.

Web: http://www.wolfram.com
Email: info@wolfram.com
Phone: 1-800-441-6284 or 217-398-0700
Fax: 1-217-398-0747

If *Mathematica* does not meet your requirements, just return it to us within 30 days and you will be issued full credit.

Mathematica 3.0 is being released for Microsoft Windows, Macintosh, and over twenty Unix and other platforms. Special pricing is available for academic institutions, site licenses, Mathematica teaching labs, and the student version of Mathematica.

* This offer is valid for direct orders through Wolfram Research in the U.S. and Canada only. Outside the U.S. and Canada, evaluation copies of Mathematica are available through your local reseller. For a list of resellers in your area, contact Wolfram Research.

http://www.wolfram.com

WOLFRAM RESEARCH

Wolfram Research, Inc.: http://www.wolfram.com; info@wolfram.com; +1–217–398–0700. **Wolfram Research Europe Ltd.:** http://www.wolfram.co.uk; info@wolfram.co.uk; +44–(0)1993–883400. **Wolfram Research Asia Ltd.:** http://www.wolfram.co.jp; info@wolfram.co.jp; +81–(0)3–5276–0506.

MATHEMATICA
APPLICATIONS
LIBRARY

The range of fields represented by the one million *Mathematica* users is striking. To help meet the specific needs of each of these users, Wolfram Research is constantly developing specialized applications packages. You'll find exactly the tools you need to quickly handle specific tasks in engineering, mechanical systems, finance, time series, optics, fuzzy logic, visualization, and many other areas. Plus, ready-to-use links provide seamless connections between *Mathematica* and popular applications such as Microsoft Word and Excel, and more.

All application packages are written entirely in *Mathematica* so you can seamlessly integrate them with your own existing *Mathematica* work or customize the functions just as you would any *Mathematica* function. By giving you direct access to the system's powerful programming language and expansive collection of computational commands, each package is an incredibly rich and flexible working environment.

For prices and more information

http://www.wolfram.com

or call 1-800-441-MATH (6284) • 1-217-398-0700

WOLFRAM
RESEARCH

Wolfram Research, Inc.: http://www.wolfram.com; info@wolfram.com; +1-217-398-0700. **Wolfram Research Europe Ltd.:** http://www.wolfram.co.uk; info@wolfram.co.uk; +44-(0)1993-883400. **Wolfram Research Asia Ltd.:** http://www.wolfram.co.jp; info@wolfram.co.jp; +81-(0)3-5276-0506.